CATALYST HANDBOOK

SECOND EDITION

Edited by

Martyn V. Twigg

BSc, PhD, C.Chem., FRSC
ICI Chemicals & Polymers Ltd
Research Technology Department
Catalysis Centre, Billingham
Cleveland, England

Wolfe Publishing Ltd

Many of the topics contained in this book are the subject of patents and therefore nothing contained herein should be construed as granting a licence or permission to use any invention or process which may be protected by a patent.

Reasonable care has been used in compiling the contents of this book but the editor, the contributors, and the publisher accept no liability with respect to the accuracy of the contents. Any person using information, equipment, methods or procedures described in this book should only do so after seeking the advice of a suitably qualified person.

Copyright © M.V. Twigg, 1989
Published by Wolfe Publishing Ltd, 1989
Printed by Butler & Tanner, Frome, England
ISBN 0 7234 0857 2

All rights reserved. No reproduction, copy or transmission of this publication may be made without written permission.

No part of this publication may be reproduced, copied or transmitted save with written permission or in accordance with the provisions of the Copyright Act 1956 (as ammended), or under the terms of any licence permitting limited copying issued by the Copyright Licensing Agency, 33-34 Alfred Place, London, WC1E 7DP.

Any person who does any unauthorised act in relation to this publication may be liable to criminal prosecution and civil claims for damages.

A CIP catalogue record for this book is available from the British Library.

For a full list of Wolfe Science Atlases, plus forthcoming titles and details of our other Atlases, please write to Wolfe Publishing Ltd, 2-16 Torrington Place, London WC1E 7LT, England.

Contributors

G.W. Bridger BSc, MRSC

P.J.H. Carnell BSc, PhD

P. Davies BSc

R.T. Donald

D.R. Goodman BA

N.H. Harbord BSc, PhD

J.R. Jennings BSc, PhD

L. Lloyd BSc, PhD

W.J. Lywood BSc, ACGI

B.B. Pearce BSc, PhD

D.E. Ridler BSc

M.S. Spencer BSc, PhD, C.Chem., FRSC

M.V. Twigg BSc, PhD, C.Chem., FRSC

S.A. Ward BSc

C. Woodward BSc, PhD

Contents

Preface 16

Chapter 1.

Fundamental Principles

M.S. Spencer

1.1. Fundamentals of Heterogeneous Catalysis 17
    1.1.1. Introduction 17
    1.1.2. The Role of Catalysis 18
        1.1.2.1. Ammonia Synthesis 18
        1.1.2.2. Ammonia Oxidation 19
    1.1.3. The Nature of the Catalytic Process 23
    1.1.4. Catalyst Activity 24
    1.1.5. Catalyst Selectivity 26
    1.1.6. Steps in the Catalytic Process 27
    1.1.7. Adsorption and Desorption 29
    1.1.8. Catalyst Design 32

1.2. Catalyst Manufacture 34
    1.2.1. Introduction 34
    1.2.2. Unsupported Metals 34
    1.2.3. Fused Catalysts 35
    1.2.4. Wet Methods of Catalyst Manufacture 37
    1.2.5. Fundamentals of Precipitation Processes 38
    1.2.6. Catalyst Manufacture by Precipitation Processes 40
    1.2.7. Impregnation Processes 41
    1.2.8. Forming Stages 43

1.3. Catalyst Testing 48
    1.3.1. Introduction 48
    1.3.2. Chemical and Physical Properties 49
    1.3.3. Bulk Chemical Properties 49
    1.3.4. Surface Chemical Properties 50
    1.3.5. Physical Properties 52
    1.3.6. Catalyst Performance 55
    1.3.7. Coarse Laboratory Screening 56
    1.3.8. Fine Laboratory Screening 58
    1.3.9. Semi-technical Catalyst Testing 60
    1.3.10. Reaction Kinetics 61
    1.3.11. Catalyst Ageing 65
    1.3.12. Mechanism of the Catalytic Reaction 66

   1.3.12.1. Ammonia Synthesis 67
   1.3.12.2. Methanol Synthesis 68
1.4. Catalyst in Use 69
 1.4.1. Introduction 69
 1.4.2. Pretreatment and Activation 69
 1.4.3. Loss of Catalyst Performance 73
 1.4.4. Physical Causes of Decay 76
 1.4.5. Poisoning by Impurities in Feeds or Catalysts 77
 1.4.6. Poisoning by Reactants or Products 81
 1.4.7. Interactions in Catalyst Deactivation 82

Chapter 2.

Process Design, Rating and Performance

W.J. Lywood

2.1. Design of Catalytic Reactors 85
 2.1.1. Operating Temperature and Pressure 87
  2.1.1.1. Desulphurization Reactor 87
  2.1.1.2. Steam Reformers 87
  2.1.1.3. Water-gas Shift Reactors 88
  2.1.1.4. Methanation Reactor 89
  2.1.1.5. Ammonia and Methanol Synthesis Reactors 89
 2.1.2. Converter Types 90
  2.1.2.1. Single Adiabatic Bed 90
  2.1.2.2. Quench Converter 91
  2.1.2.3. Inter-bed Cooling 91
  2.1.2.4. ICI High-conversion Reactor 96
  2.1.2.5. Tube-cooled Reactor 96
  2.1.2.6. Steam-raising Reactor 96
 2.1.3. Catalyst Life 96
 2.1.4. Optimum Catalyst Size and Shape 101
  2.1.4.1. Voidage 101
  2.1.4.2. Catalyst Particle Size 103
 2.1.5. Design Conversion of Reactor 105
 2.1.6. Calculation of Catalyst Volume 107
  2.1.6.1. Catalyst Volume for Low concentration Reactant
      Being Removed 107
  2.1.6.2. Catalyst Volume for Low concentration Product Being
      Formed 108
  2.1.6.3. Equilibrium Concentrations 109
  2.1.6.4. Rate Constants 119
 2.1.7. Vessel Dimensions 119
2.2. Reactor Rating 121

2.2.1. Optimum Operating Temperature	121
2.3. Catalyst Performance	123
2.3.1. Fall in Apparent Catalyst Activity	123
2.3.1.1. Poisoning/Sintering	123
2.3.1.2. Poor Gas Distribution	124
2.3.1.3. Poor Mixing of Reactants	124
2.3.2. Increase in Pressure Drop	125
2.3.2.1. Breakage or Erosion of Catalyst Particles	125
2.3.2.2. Disintegration of Catalyst Particles	125
2.3.2.3. Deformation of Catalyst Particles	126
2.3.2.4. Carry-over onto Catalyst Bed	126
2.3.2.5. Collapse of Bed Support	126
2.3.3. Measurement of Performance	126
2.3.3.1. Analysis	126
2.3.3.2. Mass Balance	127
2.3.3.3. Catalyst-bed Temperature Rises	127
2.3.3.4. Catalyst-bed Temperature Profiles	127
2.3.3.5. Radioactive Tracing	127
2.3.3.6. Pressure Drop	127
2.3.4. Quantifying Catalyst Performance	127
2.3.4.1. Composition at the Exit from the Reactor	128
2.3.4.2. Approach to Equilibrium	128
2.3.4.3. Activity or Active Volume of Catalyst	128
2.3.5. Calculation of Catalyst Performance	128
2.3.5.1. Reactor Exit Composition	129
2.3.5.2. Calculation of Approach to Equilibrium	130
2.3.5.3. Calculation of Activity or Active Volume from Composition	132
2.3.5.4. Calculation of Activity or Active Volume from Temperature Profiles	133
2.3.6. Application of Methods to Ammonia and Methanol Catalysts	134
2.3.6.1. Desulphurizer	135
2.3.6.2. Primary and Secondary Reformer	135
2.3.6.3. High Temperature Shift	136
2.3.6.4. Low Temperature Shift	136
2.3.6.5. Methanator	136
2.3.6.6. Ammonia and Methanol Synthesis Converter	137
2.4. Computer Programs	137
2.4.1. Reasons for Using Computer Calculations	137
2.4.1.1. Accurate Calculations	137
2.4.1.2. Non-isothermal Reactors	138
2.4.1.3. Multiple Reactions	138
2.4.1.4. Optimization	138
2.4.1.5. Simulation	138
2.4.2. Types of Computer Programs	138

Chapter 3.

Handling and Using Catalysts in the Plant

D.R. Goodman

3.1. Introduction	140
3.2. Catalyst Storage	140
3.3. Drum Handling	141
3.4. Intermediate Bulk Containers and Socks	142
3.5. Sieving Catalyst	149
3.6. Catalyst Charging	150
3.6.1. Pre-charging Checks	150
3.6.2. Charging Vessels	151
3.6.3. Charging Ammonia Converters	155
3.6.4. Charging Reformer Tubes	156
3.7. Catalyst Reduction	161
3.7.1. Reduction of Reforming Catalyst	162
3.7.1.1. Typical Reduction with Steam and Natural Gas	163
3.7.1.2. Reduction with Gas Recirculation	164
3.7.2. Reduction of High-temperature Shift Catalyst	165
3.7.2.1. Typical Reduction of High-temperature Shift Catalyst	166
3.7.3. Reduction of Low-temperature Shift Catalyst	166
3.7.3.1. Typical Reduction of Low-temperature Shift Catalyst	170
3.7.4. Reduction of Methanation Catalyst	171
3.7.5. Reduction of Ammonia Synthesis Catalyst	171
3.7.5.1. Typical Reduction of a Tube-cooled Ammonia Converter	173
3.7.5.2. Typical Reduction of a Multibed Quench Converter	174
3.8. Catalyst Shutdown and Restarts	175
3.9. Catalyst Regeneration	176
3.9.1. Regeneration of Reforming Catalyst	176
3.9.2. Regeneration of High-temperature Shift Catalyst	177
3.9.3. Regeneration of Low-temperature Shift Catalyst	177
3.9.4. Washing of Methanation Catalyst	177
3.9.5. Regeneration of Ammonia Synthesis Catalyst	178
3.10. Blanketing of Reduced Catalyst	178
3.11. Catalyst Stabilization	179
3.11.1. Stabilization of Reforming Catalyst	180
3.11.2. Stabilization of High-temperature Shift Catalyst	180

3.11.3. Stabilization of Low-temperature Shift Catalyst 181
3.11.4. Stabilization of Methanation Catalyst 182
3.11.5. Stabilization of Ammonia Synthesis Catalyst.................... 183

3.12. Catalyst Discharge... 183
   3.12.1. General 183
   3.12.2. Discharge of Pyrophoric Catalyst 184
   3.12.3. Top Discharge... 185
   3.12.4. Blanketing Pyrophoric Catalyst During Vacuum Extraction .. 186
   3.12.5. Discharge of Ammonia Synthesis Catalyst.................. 186

3.13. Re-use of Discharged Catalyst.. 187

3.14. Disposal of Used Catalyst... 188

3.15. Safety Precautions 188

Chapter 4.

Feedstock Purification

P.J.H. Carnell

4.1. Introduction 191

4.2. Feedstocks for Ammonia, Methanol and Hydrogen Production 192
   4.2.1. Natural Gas 192
   4.2.2. Associated Gas, Natural Gas Condensates and LPG 193
   4.2.3. Naphtha 194
   4.2.4. Refinery Off Gases and Electrolytic Hydrogen 194
   4.2.5. Coal Gasification and Coke Oven Gas 194
   4.2.6. Mixed Feeds.. 195

4.3. Desulphurization 196
   4.3.1. Processes for Single-stage Sulphur Removal 196
   4.3.2. Processes for Two-stage Sulphur Removal 198

4.4. Thermal Dissociation of Sulphur Compounds....................... 199

4.5. Hydrogenolysis of Sulphur Compounds 200

4.6. Carbonyl Sulphide 203

4.7. Cobalt Molybdate Catalysts 204
   4.7.1. Presulphiding Cobalt Molybdate Catalyst 205
   4.7.2. Other Reactions Over Cobalt Molybdate Catalyst 206

4.8. Nickel Molybdate Catalysts... 207

4.9. Physical Form of Cobalt and Nickel Molybdate Catalysts 207

4.10. Replacement and Discharging of Cobalt and Nickel Molybdate Catalysts... 208

4.11. Zinc Oxide 209
    4.11.1. Background to Zinc Oxide Absorbents 209
    4.11.2. Thermodynamics and Reaction Kinetics 209
    4.11.3. Formulation of Commercial Zinc Oxide 211
    4.11.4. Use of Test Reactors to Assess Zinc Oxide Absorbents 211
    4.11.5. Effect of Temperature, Pressure and Space Velocity on
            Efficiency of Zinc Oxide Absorbents 213
    4.11.6. Effect of Gas Composition 216
    4.11.7. Effect of Reactor Design 217
    4.11.8. Other Desulphurization Uses for Zinc Oxide 219
    4.11.9. Impurities in Zinc Oxide 220

4.12. Dechlorination 220
    4.12.1. Chloride Sources and Absorbents 220
    4.12.2. Operating Conditions 222

4.13. Removal of Silica and Fluoride 223

4.14. Demetallization 223

4.15. Denitrification 224

Chapter 5.

Steam Reforming

D.E. Ridler, M.V. Twigg

5.1. History 225

5.2. Feedstock and Feedstock Pretreatment 226
    5.2.1. Natural Gas 227
    5.2.2. Naphthas 228

5.3. Chemistry of Steam Reforming 230
    5.3.1. Thermodynamics 230
    5.3.2. Kinetics 239

5.4. Design of Steam Reforming Catalysts 244
    5.4.1. Selectivity 244
    5.4.2. Thermal Stability 244
    5.4.3. Physical Properties 244
    5.4.4. Nickel as a Steam Reforming Catalyst 244
    5.4.5. Supports for Nickel Steam Reforming Catalysts 249
    5.4.6. Carbon Formation on Reforming Catalysts 250

5.5. Secondary Reforming 253

5.6. Catalyst Dimensions 254

5.7. Uses of Catalytic Steam Reforming 256

    5.7.1. Ammonia Synthesis 256
    5.7.2. Methanol Synthesis 258
    5.7.3. Oxo Synthesis Gas 259
    5.7.4. Reducing Gas 260
    5.7.5. Town Gas 261
    5.7.6. Substitute Natural Gas (SNG) 263

5.8. Practical Aspects of Steam Reformers 264
    5.8.1. Containing the Catalyst 267
    5.8.2. Reactant Gas Distribution 269
    5.8.3. Firing the Reformer 270
    5.8.4. Expansion and Contraction of Reformer Tubes 271
    5.8.5. Facilities to Charge and Discharge Catalyst 273
    5.8.6. Designing a Reformer for Efficient Operation 274
    5.8.7. Catalyst Reduction 274
        5.8.7.1. Reduction with Hydrogen 275
        5.8.7.2. Reduction with Ammonia 275
        5.8.7.3. Reduction with Methanol 276
        5.8.7.4. Reduction with Natural Gas 276
        5.8.7.5. Reduction with Other Hydrocarbons 276
        5.8.7.6. Reduction After Shutdown 277

5.9. Factors Affecting the Life of Reforming Catalyst 277

5.10. Catalyst Poisons 278
    5.10.1. Sulphur 278
    5.10.2. Arsenic 278

5.11. Hot Bands in Natural Gas Reformers 280

Chapter 6.

The Water-gas Shift Reaction

L. Lloyd, D.E. Ridler, M.V. Twigg

6.1. Introduction 283

6.2. Thermodynamics 285

6.3. Kinetics and Mechanism 288
    6.3.1. Kinetics Over HT Shift Catalyst 288
    6.3.2. Kinetics Over LT Shift Catalyst 289
    6.3.3. Mechanism of the Catalytic Water-gas Shift Reaction 290

6.4. Converter Design 291

6.5. High-temperature Shift 293
    6.5.1. High-temperature Shift Catalyst Formulation 293
    6.5.2. Diffusion Effects and Pellet Size 295

6.5.2.1. Effect of Pellet Size on Activity 295
6.5.2.2. Effect of Pellet Size on Pressure Drop 296
6.5.3. Reduction of HT Shift Catalyst 298
6.5.4. Operation of HT Shift Catalyst 302
6.5.5. Poisoning and Deactivation 304
6.5.6. Reoxidation and Discharge 306

6.6. Low-temperature Shift 308
6.6.1. General 308
6.6.2. Low Temperature Shift Catalyst formulation 309
6.6.3. Diffusion Effects and Pellet Size 312
6.6.4. Reduction of LT Shift Catalyst 314
6.6.4.1. General Considerations 314
6.6.4.2. Once-through Reductions 317
6.6.4.3. Recycle Reduction Systems 318
6.6.4.4. Commissioning Reduced Catalyst 319
6.6.5. Operation and Monitoring Performance 320
6.6.6. Deactivation and Poisoning 324
6.6.6.1. Deactivation 324
6.6.6.2. Sulphur Poisoning 326
6.6.6.3. Chloride and Other Poisons 328
6.6.7. Oxidation and Discharge 330
6.6.8. Guard Beds 331
6.6.9. Economics of Operation 335

6.7. Recent Developments 335
6.7.1. Sulphur-tolerant Shift Catalysts 335
6.7.2. Operation at Very Low Steam Ratios 338

Chapter 7.

Methanation

B.B. Pearce, M.V. Twigg, C. Woodward

7.1. Introduction 340
7.2. Methanation in Ammonia and Hydrogen Plants 341
7.2.1. Methanation Equlibria 344
7.2.2. Kinetics and Mechanisms 347
7.2.3. Catalyst Formulation 352
7.2.4. Physical Properties of Methanation Catalysts 358
7.2.5. Catalyst Reduction 359
7.2.6. Catalyst Poisons 360
7.2.7. Prediction of Catalyst Life 362
7.2.8. Operating Experience 365

7.3. Methanation in Hydrogen Streams for Olefin Plants 367

7.4. Substitute Natural Gas (SNG) 368
    7.4.1. Oil-based Routes to SNG 368
    7.4.2. Coal-based Routes to SNG 372
        7.4.2.1. Lurgi Coal/SNG Process 373
        7.4.2.2. HICOM Coal/SNG Process 374
        7.4.2.3. Other Developments 376
7.5. Heat Transfer Applications 378
    7.5.1. The EVA–ADAM Project 378

Chapter 8.

Ammonia Synthesis

J.R. Jennings, S.A. Ward

8.1. Introduction 384
8.2. Thermodynamics of Ammonia Synthesis 388
    8.2.1. Theoretical Aspects 388
    8.2.2. Process Consequences 390
    8.2.3. The Synthesis Loop 391
8.3. Ammonia Synthesis Catalysts 393
    8.3.1. The Iron Component 394
    8.3.2. Promoters 395
        8.3.2.1. Structural Promoters 395
        8.3.2.2. Electronic Promoters 398
8.4. Catalyst Reduction 400
    8.4.1. Typical Plant Procedure 400
    8.4.2. Prereduced Catalysts 402
    8.4.3. Economics of Prereduced Catalyst 404
8.5. Poisoning and Deactivation 404
    8.5.1. Introduction 404
    8.5.2. Temporary Poisoning in Ammonia Converters 406
    8.5.3. Permanent Poisoning in Ammonia Converters 407
8.6. Kinetics and Mechanism 409
    8.6.1. Temkin Kinetics 409
    8.6.2. Effect of Catalyst Size 411
    8.6.3. Implications on Process Design 412
    8.6.4. Reaction Mechanism 413
8.7. Plant Operation 415
    8.7.1. General Considerations 415
    8.7.2. Circulation 418
    8.7.3. Hydrogen/Nitrogen Ratio 420
    8.7.4. Influence of Inert Gas Concentration and Purge Rate 420

8.8. Commercial Ammonia Converters.. 423
    8.8.1. General Considerations 423
        8.8.1.1. Flow Type.. 424
        8.8.1.2. Temperature Control and Heat Recovery 425
    8.8.2. Quench Converter... 426
    8.8.3. Indirectly Cooled Multi-bed Converter 433
    8.8.4. Tube-cooled Converter... 433
8.9. The Future 439

Chapter 9.

Methanol Synthesis

G.W. Bridger, M.S. Spencer

9.1. Introduction 441
9.2. Thermodynamic Aspects 442
    9.2.1. Methanol Formation 442
    9.2.2. Selectivity... 444
9.3. The Methanol Synthesis Process 446
    9.3.1. The Synthesis Loop 446
    9.3.2. Make-up Gas Composition 452
9.4. Methanol Synthesis Catalysts... 453
    9.4.1. High-pressure Catalysts 453
    9.4.2. Low-pressure Catalysts 453
9.5. Selectivity and Poisons 460
9.6. Mechanisms and Kinetics... 462
    9.6.1. Reaction Mechanism 462
    9.6.2. Kinetics... 467
9.7. Recent Developments... 467

Chapter 10.

Catalytic Oxidations

P. Davies, R.T. Donald, N.H. Harbord

10.1. Introduction 469

10.2. Ammonia Oxidation	470
10.2.1. History of Nitric Acid Production	470
10.2.1.1. Routes from Atmospheric Nitrogen	470
10.2.1.2. Ammonia Oxidation	471
10.2.2. Chemistry of the Modern Process	477
10.2.3. The Chemistry of Absorption	477
10.2.4. Nitric Oxide Oxidation Chemistry	478
10.2.5. Ammonia Oxidation Chemistry	479
10.2.6. Modern Plants	482
10.2.7. The Burner Gauze—Platinum/Rhodium Catalyst	484
10.2.7.1. Gauze Activation	484
10.2.7.2. Gauze Deactivation and Cleaning	488
10.2.7.3. Metal Recovery	489
10.3. Methanol Oxidation	490
10.3.1. Introduction	490
10.3.2. The Silver-catalysed Process	490
10.3.2.1. Silver-catalysed Reactions	493
10.3.2.2. Selectivity	494
10.3.2.3. Poisoning	494
10.3.2.4. Composition of Reaction Gases	497
10.3.3. The Metal Oxide-catalysed Process	499
10.3.3.1. Metal Oxide-catalysed Reactions	501
10.3.3.2. Composition of Reaction Gases	502
10.3.4. Future Process Developments	502
10.4. Sulphur Dioxide Oxidation	503
10.4.1. Introduction	503
10.4.2. Thermodynamics	503
10.4.2.1. Equilibrium Calculations	503
10.4.2.2. Application to the Contact Process	506
10.4.3. The Contact Process	507
10.4.3.1. Vanadium Catalysts	508
10.4.3.2. The Modern Sulphuric Acid Plant	510
10.4.4. Mechanisms and Kinetics	511
10.4.5. Catalyst Poisoning	514
10.4.6. Disposal of Used Vanadium Catalysts	517
10.4.7. Possible Further Developments	517

Appendices

1. Further Reading	519
2. Numerical Examples of the Use of Equations Derived in Chapter 2	525
3. ICI Catalysts for the Production of Hydrogen, Ammonia and Methanol	528
4. Pigtail Nipping	530
5. ICI Technical Publications	532
6. Equilibrium Constants: for the Methane-Steam Reaction at Various Temperatures	537
7. Equilibrium Constants: for the CO Conversion Reaction (Shift) at Various Temperatures	543

8. Nomograph of Selected Properties of Ammonia 549
9. Thermodynamic Properties of Elements and Compounds at 298.15K ... 550
10. Physical Properties of Methanol 553
11. Approximate Boiling Ranges of Hydrocarbon Feedstocks............... 554
12. Monitoring Steam Reformer Tube Wall Temperature 555
13. Heat Released During Catalyst Reduction................................. 557
14. Heat Released During Catalyst Oxidation 558
15. Temperature Conversions 559
16. Specific Heats of Catalysts... 561
17. Atomic Weights of the Common Elements 562
18. Measurement of Pressure Drop Across Steam Reformer Tubes........ 565
19. Charging Primary Steam Reformer Catalyst—a Case Study 568
20. Equilibrium Constants for the Reaction for Zinc Oxide
    with Hydrogen Sulphide 572
21. Temperature Measurements in Catalyst Bed 574

References 576

Index 591

Preface

The first edition of *Catalyst Handbook* was published twenty years ago. It contained fundamental information on heterogeneous catalysis and practical details about the catalysts and processes employed in the production of hydrogen and ammonia via the stream reforming of hydrocarbons. It was used extensively by industrialists, and also by those working in research and teaching institutions, who found it valuable because it was one of the very few easily accessible authoritative sources of information about industrial catalysis and the operation of catalytic processes.

Since the publication of the first edition, there have been significant advances in areas of catalysis and the technologies it covered, many of which originated from ICI's operations at Billingham, in the North East of England. As a result there have been numerous requests for an updated edition incorporating the new developments.

This second edition is very different from the first edition. The sections concerned with the catalysts and catalytic processes employed in the production of synthesis gas for making ammonia and hydrogen have been completely rewritten. In addition to these broadened main sections, there is a chapter on methanol synthesis as the technology is in many ways similar to that used in ammonia and hydrogen plants, having as it does the same initial steps for the production of synthesis gas from hydrocarbons. Conversion of methanol to formaldehyde is the largest outlet for methanol, and as it is often operated with methanol synthesis on the same site it is appropriate for it to be covered here. Similarly, sulphuric acid and nitric acid plants have long been associated with ammonia for the manufacture of nitrogen and compound fertilizers, and so brief sections dealing with them are included.

The production of this book was truly multi-disciplinary involving specialists with extensive experience of catalyst development, catalyst manufacture, plant design, and plant operation. This brought together a team of research workers and technologists comprising chemists, physicists, metallurgists, chemical engineers and others. The result is a handbook that is intended primarily for people working in the chemical industry. However, it has been written with other readers in mind, so that it can be used as a reference source for research and teaching purposes in universities.

In addition to the named contributors, many other people at ICI both at Billingham and at Chicago, USA, helped in compiling this book, and the editor would like to extend his sincere thanks to them.

Chapter 1

Fundamental Principles

1.1. Fundamentals of Heterogeneous Catalysis

1.1.1. Introduction

The word "catalysis" is used to describe many phenomena but in all instances an agent (the catalyst) exerts a more-than-proportional influence over some change. Indeed, the word *catalysis*, coined by Berzelius in 1836 to describe some enhanced chemical reactions,[1] is now used popularly in a non-technical way. In this book we shall restrict use of the term "catalysis" to the heterogeneous catalysis of gas-phase reactions; that is, the promotion of reactions between gases by the use of solid catalysts. We shall be concerned mainly with a few reactions of great importance in the heavy inorganic chemicals industry. The simplicity of the overall chemistry of these reactions usually hides a bewildering complexity of separate reaction steps on the catalyst surface. The discussion in this chapter is limited to a few general principles, which are illustrated by examples from the processes described in later chapters. Further study, essential for any full understanding of heterogeneous catalysis, can be undertaken using the books listed in Appendix 1. All of the topics of this chapter are treated comprehensively in a recent book by Richardson[1a].

Catalysts are frequently defined as materials which accelerate chemical reactions without themselves undergoing change. As the manager of any plant using a catalytic process knows, this is too optimistic a definition: the properties of all real catalysts do change with use. The definition is also unsatisfactory in a more general way, for it implies that the acceleration is brought about without direct involvement of the catalyst in the process. Rather than attempting to produce an alternative succinct definition, we can list the essential features of the heterogenous catalysis of gas-phase reactions.

1. The presence of a solid material (the heterogeneous catalyst) changes the rate of an overall chemical reaction. The term "solid" is subject to some qualification, as the sulphuric acid catalyst (see Chapter 10) is a melt held in a porous solid and some other catalysts also contain mobile phases, e.g. some potassium salts in steam reforming catalysts (see Chapter 5). The overall chemical reaction

concerns the gas-phase species only, but this does not preclude the involvement of the solid in the formation of intermediate species—indeed, this is essential for heterogeneous catalysis.

2. The products of the catalysed reaction can, at least in principle, be obtained from an uncatalysed reaction under the same conditions. There is, therefore, no way of using catalysis to "cheat" equilibrium. In practice, however, the uncatalysed reaction may be immeasurably slow or may give very different products.

3. Any useful catalyst must have a high productivity. Thus, we expect 1 tonne of catalyst to "make" many tonnes of product. Alternatively, to look at catalysis on the atomic scale, the catalysed reaction steps must occur many times at the reaction site on the catalyst surface before catalytic activity is lost.

4. The catalysed reaction steps take place very close to the solid surface. These steps may be between gas molecules adsorbed on the catalyst surface, or extensive reaction can take place involving the topmost atomic layers of the catalysts. The influence of the solid does not effectively extend more than an atomic diameter into the gas phase, and the direct involvement of atoms below the topmost layers is not usually possible.

1.1.2. The Role of Catalysis

The synthesis of ammonia and its subsequent oxidation to nitric oxide give illustrations of two ways in which catalysts are essential to the chemical industry.

1.1.2.1. Ammonia synthesis

Ammonia synthesis from nitrogen and hydrogen occurs by the overall reaction shown in equation (1).

$$N_2 + 3H_2 \rightleftharpoons 2NH_3 \qquad (1)$$

No reaction takes place with the reactants at ambient temperatures, and even with an active catalyst a temperature of some 400°C is needed to obtain commercially useful rates of reaction. Nothing happens in the system without a catalyst as the temperature is raised until, at temperatures higher than 1000°C, a significant proportion of the hydrogen molecules are dissociated into atoms, as shown in equation (2). For example, at 1430°C, with a pressure of molecular hydrogen of 150 bar, the partial pressure of atomic hydrogen would be 0.1% of the H_2 pressure, i.e. 0.15 bar. Even this does not provide a mechanism for fixing nitrogen, for the reaction of hydrogen atoms with nitrogen molecules is very slow.[2] Only above 3000°C, where the even more strongly-bound nitrogen molecules start dissociating to atoms, equation (3), does nitrogen fixation become possible.

$$H_2 \rightleftharpoons H + H \qquad (2)$$

$$N_2 \rightleftharpoons N + N \qquad (3)$$

If each nitrogen atom formed in this way gave a molecule of ammonia by subsequent reactions with hydrogen molecules or atoms, then the rate of ammonia formation can be calculated from the rate of reaction (3).[2] A 100 m^3 reactor containing a 3:1 hydrogen/nitrogen mixture at a total pressure of 200 bar and a temperature of 3150°C would give 1300 tonnes of ammonia per day.

However, this simple analysis ignores the reverse reaction. Reaction (1) is an equilibrium reaction, and ammonia synthesis is favoured by low temperatures and high pressures. Thermodynamic data[3] can be used to show that the partial pressure of ammonia under the conditions above cannot exceed 0.07 bar, so with a gas space velocity of 10^4 h^{-1}, the make rate of ammonia would be only 6 tonnes day^{-1}.

The role of the catalyst in ammonia synthesis is therefore that of making the reaction go sufficiently fast (by facilitating the dissociation of molecular nitrogen) so that significant rates are obtained under conditions where the equilibrium conversion is large enough to be useful (Figure 1.1).

1.1.2.2. Ammonia oxidation

The question of selectivity does not arise in ammonia synthesis, because there is only one stable product that can be made from nitrogen and hydrogen. Ammonia oxidation, by contrast, can give various products depending on reaction conditions. These include molecular nitrogen (N_2), nitric oxide (NO), nitrogen dioxide (NO_2), nitrous oxide (N_2O), hydrazine (N_2H_4), nitrous acid (HNO_2) and nitric acid (HNO_3). For nitric acid manufacture, the selective formation of nitric oxide as an intermediate is essential (see Chapter 10). The catalyst in this process is required, not to accelerate ammonia conversion (the simplest way of oxidizing ammonia is to burn it in a flame), but to give selectivity in the oxidation to the wanted product, nitric oxide (Figure 1.2):

$$2NH_3 + {}^5/_2O_2 \rightarrow 2NO + 3H_2O \qquad (4)$$

$$2NH_3 + {}^3/_2O_2 \rightarrow N_2 + 3H_2O \qquad (5)$$

This selectivity is gained by an increase in the rate of reaction (4), relative to the unselective reaction (5). The difficulty of achieving this aim by means other than the subtleties of catalysis becomes apparent

Chapter 1. Fundamental Principles

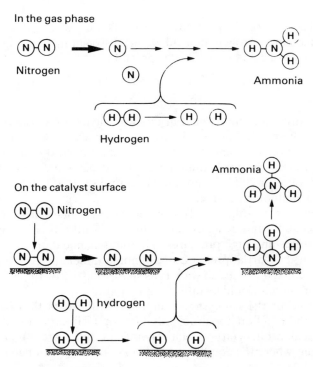

Figure 1.1. Comparison of the reaction steps in the synthesis of ammonia by homogeneous and heterogenous (catalytic) routes. Suitable catalysts make the heterogeneous route fast enough to be useful.

when it is realized that reaction (5), giving the more stable products, nitrogen and water, has the stronger thermodynamic driving force. Moreover, it can be seen from the stoichiometries of the reactions that reaction (4) requires more oxygen than the unselective reaction. Selectivity cannot therefore be achieved by limiting the extent of oxidation.

Despite the wide range of chemical processes in use which would not be possible without catalysis, catalysts are not quite the modern equivalents of the philosophers' stone. There are a number of transformations which would be desirable industrial processes if only one could get them to go, or, in some cases, get them to go faster or give different products. This can be demonstrated by three examples, all related to nitric acid manufacture (Figure 1.2).

The direct oxidation of ammonia to nitric acid is theoretically possible, since reaction (6) lies well on the right-hand side. In practice,

$$NH_3 + 2O_2 \rightleftharpoons HNO_3 + H_2O \qquad (6)$$

1.1. Fundamentals of Heterogeneous Catalysis

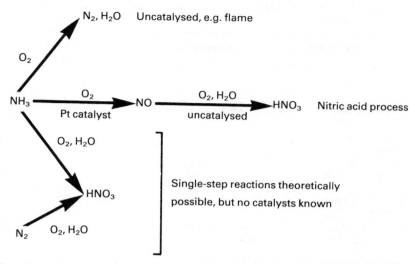

Figure 1.2. Various process routes, all theoretically possible, to make nitric acid from ammonia or molecular nitrogen. Suitable catalysts achieve the necessary selectivity in the conversion of ammonia but none has yet been found for the nitrogen-based reactions.

no suitable catalyst has been found for reaction (6), which would need to be operated at low temperatures to avoid decomposition of nitric acid. The principles of the nitric acid process are described in detail in Chapter 10. The commercialization of reaction (6) awaits the discovery of a new ammonia oxidation catalyst which is much more active than currently available ones.

The conventional nitric acid process provides the second example of the "limits" of catalysis. It is interesting that in the series of process steps by which natural gas is converted first to ammonia and then to nitric acid, a catalyst is used to bring about each chemical change up to and including nitric oxide formation (Table 1.1). The subsequent reactions, which also involve simultaneous absorption in water, consist of the oxidation to NO_2 and the reaction with water and oxygen to give nitrous acid and, finally, nitric acid. No catalysts are used in these stages (Chapter 10). Reaction (7) is a gas-phase process and requires no energy of activation:[2] its slowness is due to the infrequent occurrence at ambient temperatures of the three-body (i.e. $NO + NO + O_2$) collisions which are necessary for the reaction. Here there is no energy barrier to be decreased.

$$2NO + O_2 \rightarrow 2NO_2 \qquad (7)$$

Chapter 1. Fundamental Principles

Table 1.1. Process steps in the manufacture of ammonia and nitric acid from natural gas

Process step	Heterogeneous catalyst	Catalytic properties required
Hydrodesulphurization $CH_3SH + H_2$ $\rightarrow CH_4 + H_2S$	Sulphided Co/Mo/Al_2O_3 + ZnO (ICI Catalysts 41-6 and 61-1)	Activity, life
Primary steam reforming $CH_4 + H_2O$ $\rightarrow CO, CO_2 + H_2$	Ni/Al_2O_3 (ICI Catalysts 46-1, 46-4, 46-9 and 57-3)	Activity, life
Secondary steam reforming $CH_4 + O_2 + H_2O$ $\rightarrow CO, CO_2 + H_2$	Ni/Al_2O_3 (ICI Catalysts 54-3 and 54-4)	Activity, life
High temperature water-gas shift $CO + H_2O$ $\rightarrow CO_2 + H_2$	Fe_3O_4/Cr_2O_3 (ICI Catalysts 15-4 and 15-5)	Activity, selectivity, life
Low temperature water-gas shift $CO + H_2O$ $\rightarrow CO_2 + H_2$	Cu/ZnO/Al_2O_3 (ICI Catalyst 53-1)	Activity, selectivity, life
CO_2 removal	None	
Methanation $CO, CO_2 + H_2$ $\rightarrow CH_4 + H_2O$	Ni/Al_2O_3 (ICI Catalyst 11-3)	Activity, life
Ammonia synthesis $N_2 + 3H_2$ $\rightarrow 2NH_3$	Fe, promoted by K, Ca, Al_2O_3 (ICI Catalysts 35-4 and 35-8)	Activity, life
Ammonia oxidation $2NH_3 + {}^5/_2 O_2$ $\rightarrow 2NO + 3H_2O$	Pt/Rh	Activity, selectivity, life
Nitric acid formation $2NO + H_2O + {}^3/_2 O_2$ $\rightarrow 2HNO_3$	None	

Sometimes, however, the failure to achieve any catalysis is a result of the lack of reactivity of the raw materials, and then the possibility of achieving a new process persists as a gleam in the catalyst researcher's eye. In 1923, Lewis and Randall[4] calculated from the energetics of reaction (8) that starting with water and air it should be possible to form

$$H_2O + N_2 + 5/2 O_2 \rightarrow 2HNO_3 \qquad (8)$$

nitric acid until it reaches a concentration of about 0.1 mol l^{-1}. Clearly, this reaction could form the basis of the cheapest nitric acid process possible. However, nitrogen is unreactive, and those few catalysts which can activate nitrogen are poisoned by water and oxygen, so it is hardly surprising that no-one in the past 60 years has succeeded in making any nitric acid by reaction (8). Finally "it is to be hoped that nature will not discover a catalyst for this reaction, which would permit all the oxygen and part of the nitrogen of the air to turn the oceans into dilute nitric acid".[4]

1.1.3. The Nature of the Catalytic Process

The good catalyst has three cardinal virtues; those of *activity*, *selectivity* and *life*. While the general meaning of these terms is obvious, it is useful to define them a little more closely, especially with respect to plant performance (Table 1.2).

Table 1.2. Essential catalyst properties

Property	Definition
Activity	Ability of the catalyst to convert feedstock to (various) products, e.g. kg reactant converted kg^{-1} (or l^{-1}) catalyst h^{-1} percentage change in reactant (conversion) number of molecules reacting s^{-1}/active site (turnover number, TON, or turnover frequency, TOF) NB: All measurements referred to standard conditions
Selectivity	Ability of the catalyst to give the desired product, out of all possible products, e.g. product as percentage of reactant converted (yield)
Life	Time for which the catalyst keeps a sufficient level of activity and/or selectivity

1.1.4. Catalyst Activity

The "space" available for heterogeneous catalysis is smaller than the total volume of the reactor. Although catalysts are designed with a large surface area, sometimes as high as 300 $m^2\ g^{-1}$ of catalyst, the influence of the catalytic surface extends only about one atom diameter from the surface. For example, a 100 m^3 reactor containing a typical catalyst with surface area of 50 $m^2\ g^{-1}$ will contain, in a minutely convoluted form, some 5000 km^2 of catalyst surface. That is equivalent to the areas of Oxfordshire and Cambridgeshire combined, or of the State of Delaware. If the reaction space is all within 0.5 nm of this surface, then the volume of this space is only 2.5 m^3, or 2.5% of the reactor volume. The available reaction space in the gauze catalysts used for ammonia oxidation is some orders of magnitude smaller.

Although some concentration of reactants takes place in the adsorbed phase on the catalyst surface, the increase in reaction rate which this brings about rarely does much more than compensate for the decrease in reaction space. The large increases in the reaction rate given by active catalysts are almost entirely due to the reduction of energy barriers. The rates of catalysed reactions, like those of other chemical reactions, are exponential functions of the energy barriers.

This effect can be demonstrated for ammonia synthesis. First of all, we show how effective the catalyst is from a comparison of the catalysed and uncatalysed rates under comparable conditions. The rate of ammonia synthesis over the face of a single iron crystal has been carefully measured[5] with a 1:3 N_2/H_2 mixture at 20 bar and 525°C. If we assume (Chapter 8) that the reactor is filled with a typical ammonia synthesis catalyst that has about 5 m^2 of active catalyst surface per cubic centimetre of reactor space, then the rate of formation of ammonia is about 5×10^{-3} mol $NH_3\ cm^{-3}\ s^{-1}$. The uncatalysed reaction requires the dissociation of nitrogen molecules, and the rate of uncatalysed ammonia synthesis can then be shown to be about 10^{-52} mol $NH_3\ cm^{-3}\ s^{-1}$, i.e. some 5×10^{-49} times slower than the catalysed reaction. The extreme slowness of uncatalysed ammonia synthesis can be shown in the following way. A typical reactor in an ammonia plant with a volume of about 100 m^3 will make some 1500 tonnes of NH_3 per day. If we imagine a reactor of the size of the solar system (assumed to be a sphere of radius equal to the mean distance of Pluto from the Sun) in which ammonia is being made by the uncatalysed reaction, under typical synthesis conditions of 200 bar and 500°C, the rate of uncatalysed synthesis would be only about 0.1 g $NH_3\ day^{-1}$.

The reaction profiles for the reaction and the uncatalysed gas reaction are shown in Figure 1.3, largely derived from the fundamental studies of

1.1. Fundamentals of Heterogeneous Catalysis

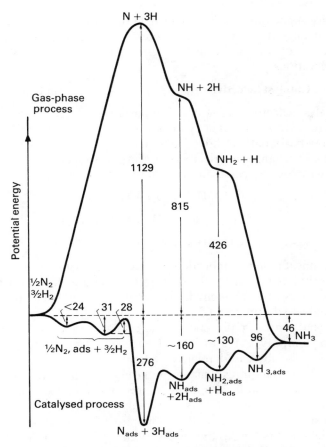

Figure 1.3. Energy profiles for the series of reaction steps to make ammonia from nitrogen and hydrogen by both homogeneous gas-phase and iron-catalysed reactions. The role of the catalyst in decreasing the energy barrier to reaction can be seen (numerical values are kJ mol^{-1}).

Ertl and his co-workers.[6–9] The energy barrier for reaction is reduced by the catalyst from 942 kJ mol^{-1}, the energy required to dissociate the N_2 molecule in the gas phase, to about 13 kJ mol^{-1}. This decrease should give an increase in reaction rate of about 10^{60} times, from the exponential dependence of reaction rate on activation energy. This factor is larger than the ratio of catalysed to uncatalysed reaction rates, and the explanation is that the ammonia synthesis catalyst has a "built-in inefficiency": only about one N_2 molecule in 10^6 with sufficient energy to dissociate on the iron surface actually does so. In this comparison of the ratio of catalysed and uncatalysed ammonia synthesis we have taken a very simple approach, using single reactions. A computer model[10] of the catalysed reaction under working conditions, derived from the kinetics of the individual reaction steps, agrees well with plant observations.

1.1.5. Catalyst Selectivity

Catalyst selectivity is of equal importance to activity in most catalytic reactions (not ammonia synthesis, of course), and smaller changes in energy barriers than those found in ammonia synthesis are sufficient to give large changes in selectivity. As an oversimplification (Chapter 10) we can write two processes in which methanol is oxidized:

$$CH_3OH + \tfrac{1}{2}O_2 \rightarrow HCHO + H_2O \qquad (9a)$$

$$CH_3OH \rightarrow HCHO + H_2 \qquad (9b)$$

$$CH_3OH + \tfrac{3}{2}O_2 \rightarrow CO_2 + 2H_2O \qquad (10)$$

The uncatalysed combustion of methanol with air at about 500°C corresponds broadly with reaction (*10*), although some CO, together with traces of formaldehyde, is formed. The silver catalyst used in the formaldehyde process (Chapter 10) gives formaldehyde, by reactions (*9a*) and (*9b*). The presence of the catalyst increases the formaldehyde yield from, say, 0.1% to 95%. To achieve this the rate of reactions (*9*) have to be increased relative to reaction (*10*) by a factor of about 2×10^4. This requires a change in the energy barrier to reaction of about 60 kJ mol^{-1}—well within the range of decreases in energy barriers brought about by catalysts. A catalyst may change the rate of the undesired reaction as well as the rate of production of the wanted product, but as long as the necessary change in relative rates is achieved, then the catalyst will be selective.

As energy barriers are dominant in determining catalyst selectivities, various general patterns of selectivity can be deduced. By-products are almost always the products of reactions with higher energy barriers than those of the main reaction, so accounting for their relatively slow formation. It follows that an increase in temperature usually accelerates by-product formation more than main product formation, i.e. product selectivity decreases. A comparison of the ICI low-pressure methanol process with the older, high-pressure process illustrates this point.

Methanol synthesis can be described by the two overall equations (*11*) and (*12*).

$$CO/CO_2 + H_2 \rightarrow CH_3OH \qquad (11)$$

$$CO/CO_2 + H_2 \rightarrow \text{by-products} \qquad (12)$$
$$(CH_4, HCOOCH_3, C_2H_5OH, \text{etc.})$$

1.1. Fundamentals of Heterogeneous Catalysis

The Cu/ZnO/Al$_2$O$_3$ catalyst developed by ICI has a very much higher activity for reaction (*11*) than the earlier ZnO/Cr$_2$O$_3$ catalyst had, and as the rate of reaction (*12*) was not accelerated as much as that of reaction (*11*) there was an immediate improvement in selectivity (Chapter 9). However, the high activity allowed operation at lower temperatures and pressures, so this in turn brought about a further improvement in selectivity.

The relationship between activity and selectivity can be more complex than these examples show. In ammonia oxidation high selectivity depends on high activity, but for reasons which are different from those in methanol synthesis (Chapter 10). In some other reactions, especially organic oxidations, high activity is frequently associated with low selectivity. We can again take methanol oxidation as an illustration, and add the reaction

$$HCHO + O_2 \rightarrow CO, CO_2, H_2O \qquad (13)$$

to reactions (*9*) and (*10*). Thus, the requirement for high selectivity is that reactions (*9*) should be faster than reactions (*10*) and (*13*) (Figure 1.4). A very active oxidation catalyst, such as platinum, has low energy barriers for all three reactions, so selectivty is poor and oxidation proceeds to CO, CO$_2$ and H$_2$O. Less-active oxidation catalysts—for instance, silver or Fe/Mo oxides—are used in formaldehyde processes, and with these catalysts the differences in energy barriers are high enough to ensure selectivity for formaldehyde.

1.1.6. Steps in the Catalytic Process

So far we have made the implicit assumption that process rates are controlled solely by the rates of key reactions on the catalyst surface, but this is not always so. For example, suppose the addition of a new promoter to the iron catalyst for ammonia synthesis greatly increased the proportion of nitrogen molecules which dissociated on colliding with the surface. If the proportion rose from 1 in 10^6 to almost all, would this give a million-fold increase in the rate of ammonia synthesis—and a reduction in the size of plant converter from 100m^3 to, say, 100 cm^3? Even apart from the problems of heat transfer and fluid flow in trying to make 1500 tonnes of NH$_3$ per day in a reactor the size of a wine glass, the answer is no. The separate stages of a heterogeneously catalysed gas-phase reaction are shown in Table 1.3.

Any of these stages, if slow, may limit the overall rate of a catalytic reaction. Distinctions are often drawn between catalysts which are "film-diffusion controlled" (i.e. limited by stages 1 and/or 7), "pore-diffusion controlled" and "reaction controlled" (i.e. limited by

Chapter 1. Fundamental Principles

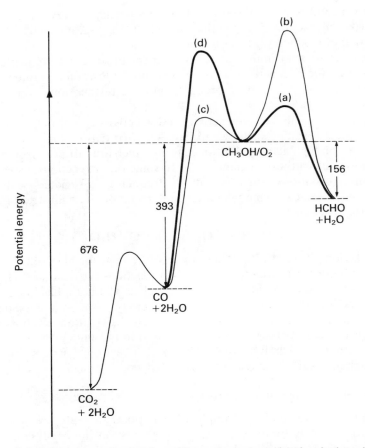

Figure 1.4. Schematic representation of the various energy barriers in the selective and unselective oxidation of methanol. A suitable choice of catalyst (for example silver) with appropriate energy barriers (a) and (b) can give high selectivity to formaldehyde. Numerical values are kJ mol^{-1}.

stages 3, 4 and/or 5). However, the development of the best industrial catalysts usually leads, either by design or empiricism, to the elimination of any strongly-limiting constraint on catalyst performance, so these simple distinctions rarely apply in practice. The concentration profiles of reactants and products in and around a typical catalyst pellet are shown in Figure 1.5. The combined effects of diffusion and reaction give significant concentration gradients, both within and outside the catalyst pellets.

We can now see why a large increase in the rate of dissociation of nitrogen molecules on an ammonia synthesis catalyst would not give a

1.1. Fundamentals of Heterogeneous Catalysis

Table 1.3. Sequence of stages in the catalysis of a gas-phase reaction by a heterogeneous catalyst

1	Transport of reactants through the gas phase to the exterior of the catalyst pellet
2	Transport of reactants through the pore system of the catalyst pellet to a catalytically-active site
3	Adsorption of reactants at the catalytically-active site
4	Chemical reactions between reactants at the catalytically-active site (frequently several steps)
5	Desorption of products from the catalytically-active site
6	Transport of products through the catalyst pore system from the catalytically-active site to the exterior of the catalyst pellet
7	Transport of products into the gas phase from the exterior of the catalyst pellet

Notes

1. In this description the catalyst is assumed to be in the form of a pellet. The same stages apply to other types of catalyst particle (e.g. spheres made by granulation, or extrusions), but stages 2 and 6 clearly do not take place in catalysts made of wire gauzes or non-porous metal crystals, unless surface roughening becomes so advanced that in effect a pore system is formed.
2. Transport through the catalyst pore system can be either through the gas phase or across the interior surface of the catalyst surface.
3. Several different catalytically-active sites may be involved. Adsorption, possibly followed by reaction, may occur at one site, followed by transport of an intermediate product to a different site for further reactions. Transport of the intermediate product can be either through the gas phase, if the intermediate is a stable molecule, or across the catalyst surface (the "spillover" effect).

proportional increase in overall rate: other stages would become of more significance in controlling the process. Indeed, even with the present catalyst there is experimental evidence that under some industrial conditions this occurs. There is a complex interaction between the relative importance of these different stages and the resulting selectivity when several products are formed.

1.1.7. Adsorption and Desorption

Table 1.3 shows that adsorption and desorption are both essential and critical stages of the overall catalytic process. In the act of adsorption, a molecule approaches the solid surface from the gas phase and is held close to (or in) the surface. This is different from the collision and rebound that occurs at all solid surfaces in contact with gases. Thus, a molecule which stays on the surface for a time longer than that of a collision is said to be adsorbed. The removal of an adsorbed molecule is known as desorption. A distinction is drawn between adsorption, where the adsorbed material stays on (or at least close to) the solid surface,

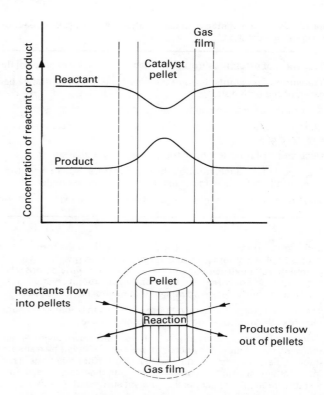

Figure 1.5. Concentration profiles of reactants and products around and in a typical catalyst pellet under reaction conditions where mass-transfer control is significant.

and absorption, where the absorbed material "soaks" into the bulk of the solid.

Different types of adsorption are listed in Table 1.4 and are shown for carbon monoxide in Figure 1.6. We can see the differences by considering the adsorption of carbon monoxide on α-alumina, copper metal and nickel metal surfaces. There is no special bonding between CO and an α-Al_2O_3 surface, so physical adsorption takes place. The energy of adsorption is of the same order as that between CO molecules in liquid CO, so the physisorption is extensive at low temperatures only. Physisorption is often important in catalysis as a precursor to chemisorption. Associative chemisorption occurs with CO on copper because the energy of interaction is greater than that of physisorption. Various adsorbed states have been identified, with different adsorption energies, and although the C–O bond strength is weakened no breaking of this bond takes place on copper surfaces. This is significant in the hydrogenation of CO to methanol, when dissociation of the C–O bond

1.1. Fundamentals of Heterogeneous Catalysis

Table 1.4. Types of adsorption

Type of adsorption	Characteristics
1 Physical adsorption (physisorption)	Unselective. Low energy of adsorption. Extent of adsorption related to boiling point of gas, not nature of solid surface. No breaking of bonds in molecules and negligible changes in bond energies
2 Associative chemical adsorption (chemisorption)	Selective, strongly dependent on both gas and solid surface. Higher energies of adsorption than those of physisorption. Bonds in the adsorbed molecules are changed in strength but not broken, i.e. molecule adsorbed whole
3 Dissociative chemical adsorption (chemisorption)	Selective, strongly dependent on both gas and solid surface. Higher energies of adsorption than those of physisorption. Bonds in the adsorbed molecules are broken, i.e. molecule adsorbed as two or more molecular fragments

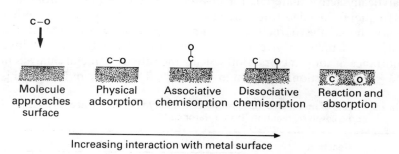

Figure 1.6. Schematic representation of the adsorption, and possible subsequent reaction, of carbon monoxide on various solid surfaces.

could lead to the formation of methane and water (Chapter 9). The removal of CO in the methanation reaction (Chapter 7) does require its conversion to methane and water as in equation (*14*)

$$CO + 3H_2 \rightarrow CH_4 + H_2O \qquad (14)$$

which is catalysed by nickel. Here dissociative chemisorption of CO occurs and the CO molecule is broken on adsorption as in equation (*15*) thus facilitating the methanation reaction.

$$CO \rightarrow C_{(a)} + O_{(a)} \qquad (15)$$

1.1.8. Catalyst Design

In the previous section we saw how the patterns of adsorption on different solids led to the choice of appropriate metals as catalysts for different reactions. The design of a catalyst covers all aspects from the choice of catalytically-active material to the method of forming particles. This exercise can be detailed, rigorous and extensive, as described by Dowden,[11] starting from fundamentals to get the best catalyst for a new process; but sometimes the design of a new catalyst may be only a minor modification of an existing industrial catalyst. The optimum catalyst is the one that provides the necessary combination of properties, including activity, selectivity and life, at an acceptable cost. These requirements always put conflicting demands on the catalyst designer, and much of the designer's art consists of the achievement of a suitable compromise[1a].

At the start of catalyst design a consideration of both desirable and undesirable reactions in the overall process leads to a choice of catalytically-active materials for possible catalysts. Suitable promoters are frequently needed to get adequate performance: these may either modify the catalyst structure, so improving stability, or enhance the catalytic reactions to give better activity or selectivity. The relative importance of some factors[12] influencing the activity of metal catalysts in some simple reactions is shown in Table 1.5. The nature of the metal is

Table 1.5. Effects of metal structure, promoters and type of metal on activity for some catalytic reaction (from reference 12)

Reaction	Effect[a] of		
	Structure	Promoters	Type of metal
$H_2 + D_2 \rightarrow 2HD$	VS	S	M
$C_2H_4 + H_2 \rightarrow C_2H_6$	VS	S	M
$Cyclo\text{-}C_3H_6 + H_2 \rightarrow C_3H_8$	VS	S	M
$C_6H_6 + 3H_2 \rightarrow C_6H_{12}$	VS	S	M
$C_2H_6 + H_2 \rightarrow 2CH_4$	S	L	VL
$N_2 + 3H_2 \rightarrow 2NH_3$	M	L	VL

[a]The order of magnitude of the effect is classified as follows:
$1 < VS < 10$, $10 < S < 100$, $100 < M < 10^4$, $10^4 < L < 10^6$, $10^6 < VL$.

always the most important factor. The dependence of intrinsic activity on the refractory support is usually small, but there are some remarkable exceptions. In most catalysts some refractory support material is needed to give adequate physical stability. This support has to be compatible with the active catalytic phases, both in the reactor and during catalyst manufacture. The maximum reaction rate is obtained by maximizing the available surface area of the active phase. The actual size and shape of the fabricated catalyst is influenced by considerations such as heat and mass transfer, and reactor pressure drop.

There is insufficient space here to go through the full design exercise for even one catalyst, but to give some impression of the factors involved we will look at catalysts for the high- and low-temperature water-gas shift processes. These processes are very similar, as they both involve reaction (*16*) but the two catalysts used have no component in common. In the high-temperature stage of the water-gas shift process (Chapter 6), both the temperature range (350–450°C) and the high partial pressure of CO (typically 1–3 bar) allow the use of catalysts of only moderate activity. The main critical requirement is selectivity—the avoidance of any measurable methanation [reactions (*14*) and (*17*)].

$$H_2O + CO \rightarrow H_2 + CO_2 \qquad (16)$$

$$CO_2 + 4H_2 \rightarrow CH_4 + 2H_2O \qquad (17)$$

Many metals catalyse the shift reaction, but without the necessary selectivity. Magnetite, Fe_3O_4, is sufficiently active for process purposes, has negligible activity for methanation and it is the chemically stable form of iron under the reaction conditions. It is also quite resistant to various poisons such as H_2S (which may be present in the process gas) because the sulphided form has significant activity. However, pure Fe_3O_4 of adequate surface area is not sufficiently refractory: the crystallites would sinter rapidly in use, with consequent loss of surface area and, hence, activity. A more refractory material has to be combined with the magnetite to inhibit sintering, and chromia, Cr_2O_3, has been found to be both sufficiently refractory and compatible. Some sintering of the active Fe_3O_4 phase still occurs, but it is sufficiently slow for the Fe_3O_4/Cr_2O_3 catalyst to have a useful plant life of several years.[13] The moderate activity of the Fe_3O_4/Cr_2O_3 catalyst allows the use of medium to large pellets (ICI Catalyst 15-4, 8.5 mm × 11.0 mm; ICI Catalyst 15-5, 5.4 mm × 3.6 mm) without too great a loss of activity from pore diffusion. The large pellets are beneficial in giving physically strong catalysts and a low reactor pressure drop.

Much higher catalyst activity is required for the low-temperature water-gas shift process because both temperature (200–250°C) and CO partial pressure are much lower. Magnetite is inadequate and a metal active phase has to be used. The need to avoid methanation restricts the choice among the more common metals to copper.[14] The disadvantages of copper as the active phase lie in the ease with which it both sinters and is poisoned, especially by chlorine and sulphur compounds. The use of zinc oxide and alumina as refractory supports greatly limits the sintering of the copper crystallites, and it gives two further benefits. The precipitation of mixed copper/zinc carbonates (Section 1.2.6) leads to the eventual formation of very small and well dispersed copper crystallites.

Zinc oxide also acts as an absorbant for catalyst poisons, thus further extending catalyst life. The optimum proportion of copper depends on the copper and support crystallite sizes.[15] A typical reduced LT shift catalyst contains about 20% Cu by volume, with both Cu and ZnO crystallites about 5 nm in size. A catalyst with about 40% Cu by volume would need the support crystallites to be <3 nm (impossible in practice) to give a greater copper area and hence greater activity. Medium-sized catalyst pellets are used (ICI Catalyst 53-1, 5.4 mm × 3.6 mm), partly to minimize the effect of pore diffusion, which is more significant than with a high-temperature shift catalyst because of the higher intrinsic activity, and partly to ensure that poisons are held in a concentrated zone next to the reactor inlet (see also Section 1.4).

1.2. Catalyst Manufacture

1.2.1. Introduction

Successful catalyst production necessarily involves the making of catalysts which fulfil all of the process requirements described in Section 1.1.8 and which can also be transported, charged to the reactor and activated. This has to be done reproducibly and at reasonable cost.

Catalyst makers achieve their aims by a judicious use of practical skills, knowledge of fundamental principles and many years' experience.[16] It is therefore not surprising that there is much less in the open literature on catalyst manufacture than there is on catalyst use, testing or fundamentals (but see references 17–19). Attention to detail in catalyst manufacture is crucial to success, because small changes in procedure can have large effects on catalyst performance. Some of the ways in which industrial catalysts are made are outlined below.

1.2.2. Unsupported Metals

In most processes which need metal catalysts a prime requirement is the use of a large metal surface area, and this area can be maintained only by the spreading of the metal across a refractory support (Section 1.1). Accordingly, few industrial catalysts are used in the form of unsupported metals, but there are some instances where the rate of the catalysed reaction is so fast that a small metal area is capable of supplying enough product, and so a support is not necessary. The other functions of a support (such as the absorption of poisons) are also necessarily absent, so this may impose operating conditions such as the high purity of methanol feed needed for a formaldehyde plant. Poisoning of platinum gauzes was one of the reasons for their replacement by the now conventional vanadium-based catalysts in sulphuric acid plants (Chapter 10).

One of the most widely used unsupported metal catalysts is the precious metal gauze as used, for example, in the oxidation of ammonia to nitric oxide in nitric acid plants. A very fast rate is needed to obtain the necessary selectivity to nitric oxide, so a low metal surface area and short contact time are used. These gauzes are woven from fine wires (a few thousandths of an inch in diameter) of platinum alloy, usually platinum–rhodium. Several layers of these gauzes, which may be up to 3 m in diameter, are used, often with "getter" gauzes to catch the platinum lost from the gauze in use (Chapter 10). Methanol oxidation to formaldehyde is another process in which an unsupported metal catalyst is used, but here the metal is normally in the form of a bed of granules, although gauzes are also used (Chapter 10). For optimum catalyst efficiency the silver is made as granules of agglomerated dendritic crystals.

Raney nickel catalysts (Section 1.2.3) (used in some organic hydrogenations) are sometimes regarded as unsupported metal catalysts, but their high metal areas are probably maintained by traces of residual alumina acting as a support.

1.2.3. Fused Catalysts

The manufacture of catalysts by fusion of either metals or oxides is not extensively used. Although clearly the technique can give very good mixing of all of the catalyst components (unless phase separation occurs during melting or subsequent cooling) and catalysts of high density can be obtained, problems arise in forming and activation. The product of fusion obviously has negligible surface area, so the internal areas necessary for catalysis can be developed only during activation; i.e. by the removal of some component of the catalyst, or some other method of generating porosity. However, in many cases where fusion routes might be potentially useful the removal of oxygen (or some other component) from the fused catalyst either occurs far too slowly or leads to collapse of the structure.

The most important catalyst made in this way is the promoted iron ammonia synthesis catalyst[20] (Chapter 8). The key feature which allows the use of a fusion method is the way in which the iron oxide, Fe_3O_4, is reduced by hydrogen. On reduction iron metal is formed as a porous solid of the same overall dimensions as the original iron oxide. For pure Fe_3O_4 the removal of oxygen corresponds to a weight loss of 28%, but because metallic iron is denser than Fe_3O_4 the resulting theoretical porosity is 52%. In practice, ammonia synthesis catalysts contain small amounts of other solids, so the weight loss and porosity achieved on reduction are less than theoretical.

Ammonia synthesis catalysts are made[16] by the fusion (in an electric furnace at ~1600°C) of magnetite, Fe_3O_4, of a suitably pure grade, together with the promoters, typically calcium, potassium and alumina, in a "triply-promoted" catalyst. The appropriate purity of the raw materials is of critical importance in fused catalysts because there is no subsequent stage in catalyst manufacture during which poisons can be removed (as can be done with precipitated catalysts). The only opportunity for the removal of poisons is during catalyst activation (Section 1.4). Conventional forming techniques (pelleting, extrusion and granulation) are not practical with fused catalysts, so the common practice is cooling of the melt, followed by crushing and size grading, to give particles of the required dimension ranges (Chapter 8). A flow sheet for the process is shown in Figure 1.7. Very good mixing of the components is attained in the fusion stage, but some segregation, on a micro-scale, occurs in the cooled solid.[21] This separation of components is not sufficient to affect the final catalysts, since further migration to the optimum distribution for activity occurs on activation/reduction.[21]

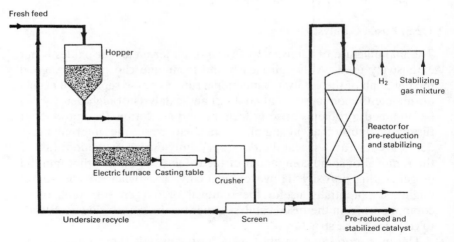

Figure 1.7. Flow sheet for a plant to make an ammonium synthesis catalyst by the fusion route.

Raney nickel[22] is prepared as a nickel–aluminium alloy by conventional metallurgical techniques. The aluminium is removed with aqueous sodium hydroxide, to leave porous nickel (of high surface area) which contains a small amount of stabilizing alumina, formed in the extraction process. The resulting catalyst is of high area (e.g. 100 m^2 g^{-1}) and activity, and is useful for mild, low-temperature hydrogenations, but it is sensitive to poisons and is rapidly deactivated at higher temperatures. Raney cobalt and Raney copper catalysts can also be made by the same techniques.

1.2.4. Wet Methods of Catalyst Manufacture

Most industrial catalysts are made (Figure 1.8) either by precipitation, when active phase and support are made together, or by the impregnation of an active phase on to a preformed support.[16–19, 23] The process used depends on many factors, such as the chemistry of the catalyst components and their possible precursors, the concentrations of different components required, physical strength required, reaction conditions of catalyst in use and the need and ease of removing contaminants. It is usually easier to get a high concentration of the catalytically active phase by precipitation than by impregnation processes, but the development of adequate strength can be more difficult. Sometimes the processes are combined, as when a precipitated catalyst is impregnated with potassium hydroxide solution to give the required level of alkali doping.

Figure 1.8. Aerial view of the ICI factory at Clitheroe (Lancashire, England) for the manufacture of catalysts.

1.2.5. Fundamentals of Precipitation Processes

Precipitated catalysts are generally prepared by rapid mixing of concentrated solutions of metal salts, and a typical flowsheet is shown in Figure 1.9. The product precipitates in a finely divided form of high surface area. Precipitated mixed hydroxides or carbonates are most frequently prepared. The reaction

$$2Ni(NO_3)_2 + 2NaOH + Na_2CO_3 \rightarrow Ni_2(OH)_2CO_3 + 4NaNO_3 \quad (18)$$

is typical, but the precipitates are usually non-stoichiometric and often amorphous. After filtration and washing the precipitate is dried and heated to decompose the hydroxides/carbonates to the corresponding oxides. The final size of crystallites present in precipitated catalyst are typically in the range 3–15 nm, while overall surface areas can be 50–200 $m^2 g^{-1}$ or more. These values can change markedly in use (Section 1.4), but the activity of precipitated catalyst is nevertheless usually high.

The main aims in the use of a precipitation process for catalyst manufacture are the intimate mixing of the catalyst components and the formation of very small particles to give a high surface area. The necessary degree of mixing can be achieved either by the formation of very small crystallites, in close proximity, of the different components or by the formation of mixed crystallites containing the catalyst constituents. Hydroxides, carbonates or basic carbonates are the favoured precipitated intermediates for the following reasons:

1. The solubilities of these salts of transition metals and other catalytic components are very low. Conseqently, very high supersaturations, leading to very small precipitate particle sizes, can be reached.

2. The solubilities of the precursors, typically metal nitrates and sodium hydroxide or carbonate, are high, so concentrated solutions can be used, again giving high supersaturations.

3. Hydroxides and carbonates are readily decomposed, by heat, to oxides of high area without leaving catalyst poisons as, for example, sulphur residues from the calcination of sulphates.

4. Many mixed hydroxides, carbonates and hydroxycarbonates are known, so there is a good chance of getting a mixed compound of the required composition for given components.

5. Environmental difficulties arising from the calcination of hydroxides and carbonates are minimal.

1.2. Catalyst Manufacture

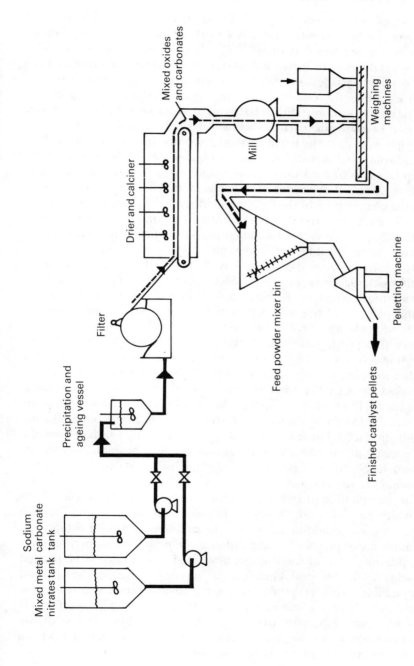

Figure 1.9. A flow sheet for a plant to make catalysts by the precipitation route.

1.2.6. Catalyst Manufacture by Precipitation Processes

An example of the complexity of development work with precipitated catalysts can be taken from the ICI work on the copper/zinc/alumina catalysts for methanol synthesis.[24, 25] Earlier work had shown that copper catalysts were active but unstable, so activity was rapidly lost. The copper particles in these catalysts are the essential catalytic material, while the other components are the stabilizers. In earlier catalysts made by conventional batch methods these stabilizers were not effective because the particles were not small enough and were not uniformly mixed. In the batch method involving the addition of alkali to an acid solution, the catalyst which is obtained at the beginning of the precipitation is formed under acid conditions, whereas later, as more alkali is added, it is produced under alkaline conditions. This affects both the composition of the catalyst and the size of the particles (Figure 1.10). The catalyst produced under the acid conditions at the beginning of the precipitation is rich in aluminium and deficient in zinc, while at the end, under alkaline conditions, the catalyst is low in copper. The catalyst particle size is also affected by the pH: under acidic and alkaline conditions large particles are produced, but under neutral conditions they are much smaller. The best catalyst is obtained by precipitation at around pH 7, and this was achieved by a precipitation procedure in which the acidic and alkaline solutions were mixed continuously.

Very fine precipitates can, however, cause severe problems in the later stages of processing and much skill and empirical development is needed to reach the optimum process. If the precipitate is too fine it may pass through the filter or block the filter pores. Even when a satisfactory filter cake is made, the removal of sodium and other unwanted ions can be difficult. Salt solutions held in the filter cake are readily displaced, but salts can be *adsorbed* on the high surface area of a fine precipitate. Washing with water may not then be enough to reduce sodium levels to the specification for the final catalyst, and special techniques have to be used.

The precipitation process outlined above represents the simplest form in which it is used, many catalyst production routes involve more complex or additional steps. The addition of a solid component either during or just after precipitation is common. Ageing periods after precipitation, or after washing out unwanted materials, may be needed to achieve the desired chemical or physical state of the catalyst intermediate. This may not always be advantageous because some crystal growth may occur in a precipitate in contact with the filtrate. Crystallization, especially from concentrated solutions, is one of the less-well understood chemical processes, so scale-up (Figure 1.11) from laboratory experiments has to be done with care.

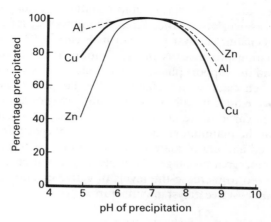

Figure 1.10. Variation with pH of the properties of copper/zinc oxide catalysts prepared by precipitation. Deviation from the optimum value of pH affects composition of the catalyst, and the size of the particles, and other physical properties.

Figure 1.11. Semi-technical unit for catalyst preparation, an intermediate stage in scale-up from laboratory to full-scale manufacture.

1.2.7. Impregnation Processes

As indicated in Section 1.2.4, the choice of a precipitation or an impregnation route for catalyst preparation is determined by a complex interplay between many different factors.[16, 25] In Table 1.6 several

Chapter 1. Fundamental Principles

examples are given where impregnation techniques have marked advantages over other procedures. The flowsheet for a plant for making an impregnated catalyst is shown in Figure 1.12. The great virtue of impregnation processes is the separation of the making of the active phase and the support phase, which is clearly not possible in the case of precipitated catalyst manufacture. The support is normally a porous refractory oxide, usually fired at a high temperature to give stability. High-area carbons are used for some impregnated catalysts, especially metals of the platinum group. The disadvantage of impregnation lies in the limited amount of material which can be incorporated in a support by an impregnation stage. Multiple impregnation/drying/firing cycles can be used to increase the levels of active components, but only at a significant cost in catalyst manufacture.

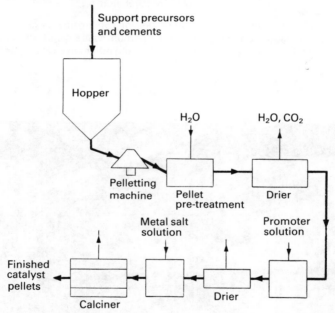

Figure 1.12. Flow sheet for a plant for the manufacture of catalysts by the impregnation of pre-formed supports with a compound of the active metal.

In the absence of specific interactions between the preformed support and the components of the impregnating solution, the impregnation process stages can be described quite simply. A solution is made up containing the component to be put on the catalyst. In the next stage either the support is dipped into a tank of this solution or the solution is sprayed on (or added to) the support—the latter procedure is sometimes

1.2. Catalyst Manufacture

Table 1.6. Catalysts made by impregnation processes

	Catalyst	Reason for use of impregnation techniques
1	Supported platinum-group metal catalysts	Good metal dispersion and no loss of metal within support phases
2	Catalysts for high-temperature processing	Support can be stabilized to resist reaction conditions before active phase incorporated
3	Catalysts with an uneven distribution of components, e.g. active phase on exterior	Precipitation methods give uniform catalysts, at least on a macro-scale: impregnation can concentrate components where required
4	Alkali-doped catalysts	Alkali would be removed at washing stages of precipitation route
5	Metal/zeolite catalysts	Usually more convenient to add metal by ion-exchange after zeolite synthesis rather than include metal in zeolite synthesis
6	Supported molten-salt catalysts	Usually more convenient to add molten salt phase (e.g. V_2O_5–K_2SO_4 for sulphuric acid catalysts) to preformed support, e.g. silica

described as the "incipient wetness" technique. In both methods the take-up of the solution is governed by the porosity of the support, so the level of active component incorporated in the catalyst is a function of the solution concentration and support porosity. After absorption of the solution into the pore system of the support, a drying stage is used to remove water. This has to be carried out so that the impregnated component remains within the support pore system and does not migrate to the exterior surface of the support. If this stage is done correctly, the support then has crystallites of the impregnated component, typically a metal nitrate, in the interstices of the pore system. Most impregnated catalysts are calcined in air after drying, thus converting the soluble salt to insoluble oxide. Calcination can also have other effects; for instance, the firing of a potash-doped catalyst for naphtha–steam reforming helps to "fix" the potassium by reaction with the refractory phases to give kalsilite, a potassium aluminosilicate.[26] This is resistant to hydrolysis by steam, and without this "anchor" the potash would be lost rapidly as volatile potassium hydroxide in steam during use.

Interactions between impregnating solution and support are commonplace, and indeed useful, in the impregnation process. Three

Chapter 1. Fundamental Principles

types of interaction can be recognized although the boundaries of the classification are somewhat arbitrary.

1. *Specific adsorption.* If the impregnating component (e.g. a metal ion) is more strongly adsorbed on the support surface than the molecules of the solvent (usually water), metal ions are removed from the impregnating solution as it passes into the support, with the result that the impregnated metal is concentrated in the outer layer of the support pellet. This "shell" catalyst can ensure that maximum use is made of the catalytic metal in reactions that are strongly limited by pore diffusion (Section 1.1). Further treatment with water and/or other solutions can give different distributions. For example, in coking or poisoning conditions there can be advantages in having the active phase in a layer below the pellet surface.

2. Some high-area solids (zeolites form the most important class) have significant *ion-exchange* capacity. Thus, metal ions in the impregnating solution will exchange with ions already present in the support surface before impregnation. Since this gives almost an atom-by-atom deposition over the surface, the metal distribution, initially at least, is very good. As with adsorption, the amount of metal taken up by the support is not simply determined by concentration and support porosity, but is a complex function of ion-exchange equilibria and diffusion rates in the liquid phase.[27] Ion-exchange, like less-specific adsorption, can be used to give non-uniform distributions of metals. Some high-area solids which are not conventional ion-exchange materials may nevertheless display ion-exchange properties; for example, the residual sodium in a precipitated alumina may be held in the surface and behave in this way.

3. A stage beyond ion-exchange is the possiblity of *reaction and precipitation* in the pores of the support. Thus, the residual sodium on the surface of a precipitated alumina may increase the pH of an impregnating solution to the point at which metal hydroxides are precipitated. While this technique can again be of use in getting a non-uniform distribution of metal, its unexpected occurrence can sabotage attempts to reach a uniform distribution.

1.2.8. Forming Stages

Precipitated catalysts and supports for impregnation need to be formed into suitably sized particles for use in the reactor. The size and shape of the catalyst pieces (Figure 1.13) is a compromise between the wish to minimize pore diffusion effects in the catalyst particles (requiring small sizes) and pressure drop across the reactor (requiring large particle sizes).

Figure 1.13. A selection of industrial catalysts, showing the variation in size and shape of catalyst particles necessary for different uses.

The three main processes used in catalyst manufacture to make conveniently sized particles from powders are pelleting (Figure 1.14), extrusion (Figure 1.15) and granulation (Figure 1.16). The products of each process are compared in general terms in Table 1.7. The choice of method depends of the size, shape and density of the catalyst particle required, on the strength required and on the properties of the starting powder. If the catalytic reaction is not pore-diffusion limited, or at least not very much so, in catalyst particles with dimensions of a few millimetres, then the maximum conversion from a reactor is obtained by filling it with as much catalyst as possible, and the catalyst particles should be of maximum density. (The usual range of densities available generally has little influence over the extent of pore-diffusion control.) It is clear from Table 1.7 that pelleting is the preferred forming technique for catalysts intended for this type of use.

On the other hand, the choice is different for processes in which the activity of the catalysts are strongly limited by pore-diffusion. Under

Chapter 1. Fundamental Principles

Figure 1.14. A typical machine used in the manufacture of catalyst pellets.

Figure 1.15. A typical machine used to extrude catalysts.

1.2. Catalyst Manufacture

Figure 1.16. A typical machine used to form catalyst particles by granulation.

these conditions only the outside layer of the catalyst particle is effectively used, so the maximum activity from a reactor is obtained with catalyst particles of maximum superficial area. As smaller particles can be made more easily by granulation or extrusion than by pelleting, these are the preferred methods of forming. Cross-sectional shapes of longer periphery than a circle (e.g. a "clover leaf") can give even larger superficial areas on extruded catalysts, and these have been used in some hydrodesulphurization catalysts. Occasionally just a decrease in pellet size can give a sufficient increase in activity under pore-diffusion limited conditions. With the rise in temperature moving down a HT water-gas shift reactor the process moves from almost pure reaction control to substantial pore-diffusion control.[13] Consequently, there are advantages in the use of large, dense pellets in the inlet regions of the reactor and small pellets in the exit region of the reactor.[28]

Table 1.7. Relative merits of different types of catalyst particle

Type of particle	Properties
Pellet	Cylindrical shape. Denser and stronger than extrusions or granules. Convenient size range from about 5 mm to about 20 mm. Raschig rings as well as cylinders can be made, but otherwise limited in shape. Two stages of compaction often needed
Extrusions	Usually long irregular cylindrical shape, but other shapes of larger external surface area (e.g. "clover-leaf") possible. Less dense than pellets. Convenient size from about 1 mm diameter upwards. Also possible to make hollow matrices (e.g. car exhaust catalysts)
Granules	Spherical shape only. Less dense than pellets. Convenient size from about 2 mm diameter upwards

The choice between extrusion and granulation for small catalyst particles depends primarily on the nature of the powder precursor: not all powders can be either extruded or granulated satisfactorily, even with the help of special additives (e.g. plasticizers or cements). The requirements of the forming process lead to constraints on earlier stages of the process.[15] Thus, in the pelleting of methanol synthesis catalysts[16] it is important to control the amount of hydroxide and carbonate remaining from the calcination stage. An excess weakens the pellets after reduction, whereas an inadequate proportion produces fragile pellets because the hydroxide and carbonate have a role in binding the oxides under pressure in the pelleting machine.

1.3. Catalyst Testing

1.3.1. Introduction

"Commit your blunders on a small scale and make your profits on a large scale." This principle, given by Baekeland[29] in his Perkin Medal address in 1916, summarizes the whole basis of catalyst testing. In catalyst testing there are two obvious features which can be emphasized.

1.3. Catalyst Testing

The first is that some simulation of what will (or could) happen to the catalyst is required for each test, but it is not possible to do all of the tests simultaneously. The results from various tests are combined in order to predict full-scale behaviour. The second point to be emphasized is that all of these figures have to be "right", or very nearly so.

All of the properties which can affect catalyst activity, selectivity and life need to be investigated, no matter whether a novel process or a new catalyst for an existing process is being examined—or, indeed, for ensuring the constant quality of a production catalyst. A range of experimental techniques is used in the small-scale simulation of the different aspects of the full-scale process, but the equipment rarely resembles a full-scale reactor. Baekeland stated that the principle quoted above "should guide everybody who enters into a new chemical enterprise, even if it taxes the patience of some men who cannot conceive that one single apparently minor detail in a chemical process may upset all the good points and lead to ruin".[29]

1.3.2. Chemical and Physical Properties

Many of the chemical and physical properties of a catalyst influence the performance of the catalyst in a full-scale reactor. A wide range of properties, classified as *bulk chemical properties*, *surface chemical properties* and *physical properties*, is determined, usually on a routine basis, as part of catalyst testing. In contrast with the various techniques developed for the assessment of catalyst performance, most of the analytical procedures used to determine other catalyst properties are standard techniques which require little modification for these applications.

1.3.3. Bulk Chemical Properties

Techniques used to determine these properties are given in Table 1.8. The first three methods—elemental analysis, X-ray diffraction and electron microscopy—are of widest application in the determination of the chemical nature of a catalyst. The advances in electron microscopy over the past decade have made this potentially the most powerful technique available for uncovering the chemical nature of catalysts.[30] Both the element analysis and crystal structure of particles identified in the field can be determined.[31] The other experimental techniques given in Table 1.8 tend to be used in special cases only. Thus, in an investigation of a poisoned catalyst, GC/MS can identify poisons present, and electron probe analysis or radiography will give the distribution of the poison across the catalyst structure.

Chapter 1. Fundamental Principles

Table 1.8. Techniques used for bulk chemical properties of catalysts

Technique	Properties determined
Element analysis (qualitative and quantitative)	Bulk elemental composition
X-ray diffraction	Crystalline phases present, crystallite sizes
Electron microscopy and associated X-ray fluorescence and electron diffraction	Particle shapes and sizes, particle compositions, particle crystal structures
Micro-probe analysis	Variation in composition across pellet
Radiography with radioactive isotopes	Distribution of radioactive component
Nuclear magnetic resonance (NMR)	Chemical environment of element
Mössbauer spectroscopy	Chemical environment of element
IR, Visible, UV spectroscopy	Types of chemical bond present
Extended X-ray adsorption fine-structure analysis (EXAFS)	Types of chemical bond present
Thermal analysis (DTA, TGA)	Phase changes, weight changes on heating
Temperature-programmed reduction (TPR)	Size and temperature range of reduction stages
Combined gas chromatography and mass spectrometry (GC/MS)	Analysis of volatile components

1.3.4. Surface Chemical Properties

As heterogenous catalysis is a surface phenomenon, the determination of surface properties (Table 1.9) plays a large part in catalyst characterization. Some techniques (e.g. LEED) can be used only with single crystal surfaces, and not with practical catalysts, and are not included in Table 1.9 (for further details see, for example, reference 32). No single technique can give a full description of the nature of the surface, so the use of several techniques is essential. As different equipment is used under various conditions, it is also essential to ensure that it is really the same catalyst surface which is being examined in each experiment. Some of the modern techniques of surface physics, listed at

Table 1.9. Some techniques used for determination of surface properties of catalysts

Technique	Properties determined
Photoelectron spectroscopy (UPS, XPS)	Chemical identity of surface layers
Auger spectroscopy	Chemical identity of surface layers
Secondary ion mass spectrometry (SIMS)	Chemical identity of surface layers
Temperature-programmed desorption, flash desorption	Chemical identity of adsorbed surface species
Physisorption of gases (e.g. N_2)	Total surface area
Chemisorption of CO, H_2 or O_2	Surface area of metal components
Surface reaction of N_2O	Surface area of metal components
Chemisorption of bases (e.g. NH_3, pyridine)	Surface concentration of acidic sites
Chemisorption of acidic gases (e.g. CO_2)	Surface concentration of basic sites
IR, Visible, UV spectroscopy	Types of chemical bond present
High-resolution electron energy-loss spectroscopy (HREELS)	Type of chemical bond present
Extended X-ray absorption fine-structure analysis (EXAFS)	Atomic structure of surfaces and adsorbates
Work function determination	Surface ionization
NMR	Chemical environment of element
Electron microscopy	Chemical identity and structure of surface layers
Isotopic labelling of reagents or catalysts	Chemical origin of adsorbed species

the top of Table 1.9, provide a wider range of chemical surface information, especially concerning minor components such as poisons, than could be obtained by the older, selective chemisorption methods.[33–36]

For example, in recent work[21] on the surface composition and topography of both unreduced and reduced ammonia synthesis catalysts, scanning Auger electron spectroscopy, X-ray photoelectron spectroscopy and scanning electron microscopy were used. The application of these techniques has given much detailed information on the microstructure of the reduced iron component, and the distribution, on a micro-scale, of the catalyst promoters. Nevertheless, Emmett and his co-workers,[37] with the few, somewhat primitive techniques available some 50 years ago, achieved a broad understanding both of the nature of the promoted-iron ammonia synthesis catalyst, and of how it worked, which still holds today.

1.3.5. Physical Properties

A relatively small range of techniques is used to monitor physical properties[38] of catalysts, and these are listed in Table 1.10. The *strength of catalyst particles* is needed to assess the possiblity of failure in use. Apart from chemical causes such as pellet rupture upon excessive coking, two different physical stresses are imposed on industrial catalysts: first, in the loading of a catalyst into a reactor (Chapter 3) and second, when the catalyst bed is installed in the reactor. Even under steady operating conditions catalyst particles are stressed by the weight of the catalyst bed above, and by the pressure drop across the catalyst bed due to gas flow. Catalyst break-up during loading is simulated (Figure 1.17) by the use of a standard tumbling test to determine attrition loss,[39] but all estimates are done by comparison with standard

Table 1.10. Selected techniques used for determination of physical properties of catalysts

Technique	Properties determined
Pellet crushing	Compression strength
Pellet tumbling	Attrition loss
Physisorption isotherms	Textural properties
Porosimetry	Pore size distribution
Controlled packing	Bulk density and packing characteristics
Inert gas flow through catalyst bed	Fluid flow properties, including pressure drop
Single-pellet reactor	Pore-diffusion coefficient
Non-steady state gas flow through catalyst bed	Diffusion characteristics

1.3. Catalyst Testing

Figure 1.17. Typical apparatus used to test pellet strength by measuring forces necessary to crush them. The photograph (right) shows a ring placed on the plattern before it is correctly aligned and crushed.

Chapter 1. Fundamental Principles

materials, since there is no direct relationship between the test and losses in plants. Similarly, comparative measurements are used in the crushing tests which simulate static loads in a reactor. Obviously, the way in which a crushing strength is determined depends on the shape of the catalyst particle. Spheres and long extrudated cylinders can be crushed only along a diameter, but pellets can be crushed either along the axis or along a diameter of the cylinder. The latter, the horizontal crushing strength, correlates better with behaviour in the plant.

Many of the physical properties required can be described as *properties of the texture of the catalyst* or properties (e.g. pore diffusion coefficients) which are determined by catalyst texture. Average properties are generally used (a complete description of the texture of a catalyst pellet is not possible):

(a) specific surface area (i.e. total accessible area);

(b) specific porosity (i.e. total accessible pore volume);

(c) pore size distribution (in distribution of pore volume as a function of pore radius);

(d) mean pore radius and

(e) particle size distribution.

Pores are classified as macropores (>30–35 nm), micropores (<2 nm) or mesopores (intermediate size). Mercury porosimetry and the analysis of physical adsorption isotherms are the two principal means of obtaining a description of the texture of a catalyst. These can be combined with information on particle size and shape from X-ray

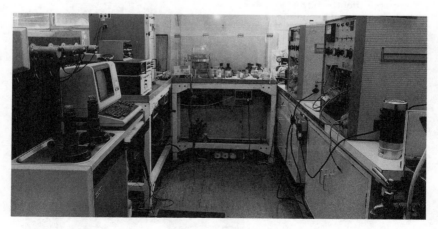

Figure 1.18. A modern laboratory for the determination of catalyst micromeritics.

diffraction work and electron microscopy. The pore shape can be deduced, but rarely without ambiguity, from physical adsorption isotherms: the only pore shapes known with accuracy have been determined by X-ray diffraction of a crystal structure (e.g. zeolites). Standard methods are available.[40-44] In recent years these methods have been automated, and a modern laboratory is shown in Figure 1.18. Diffusion characteristics of catalyst particles can be estimated by analysis of transient flows, but a more direct method involves the use of a single-pellet reactor[44] in which diffusion and reaction are measured simultaneously.

1.3.6. Catalyst Performance

To assess catalytic performance a reactor of some form is needed. The type of experimental reactor, and how it is operated, has to be chosen carefully and it depends on the nature of the information required.[45-50]

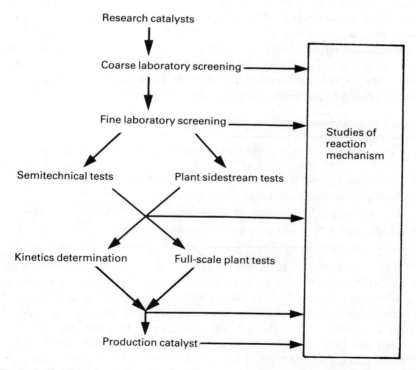

Figure 1.19. Flow sheet showing the stages of catalyst development from exploratory work through to catalyst production.

Chapter 1. Fundamental Principles

The flow sheet in Figure 1.19 shows the stages of catalyst development from exploratory work through to catalyst production. Few research programmes follow the full sequence: most speculative research projects fail and many development projects start from a considerable body of knowledge. Studies of the reaction mechanism (considered below) can be carried out at any stage. For example, mechanistic studies may be used to interpret unexpected results in the coarse screening at the start of catalyst development, but work is still being done on the mechanism of ammonia synthesis some 70 years after the successful development of a commercial catalyst.

1.3.7. Coarse Laboratory Screening

At each stage of development various catalyst formulations are rejected as unsuitable, and this rejected fraction is largest at the initial laboratory stage, which may be considered "coarse screening". The catalysts tested here are either speculative catalysts for an existing, possibly even well-established, industrial process, or a range of plausible catalysts for a new process. The primary aim in coarse testing is the separation of the promising catalysts from the useless catalysts. It is important at this stage that, as far as possible, no promising formulation is rejected. The experimental requirements to achieve coarse screening at a reasonable cost are given in Table 1.11. Although the criteria for acceptance at this stage are coarse, i.e. semiquantitative or even qualitative, successful coarse testing required much experimental finesse.

Table 1.11. Experimental requirements for coarse screening of catalysts

1	Simple criteria (e.g. minimum activity or selectivity to desired product) to reject useless formulations
2	Rapid tests, preferably with several catalysts simultaneously
3	Small catalyst samples only needed for evaluation
4	Determination of intrinsic catalyst performance
5	Able to cope with as wide a variety of catalysts as possible

The requirements in Table 1.11 are most easily achieved with microreactor systems. A modern installation with full computer control of operation and analysis is shown in Figure 1.20. Most microreactor systems consist of continuous-flow reactors, and the use of crushed catalyst minimizes any disguise of true activity by pore diffusion and inadequate heat transfer. Sometimes, however, a pulse microreactor (Figure 1.21), the first type of microreactor to be developed,[51] can give results which are unobtainable by other techniques.[52] As well as being of

1.3. Catalyst Testing

Figure 1.20. A modern laboratory installation of microreactor systems for catalyst testing.

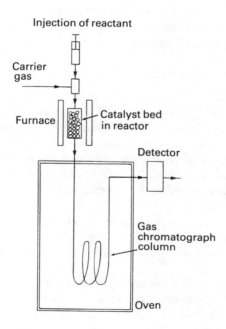

Figure 1.21. Flow sheet for a pulse microreactor system.

value for discovering the intrinsic performance of experimental catalysts, pulse microreactor systems can also be very useful in studies of catalyst poisoning.[53]

So far, it has been tacitly assumed that the reaction used for the tests is the same as the reaction occurring in the full-scale reactor. The ideal test (of more importance in the fine screening) should encompass:

(a) the real catalyst

(b) the real reaction

(c) the real conditions.

Any departure from this ideal must be regarded as a potential source of error in the testing. For example, some attempts were made[45] to use methanol decomposition, rather than the synthesis reaction, as a screening test for methanol synthesis catalysts because of experimental simplicity (e.g. operation at atmospheric pressure). While a broad correlation was found between test performance and activity in methanol synthesis, there were too many exceptions for acceptance of methanol decomposition activity as a standard test. Those discrepancies can be understood[54] as a consequence of catalyst variability: the nature of the surface of many catalysts, including methanol synthesis catalysts, is markedly dependent on the gases with which they are in contact.

1.3.8. Fine Laboratory Screening

In the fine screening of catalysts an accurate quantitative assessment of activity and selectivity is required, so the simulation of full-scale practice is closer than for coarse testing. Microreactor systems are widely used[45–47] for this testing, and automation allows high productivity in their use. As in most activity testing, experimental formulations should be compared with a "standard" catalyst, although with sufficient care and skill it is now possible to operate catalyst test units for most heterogeneous catalytic processes so that a test catalyst gives an accurately reproducible performance. Where such good reproducibility has not been attained the causes are usually inadequate temperature control, trace poisons in reactants, non-steady state conditions or poor reactor design.

Activities of catalysts are sometimes based on conversions achieved in tests at constant space velocity. This procedure should not be used, because a quantitative comparison is not possible without a full knowledge of the kinetics. This also applies to selectivity determinations because selectivity is defined in terms of conversions to the various possible products. The procedure can also give false answers for reactions which are near equilibrium. The correct procedure involves more experimental work, but with modern computer control this is not too arduous. Instead of a single run at fixed space velocity, a series of runs is carried out under constant conditions except for space velocity, which is varied. A comparison of the space velocities required for different catalysts to give the same conversion (from interpolation of experimental results) gives directly the relative amounts of catalyst needed to obtain the same conversion. This gives the ratio of catalyst activities directly. For similar reasons, comparisons of selectivity should always use values obtained at constant conversion.

Catalyst poisoning can be much more troublesome in small-scale testing than in full-scale practice.

1. It can be difficult to obtain laboratory gases that are as pure as those in a plant.

2. Space velocities used in catalyst testing are higher than in plant practice, so more poison is fed to the catalyst.

3. In plants with recycle (ammonia and methanol syntheses), only the make-up gas carries fresh poison to the catalyst.

As a result, even with the same poison content in the feed gas, the rate of poisoning in a test reactor can be 50 times that in a plant.

Other difficulties arise from "poisoning profiles". In a large reactor a poison front often develops, deactivating catalyst near the inlet. Although the shape of the poison profile is a complex function of mass and heat transfer, and poisoning and reaction kinetics, the poison front is generally sharper and narrower in a long reactor than in a short reactor. Thus an extent of poisoning which, in a full-scale reactor may deactivate only 1% or 2% of the catalyst bed at the inlet (i.e. a barely perceptible effect), can affect all of the catalyst in a very small laboratory reactor.

In both coarse and fine catalyst testing the measured catalyst activities must be those of the catalyst itself, free from any intrusion by mass-transfer or heat-transfer phenomena. Many tests for these effects have been proposed in the literature, but most are inadequate. Madon and Boudart[55] examined these problems in some detail, and concluded that a diagnostic test in which the concentration of active sites in the catalyst is varied gives the most reliable assurance of freedom from mass- and heat-transfer effects. If the observed activity of the catalyst is directly proportional to the concentration of active sites, then the experimenter can be reasonably sure that catalyst activity is really being determined. Any different variation of observed catalyst activity with active site concentration can be analysed to determine the nature of the interfering artefact. Most of the literature on this diagnostic test concerns supported metal catalysts, but it can also be applied to oxide catalysts, e.g. Fe_3O_4/Cr_2O_3, used for the high-temperature water-gas shift reaction.[13]

Sometimes the special features of a process lead to the development of special methods of fine catalyst testing. In one example[56] it was necessary to compare accurately the activity of various cobalt oxide catalysts for ammonia oxidation. A reactor was devised (Figure 1.22) in which a single pellet was mounted on a thermocouple and the pellet temperature was measured as a function of the gas rate. The pellet, together with the surrounding gas and apparatus, was first heated to reaction ignition temperature, when the pellet temperature rose because of the exothermic reaction. The gas flow rate was then gradually increased until, quite suddenly, the pellet temperature fell to virtually the gas temperature as ignition was lost. The gas rate at which extinction of the oxidation process occurred was taken as a measure of the catalyst activity.

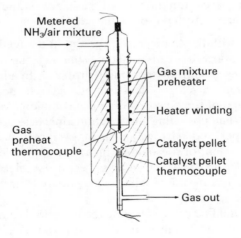

Figure 1.22. Special reactor, using a single catalyst pellet, for the testing of cobalt oxide catalysts in ammonia oxidation.

1.3.9. Semi-technical Catalyst Testing

Semi-technical testing with integral reactors is the next scale-up from laboratory screening, with typically between 100 ml and 1 l of catalyst in the reactor (Figure 1.23). Once widely used for catalyst testing, smaller-scale testing has taken over much of the work, mainly for

1.3. Catalyst Testing

Figure 1.23. A modern building for semi-technical catalyst test units, with full computer control and data-logging.

reasons of both convenience and accuracy. Semi-technical and pilot plants are, however, still of value in two broad areas: the larger-scale operation, especially if continued for weeks or months, provides an "insurance policy" by indicating whether some type of catastophic failure is likely. This is especially necessary if the catalyst has some hitherto untried features. The other main use lies in product quality, etc.; for example, operation close to plant conditions gives reliable information on by-product formation. Multiple sampling points along an integral reactor have been used[57] to obtain more information in each run.

1.3.10. Reaction Kinetics

The prediction of reactor performance, either for plant design or for the optimization of the performance of an existing plant, requires a knowledge of the reaction kinetics. Kinetic expressions are sometimes determined in an attempt to deduce catalytic mechanism, but that exercise is quite different from the development of a semi-empirical rate equation which is required only to give the rate of reaction, to whatever desired accuracy, over the range of temperature, pressure and composition found in the plant reactor. There are therefore three broad requirements for kinetics measurement.

1. Use of the real plant catalyst (or at least one in the later stages of development, so that prediction of plant performance is required) in the form of pellets, etc., used in the reactor. Crushed catalyst should not be used, because the performance of whole pellets is needed.

2. Experiments should cover all plant conditions, so that all uses of the resulting kinetic equation represent an interpolation from experimental values rather than an extrapolation.

3. The rate of reaction must be determined to sufficient accuracy as a function of temperature, pressure and composition of the reacting gas.

Continuous stirred tank reactors (CSTRs) are customarily recommended for kinetics measurement because of the mathematical simplicity of treating the results. In a CSTR the conditions are constant throughout, so the measurement of the rate of formation of the products (from their concentrations in the reactor exit stream and the flow rate) gives the rate directly for a given set of conditions. In an integral or plug flow reactor (PFR) conditions vary along the reactor, so a different treatment of results is needed. Some form of kinetic equation has to be postulated and then integrated to fit inlet and exit conditions. However, with current computing facilities this exercise can be done readily, so the choice of reactor type is no longer as clear-cut as it used to be. Moreover, the much better temperature control reached with microreactor systems ensures better data for equation fitting. However, CSTRs do still have two significant advantages.

1. Unless the catalyst particles are quite small (e.g. from a fluidized bed reactor), it is much easier to accommodate full-sized particles (pellets, granules or extrudates) in a CSTR than in a microreactor (PFR).

2. The direct determination of reaction rate means that its variation with experimental parameters is obvious to the experimenter during the course of the work, so a much better "feel" for what is taking place in the system can be obtained.

1.3. Catalyst Testing

There are three broad types of CSTRs which can be used for the measurement of catalyst kinetics (Table 1.12 and Figure 1.24). All three systems work well at moderate temperatures and pressures close to 1 bar. Difficulties arise when measurements are required at elevated temperatures and/or pressures. The Boreskov system is now little used: the available pumps for severe conditions are less suitable, temperature control in the reactor often proves to be difficult and the gas volume/catalyst volume ratio is much larger than in plant reactors. Both Carberry/Brisk and Berty systems have been used successfully for high-pressure/high-temperature systems, but the Berty reactor has several fundamental advantages over the earlier systems (Table 1.13). The relative merits of these reactors remains the subject of some controversy.[58]

The prediction of catalyst performance in a full-scale reactor must include the changes which occur during catalyst life (Section 1.4), and the way this is done depends on the mode of deactivation. When a poison front (as for example in LT shift beds) or a coking front (as in

Table 1.12. Type of continuous stirred tank reactor (CSTR) used in the determination of catalyst kinetics

Type of reactor	Method of mixing
Boreskov	External recycle of products and reactants. Recycle ratio high enough for near-constant conditions in reactor
Carberry/Brisk	High-speed rotation of catalyst basket in reactor
Berty	Internal recycle of products and reagents within reactors Recycle ratio high enough for near-constant conditions in reactor

Table 1.13. The advantages of the Berty-type reactors over Carberry–Brisk reactors for catalyst kinetics measurement

1. The gas flow through the catalyst bed is well-defined and calculable
2. As the catalyst bed is stationary, its properties (e.g. temperature) can be measured directly
3. Little if any catalyst breakage
4. As the rotor is an impeller rather than a catalyst basket it is smaller and can be balanced accurately
5. The gas volume/catalyst volume ratio can, if necessary, be kept closer to the plant value

Chapter 1. Fundamental Principles

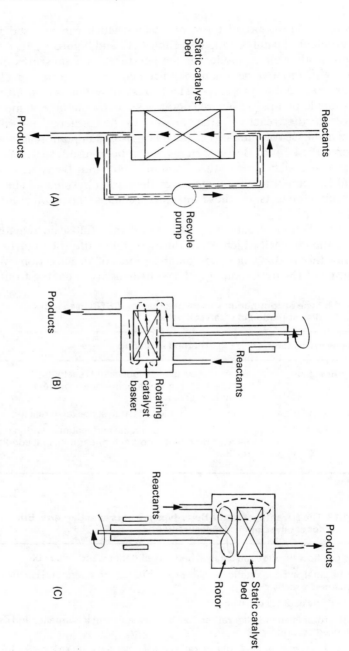

Figure 1.24. Three types of continous-stirred tank reactor (CSTR) used for the measurement of catalyst reaction kinetics.

substitute natural gas production by naphtha–steam reforming) passes down the bed, giving complete deactivation, then the change is solely a decrease in effective catalyst volume, and so is readily accommodated in reactor modelling. In other cases (e.g. HT shift and ammonia synthesis) when the loss in activity occurs throughout the bed largely due to slow sintering of the active phase of the catalyst, a change in the kinetic equation is necessary for correct reactor modelling. This is commonly done by the use of a factor, called "relative catalyst activity", by which the whole kinetic expression is multiplied. This can be justified for catalysts, such as HT shift catalyst,[13] in which decay involves a loss of the number of active sites but no change in the site properties.[59, 60] However, a change in the characteristics of the active phase or in the balance between pore-diffusion and reaction rates can cause more extensive changes in form and constants of the kinetic equation.[61, 62] The simpler procedure of using an overall factor for activity is almost always used in practice, since the errors introduced are usually acceptable. Operation at constant exit concentration by adjustment of feed rate to compensate for decay is recommended.[49]

Catalyst activation for kinetic studies presents fewer problems. As long as feed and reduction gases are as pure as plant gases (to avoid poisoning during activation), catalyst activation by a well-defined and reproducible procedure based on full-scale practice gives catalysts of a constant performance which is within the range found in plants.

1.3.11. Catalyst Ageing

The simulation of catalyst ageing is the most difficult of all the simulations of catalyst performance. What we need to achieve is a catalyst treatment which is equivalent to some months or years in the full-scale reactor, so that all of the tests described above for fresh catalysts can be repeated with aged catalysts. There are two broad aims in this work: to find out how far the catalyst performance has deteriorated (i.e. the changes in catalyst activity and selectivity) and why this has happened. The simplest technique is that of using catalysts discharged from the full-scale plant. Except for fluidized-bed processes, where catalysts are frequently added and removed, it is usually difficult to obtain catalyst samples other than at a plant shutdown, so the catalyst decay cannot be followed. Also, plant-discharged catalyst may not be in the same state as it was in the reactor.

An alternative procedure is the use of a laboratory accelerated-ageing procedure. This can give useful qualitative information, but its limitations can be shown by two examples. If catalyst decay is due to poisoning at a level of 1 part in 10^9, then 1 year of plant poisoning could

Chapter 1. Fundamental Principles

be simulated by the use of 50 ppm of poison for 10 min. However, the adsorption and subsequent reactions of the poison would be quite different under the two sets of conditions. Similar problems arise with accelerated sintering. Suppose a catalyst suffers a 50% loss of area in 1 year of plant duty at 250°C. If this follows typical sintering kinetics with an activation energy of 167 kJ mol^{-1}, then treatment at 400°C for 100 min will also give a 50% loss of area, i.e. accelerated ageing. However, if another phase of the catalyst, with equivalent sintering kinetics but an activation energy of 251 kJ mol^{-1}, also loses 50% surface area at 250°C, then the same accelerated ageing will cause a 79% loss of area.

The third technique for monitoring catalyst ageing is the use of a sidestream reactor (i.e. an experimental reactor attached to a full-scale plant using actual plant feed). This has the advantage of operating in real plant time, so the problems of accelerated ageing do not arise. The catalyst undergoes treatment that is very similar to that experienced by catalyst in the plant reactor, but at the same time its performance can be monitored readily. The disadvantages of a sidestream reactor are that it requires large quantities of catalyst, and is restricted to plant gas composition and operation.

1.3.12. Mechanism of the Catalytic Reaction

The first stages in unravelling the mechanism of catalytic reaction are the identification of the reaction steps, the sequence of which consitutes the overall process, and the parts of the catalyst on which these reaction steps occur. Next we wish to know the rates of the individual reaction steps and which, if any, are in quasi-equilibrium and which effectively limit the rate of the overall process. We also need to know the energetics of these steps and the energies of adsorption of the intermediates. When the intermediates have been identified, their structures (bond lengths and energies) and their method of bonding to the catalyst surface may be probed. Catalysts decay in practice, so we also want to know about the mechanism of decay—what happens in the catalyst to change its properties, and how does this affect all of the details of the mechanism of the catalytic reaction itself?

The books and other references for further reading given in Appendix 1 provide descriptions of the many methods used to unravel the mechanisms of catalytic reaction, together with examples from all categories of reaction. Here we shall look briefly at some work done towards elucidating the mechanisms of two catalytic syntheses, those of ammonia and methanol.

1.3.12.1. Ammonia synthesis

Some of the studies of the nature of the surface of the ammonia synthesis catalyst including the early work by Emmett, and more recent work by Ertl and Somorjai with the new techniques, was described in Section 1.1.4. The surface consists essentially of iron, almost covered with a potassium compound, the nature of which is still the subject of speculation,[6-8,21,63,63a] although the overall bulk concentration of potasium is only about 1%. Metallic iron is the only component capable of catalysing ammonia synthesis alone. Experiments with catalysts of different compositions showed that alumina and other refractory components do not affect catalysis directly, but play a structural role while potassium actually increases the rate of synthesis on a given iron catalyst. Potassium may also have a structural function[63a]. The intermediates in the synthesis, adsorbed on the metallic iron surface in the catalyst, are shown by the equations (19) and (20) and the energetic relationships between them and the equivalent gas-phase species are shown in Figure 1.3. These relationships were established in studies[6-9] of the separate adsorption desorption and reactions of N_2, H_2 and NH_3 on various faces of iron single crystals.

$$N_2 \rightarrow N \rightarrow NH \rightarrow NH_2 \rightarrow NH_3 \qquad (19)$$

$$H_2 \rightarrow H \qquad (20)$$

In a study[5] with single crystal iron faces at 525°C and 20 bar the relative rates of ammonia formation were found to be 418 : 25 : 1 for Fe(111), Fe(100) and Fe(110) faces, respectively. This sensitivity of synthesis rate to crystal structure had been indicated indirectly in earlier work[64] with supported iron catalysts of different crystallite sizes. Measurements of the concentration of adsorbed nitrogen atoms by Auger spectroscopy were used to determine the rate-limiting step. Ertl and co-workers were able to show that ammonia was formed through the reaction of N(ad) as in equation (21) rather than involving adsorbed dinitrogen, as in equation (22) and under most conditions the rate-limiting step is dissociative nitrogen chemisorption.

$$N(ad) + 3H(ad) \rightarrow NH_3 \qquad (21)$$

$$N_2(ad) + 6H(ad) \rightarrow 2NH_3 \qquad (22)$$

In a study on the role of potassium in promoting ammonia synthesis, metallic potassium was evaporated on to a surface, the (100) surface, of a single iron crystal. It was found that the initial rate of formation of adsorbed nitrogen atoms at 160°C was increased by more than two orders of magnitude and the effective energy barrier to the dissociation of molecules dropped to close to zero. Much smaller effects were found with other iron crystal faces. Thermal desorption experiments have shown that the potassium is partially oxidized in actual synthesis catalysts, and the exact mechanism of the promoter action remains unresolved, although it is clearly related to the phenomena observed in model studies. Confirmation of the correctness of the mechanism has been obtained[10] in a computer model of ammonia synthesis under industrial conditions which uses the mechanism and energies from single-crystal studies.

1.3.12.2. Methanol Synthesis

The second example is methanol synthesis over a $Cu/ZnO/Al_2O_3$ catalyst (Chapter 9). Several features of the mechanism are understood, but as with ammonia synthesis there are details yet to be exposed. Commercial methanol plants operate with $CO/CO_2/H_2$ feeds, and the first mechanistic question to be asked is whether methanol is made from CO or CO_2. The shift reaction, shown by the equation (23) allows CO/CO_2 interchange, so that eventually both sources of carbon are used. Standard methanol catalysts work with both CO/H_2 and CO_2/H_2 mixtures, so a more subtle technique must be used.[65] A standard $CO/CO_2/H_2$ mixture was made up in which some of the CO_2 contained the radioactive isotope ^{14}C. The methanol formed under normal synthesis temperature and pressure initially had the same level of radioactivity as the feed CO_2, showing that essentially all the methanol is formed from CO_2 by reaction (24), and very little, if any, by reaction (25).

$$CO + H_2O \rightleftharpoons CO_2 + H_2 \qquad (23)$$

$$CO_2 + 3H_2 \rightarrow CH_3OH + H_2O \qquad (24)$$

$$CO + 2H_2 \rightarrow CH_3OH \qquad (25)$$

The standard ICI Catalyst 51-2 contains metallic copper, zinc oxide and a zinc oxide/alumina spinel-type phase. The copper surface area and synthesis activities of a wide range of $Cu/ZnO/Al_2O_3$ catalysts were measured.[65] The copper surface areas of the catalysts were determined by reaction with N_2O as in the equation (26). Activity was found to be directly proportional to the copper surface area, showing that the critical reaction step occurred on this phase. Further work[65, 66] also showed that the copper surface of the working catalyst was partially oxidized. So far, none of the work has indicated which adsorbed intermediates took part in the overall conversion of CO_2 to CH_3OH. Other phases of the catalyst may play some part in their formation or reaction. Temperature-programmed desorption with ZnO catalysts have demonstrated[67] the presence of a formate intermediate on the catalyst surface under synthesis conditions. Similar experiments with the standard Cu/ZnO/Al$_2$O$_3$ catalysts, and comparison with other supported catalysts, showed[65] similar formate (desorption temperature 580 K) to be present on the ZnO component and a more weakly bonded formate (desorption temperature 440 K) on the copper surface. Hydrogenolysis of the latter formate is probably the rate-determining step in methanol synthesis.

$$N_2O\ (+\ Cu) \rightarrow N_2 + O(ad) \tag{26}$$

1.4. Catalyst In Use

1.4.1. Introduction

In this section the ways in which catalysts are activated, i.e. brought into the working state in the reactor, are discussed, and then the various ways in which the performance of a catalyst can change during its working life are considered.

A recurrent theme in this chapter is the necessity for compromise. Any practical catalyst has many conflicting requirements, and the optimum catalyst is formed by a judicious blend of its various properties. This is also the case for the conditions of use of the catalyst. To achieve the maximum possible economic benefit from a catalyst charge a fine balance has to be kept between all of the variables of plant operation.

1.4.2. Pretreatment and Activation

In some ways catalyst activation[25, 68] can be regarded as the final stage of catalyst manufacture, so the first question to be asked is *why do it in the plant?* While occasionally some part of the activation process is carried out before the catalyst is put into the reactor (e.g. with the pre-reduction and stabilization of some supported metal crystals), there are three general reasons for activation in the plant reactor.

1. Reduced metal catalysts are very sensitive to air, so they are not readily transported or loaded into converters.

2. Correct activation is essential to achieve the proper catalyst structure and thus the optimum performance.

3. It is very difficult to avoid some contamination of catalysts during forming, transport and charging the reactor, and activation in the reactor gives a final opportunity to remove adventitious poisons.

Chapter 1. Fundamental Principles

The techniques used for catalyst activation depend primarily on the nature of the catalyst, but also on the process for which the catalyst is to be used.[68] Most catalysts used in hydrogen, ammonia and methanol plants are supported metals, and the activation stage consists essentially of reduction of the precursor metal oxide to fine crystallites of the metal. Hydrogen is most commonly used as the reducing agent. To get the maximum surface area of metal the concentration of the product of the reduction process (steam) and the temperature of reduction needs to be controlled and low. Excessive temperature during reduction causes sintering, i.e. premature catalyst ageing. The effect of steam is more complex.[25, 68] Steam, even at low concentrations of 0.1 bar or less, is very effective at promoting the sintering of oxides, so although the metal crystallites may not themselves be affected by steam, the removal of the refractory "hedges" allows metal aggregation (Figure 1.25). However, there is a further deleterious effect arising from the P_{H_2}/P_{H_2O} ratio,[16, 25] which has been investigated in detail for the reduction of ammonia synthesis catalysts (Figure 1.26). A high P_{H_2}/P_{H_2O} ratio gives a high rate of reduction, leading to fine crystallites. Other types of catalyst require different activation procedures. Cobalt/molybdenum/alumina catalysts, used for hydrodesulphurization (Chapter 4), need partial sulphiding to develop full activity.

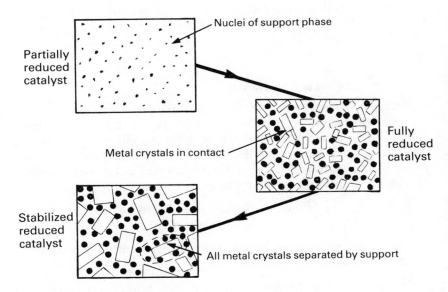

Figure 1.25. Reductive activation of supported metal catalysts, showing the function of the refractory support in prevention of metal aggregation.

1.4. Catalyst in Use

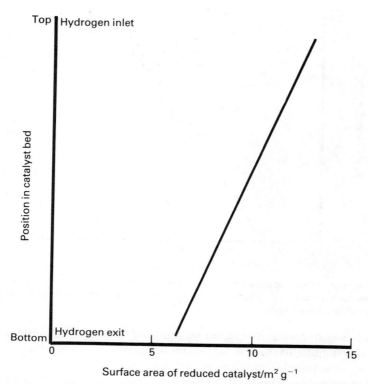

Figure 1.26. Variation in total surface area of ammonia synthesis catalysts as a function of position in the catalyst bed. Water, produced by the reduction of catalyst in the upper part of the bed, causes the decrease in surface area found in the lower part of the bed.

The high-temperature water-gas shift catalyst (Chapter 6) initially contains ferric oxide, Fe_2O_3, with a chromia support, Cr_2O_3, which remains essentially unchanged during catalyst life.[13] The active phase of the catalyst is magnetite, Fe_3O_4, so partial reduction is needed. If carried out under the conditions necessary for the reduction of ammonia catalysts, the HT shift catalyst would be over-reduced to metallic iron. This must be avoided since metallic iron catalyses methanation reactions rather than the desired shift reaction. The high selectivity of shift catalysts depends on correct reduction procedures. Figure 1.27 shows the range of P_{H_2}/P_{H_2O} ratios for which Fe_3O_4 is the stable phase: these correspond to normal plant operation, and are used for activation.

Different activation procedures are used in other types of plant, e.g. those for oil refining.[70, 71]

As successful reduction of supported metal catalyst in the plant

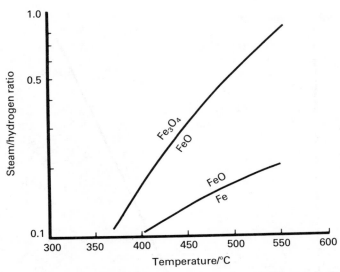

Figure 1.27. Stable phases of iron and iron oxides at different hydrogen/steam ratios. Normal plant operation with the high temperature shift catalyst is within the range of Fe_3O_4 stability.

requires both care and time, *prereduction of catalysts* is sometimes done before charging to the reactor.[68] The reduction of a plant charge of 100 tonnes of ammonia synthesis catalyst gives more than 20 tonnes of water and prereduction avoids the problems associated with this. Prereduction may also be necessary when the temperature required for activation cannot be attained in the plant. For example, in the liquid phase organic reductions which are operated at temperatures up to 150–200°C, whereas reduction of a supported nickel catalyst may need 300–400°C. In prereduction the catalyst is reduced to the desired activated state, but, because of the problem of transporting and charging a pyrophoric solid, it is then "stabilized". This usually involves a careful partial oxidation which prevents gross oxidation on exposure to air. Ideally a monolayer of adsorbed oxygen should be sufficient, but in practice a greater degree of oxidation may be necessary. Even then, plant start-up times are decreased and the temperatures needed for activation of prereduced catalysts are lower than with unreduced catalysts. The economic balance of using prereduced catalysts has to be assessed for each case: the prereduction and stabilization procedure inevitably adds to catalyst cost, and this has to be set against the gain from faster plant start-up.

The *reactivation of catalysts* is sometimes effective.[68] Most frequent is the removal of "coke" or other carbonaceous deposits by air oxidation or steaming, and this can be done continuously, intermittently on a regular basis or intermittently as required by, for example, a plant mishap. Reactivation after decay or from other causes, such as poisoning or metal sintering, is sometimes also possible, but as recovered activity can be much less than initial activity it is usually not economic in practice, so a fresh catalyst charge is needed.[14]

Pre-treatment and activation not only produce the active phase but can also cause other considerable changes within the catalyst. For example, a large volume change may accompany reduction (e.g. more than 40% for bulk cupric oxide). If not controlled, such transformations may therefore cause subsequent structural changes, including decrease in pellet strength and enhanced sintering of the active phase.

1.4.3. Loss of Catalyst Performance

A loss of either activity or selectivity can be sufficient to necessitate catalyst discharge or reactivation. The various aspects of catalyst deactivation are well covered in the literature, both general treatments[25, 68, 72–78] and many case studies[13, 16, 24, 31, 52, 56, 69] (also see Appendix 1) with much detailed work on the causes of decay and modelling of catalyst performance during life. It is useful to distinguish between temporary and permanent loss of performance; that is, between cases where catalyst performance can be recovered (usually by reactivation in the plant) and those where it cannot. Wide variations in catalyst life occur in practice, dependent on the process (see Tables 1.14 and 1.15, expanded from reference 76) and, including temporary and permanent lives together, the range of catalyst lives covers more than 10^8 to 1. It will be seen that the catalysts used in ammonia and methanol synthesis trains are among the longest-lived.

Among the techniques for coping with declining catalyst activity, temperature-ramping to maintain constant conversion (or reactor output) is widely used. For example, an increase in reactor temperature from 300°C to 310°C compensates for a 25% loss of catalyst activity in a reaction with an activation energy of 84 kJ mol^{-1}. Clearly, the method cannot be used for drastic falls in activity. Theoretical aspects have been described.[59, 79]

The three different causes of catalyst decay—physical causes, poisoning by impurities, and poisoning by reactants or products—are discussed in the following sections.

Chapter 1. Fundamental Principles

Table 1.14. Typical lives of some catalysts used in the manufacture of ammonia, nitric and sulphuric acids, methanol and formaldehyde

Process and typical conditions	Typical catalyst	Life/ years	Common cause of decay and catalyst property affected
Ammonia synthesis: $N_2 + 3H_2 \rightarrow 2NH_3$ 450–550°C 200–500 bar	Fe/Al$_2$O$_3$/CaO/K granules	5–10	Slow sintering: activity
Methanation: CO/CO$_2$ + H$_2$ $\rightarrow$ CH$_4$/H$_2$O 250–350°C 30 bar	Supported Ni pellets	5–10	Slow poisoning (e.g. S, As): activity
Sulphuric acid: $2SO_2 + O_2 \rightarrow 2SO_3$ 400–600°C 1 bar	V/K sulphates/SiO$_2$ extrusions	5–10	Slow physical deterioration (dust): pressure drop, activity
Methanol synthesis: $CO + 2H_2 \rightarrow CH_3OH$ 200–300°C 50–100 bar	Cu/ZnO/Al$_2$O$_3$ pellets	2–8	Slow sintering: activity
Low temperature carbon monoxide shift: CO + H$_2$O $\rightarrow$ CO$_2$ + H$_2$ 200–250°C 30 bar	Cu/ZnO/Al$_2$O$_3$ pellets	2–6	Slow poisoning; sintering accelerated by poisons: activity
Natural gas, naphtha hydrodesulphurization: $R_2S + 2H_2 \rightarrow 2RH + H_2S$ 300–400°C 30 bar	Co/Mo sulphides/ Al$_2$O$_3$ extrusions	2–8	Slow coking: activity, pressure drop
High temperature carbon monoxide shift CO + H$_2$O $\rightarrow$ CO$_2$ + H$_2$ 350–500°C 30 bar	Fe$_3$O$_4$/Cr$_2$O$_3$ pellets	2–4	Slow sintering, pellet breakage: activity, pressure drop
Steam reforming, natural gas CH$_4$ + H$_2$O $\rightarrow$ CO + 3H$_2$ 500–850°C 30 bar	Ni/CaO/Al$_2$O$_3$ rings	2–4	Sintering, occasionally C formation, ring breakage: activity, pressure drop

Table 1.14. continued

Process	Catalyst	Life/years	Common cause of decay
Oxidation of methanol to formaldehyde: $CH_3OH \rightarrow HCHO + H_2$ $CH_3OH + \tfrac{1}{2}O_2 \rightarrow HCHO + H_2O$ 500–600°C 1 bar	Silver granules	0.3–1	Poisoning (e.g. iron), coking (poison promoted): selectivity
Ammonia oxidation: $2NH_3 + \tfrac{5}{2}O_2 \rightarrow 2NO + 3H_2O$ 800–1300°C 1–10 bar	Pt alloy gauze	0.1–0.5	Surface roughness, loss of Pt, poisons: selectivity

Table 1.15. Typical lives of some catalysts used in oil refining and the petrochemical industry

Process and typical conditions	Typical catalyst	Life[a]/ years	Common cause of decay and catalyst property affected
Distillate oil hydrodesulphurization: 300–400°C 5–100 bar	Co/Mo sulphides/ Al_2O_3 extrusions	T 1–2 P 2–8	Coking: activity, pressure drop
Residual oil hydrodesulphurization: 250–400°C 100 bar	Co/Mo sulphides/ Al_2O_3 extrusions	P 0.5–1	Coking and metal (V, Ni) deposition: activity, selectivity, pressure drop
Hydrocracking heavy oil: 250–400°C 100–200 bar	Ni/W sulphides/ $Al_2O_3 + SiO_2/Al_2O_3$ (Pd/zeolite) extrusions	T 1–5 P 1–10	Coking, metal deposition: activity, selectivity, pressure drop
Catalytic reforming: 450–550°C 5–50 bar	$Pt/Re/Al_2O_3/Cl^-$ spheres, extrusions	T 0.01–0.5 P 2–15	Coking, Pt sintering: activity, selectivity
Fluid catalytic cracking: 500–600°C 1–3 bar	Zeolite/SiO_2–Al_2O_3 matrix microspheroids	T 10^{-8}–10^{-9} P 0.1	Coking, metal deposition, loss of zeolite structure: activity, selectivity
Ethylene oxidation: $C_2H_4 + \tfrac{1}{2}O_2 \rightarrow C_2H_4O$ 200–300°C 10–20 bar	Ag/α-Al_2O_3/ promoters rings	P 1–4	Slow sintering: activity, pressure drop

Chapter 1. Fundamental Principles

Table 1.15. continued

Benzene oxidation to maleic anhydride $C_6H_6/O_2 \rightarrow C_4H_2O_3$ 350°C 1 bar	V/Mo oxides/ promoters/α-Al_2O_3 rings	P 1–2	Formation of inactive V phase: activity, selectivity
Reduction of aldehydes to alcohols: $2RCHO + H_2 \rightarrow RCH_2OH$ 250–400°C 100–300 bar	Cu/ZnO pellets	T 0.5–1 P 2–8	Slow sintering, pellet breakage: activity or pressure drop
Acetylene hydrogenation: $C_2H_2 + H_2 \rightarrow C_2H_4$ 30–100°C 50 bar	Pd/support	T 0.1–0.5 P 5–10	Coke, oil formation: activity
Ethylene oxychlorination: $C_2H_4 + 2HCl + \frac{1}{2}O_2 \rightarrow C_2H_4Cl_2 + H_2O$ 250°C 1–10 bar	Cu chlorides/ Al_2O_3 microspheroids	P 0.2–0.5	Attrition, other causes from plant upsets: fluidized state, activity

[a] P = permanent loss of catalyst performance; T = temporary loss of catalyst performance, recoverable (at least partly) by regeneration.

1.4.4. Physical Causes of Decay

Physical changes on a micro- or macro-scale in a catalyst can lead to decay in performance. Agglomeration of the crystallites of the active phase leads to loss of active surface and, consequently, a decrease in activity. On a larger scale, the break-up of catalyst pellets will hinder gas flow through the catalyst bed, with greater pressure drop, and so will also give a decreased output from the reactor. Either type of failure, if severe enough, may limit plant performance so much that shutdown and a catalyst change may become imperative.

The sintering of supported metals has been studied extensively.[68, 73, 74] Commercial catalysts are designed for stable performance, so the rate of sintering is normally very slow.[14, 69] Similar patterns are seen with oxide catalysts. For example, HT shift catalysts,[13] where a fast, initial loss of activity was attributed to the agglomeration of unstabilized Fe_3O_4, formed in reduction, whereas Fe_3O_4 stabilized with the more refractory chromia, Cr_2O_3, sinters slowly, giving a long plant life (Table 1.14). The actual rate and extent of sintering depends on many factors, including the metal concerned, the metal content, initial crystallite size and size distribution, the dispersion of the metal across the support, the nature of the support material and the operating conditions.

1.4. Catalyst in Use

The most important factor is the method of catalyst manufacture (Section 1.2). The aggregation of metal crystallites necessarily involves the transport of metal within the catalyst, although it is often difficult to determine whether this occurs by the transfer of metal atoms or clusters, across the surface of the support, by atoms through the gas phase (usually promoted by poisons which form compounds with the metal) or by the movement of whole crystallites.[68] In most catalytic processes the temperatures and the size of the metal crystallites are such that extensive agglomeration would occur in seconds without the presence of a support. The effectiveness of the support in hindering metal movement and the movement of the support itself, factors controlled primarily by catalyst formulation, are thus of prime importance.[25]

Catalyst failure due to macro-scale physical causes are almost always due to plant maloperation. Catalysts are made strong enough to withstand the stresses of normal plant operation for, at least, the normal catalyst life, but the production of much stronger catalysts would entail the loss of other necessary properties such as activity. The nature of plant troubles which lead to gross catalyst failure can be purely physical (e.g. unsteady-state running, giving "bumping" of the catalyst bed) or chemical (e.g. moving into the temperature/composition regime where carbon deposition is so extensive that catalyst pellets are burst).

1.4.5. Poisoning by Impurities in Feeds or Catalysts

Impurities (in feed or catalyst) can affect catalyst performance when their interaction with the catalyst is stronger than that of the feed.[68, 77, 80–82] The active site of the catalytic reaction or, less often, the pore structure giving access to the active site, is modified in some way so that catalyst performance is altered—for the worse, as is implied by the use of the word "poison". Overall catalyst activity may be decreased without affecting selectivity when some of the sites are totally deactivated while others are unaffected. If, however, some active sites are modified without losing all activity, then the relative rates of different reactions may change to give a different catalyst selectivity.

Occasionally plant maloperation can cause catalyst fouling, e.g. carry-over of CO_2-removal liquor into the methanation reactor,[75] with consequent loss of catalyst performance. This is distinct from poisoning because the transient impurity levels are much higher. The initial mode of action of most poisons is that of strong adsorption on the active site. Thus, the poisoning effect may be essentially due to a blocking action. The poison, by being more strongly adsorbed, prevents the reactants from being adsorbed and converting to products. More serious effects

Chapter 1. Fundamental Principles

are found when reconstruction of the active site follows adsorption of the poison.

The types of impurities which can act as poisons depend on the chemical nature of impurity and active site, since strong chemisorption is the main feature of the poisoning action.[81] Significant poisons for some industrial catalysts are given in Table 1.16. The list is not complete, but nevertheless the pattern is clear: strong interaction, usually related directly to the chemistry of the bulk states, between poison and catalyst is present. This strong interaction explains why poisons are effective at such low levels, e.g. 1 ppm H_2S in the feed to a nickel catalyst for steam reforming or $\ll 1$ ppm HCl to a copper catalyst. The bulk analysis of poison in a fully deactivated catalyst can also be very low, e.g. 0.1% or less, so analyses of deactivated catalyst need to be supplemented by other data, such as surface analysis, element distribution across the pellet or information about plant operation. Most poisons get to the catalyst surface by being in the feedstock entering the reactor, but occasionally the catalyst itself can be the source. Catalyst manufacturers take great trouble (Section 1.2) to ensure that, for example, residual traces of chloride and sulphate are at very low levels in copper catalysts. One of the functions of the support in many catalysts is that of holding poisons to prevent them from reaching the active sites. Zinc oxide fulfils this role in LT shift catalysts by reacting with H_2S and HCl.[14, 69]

The distribution of poison both within the reactor and across the catalyst particles is determined by the kinetics of the poisoning reaction and the mobility of the poison.[68, 77, 80, 83] As poison concentrations in the feed are usually low and poisoning reactions fast, most poisoning reactions are strongly diffusion-limited, with consequent poison deposition near the outside of the catalyst particle. Copper catalysts, when poisoned by chlorine compounds, form copper chlorides which, because of their low melting points, are sufficiently mobile to move from the exterior to the middle of the catalyst pellets. It is frequently found that poisons are taken up preferentially near the inlet of the reactor (where this occurs it allows the use of small guard beds to remove poisons), giving a "front" which slowly progresses through the reactor (Figures 1.28 and 1.29).

The most direct way of dealing with catalyst poisoning is the purification of reactor feeds and the catalysts themselves. Special process stages are used specifically to achieve this, e.g. hydrodesulphurization to protect nickel catalysts in steam reforming. Other process stages may also do this; for example, the methanation stage, which is designed to convert CO and CO_2 to methane to prevent

1.4. Catalyst in Use

Table 1.16. Some typical poisons of industrial catalysts

Poison	Catalysts	Processes
CO, CO_2, H_2O, C_2H_2	Fe	Ammonia synthesis
CO	Pd, Pt	Hydrogenation
H_2S, AsH_3, PH_3	Co, Ni, Pd, Pt, Cu	Hydrogenation, steam reforming Catalytic reforming Methanol synthesis LT CO shift
H_2S (high levels)	Fe_3O_4	HT CO shift
HCl	Cu	Methanol synthesis, LT CO shift
Na, NH_3, organic bases	SiO_2/Al_2O_3, zeolites	Catalytic cracking, hydrocracking
Transition metals (V, Ni)	SiO_2/Al_2O_3, zeolites, Co/Mo sulphides	Catalytic cracking, hydrocracking, hydrodesulphurization
Fe, Ni carbonyls	Ag	Methanol oxidation to formaldehyde
Pb, Hg, Zn	Most transition metals	Many

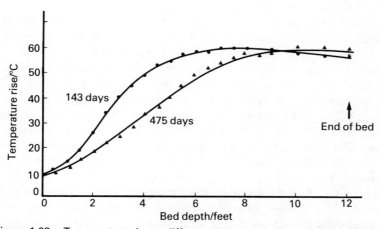

Figure 1.28. Temperature rise at different times in a bed of catalyst that is subject to sintering.

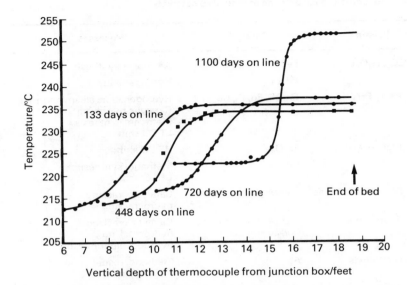

Figure 1.29. Variation of temperature profile with time for the poisoning of low temperature shift catalyst. A front slowly progresses through the reactor.

poisoning of ammonia catalyst, will also remove any traces of H_2S reaching this stage.

When the poison is not too strongly adsorbed and no reconstruction of the active site has taken place, then regeneration of the catalyst by poison removal may be possible. In this case the poison is described as "reversible". The poisoning caused by traces of water or carbon dioxide in ammonia synthesis is due to adsorbed oxygen on the iron surface,[75] which prevents nitrogen adsorption. Reduction with hydrogen removes the adsorbed oxygen to leave the iron surface as it was before. Gross oxidation with oxygen gives bulk changes which are not readily reversed. The poisoning of nickel catalysts with arsenic and of copper catalysts with chlorine are examples of strong chemisorption, or "irreversible" poisoning.

In some cases of irreversible poisoning where removal of the poison is not practicable, a change of catalyst type may be the only solution. The original catalysts for SO_2 oxidation to SO_3 were unsupported-platinum, which worked satisfactorily except that arsenic compounds (found in the products of sulphur burning) poisoned the platinum. As the necessary gas purity was unattainable, the catalyst was changed to potassium-promoted V_2O_5 (Chapter 10) which is sufficiently resistant to arsenic poisoning. The susceptibility of Co_3O_4-based catalysts to poisoning is one reason[68] why they are not used for ammonia oxidation.

1.4.6. Poisoning by Reactants or Products

Most catalytic processes that use organic reactants form carbonaceous deposits on the catalysts. These are usually described as "coke", even though they are never pure carbon and always contain hydrogen, typically $C_1H_{0.4}$ to C_1H_1, and possibly other elements. "Coking" is by far the most common type of poisoning caused by reactants or products, but other cases are known. Methanol synthesis catalysts are carefully formulated[69] to avoid the possiblity of impurities; e.g. iron, which would promote Fischer–Tropsch reactions that would lead to blockage by waxes. Organometallic compounds in hydrocarbon feeds can react with H_2S to deposit metal sulphides on hydrodesulphurization catalysts.

Catalyst "coking" has been extensively studied,[68,77,80] especially in relation to oil-refinery processes. With hydrocarbon feeds (e.g. in naptha–steam reforming as well as oil-refinery processes), the coke deposits are built up by complex dehydrogenation/polymerization reactions. This route to coke is controlled in practice by using an optimized catalyst and an appropriate combination of process conditions, such as the use of a high pressure of hydrogen to decrease the tendency for dehydrogenation.

In ICI naphtha–steam reforming,[26] coke formation from the naphtha is controlled partly by hydrogen addition, which hydrogenates unsaturated coke precursors on the nickel component of the catalyst, and partly by the incorporation of potassium in the catalyst. Potassium has several roles: it neutralizes acidic sites on the support which would catalyse polymerization of olefins, and it limits coke formation on the nickel surface. If coke is formed, potassium compounds catalyse its gasification by steam as in the equation (27) providing an *in situ* route for catalyst regeneration.

$$C + H_2O \rightarrow CO + H_2 \qquad (27)$$

A different cause of carbon formation is present in ammonia and methanol production plants. Carbon monoxide can disproportionate under some conditions to give carbon and carbon dioxide according to the equation (28).

$$2CO \rightarrow C + CO_2 \qquad (28)$$

The equilibrium constants for carbon-forming reactions are shown in Figure 1.30. This route to carbon is controlled in two ways. As far as possible, plants are designed so that the gas mixtures are kept well away

Chapter 1. Fundamental Principles

from conditions under which carbon can be thermodynamically formed. This is essential for gases at high temperatures, but at low temperatures, e.g. below 200°C, carbon formation can be prevented by ensuring that there is no material present which will catalyse reaction (*28*).

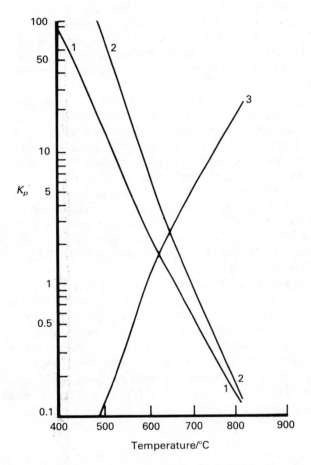

Figure 1.30. Equilibrium constants for the various carbon-forming reactions which can occur in ammonia and methanol plants. (1) $CO + H_2 \rightleftharpoons C + H_2O$, $K_p = P_{H_2O}/P_{CO}P_{H_2}$; (2) $2CO \rightleftharpoons C + CO_2$, $K_p = P_{CO_2}/P^2_{CO}$; (3) $CH_4 \rightleftharpoons C + 2H_2$, $K_p = P^2_{H_2}/P_{CH_4}$

1.4.7. Interactions in Catalyst Deactivation

The examples given above show that catalyst decay is a complex phenomenon and its diagnosis and cure is frequently difficult. Thus, the

1.4. Catalyst in Use

chloride poisoning of copper catalysts (methanol synthesis, LT shift) gives relatively mobile copper chlorides which poison the catalytic sites on the copper crystallites. These chlorides also act as "micro-fluxes" to promote sintering of copper crystallites, decreasing catalyst activity still further.[16] LT shift catalyst discharged from a plant after several years was examined[76] and the catalyst activity (laboratory test), Cu-crystallite size and poison levels (Cl and S) are plotted against the reactor bed depth in Figure 1.31, together with the appropriate temperature profile.

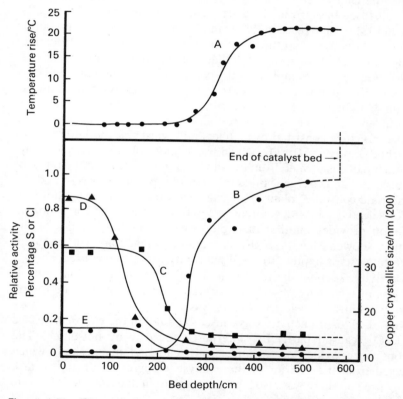

Figure 1.31. The variation of catalyst activity (from laboratory tests), copper crystallite size and poison levels (Cl and S) with reactor depth for an old charge of low temperature shift catalyst in a commercial reactor. A: temperature profile immediately before discharge; B: relative activity; C: copper crystallite size; D: sulphur content; E: chloride content.

The results show good correlation of loss of activity with poisoning. Chloride absorption levels as low as 0.05% can cause loss of activity. From the analytical data the HCl level in the gas phase was calculated to be no more than 0.001–0.003 ppm. Most of the loss in activity could be accounted for by the decrease in the copper surface area.

Another relevant example comes from the poisoning of naphtha–steam reforming catalysts with arsenic or sulphur compounds. These poisons deactivate the nickel crystallites, with the result that olefins and other unsaturated intermediates from the cracking of naphtha are not hydrogenated quickly enough to avoid further dehydrogenation/polymerization reactions to give coke. If the coking is severe, the catalyst rings burst open by coke deposition in the pores. Thus, all three basic forms of decay (corresponding to Sections 1.4.4, 1.4.5 and 1.4.6) take place in this example, and in a cursory examination of the discharged catalyst the loss of performance could be attributed to coking and catalyst collapse rather than to the primary cause, poisoning.

The final two examples both come from ammonia oxidation processes. Anomalously low yields of NO were occasionally found in medium- and high-pressure plants using Pt/Rh gauze catalysts. Detailed physical and chemical analysis revealed complex phenomena.[84] The surface structure consisted of platelets and whiskers, the latter almost pure rhodium oxide, Rh_2O_3, which was responsible for catalyst decay. Rhodium oxide is formed from Pt/Rh alloy under only a limited range of temperature and oxygen pressure, so the problem was cured by appropriate control of plant operating conditions. An investigation[56] of loss of selectivity in a development cobalt oxide catalyst for ammonia oxidation provides another example. Analysis of the kinetics of the reaction showed that, like the conventional reaction over platinum gauze catalysts (Chapter 10), high selectivity to NO is a consequence of high activity, keeping the process in a region of film-diffusion control. The loss in selectivity observed with trial catalysts was not due to any change in intrinsic selectivity, but was a consequence of decreased activity. The loss in activity was caused by a fall in the exposed surface of Co_3O_4, caused partly by sintering and partly by poisoning. The poisons were oxides of calcium, iron, lead and magnesium, which emerged from the interior of the catalyst to segregate on the surface during use and there was also a partial sulphation due to traces of SO_2 in the feed air stream.

Chapter 2

Process design, rating and performance

There are three basic types of question that can be asked about a catalytic reactor.

1. What size (and possibly what kind) of reactor is needed for a particular duty?
2. Given a particular reactor, how will it behave under a given set of operating conditions?
3. Given a particular reactor, and its behaviour under a set of operating conditions, what is the activity of the catalyst?

These three classes of problem are embraced by the terms of design, rating and calculation of performance. In order to deal with these problems numerically it is necessary to ascertain the kinetics for the catalyst to be used in the reactor, as discussed in Chapter 1. This chapter outlines the steps that are taken to do design, rating and performance calculations with reference to the reactions in the production of hydrogen, ammonia and methanol.

2.1
Design of Catalytic Reactors

There are a number of steps in the design of a catalytic reactor once the catalyst kinetics and the catalyst performance with time have been established. The calculation steps and their interdependence are shown in Figure 2.1. The calculation steps are not all done for every catalytic reactor design. Often the catalyst size and shape will be determined by the catalyst vendor, who will offer a limited range of catalysts appropriate to a given duty. When a complex converter is needed, proprietary designs will be offered by process licensors. The other calculation steps will then be done by the process licensor or engineering contractor, and the calculated size of the reactor will be checked by the catalyst vendor.

The procedure for calculating the process design parameters in Figure 2.1 will depend on what information is already known. If starting from scratch with a new catalyst, some preliminary iterations are needed around the steps in Figure 2.1. A rough idea of the catalyst volume is

Chapter 2. Process Design, Rating and Performance

needed to decide what type of converter should be used, to calculate catalyst size and shape, and the catalyst design life and activity. This preliminary work is usually done by a catalyst manufacturer or licensor, so it is assumed in this section that catalyst users have a rough idea of the volume of catalyst needed. The methods of calculating the process design parameters outlined in Figure 2.1 are set out in the rest of this section.

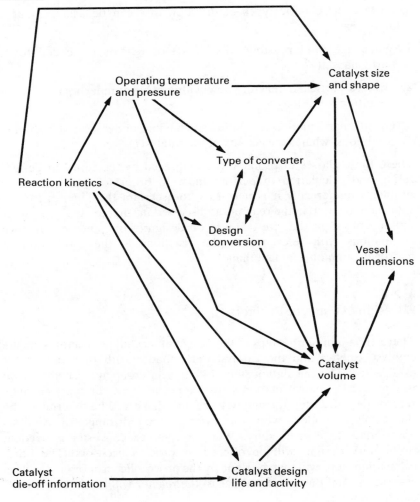

Figure 2.1. Schematic representation of the process steps for design of catalytic reactors.

2.1.1. Operating Temperature and Pressure

When a flowsheet is being generated for a new plant, the major decisions that have to be made are the operating temperatures and pressures of the reactors. The correct choice of operating conditions often involves a large amount of work evaluating the cost variations of the plant by changing these conditions. This is usually done using complex computer programs. However, it is possible to give some general guidelines for determining the operating conditions of particular catalysts and to explain the main factors that are taken into consideration in choosing them.

2.1.1.1. Desulphurization Reactor

The pressure used for desulphurization is set either by the available natural gas pressure or by the operating pressure of the primary reformer. The operating temperature for hydrodesulphurization depends on what sulphur compounds are present in the natural gas and whether there is a natural gas saturator in the flowsheet. If the natural gas contains compounds that are easily hydrogenolysed, then the hydrodesulphurizer can be operated at 250°C. However, for compounds where hydrogenolysis is more difficult a higher temperature in the region of 340—400°C must be used. The higher the operating temperature, the smaller the volume of catalyst necessary so, other things being equal, the desulphurizer is usually run at about 400°C regardless of the sulphur compounds present. However, with flowsheets that incorporate a natural gas saturator after the desulphurizer the gas is cooled in the saturator, so it can be economical to operate the desulphurizer at about 250°C. The operating temperature of zinc oxide for removal of hydrogen sulphide will be the same as the temperature chosen for hydrodesulphurization.

2.1.1.2. Steam Reformers

It is generally economical to run the primary reformer at as high a temperature and/or pressure as possible. Because of the creep limit of the reformer tubes (see Chapter 5, Figure 5.20), there is a limit to the operating temperature for a given pressure. The higher the temperature of the tube, the lower the pressure must be in order to prevent creep of the tubes. The operating temperature and pressure of a particular plant is determined from a trade-off between methane slip from the reformer, gas compression cost to the synthesis loop, raw natural gas compression cost and power obtained from letting down process steam from a higher pressure. With a methanol plant or hydrogen plant there is no secondary reformer, so a high temperature and a low pressure are needed on the primary reformer to reduce the methane slip.

Chapter 2. Process Design, Rating and Performance

On a conventional ammonia plant the secondary reformer raises the temperature of the process stream so high a very low methane slip results. Because of this effect it is possible to use a higher pressure and a relatively low exit temperature from the primary reformer.

In the ICI AMV ammonia process, methane slip from the reformers costs less than on a conventional plant, so the optimum secondary reformer temperature is lower. Also, because excess air is added to the secondary reformer, the primary reformer exit gas temperature is lower for a given secondary reformer exit temperature. These two effects mean that the operating conditions of an AMV reformer are at a higher pressure and a lower temperature than for a conventional ammonia plant. Typical primary reformer operating conditions for modern plants are shown in Table 2.1.

Table 2.1. Typical primary reformer operating conditions for modern plants

	Exit temperature/°C	Exit pressure/bar
Methanol or hydrogen plant	880	20
Conventional ammonia plant	820	30
ICI AMV ammonia plant	789	40

2.1.1.3. Water-gas Shift Reactors

The operating pressure of the high and low temperature (HT and LT) shift reactors in a hydrogen or an ammonia plant is determined by the operating pressure chosen for the reforming section. In hydrocarbon-based plants the HT shift reactor normally follows a high-pressure boiler or steam superheater. The design operating temperature of the HT shift reactor is therefore a trade-off between the cost of the upstream equipment and the cost of carbon monoxide slip from the HT shift reactor during operation.

As the operating temperature is lowered, the equilibrium carbon monoxide slip is reduced, but the temperature difference between the gas inlet the HT shift reactor and the saturation temperature of the high-pressure steam is also reduced, and this gives higher heat-exchanger costs. Typical temperatures inlet into the HT shift reactor are about 350°C for a 100-bar steam system and about 370°C for a 120-bar steam system. In coal-based plants there are two or three stages of HT shift, with heat interchange or quenching between them. The operating temperatures depend on the number of stages and on the method of cooling between stages.

The operating temperature of the low-temperature shift reactor is determined by a compromise between the size of the catalyst bed and the cost of carbon monoxide slip. The lower the operating temperature, the larger the volume of catalyst and vessel required, because of the lower reaction rate, but the carbon monoxide slip is lower and the overall plant efficiency is better. Further information concerning the derivation of the optimum LT shift reactor temperature is given in Section 2.1.6.

2.1.1.4. Methanation Reactor

If the operating temperature of the carbon dioxide removal system is at ambient or higher temperature (e.g. a hot carbonate system), then the methanator directly follows carbon dioxide removal and is at a similar pressure to the reformers. However, if a refrigerated carbon dioxide removal system (e.g. Selexol) is used, it is better to compress the cold gas, so saving on compression power. Methanation is then carried out at the compressor delivery pressure. The gas temperature rise during compression also means that less additional heat is needed to achieve the necessary methanator inlet temperature.

The operating temperature of the methanator is determined by a trade-off between the size of reactor needed and the cost of heating the gas to the operating temperature. The lower the operating temperature, the larger the methanator size needed because of the lower reaction rate, but less capital cost is needed to provide the equipment to heat and cool the gas before and after methanation.

2.1.1.5. Ammonia and Methanol Synthesis Reactors

The operating pressures of ammonia and methanol synthesis loops are determined by balancing capital costs of the synthesis loop against the necessary compression power and the refrigeration equipment for ammonia loops. Where energy is expensive, the tendency is to design for low-pressure operation in order to save on compression power, but the synthesis loop is then more expensive. The synthesis pressure can be kept low if a high-activity catalyst is used, or if a high-conversion reactor is used (see Section 2.1.2), or if it is not necessary to achieve a high hydrogen (ammonia) or high carbon (methanol) conversion efficiency. For example, the ICI low-pressure methanol process was made possible by having a high-activity catalyst, while the ICI AMV ammonia process has a loop pressure of 80–100 bar instead of the conventional 150–350 bar, due to a combination of all three of these reasons. The hydrogen conversion efficiency of the AMV loop is only 80%, but using a hydrogen recovery system the overall hydrogen efficiency of the synthesis section is raised to 98%.

The operating temperatures of methanol and ammonia synthesis catalysts depend on the pressure, the type of converter and the heat recovery required for the synthesis loop. The optimum temperature is lower at lower pressures because of equilibrium constraints.

2.1.2. Converter Types

When the operating temperature and pressure of a catalyst charge have been ascertained, the type of converter must be decided before calculating the optimum conversion and the catalyst volume required. This is because the type of converter will affect the conversion that can be obtained from a given volume of catalyst. In this section single adiabatic beds are considered first before discussing more complex arrangements.

2.1.2.1. Single Adiabatic Bed

When possible, it is normal simply to use a converter with a single adiabatic bed. This is possible when the conversion required will give an adiabatic temperature rise within the operating temperature range of the catalyst. When the converter is not equilibrium limited (e.g. the methanator) or when the temperature rise is very small, there is no need to consider anything other than a single adiabatic bed.

When the conversion is equilibrium limited, as with LT shift, there is an advantage in using a more complex converter, but the extra complication does not normally justify this. When the conversion required gives an adiabatic temperature rise greater than the operating temperature range of the catalyst, a more complicated converter must be used with some form of cooling. Cases where this occurs include methanol synthesis, ammonia synthesis, HT shift conversion for coal-based plants, and methanation when making SNG. The types of more complex converter can be divided into four main groups:

(a) multibed with quench between beds;

(b) multibed with interbed cooling;

(c) tube-cooled and

(d) steam-raising.

Each of these has advantages and disadvantages, and the choice of converter type used will depend on the operating experience, reliability, complexity and cost of the converter, as well as the volume of catalyst, the heat recovery required and the cost of associated heat exchangers. There are a number of different mechanical designs for each type of converter, which will depend on the duty to be performed and on the manufacturer or licensor.

The different types of converter are compared in Table 2.2 and Figures 2.2–2.5 for a typical duty. In the example given in Table 2.2 the conversion required will give an adiabatic temperature rise of 130°C, and since the catalyst should be operated between 240°C and 290°C, a

2.1.2. Converter Types

single adiabatic bed is inappropriate. The relative volumes of catalyst given for the different types of converter are only approximate, and will depend on the kinetics of the reaction. A minimum approach on converter feed/effluent interchangers of 20°C is assumed. This will mean that the amount of heat recovered will be exactly the same for each type of converter.

Table 2.2. Comparison of different types of converter

	Type of converter			
	Multibed		Tube cooled	Steam raising
	Quench	Intercooled		
Number of beds	6	3	1	1
Relative volume	1	0.95	0.9	0.95
Relative interchanger area	0.8	1.7	0.75	3.5
Heat recovery	270–120	270–160	260–150	360
Type of heat recovery	←———————— Any ————————→			Steam only
Type of flow				
axial	✓	✓	✓	✓
radial	✓	✓	X	✓
crossflow	X	X	X	✓

2.1.2.2. Quench Converter

In a quench converter cooling is achieved by mixing cold feed gas with the hot, partially reacted gas, from one catalyst bed before it enters another bed of catalyst as shown in Figure 2.2. There is a trade-off between the number of beds of catalyst and the temperature at which heat can be recovered. The higher the temperature of the quench stream, the higher the temperature at which heat can be recovered, but the greater the number of beds required for a given conversion. For some duties, to keep to a reasonable number of beds, the quench stream temperature may be so low that not all of the reaction heat can usefully be recovered.

2.1.2.3. Inter-bed Cooling

Inter-bed cooling can either be by interchange with the feed gas to the converter or by steam raising. When using feed gas interchange for cooling, it is normal to have the interchangers inside the same vessel as the catalyst as shown in Figure 2.3, but with steam raising between beds it is normal to have intercoolers in separate vessels in order to avoid mechanical complexity. However, the arrangement of the converter can become complicated when there is more than one inter-bed cooler when using feed gas interchange.

Chapter 2. Process Design, Rating and Performance

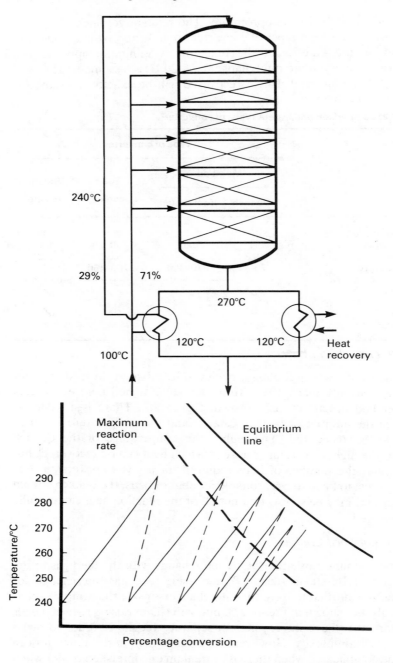

Figure 2.2. Arrangement and temperature profile for a multibed quench converter.

2.1.2. Converter Types

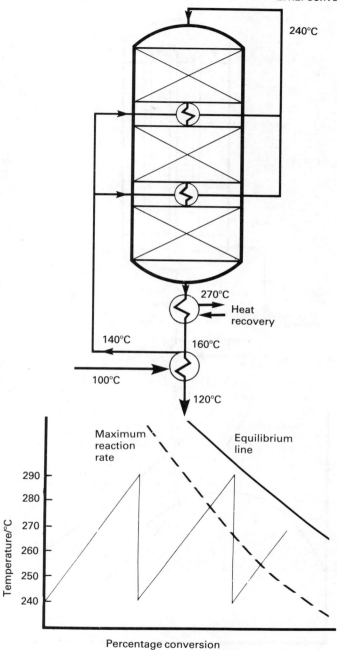

Figure 2.3. Arrangement and temperature profile for a multibed quench converter with intercooling between beds.

Chapter 2. Process Design, Rating and Performance

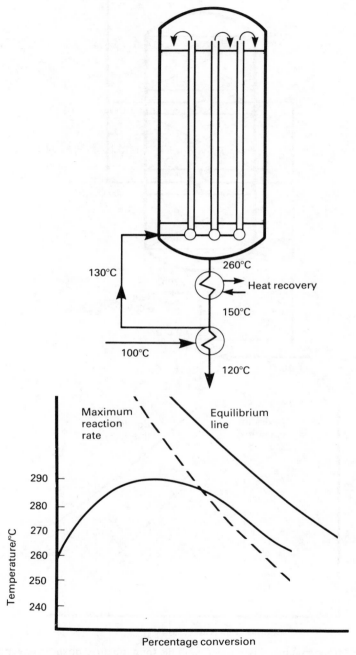

Figure 2.4. Arrangement and temperature profile for a tube-cooled converter.

2.1.2. Converter Types

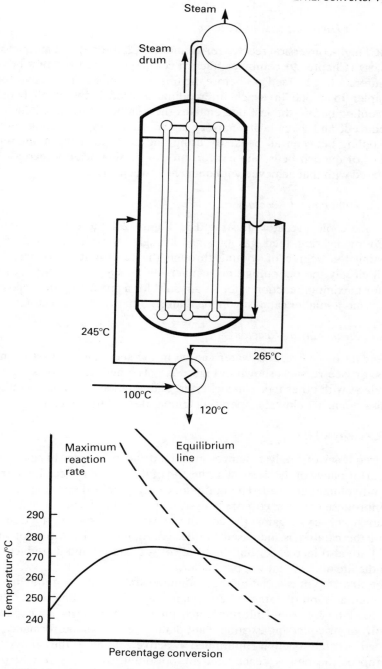

Figure 2.5. Arrangement and temperature profile for a steam-raising converter.

95

2.1.2.4. ICI High-conversion Reactor

The ICI high-conversion reactor for ammonia (Chapter 8) and methanol synthesis (Chapter 9) combines quench cooling between the first beds with intercooling by feed gas before the final catalyst bed. The converter is simpler than one in which intercooling is used between all beds. Intercooling before the last bed enables the number of quench beds to be reduced, and gives heat recovery at the same temperature as with intercooling between all beds. By using one more than the minimum number of quench beds, the conversion can be significantly increased compared with that achieved with quench between all beds.

2.1.2.5. Tube-cooled Reactor

In a tube-cooled reactor illustrated in Figure 2.4 the catalyst bed is cooled by heating feed gas passing through tubes in the bed. By adjusting the degree of cooling through the reactor it is possible to match closely the operating line (conversion versus temperature) with that for maximum reaction rate. As a result, for a given duty this type of reactor has a smaller catalyst volume than do other types of reactor.

2.1.2.6. Steam-raising Reactor

In a steam-raising reactor water can be in the shell and catalyst in the tubes, or vice versa as shown in Figure 2.5. The heat recovery is not as flexible as with other types of converter, and reaction heat must be used to raise steam at below the operating temperature of the reactor.

2.1.3. Catalyst Life

The frequency of catalyst changes must be decided in order to calculate the performance of the catalyst at the end of its life and hence obtain the catalyst volume required. The optimum catalyst life is determined from a compromise between cost of catalyst, plus cost of reducing the catalyst (where necessary), against the cost of the catalyst vessel. As the design life of the catalyst is increased the design volume and the cost of the vessel are also increased, but the total cost of catalyst and its reduction over the life of the plant will be smaller.

The first step in calculating the optimum catalyst change frequency is to obtain a graph of catalyst performance against operating time. This can be done by using information from laboratory tests on catalyst die-off, or by collecting existing plant data. The catalyst performance is defined as catalyst activity multiplied by active volume. If the catalyst at the top of the bed is deactivated by poisoning while the rest of the catalyst bed remains active, as effectively happens with zinc oxide in the

2.1.3. Catalyst Life

desulphurization section and LT shift catalysts, then the performance will decrease with time as shown in Figure 2.6. If all of the catalyst charge is gradually poisoned or sintered, then the performance of the charge decreases with time as in Figure 2.7. Examples of this kind of behaviour include HT shift, ammonia synthesis and methanol synthesis catalysts.

Using the kinetics associated with a particular catalyst, its activity and die-off behaviour, the plant output and/or plant efficiency can be calculated from the installed volume of catalyst. This is done for different installed volumes at different times during catalyst life. Typical graphs so produced are shown in Figures 2.8 and 2.9. If the catalysed reaction is virtually complete, then it is not possible to increase plant output when the catalyst is new, so the graph of output against operating time for different volumes of catalyst will be as in Figure 2.8, and this is the case for desulphurization. If the reaction is equilibrium-limited, then it will always be possible to obtain a higher plant output with new catalyst by optimizing the operating temperature (see Section 2.2.1). The graph of plant output against time will then be as shown in Figure 2.9. Examples of this situation include HT shift, ammonia synthesis and methanol synthesis catalysts.

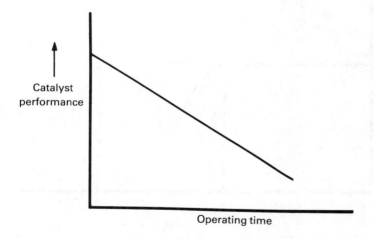

Figure 2.6. Graph of catalyst performance against operating time for catalyst where active volume decreases.

Chapter 2. Process Design, Rating and Performance

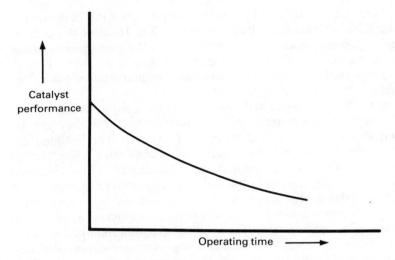

Figure 2.7. Graph of catalyst performance against operating time for catalyst where activity falls off throughout the bed.

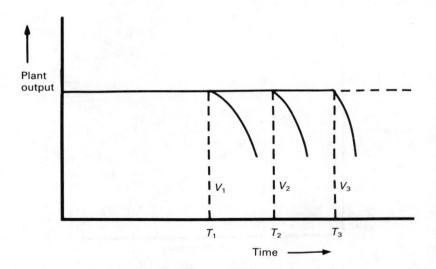

Figure 2.8. Variation of plant output with time for catalyst giving complete conversion. T_1, T_2 and T_3 are the catalyst lives for installed volumes of catalyst V_1, V_2 and V_3, respectively.

2.1.3. Catalyst Life

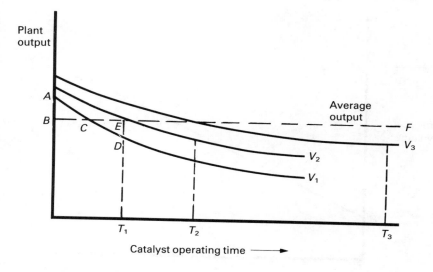

Figure 2.9. Variation of plant output with time for catalyst where conversion is equilibrium limited. T_1, T_2 and T_3, are the catalyst lives to give an average plant output B for installed volumes of catalyst V_1, V_2 and V_3, respectively. Area ABC = area CDE.

Point B in Figure 2.9 represents the average output required from the plant. For each volume of catalyst (V_n) for which the plant output versus time curve crosses the average output line BF, there is a catalyst life (T_n) for which the average plant output will be equal to the required output B. For example, if the catalyst volume is V_1, then at the beginning of life the plant output A is greater than the average output required, and after some time it becomes less than the average required. In order for the average output with volume V_1 to be equal to the average output, the catalyst should be changed at a time interval T_1 such that area ABC = area CDE. From Figures 2.8 and 2.9, additional graphs are drawn to give the catalyst volume needed for different design lives of the catalyst. The plant output against operating time graph in Figure 2.8 will give a catalyst volume against design life graph shown in Figure 2.10. Similarly, the graph in Figure 2.9 will translate into the graph in Figure 2.11. The graph in Figure 2.11 must be linearized in the region of interest by drawing a line GH at a tangent to the curve. In this region $V = H + gT$, where H is the intercept on the y-axis and g is the gradient of the line HG, and is equal to dV/dT.

The optimum catalyst life T_{opt} is calculated from equation (1)

$$T_{\text{opt}} = \left(\frac{(H \times C_c) + R}{g \times C_v \times \text{CCF}}\right)^{1/2} \qquad (1)$$

Chapter 2. Process Design, Rating and Performance

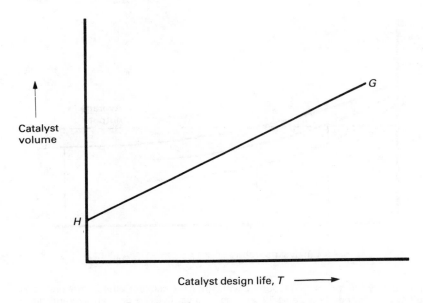

Figure 2.10. Catalyst volume needed as a function of catalyst design life for catalyst giving complete conversion.

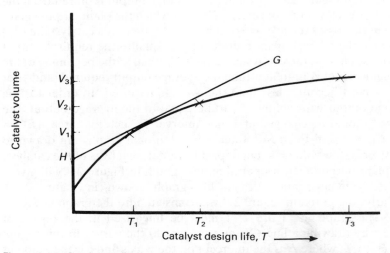

Figure 2.11. Catalyst volume needed as a function of catalyst design life for catalyst where conversion is equilibrium-limited. V_1, V_2 and V_3 are volumes of catalyst needed for catalyst design lives T_1, T_2 and T_3, respectively. HG is a tangent to the curve.

where C_c is the catalyst cost (£ m^{-3}), C_v is the marginal catalyst vessel capital cost (£ m^{-3}), R is the total cost of reducing the catalyst (£ per occasion) and CCF is the annual capital charge factor to relate capital costs to annual running costs.

If the value of T_{opt} calculated is different from the point at which the tangent to the graph is drawn, then a new tangent can be drawn at the calculated value of T_{opt} and a more accurate design life can be derived. When H is small compared with g, as for example with the desulphurization system with a high sulphur feed gas, it can be seen from the equation that the optimum zinc oxide life will be short. This could be a fraction of a year, and in this case the volume will be determined by the need to avoid changing the charge between planned plant overhauls. In cases where T_{opt} is not large compared with the period between planned overhauls, the design life of the catalyst will be adjusted so that it is equal to a whole number of planned shutdown intervals.

When the design catalyst life has been decided the corresponding catalyst activity is obtained from Figures 2.6 and 2.7. For some catalysts, the optimum design life and activity are calculated by the catalyst manufacturer or process licensor, who will guarantee the performance of the catalyst for that period. For other catalysts, sufficient information is available for the customer to ascertain what design catalyst life is optimum.

2.1.4. Optimum Catalyst Size and Shape

When the operating conditions of a catalytic reactor have been determined, it is necessary to decide the form of catalyst that should be used. The catalyst size and shape can be characterized by two independent variables: the voidage (e) and the equivalent diameter of the catalyst particle (d_e) defined by equation (2).

$$d_e = \frac{6 \times \text{volume of particle}}{\text{surface area of particle}} \qquad (2)$$

It is assumed in the derivations below, that changing the catalyst particle size and shape does not affect the particle density, or its micromeritic properties such as pore size distribution.

2.1.4.1. Voidage

The shape of the catalyst particle will determine the voidage of the catalyst bed — the more eccentric the shape, the greater the voidage. Table 2.3 gives typical voidages for some different particle shapes. The optimum voidage is determined by a trade-off between the cost of the

Chapter 2. Process Design, Rating and Performance

containing vessel and the pressure drop through the bed. The greater the voidage, usually the larger is the vessel necessary but the smaller the resultant pressure drop. Another factor that must be taken into account, but which cannot be easily quantified, is that the more eccentric the particles are, the more likely they are to break during operation. In practice catalyst strength is an important factor.

Table 2.3. Typical voidage of some catalyst shapes in vibrated catalyst beds

Particle	Length / Diameter	Hole diameter / External diameter	Voidage
Spheres			0.360
Pellets	0.5		0.405
	1.0		0.355
	1.33		0.376
Rings	1.0	0.39	0.450
	0.68	0.39	0.460

For a fixed bed catalyst vessel with axial flow operating at a pressure greater than 10 bar the installed cost can be approximated using equation (3),

$$\text{vessel cost} = C_v V_c + C_d D^3 \quad (3)$$

where V_c is the volume of vessel excluding dished ends, D is the catalyst vessel diameter, and C_v and C_d are constants which can be estimated. Assuming that no catalyst is packed in the dished ends of the vessel, V_c is equal to the bulk volume of catalyst, V_b, that is given by equation (4), where V_s is the solid volume of catalyst.

$$V_b = V_s/(1 - e) \quad (4)$$

The pressure drop in catalyst beds can be calculated by the correlation derived by Ergun.[85] In most catalytic reactors on commercial plants, the flow is fully turbulent, so the Ergun equation can be simplified to give equation (5),

$$\text{pressure drop} = 1.75 \frac{(1-e)}{e^3} \frac{G^2 h}{d_e \rho} \quad (5)$$

where ρ is the gas density and h is the height of catalyst in the bed. If no catalyst is packed in the dished ends, then h is given by equation (6), and

$$h = 4V_b/\pi D^2 \quad (6)$$

2.1.4. Optimum Catalyst Size and Shape

G in equation (5) is the mass flow rate for unit cross section of bed, and is given by equation (7), M being the mass flow rate.

$$G = 4M/\pi D^2 \qquad (7)$$

During operation, beds of catalyst usually settle, so that the actual pressure drop is greater than that calculated by the Ergun equation. The pressure drop can be multiplied by a factor f to account for this. From equations (5)–(7), equation (8) can be obtained in which a is given by equation (9).

$$\text{pressure drop} = \frac{aV_b(1-e)}{D^6 e^3 d_e} \qquad (8)$$

$$a = 3.61 f M^2/\rho \qquad (9)$$

The capitalized cost of pressure drop (CCPD) in a process stream is given by equation (10) with C_p being defined in equation (11).

$$\text{CCPD} = C_p \times \text{pressure drop} \qquad (10)$$

$$C_p = tPL_p/\text{CCF} \qquad (11)$$

In equation (11) t is the running time of the plant in a year, P the cost of power and L_p the power lost per unit pressure drop through the catalyst. The capital charge factor (CCF) relates annual running costs to capital costs, and is the inverse of simple payback time. The power lost per unit of pressure drop depends on where any pressure lost in the reactor must be made up by extra compression.

Here it has been assumed that the catalyst cost is constant and independent of catalyst shape, which may not always be the case. Equation (12) incorporates the assumption that vessel dimensions are optimized (see Section 2.1.7), and gives the optimum voidage in a catalyst bed. The dimensionless n is given by equation (13).

$$\text{voidage} = \sqrt{n}/(1 + \sqrt{n}) \qquad (12)$$

$$n = \frac{1.89}{C_v} \left(\frac{aC_p}{d_e}\right)^{1/3} \left(\frac{C_d}{V_s}\right)^{2/3} \qquad (13)$$

Normally the optimum voidage ranges from 0.35 to 0.55, and is consistently higher for some reactors (e.g. secondary reformers) than for others. An example of the calculation of optimum voidage is given in Appendix 2.

2.1.4.2. Catalyst Particle Size

As the catalyst size is increased, the pressure drop through the reactor decreases. However, as the size is increased there comes a point where

Chapter 2. Process Design, Rating and Performance

the catalyst activity is limited by diffusion within the particle. The total volume of catalyst must then be increased in order to maintain the same "effective volume". The optimum catalyst particle size results from a compromise between the cost of the catalyst, the cost of the vessel and the cost of pressure drop through the bed.

In most commercial catalytic reactors, the catalyst is limited by pore diffusion rather than by film diffusion, and the effectiveness (E) of a catalyst depends on the Theile modulus ϕ, which is given by equation (14).

$$\phi = \frac{d_e}{2}\left(\frac{\text{intrinsic reaction rate}}{\text{effective diffusivity of limiting reactant or product}}\right)^{1/2} \quad (14)$$

The Theile modulus can be calculated at the average temperature in the bed and at the design activity of the catalyst and can be represented by equation (15),

$$\phi = bd_e \quad (15)$$

where b is a constant. For a spherical catalyst particle the effectiveness is given by equation (16).

$$E = \frac{3}{\phi}\left\{\frac{1}{\tanh \phi} - \frac{1}{\phi}\right\} \quad (16)$$

In practice, equation (16) is sufficiently accurate for catalyst pellets and rings, as well as for spheres. b can be calculated directly from equations (15) and (16) using data for the relative activities of two different sizes of catalyst particle under the same operating conditions. The solid volume of catalyst required is given by equation (17),

$$V_s = V_a/E \quad (17)$$

where V_a is the volume of catalyst needed for an effectiveness of unity. The capitalized cost of catalyst per unit volume (C_{cat}) is given by equation (18).

$$C_{cat} = \frac{\text{cost of catalyst per unit volume}}{\text{catalyst life} \times \text{CCF}} \quad (18)$$

It can be shown that the optimum size of catalyst is given by equation (19),

$$m = (bd_e)^{8/3}(0.4 + 0.0216[bd_e]^2) \quad (19)$$

where m is dimensionless, and is calculated from equation (20).

$$m = \frac{1.89(1-e)}{(C_v + C_{cat})e}\left(\frac{C_d}{V_a}\right)^{2/3}(C_p ab)^{1/3} \quad (20)$$

If the product $b \times d_e$ is small, then the optimum value of d_e is obtained from equation (21), whereas if the product $b \times d_e$ is greater than unity, then it is necessary to calculate an approximate value of d_e from equation (21) and then obtain an improved estimate using equation (22).

$$\text{optimum } d_e = (1/b)(2.5\,m)^{3/8} \qquad (21)$$

$$\text{optimum } d_e = \frac{1}{b}\left(\frac{m}{0.4 + 0.0216(bd_e)^2}\right)^{3/8} \qquad (22)$$

There is a small interdependence of optimum voidage and optimum catalyst particle size, but it is rarely necessary to recalculate the voidage after calculating the particle size. A typical calculation of optimum catalyst particle size is given in Appendix 2.

2.1.5. Design Conversion of Reactor

In Section 2.1.3 the optimum catalyst design activity was calculated from a compromise between reactor volume cost and catalyst cost, assuming that all other plant equipment remained the same. This section outlines how the conversion duty (and efficiency) of the catalyst is determined by an optimization of the reactor and catalyst costs with the capital cost of the rest of the plant, assuming a constant catalyst activity.

The optimum conversion in a reactor is calculated by minimizing the cost function given in equation (23).

$$\text{total cost} = \text{capitalized catalyst cost} + \text{vessel cost} + \text{other costs} \qquad (23)$$

The capitalized cost of catalyst per unit volume (C_{cat}) is given by equation (18), while the vessel cost can be calculated from equation (3). "Other costs" are the capital cost of the rest of the plant plus the capitalized variable costs (given by equation (24)) minus the value of product from the plant.

$$\text{capitalized variable cost} = \frac{\text{unit variable cost} \times \text{annual production}}{\text{CCF}} \qquad (24)$$

From equations (3), (18) and (23), the total cost equation (25) can be obtained.

$$\text{total cost} = VC_{cat} + VC_v + D^3C_d + \text{other costs} \qquad (25)$$

The problem is set up in one of two ways. Either the catalyst volume is changed and parts of the rest of the plant are changed to maintain the

Chapter 2. Process Design, Rating and Performance

same plant output, or the catalyst volume is changed, while the rest of the plant is kept the same and the plant output is varied. For example, when considering the conversion in a methanol or ammonia synthesis loop, either the circulation rate can be changed to keep the same output, or the circulation rate can be kept the same and the plant output will change.

At the minimum total cost, the differential of total cost with respect to conversion will be zero, so the optimum reactant exit concentration (A) can be found by solving equation (26).

$$\frac{dV}{dA}(C_v + C_{cat}) = \frac{d}{dA}(\text{other costs}) \qquad (26)$$

The terms on the right-hand side of this equation are normally constant over a fairly large range of conversion. For example, with low-temperature shift the carbon monoxide conversion is normally between 85% and 95%. For any small step in this range the effect of a change in conversion on the cost of the rest of the plant will be constant. The right-hand side of the equation is therefore set equal to a constant C_a, as shown in equation (27).

$$C_a = \frac{d}{dA}(\text{other costs}) \qquad (27)$$

C_a can be found by comparing flowsheets with different conversions in the reactor. Equation (26) can therefore be written as in equation (28).

$$\frac{dV}{dA} = \frac{C_a}{C_v + C_{cat}} \qquad (28)$$

For a single-bed converter, if the conversion is reasonably close to equilibrium and the reaction is first-order in the concentration of a reactant, then, from Section 2.1.6, we may write equation (29),

$$-\frac{dV}{dA} = \frac{Q}{k_r(A - A_e)} \qquad (29)$$

where k_r is the rate constant for the reaction (see Section 2.1.6.4), A is the concentration of reactant, A_e is the equilibrium concentration of reactant at operating temperature and Q is the volume flow rate of gas. Therefore, at optimum conversion we have equation (30).

$$A - A_e = \frac{Q(C_v + C_{cat})}{k_r C_a} \qquad (30)$$

Example 3 in Appendix 2 gives a typical calculation of optimum conversion in a reactor. For complicated conversion optimizations, where C_a is difficult to determine, or with multibed reactors, equation (23) is solved using an optimizing computer program.

2.1.6. Calculation of Catalyst Volume

When the catalyst operating conditions, the catalyst size and shape, the design activity and the required conversion have been determined, the catalyst volume can be calculated.

The catalyst volume is a function of a large number of variables: temperature, pressure, gas composition and reaction conversion, catalyst formulation, catalyst particle size and catalyst activity. The effect of some of these variables can be complex and, moreover, a number of the variables change through the catalyst bed. Because of this, catalyst volumes are normally calculated by computer programs that integrate through the catalyst bed. However, by making a large number of simplifying assumptions it is possible to calculate approximate catalyst volumes for some of the reactions in ammonia plants under typical conditions, using simple calculations.

The rate of reaction in a converter can be represented simply by equation (31),

$$\text{rate} = k_r \times f(\text{composition}) \times \left(1 - \frac{K}{K_p}\right) \qquad (31)$$

where K_p is the equilibrium constant and k_r is the rate constant at a particular point in the converter. The latter is obtained from the full kinetic expression for the catalyst concerned and will depend on catalyst particle size, design activity and operating temperature. K is a measure of how far the reaction is from equilibrium, and is a function of the reactant and product partial pressures. For instance, in a reaction $A + B \rightarrow C + D$, K is given by equation (32), where P_a is the partial pressure of reactant A, etc.

$$K = \frac{P_c P_d}{P_a P_b} \qquad (32)$$

Equation (31) can be used to calculate the catalyst volume for two special cases: low-concentration reactant being removed and product being formed at low concentration.

2.1.6.1. Catalyst Volume for Low-concentration Reactant being Removed

In reactions such as hydrodesulphurization, secondary reforming, carbon monoxide shift and methanation, the reactant is at a low

Chapter 2. Process Design, Rating and Performance

concentration, so as the reaction proceeds the concentrations of the other components do not change by a large proportion. The f(composition) in equation (31) only changes significantly due to the reduction of the reactant concentration which is being consumed, so equation (31) can be rewritten as shown in equation (33),

$$\text{reaction rate} = Q\frac{dA}{dV} = -k_r A\left(1 - \frac{K}{K_p}\right) \quad (33)$$

where A is the concentration of reactant A, Q is the volume flow rate of gas and V is the catalyst volume. Because of small changes in other reactants and products, the K/K_p term in equation (33) may be written as in equation (34), and the simple rate equation (35) results.

$$\frac{K}{K_p} \simeq \frac{P_{ae}}{P_a} = \frac{A_e}{A} \quad (34)$$

Here P_{ae} is the equilibrium partial pressure of reactant A, and A_e is the equilibrium concentration of reactant A, so we can write equation (35), in which both k_r and A_e are temperature-dependent.

$$\text{rate} = Q\frac{dA}{dV} = -k_r A\left(1 - \frac{A_e}{A}\right) = -k_r(A - A_e) \quad (35)$$

However, most of the catalyst volume is close to the bed exit temperature, because the reaction rate reduces as the reactant concentration reduces and equilibrium is approached, so there is little error in assuming that the reaction is isothermal at the bed exit temperature. Integrating the rate equation (35) gives equation (36),

$$V = \frac{Q}{k_r}\ln\left(\frac{A_i - A_e}{A_o - A_e}\right) \quad (36)$$

where A_i is the inlet concentration of reactant A, A_o is the outlet concentration of reactant A and A_e is the equilibrium concentration of the reactant A at the bed exit temperature. An example of the use of equation (36) is given in Appendix 2.

For reactions such as methanation, the equilibrium concentration of the reactant is so low that it can be ignored, and under these conditions equation (37) gives the required catalyst volume.

$$V = \frac{Q}{k_r}\ln(A_i/A_o) \quad (37)$$

2.1.6.2. Catalyst Volume for Low-concentration Product being Formed

In reactions such as ammonia synthesis or methanol synthesis, the product is produced at a low concentration in the reactant mixture — typically less than 15% — so that in a stoichiometric mixture of reactants the concentration of reactants does not change by a large proportion. The f(composition) term in equation (31) does not change significantly and it can be assumed constant, so the rate of reaction is given by equation (38),

$$\text{rate} = Q\frac{dC}{dV} = k_r\left(1 - \frac{K}{K_p}\right) \quad (38)$$

in which C is the concentration of product C. Also, equation (39) holds,

$$\frac{K}{K_p} \simeq \frac{P_c}{P_{ce}} = \frac{C}{C_e} \quad (39)$$

where P_{ce} is the equilibrium partial pressure of product C and C_e is the equilibrium concentration of product C. Therefore, the rate of reaction per unit volume is given by equation (40).

$$\frac{dC}{dV} = k_r\left(1 - \frac{C}{C_e}\right) \quad (40)$$

Assuming isothermal operation, the volume of catalyst required is given by equation (41), where C_i is the inlet concentration of product C, C_o is the outlet concentration of product C and C_e is the equilibrium concentration of product C at the bed exit temperature.

$$V = \frac{QC_e}{k_r}\ln\left(\frac{C_e - C_i}{C_e - C_o}\right) \quad (41)$$

2.1.6.3. Equilibrium Concentrations

At low pressures the equilibrium constant for a particular reaction is a function of temperature only, and is denoted here by K_{p0}. For a given reaction K_{p0} can be represented by equation (42),

$$K_{p0} = \exp(a_1 + a_2)/T + a_3/T^2 + a_4/T^3 + \cdots \quad (42)$$

where $a_1, \ldots, a_4$ are constants and T is the absolute temperature.

For ease of calculation on computers or calculators, it is normal to calculate K_{p0} from the type of relation shown in equation (43) with Z defined as in equation (44),

$$K_{p0} = \exp\left(Z(Z(Zb_4 + b_3) + b_2) + b_1\right) \quad (43)$$

$$Z = (1000/T) - 1 \quad (44)$$

and $b_1, \ldots, b_4$ are constants of similar size. Specific examples of such formulae are given in Appendices 6 and 7 together with tabulations of

Chapter 2. Process Design, Rating and Performance

K_p for different reactions.

Over a small temperature range of, say, 100°C, the equilibrium constant can be represented more simply by equation (45).

$$K_{p0} = \exp(K_{p1} + K_{p2}/T) \qquad (45)$$

For reactions at high pressure, K_p is a function of pressure and composition as well as temperature. For reactions in the reforming and hydrodesulphurization purification sections of ammonia and methanol plants up to 40 bar, pressure corrections are not needed. However, for the synthesis loops it is important to use a pressure correction. These are detailed in literature for both ammonia synthesis[86] and methanol synthesis.[87] The correction can be simplified with a small loss of accuracy by using equation (46), so combining with equation (45), the full equation for K_p is given in equation (47).

Table 2.4. Equilibrium constants for reactions in ammonia and methanol manufacture

$$K_p = \exp\left\{K_{p1} + \frac{K_{p2}}{T} + \frac{P}{T}\left[K_{p3} + \frac{K_{p4}}{T}\right]\right\}$$

Reaction	K_p	Temperature/°C	Pressure bar	K_{p1}	K_{p2}/K	$K_{p3}/$ K bar^{-1}	$K_{p4}/$ K^2 bar^{-1}
Desulphurization over zinc oxide	$\dfrac{P_{H_2O}}{P_{H_2S}}$	300–400	0–50	−0.1839	9149		
Reforming	$\dfrac{P_{CO}P^3_{H_2}}{P_{CH_4}P_{H_2O}}$	750–1050	0–50	30.345	−27 278		
Shift	$\dfrac{P_{H_2}P_{CO_2}}{P_{H_2O}P_{CO}}$	750–1050	0–50	−3.670	3971		
HT shift		~440	0–50	−4.2939	4546		
LT shift		~250	0–50	−4.3701	4604		
Methanation	$\dfrac{P_{CH_4}P_{H_2O}}{P_{CO}P^3_{H_2}}$	~350	0–50	−29.254	26 251		
	$\dfrac{P_{CH_4}P^2_{H_2O}}{P_{CO_2}P^4_{H_2}}$	~350	0–50	−24.845	21 627		
Ammonia synthesis	$\dfrac{P^2_{NH_3}}{P_{N_2}P^3_{H_2}}$	350–550	50–300	−27.366	12 500	−1.42	2100
Methanol synthesis	$\dfrac{P_{CH_3OH}}{P^2_{H_2}P_{CO}}$	240–300	0–100	−29.06	11 900	−0.0150	10.8

2.1.6. Calculation of Catalyst Volume

$$K_p = K_{p0} \exp \frac{P}{T}\left(K_{p3} + \frac{K_{p4}}{T}\right) \qquad (46)$$

$$K_p = \exp\left[K_{p1} + \frac{K_{p2}}{T} + \frac{P}{T}\left(K_{p3} + \frac{K_{p4}}{T}\right)\right] \qquad (47)$$

Values of $K_{p1}, \ldots K_{p4}$ are given in Table 2.4 for normal reaction conditions in methanol and ammonia plants.

The equilibrium concentration of a reactant A_e or product C_e is calculated from the equilibrium constant and inlet concentrations. For the reaction $A + B \rightarrow C + D$, equation (48), in which the subscript o denotes exit concentrations, applies.

$$K_p = \frac{C_o D_o}{A_e B_o} = \frac{C_e D_o}{A_o B_o} \qquad (48)$$

If we denote the differences between actual and equilibrium concentrations by ΔA and ΔC, according to equations (49) and (50) in which the subscript i denotes inlet concentrations, A_e and C_e can then be found by solving equations (51) and (52).

$$A_i - A_e = \Delta A \qquad (49)$$

$$C_e - C_i = \Delta C \qquad (50)$$

$$A_e = \frac{(C_i + \Delta A)(D_i + \Delta A)}{K_p (B_i - \Delta A)} \qquad (51)$$

$$C_e = \frac{K_p(A_i - \Delta C)(B_i - \Delta C)}{(D_i + \Delta C)} \qquad (52)$$

Calculated equilibrium concentrations of important reactants and products for some of the reactors under typical conditions in hydrogen, ammonia and methanol plants are shown in Figures 2.12–2.18. The assumptions made in deriving these data are as follows.

1. The feed to the plant is pure methane.

2. For the secondary reformer, HT and LT shift it is assumed the plant has a methanator for final carbon monoxide and carbon dioxide removal, and the hydrogen to nitrogen ratio of the gas at the exit of the methanator is 3 : 1.

3. For the secondary reformer it is assumed that the dry carbon monoxide level in the gas from the LT shift is 0.33% and the dry carbon dioxide level in the gas from the carbon dioxide removal system is 0.1%.

4. For HT and LT shift it is assumed that the dry carbon dioxide level in

Chapter 2. Process Design, Rating and Performance

the gas from the carbon dioxide removal system is 0.1% and the methane level in the gas at the exit of the methanator is 1%.

It is now quite common with ammonia plants to install hydrogen recovery units in the purge gas stream, and recycle hydrogen to the synthesis loop (see Section 8.7). This will result in the hydrogen to

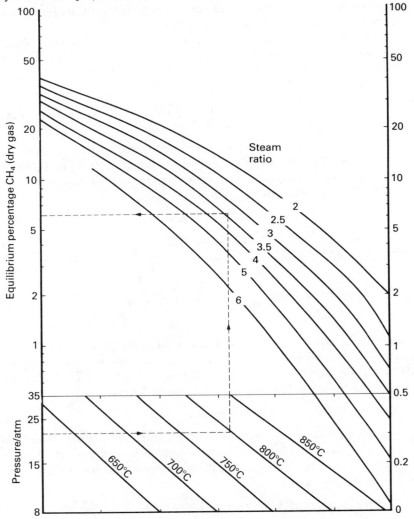

Figure 2.12. Equilibrium concentration of methane as a function of temperature, pressure and steam ratio for methane feed to a primary reformer.

2.1.6. Calculation of Catalyst Volume

nitrogen ratio from the methanator being less than 3 : 1, so corrections to the diagrams are given for lower ratios. Corrections are also given for the shift reactors when the methane level at the exit of the methanator is different from 1%.

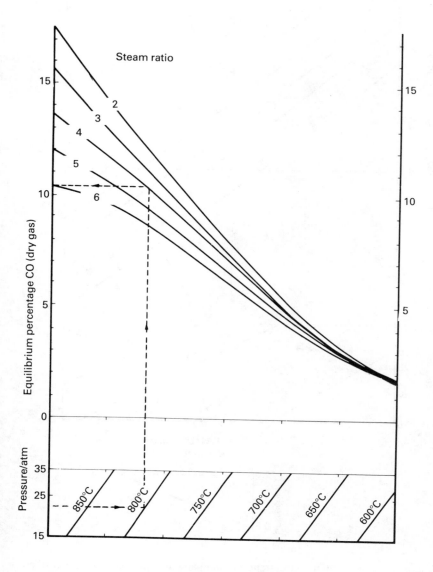

Figure 2.13. Equilibrium concentration of carbon monoxide as a function of temperature, pressure and steam ratio for methane feed to a primary reformer.

Chapter 2. Process Design, Rating and Performance

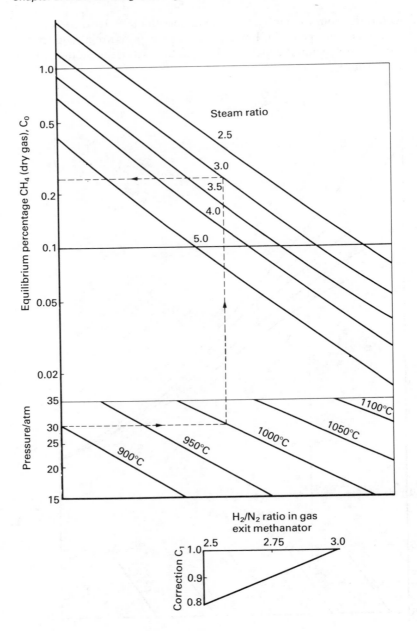

Figure 2.14. Equilibrium concentration of methane in a secondary reformer of an ammonia plant as a function of temperature, pressure and steam to carbon ratio. Feed is methane. CH_4 concentration $= C_o \times C_1$.

2.1.6. Calculation of Catalyst Volume

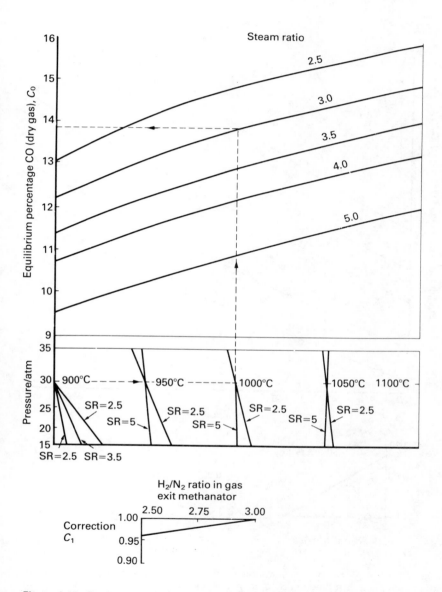

Figure 2.15. Equilibrium concentration of carbon monoxide in secondary reformer of an ammonia plant as a function of temperature, pressure and steam ratio. Feed is methane. CO concentration = $C_o \times C_1$.

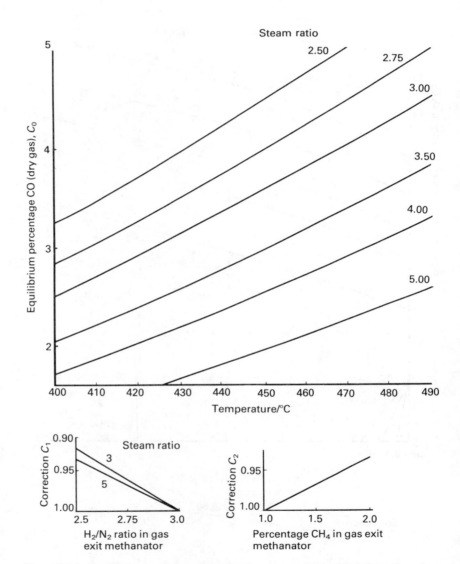

Figure 2.16. Equilibrium concentration of carbon monoxide in HT shift section of an ammonia plant as a function of temperature and gas composition at the exit from the methanator. CO concentration = $C_o \times C_1 \times C_2$.

2.1.6. Calculation of Catalyst Volume

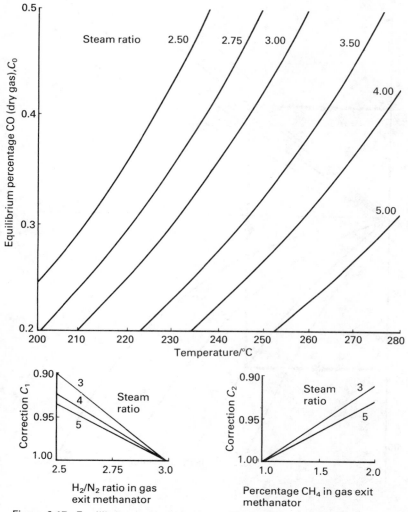

Figure 2.17. Equilibrium concentration of carbon monoxide in LT shift section of an ammonia plant as a function of temperature and gas composition at the exit from the methanator. CO concentration = $C_o \times C_1 \times C_2$.

Chapter 2. Process Design, Rating and Performance

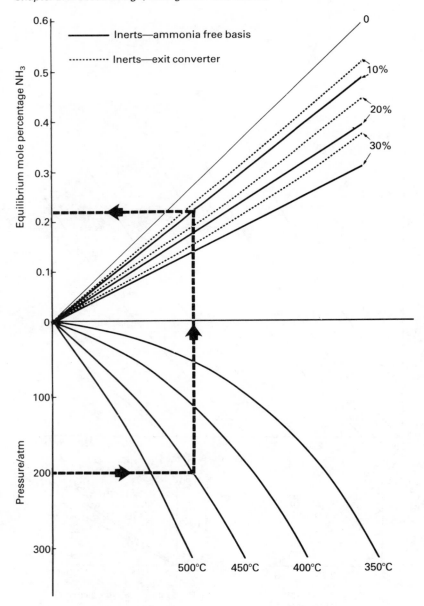

Figure 2.18. Equilibrium concentration of ammonia in synthesis converter as function of temperature, pressure and inerts level. H_2/N_2 ratio = 3 : 1.

2.1.6.4. Rate Constants

The rate constant for a particular reaction at a given pressure and composition over a small temperature range for a given catalyst size varies with temperature as represented by equation (53),

$$k_r \propto \exp(-E_a/RT) \tag{53}$$

where E_a is the activation energy of the reaction and R is the universal gas constant.

The activation energy for a given reaction may depend on the catalyst formulation, and on whether the reaction is diffusion limited. The proportionality constant will therefore depend on the activity of the catalyst and the pressure and composition of the gas stream.

For ease of calculation the rate constant can be represented as shown in equation (54).

$$k_r = \exp(k_{r1} - k_{r2}/T) \tag{54}$$

Typical design values of k_{r1} and k_{r2} for some typical commercial catalysts for conventional ammonia plant conditions are given in Table 2.5.

Table 2.5. Rate constants for commercial catalysts at typical ammonia plant conditions $k_r = \exp[k_{r1} - (k_{r2}/T)]$ (k_r in N m³ h⁻¹ wet gas m⁻³ catalyst)

Catalyst	Steam/Dry gas ratio	Pressure/bar	Operating time/years	k_{r1}	k_{r2}/K
HT shift	Any	30	2	23.0	9500
LT shift	0.30	30	New	22.8	6600
LT shift	0.45	30	New	23.0	6600
Methanator		30	2	16.4	3600

2.1.7. Vessel Dimensions

When the duty required to be done by the reactor has been decided, and when the particle size and shape have been chosen, and the bulk volume of catalyst V_b has been calculated, the vessel dimensions can be decided. The important parameters that need to be determined are the length to diameter ratio of the vessel and how much of the dished end is filled with catalyst. The optimum dimensions are a compromise between the vessel cost and the operational cost of pressure drop through the vessel.

Using equations (3), (4), (8), (9) and (11), it can be shown that the optimum diameter of the vessel using consistent units is that given by equation (55).

$$D = \left(\frac{2C_p a V_b (1-e)}{C_d e^3 d_e}\right)^{1/9} \tag{55}$$

Chapter 2. Process Design, Rating and Performance

The derivation of equation (55) assumes the reactor is adiabatic, flow is axial and that there is no bed height limitation due to crushing or channelling within the catalyst charge. It can be seen that the optimum diameter is insensitive to any of the variables except voidage because of the low exponent, so there is no need for great accuracy in calculating these. Moreover, the derivation of equation (55) assumes that the catalyst is not packed in the dished ends of the vessel. However, the size of a new catalyst vessel can be reduced if catalyst is put part way into the dished ends.

If a catalyst vessel is filled so that the minimum diameter of bed in the dished end is D_e, as illustrated in Figure 2.19, it can be shown that the optimum D_e is given by equation (56).

$$D_e = \left(\frac{C_d e^3}{C_p a} \frac{d_e}{(1-e)} + \frac{1}{D^6} \right)^{-1/6} \tag{56}$$

If the vessel diameter has been optimized according to equation (55), equation (56) can be simplified to give equation (57).

$$\frac{D_e}{D} = \left(\frac{2V_b}{D^3} + 1 \right)^{-1/6} \tag{57}$$

The height of the catalyst in a vessel when catalyst is put into dished ends will, of course, depend on their shape. Examples 5 and 6 in Appendix 2 give typical calculations of vessel shape.

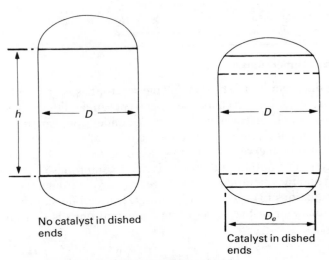

Figure 2.19. Design dimensions of catalyst vessels.

2.2 Reactor Rating

When a catalytic reactor has been designed, it is usual to calculate how it would perform under a range of operating conditions; for example, when the catalyst is new or when the gas composition or inlet temperature changes. This calculation of what the reactor will do under changed conditions is termed rating of the reactor. Basically the same equations can be used for rating simple reactors as for determining reactor volumes. For a reactor which is removing a low concentration reactant, from equation (*36*) we can write equation (*58*),

$$A_o = A_e + (A_i - A_e)\exp\left(\frac{Q}{V_a k_r k_a}\right) \quad (58)$$

for a reactor producing a product at low concentration, from equation (*41*), we can write equation (*59*),

$$C_o = C_e - (C_e - C_i)\exp\left(\frac{QC_e}{V_a k_r k_a}\right) \quad (59)$$

where k_a is the catalyst activity relative to the design activity, Q is the volume flow rate of gas, V_a is the active catalyst volume, k_r is the rate constant for the reaction at the operating pressure and reactor exit temperature at design catalyst activity (see Section 2.1.6.4), A_i, A_o and A_e are the inlet, exit and equilibrium concentrations, respectively, of a reactant, and C_i, C_o and C_e are the inlet, exit and equilibrium concentrations, respectively, of a product. For reactions which are not equilibrium-limited, e.g. methanator, then $A_e = 0$.

If the exit temperature of the reactor being rated is fixed, then equation (*59*) can be solved easily. The equilibrium concentration of the reactant or product is found using the method in Section 2.1.6.3. If the inlet temperature for the reactor rating is fixed and the exit temperature is variable, then some iteration is needed. First a guessed approximate conversion is used to calculate the reactor exit temperature (or this is found using Figures 2.21–2.23, below) then the conversion is recalculated using equation (*59*). This iteration is time-consuming and it is usually more appropriate to use a computer program — see Section 2.4.

2.2.1. Optimum Operating Temperature

Reactors are normally designed for a catalyst activity or active catalyst volume which is less than when the catalyst is new, with the result that when the catalyst is new and is operated at the design operating

Chapter 2. Process Design, Rating and Performance

temperature, the exit composition may be very close to the equilibrium value. Consequently with exothermic reactions when the catalyst is new a higher conversion can be achieved by lowering the operating temperature. In practice the inlet temperature is reduced, and the approach to equilibrium will then not be as good as for the design operating temperature, but the equilibrium conversion will be better. There will be some optimum operating temperature at which the highest conversion can be achieved. This temperature can be found by using equation (58) or (59) to calculate the exit concentration of the reactant

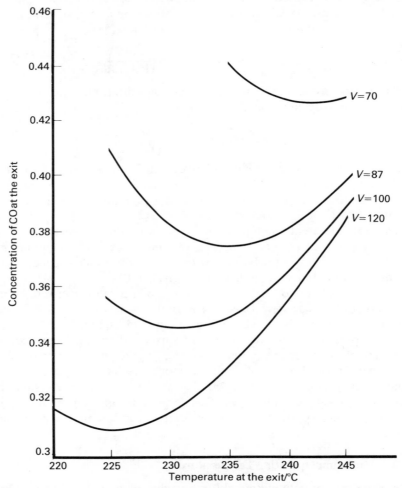

Figure 2.20. Exit concentration of CO from a LT shift reactor as a function of operating temperature and catalyst volume (V) for particular inlet gas composition and rate.

or product over a range of temperatures. This has been done for a LT shift example shown in Figure 2.20, which shows clearly how the optimum operating temperature changes with the active volume of catalyst.

2.3. Catalyst Performance

The performance of all catalyst systems gradually deteriorates over a period of time, and it is important to monitor the performance of catalysts so that the rate of performance fall-off can be assessed in order to know when the catalyst should be regenerated or replaced, or to identify an abnormal rate of deterioration due to some cause that may be rectified. Moreover, as the catalyst performance changes, the optimum operating conditions of the catalyst will change. It is therefore important at the design stage to provide adequate means for monitoring catalyst performance. There are two aspects of catalyst performance that can be monitored easily. These are the apparent activity of the catalyst and the pressure drop across the reactor.

2.3.1. Fall in Apparent Catalyst Activity

The following reasons for apparent loss of catalyst activity can be identified:

(a) poisoning/sintering

(b) poor gas distribution

(c) poor mixing of reactants

2.3.1.1. Poisoning/sintering

Catalyst deactivation is discussed in detail in Section 1.4.3, and mainly results from ingress of poison, or loss of surface area of the active phase through sintering. It is sometimes possible to regenerate catalyst by removing poisons. Well-designed catalysts only sinter very slowly at normal operating temperatures. However, sintering can occur at normal catalyst operating temperatures if particular poisons are present. For instance, halogens form low-melting surface compounds with many metals, and will significantly accelerate the rate of crystal growth. Because the catalyst structure is destroyed by sintering usually it is not possible to regenerate the catalyst. In terms of activity the effects on overall catalyst performance of poisoning and sintering are similar, and it is not always easy to distinguish which is causing the fall-off in apparent activity.

2.3.1.2. Poor Gas Distribution

Poor gas distribution can occur in large catalyst beds due to uneven packing of the bed caused by incorrect charging; for example, by pouring the catalyst into the vessel in one spot (see Chapter 3), which causes the voidage to vary across the bed. Poor gas distribution is normally only a problem in reactors which have a high conversion such as methanator and desulphurization reactors.

The following example illustrates the effect of poor gas distribution in a methanator.

Inlet CO concentration	0.5%
Design CO concentration at the exit (good distribution)	5 ppm
Voidage through one half of cross section	0.3
Voidage through other half of cross section	0.4
Actual CO concentration at the exit	20 ppm

In the half of the bed with the higher voidage, the gas velocity is much higher, so the conversion is lower than in the low-voidage half. Because more gas goes through the higher voidage section, the average concentration of carbon monoxide at the exit is higher than it would otherwise be.

In a primary reformer there can be a problem of poor gas distribution between the different tubes. This is especially a problem in methanol reformers where there is a higher conversion of methane to synthesis gas than in a primary reformer on an ammonia plant. If flow through a tube is high, the heat transferred to the process gas will not raise its temperature as much as in other tubes, so the equilibrium methane concentration will be higher. Also, because the space velocity is higher in the tube, the rate of conversion at a given temperature will be lower. These two effects will give a higher methane level at the exit from tubes with high flow, and will cause the methane level from the reformer to be higher than design. This illustrates the importance of correctly charging reforming tubes that is discussed in detail in Sections 3.6.4, 5.8.5 and Appendix 19.

2.3.1.3. Poor Mixing of Reactants

Poor mixing of reactants can be a problem in primary and secondary reformers where reactants are mixed before contacting the catalyst. In the primary reformer, if the distance between the mixing point and the split of the gases between headers and tubes is too small, then different areas of the reformer may not be supplied with feed having the same steam to carbon ratio. This will give a worse performance of the reformer than with good mixing, and in extreme cases could lead to

2.3. Catalyst Performance

carbon laydown in some tubes. In the secondary reformer air reacts with partially reformed gas in a flame before passing over the catalyst. There is a problem in some "gas-ring" type burners that parts of the burner may become overheated and break away, giving non-uniform air distribution. The poor air distribution then gives a non-uniform concentration and temperature gradient through the secondary reformer and an undesired high methane slip.

An example of the effect of poor air mixing in the secondary reformer is shown in Table 2.6, where one half of a secondary reformer gets 20% less air than it should and the other half gets 20% more air than it would if there were good mixing. This results in a 50% increase of the methane slip.

Table 2.6. Effect of poor mixing in secondary reformer

	With good mixing	With poor mixing		Average
		Half with 20% less air	Half with 20% more air	
Secondary reformer exit temperature/°C	957	902	1034	971
Methane exit concentration/%(dry)	0.62	1.89	0.13	0.97
Approach to methane/ steam equilibrium/°C	10	10	10	53
Approach to shift equilibrium/°C	0	0	0	6

2.3.2. Increase in Pressure Drop

The pressure drop through a catalyst bed can be higher for a variety of reasons, and the main ones are considered in this section.

2.3.2.1. Breakage or Erosion of Catalyst Particles

Breakage of catalyst particles can result from incorrect charging of the catalyst, or from differential expansion and contraction between the catalyst and the containing vessel due to temperature cycling at start-up and shut-down. The breakage of catalyst particles gives a smaller particle diameter, while erosion of the corners of particles gives a lower voidage due to the eroded particles packing more closely together.

2.3.2.2. Disintegration of Catalyst Particles

Disintegration of the catalyst particles can occur at the top of catalyst beds if the inlet gas distribution is poor and if the bed is not covered by a

Chapter 2. Process Design, Rating and Performance

mesh or large particles of other material — see Section 3.6.2. When the inlet gas distribution is poor, high velocities will cause catalyst particles to be thrown around the space at the top of the vessel, leading to continuous erosion and disintegration. Powder from the disintegration of the particles will percolate through the catalyst bed, clogging spaces between other particles giving a low voidage and higher pressure drop.

2.3.2.3. Deformation of Catalyst Particles

If some catalysts become wet, especially in the reduced state, then the catalyst strength decreases, leading to deformation of the particles and reduced voidage that results in increased pressure drop across the catalyst bed.

2.3.2.4. Carry-over on to Catalyst Bed

Carry-over of solution from an upstream vessel or of dust in the gas streams will cause the spaces between particles to become blocked, giving a high pressure drop.

2.3.2.5. Collapse of Bed Support

If there is a collapse of the bed support grid, or whatever device is used to hold catalyst away from the vessel exit nozzle, then the exit nozzle will be partly screened, causing a high pressure drop at the exit nozzle.

2.3.3. Measurement of Performance

The following measurement techniques are commonly used to determine data for the evaluation of catalyst performance: analysis of inlet and exit gas, catalyst bed temperature rise, catalyst bed temperature profiles, radioactive tracing and pressure drop. Each of these is discussed in this section.

2.3.3.1. Analysis

Analysis of a gas stream from a reactor is normally the primary source of information used to calculate the conversion within the reactor, and hence to determine the reactor performance. Analysis can be done by taking individual gas samples and analysing them in a laboratory, by continuous analysers on the plant or by passing a measured gas flow through a solution for a measured time, and analysing this solution. Due to errors in such plant analyses, other methods are usually also used to calculate the conversion in a reactor as detailed in the following sections.

2.3. Catalyst Performance

2.3.3.2. Mass Balance

In circulating loops, such as ammonia or methanol synthesis, a mass balance around the loop and the circulation rate are used as checks on the conversion.

2.3.3.3. Catalyst-bed Temperature Rises

The temperature rise in a catalyst bed can be measured quite accurately by thermocouples at the inlet and exit of the bed, and enables an accurate calculation of the conversion in the reactor to be made. This information can be used to check the analysis at either the inlet or the exit of the reactor.

2.3.3.4. Catalyst-bed Temperature Profiles

Catalyst-bed temperature profiles can be measured at regular intervals by passing a travelling thermocouple through a sheath in the bed and measuring the temperature at different points through the bed. Temperature profiles can be used directly to calculate the catalyst activity or the active volume of catalyst in the bed.

2.3.3.5. Radioactive Tracing

Radioactive tracing is used to check whether there is poor distribution or by-passing in the reactor. A small amount of a radioactive species is injected upstream of the reactor and the radioactivity measured at the exit from the reactor. The decay of the level of radioactivity with time at the exit of the reactor is used to calculate the residence time distribution in the reactor, and hence to establish whether by-passing is taking place.

2.3.3.6. Pressure Drop

If the pressure drop across a reactor is measured regularly, any increase of pressure drop with time on-line can be monitored. The pressure drop should always be measured by using the same pressure gauge on pressure tappings at the inlet and exit of the reactor, or by connecting a differential pressure cell (DP cell) between pressure tappings at the inlet and exit of the reactor.

2.3.4. Quantifying Catalyst Performance

Different ways of quantifying performance are applicable to different catalysts, but there are three general ways in which the performance of a reactor can be quantified, to give meaningful information about changes with time. These are:

(a) measurement of composition at the exit from the reactor

(b) determination of approach to equilibrium at the reactor exit

(c) calculation of catalyst activity or active volume of catalyst.

Each of these are discussed in the following Sections.

2.3.4.1. Composition at the Exit from the Reactor

Analysis of the product stream at the exit from a reactor is a good measurement of the catalyst performance when the operating temperature of the reactor is held constant and when the reaction over the catalyst is not equilibrium limited. An example of this is hydrodesulphurization but in practice the hydrogen sulphide level exit the zinc oxide bed is too low to measure using conventional methods.

2.3.4.2. Approach to Equilibrium

The approach to equilibrium at the exit of a catalyst bed is the difference between the gas temperature at the exit of the catalyst bed and the equilibrium temperature corresponding to the gas composition. The approach to equilibrium can be used as a good measure of the performance of the catalyst when the operating temperature of the reactor is held constant and when the reaction is equilibrium limited, such as with primary and secondary reformers. This is satisfactory of course only if the gas composition does not change during sampling. If the operating temperature of the catalyst varies with time, the approach to equilibrium by itself is not a good measure of the performance of the catalyst. For instance, a catalyst of a given activity may give an approach to equilibrium of 5°C at one operating temperature, but an approach to equilibrium of 10°C if the operating temperature is lowered by only a few degrees.

2.3.4.3. Activity or Active Volume of Catalyst

Determination of catalyst activity or the active volume of catalyst (see Sections 2.3.5.3 and 2.3.5.4) is the only meaningful measure of the performance of a catalyst when the operating temperature of the catalyst changes with time, or when there are large variations of gas composition at the inlet to the reactor. Examples include LT shift and methanol synthesis.

2.3.5. Calculation of Catalyst Performance

For all catalysts the exit composition from the reactor must first be established to quantify the performance of the catalyst. When this has been done, the approach to equilibrium and/or catalyst activity can be calculated.

2.3. Catalyst Performance

2.3.5.1. Reactor Exit Composition

The exit composition from a reactor usually can be determined directly, and can be checked from mass balances or from the temperature rise through the reactor (or perhaps by calculation from the temperature rise in an adjacent reactor). Mass balances and temperature rises enable the conversion in a reactor to be calculated, and when necessary this can be used to find the exit concentration (e.g. in an ammonia converter) or the inlet concentration of a reactant (e.g. in a LT shift). The relationships between the conversion and the adiabatic temperature rise across HT shift, LT shift and ammonia converters are given in Figures 2.21–2.23. In the methanator, the temperature rise is 60°C for 1% of CO_2 and 74°C for 1% of CO converted to methane.

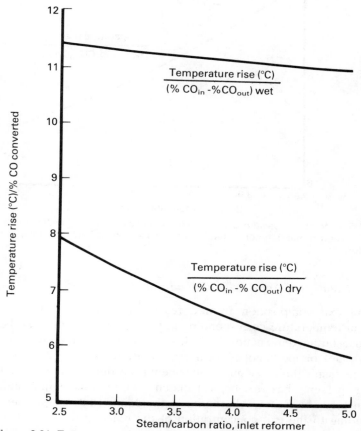

Figure 2.21. Temperature rise across a charge of HT shift catalyst as a function of the amount of carbon monoxide converted and the steam to carbon ratio inlet the primary reformer.

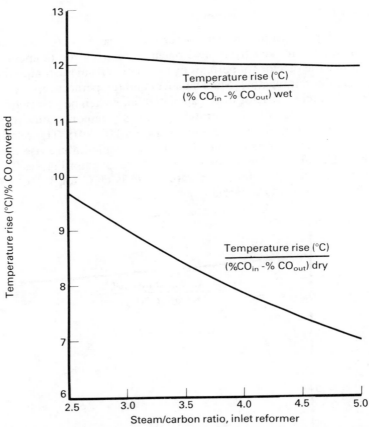

Figure 2.22. Temperature rise across a charge of LT shift catalyst as a function of the amount of carbon monoxide converted, and the steam to carbon ratio inlet the primary reformer.

2.3.5.2. Calculation of Approach to Equilibrium

If the full exit composition from a reactor is available, then the equilibrium temperature corresponding to the exit composition can be calculated as outlined in Section 2.1.6.3. Analyses in the front end of an ammonia plant or methanol plant are normally done on a dry basis because practically this is the most convenient procedure, so the amount of steam present has to be calculated before the equilibrium temperature can be found. Alternatively, the equilibrium temperature corresponding to the exit concentration of the reactant or product of interest can be estimated graphically. Diagrams for the exit concentrations of reactants or products for the primary reformer,

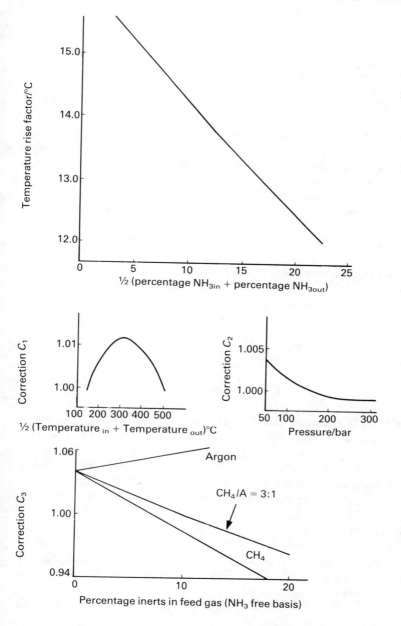

Figure 2.23. Temperature rise across an ammonia converter as a function of conversion. Temperature rise = temperature rise factor × (%NH$_{3out}$ − % NH$_{3in}$) × C_1 × C_2 × C_3.

Chapter 2. Process Design, Rating and Performance

secondary reformer, HT shift, LT shift and ammonia converters for typical operating conditions are given in Figures 2.12–2.18.

2.3.5.3. Calculation of Activity or Active Volume from Composition

For equilibrium limited reactions the equilibrium concentration must be found from the exit composition and exit temperature in order to calculate the catalyst activity or active volume. The equilibrium concentration of the reactant A_e or product C_e of interest is found using the procedure given in Section 2.1.6.3, or from the diagrams in Figures 2.12–2.18. For catalysts that are deactivated evenly throughout the bed (e.g. HT shift) the active volume can be considered to be constant and the catalyst activity will decrease over the life of the plant. For catalysts which are poisoned at the start of the bed before the rest of the bed is affected (e.g. some LT shift catalysts) the activity can be considered to be constant, but the active volume will decrease over the life of the charge.

Equations derived in Sections 2.1.6.1 and 2.1.6.2 are rearranged for calculating catalyst performance. The catalyst activity for a low concentration reactant being removed is given by equations (60) and (61) when a low-concentration product is being formed

$$k_a = \frac{Q}{Vk_r} \ln\left(\frac{A_i - A_e}{A_o - A_e}\right) \tag{60}$$

$$k_a = \frac{QC_e}{Vk_r} \ln\left(\frac{C_e - C_i}{C_e - C_o}\right) \tag{61}$$

where k_a is the catalyst activity relative to design activity, Q is the volume flow rate of gas, V is the catalyst volume, k_r is the rate constant at the operating pressure, the reactor exit temperature and the design activity (see Section 2.1.6.4), A_i, A_o and A_e are the inlet, exit and equilibrium concentrations, respectively, for a reactant, and C_i, C_o and C_e are the inlet, exit and equilibrium concentrations, respectively, for a given product.

The active volume (V_a) of a catalyst for a low-concentration reactant being removed is given by equations (62) and (63) for the situation where a low concentration product is being formed (for reactions which are not equilibrium-limited, as for methanation, $A_e = 0$):

$$V_a = \frac{Q}{k_r} \ln\left(\frac{A_i - A_e}{A_o - A_e}\right) \tag{62}$$

$$V_a = \frac{QC_e}{k_r} \ln\left(\frac{C_e - C_i}{C_e - C_o}\right) \tag{63}$$

2.3.5.4. Calculation of Activity or Active Volume from Temperature Profiles

Reactors with a catalyst that is poisoned evenly throughout the bed will show a changing temperature profile, as shown in Figure 2.24. The activity can be found by drawing a tangent to the profile curve at a temperature T_a. The slope of the curve given by the tangent is dT/dV. The catalyst activity relative to the design activity for a reaction where one of the reactants is consumed is calculated from equation (64).

$$k_a = \frac{dT}{dV} \frac{(A_i - A_o)}{(T_o - T_i)} \frac{Q}{k_r (A - A_e)} \quad (64)$$

Where k_a, Q, A_i, A_o and A_e are as in the previous section, k_r is the rate constant at temperature T_a (see Section 2.1.6.4), T_i and T_o are the inlet and outlet temperatures, respectively, and A is the concentration of reactant at temperature T_a as given by equation (65).

$$A = \frac{(T_a - T_i)(A_o - A_i)}{T_o - T_i} \quad (65)$$

A_e is the equilibrium concentration of the reactant at temperature T_a, which can be determined as in Section 2.1.6.3 or from Figures 2.12–2.18.

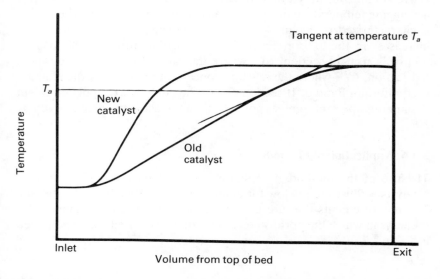

Figure 2.24. Temperature profiles for a catalyst that is poisoned evenly throughout the bed. The catalyst activity can be calculated from equation (64).

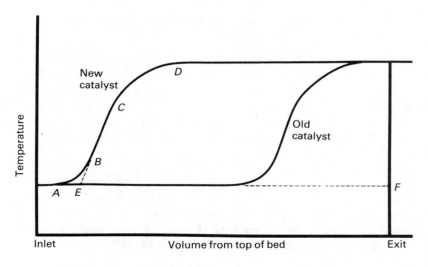

Figure 2.25. Temperature profiles for a catalyst that is only poisoned at the beginning of the bed. Active volume is given by *EF*.

Reactors with a catalyst that takes up poisons at the beginning of the bed without affecting the rest of the bed can be considered to have an active volume of catalyst that will gradually decrease. The bed will show a changing temperature profile, as illustrated in Figure 2.25. The profile is characterized by having a "take-off" section, AB, where the slope increases as the poison level decreases going through the bed, a relatively straight section, BC, where the catalyst is unpoisoned, and a tail section, CD, where the reactant concentration is low or the reaction is equilibrium limited. If section BC of the curve is extrapolated to cut the inlet temperature line at point E, then the active volume is given by EF.

2.3.6. Application of Methods

The use of the methods in Section 2.3.5 to find the performance of catalysts is illustrated below for ammonia and methanol plant catalysts. The measurements that are used for monitoring the performance and the ways in which the performance is usefully expressed are summarized in Table 2.7.

2.3. Catalyst Performance

Table 2.7. Summary of performance measurements and quantification of performance for ammonia and methanol catalysts

	Catalyst						
	Desulphurizer	Primary reformer	Secondary reformer	HT shift	LT shift	Methanator	Synthesis (NH_3, CH_3OH)
Measurements							
Exit analysis	✓	✓	✓	✓	✓	✓	✓
Mass balance						✓	✓
Temperature rise				✓	✓	✓	✓
Temperature profile				✓	✓	✓	✓
How to quantify performance							
Exit analysis	✓					(✓)	
Approach to equilibrium		✓	✓				(✓)
Catalyst activity				✓			✓
Catalyst active volume					✓	✓	

2.3.6.1. Desulphurizer

The temperature change in the desulphurization unit is small, so the only way of monitoring the performance is by exit analysis. Once the zinc oxide becomes saturated with sulphur and sulphur slip begins, the slip will rise rapidly. In order to avoid premature shutdown because of high sulphur slip, an analysis point can be put part way up the bed so that when hydrogen sulphide is detected at this point the lifetime of the remaining zinc oxide can be calculated (see Chapter 4).

2.3.6.2. Primary and Secondary Reformer

The primary and secondary reformer performance is monitored by recording compositions and temperatures at the exit in order to calculate the approach to equilibrium. The calculation of the catalyst activity is complicated and is not normally done routinely, but the approach to equilibrium provides a good indication of performance. The temperature measurements at the exit from the primary and secondary reformers are often inaccurate due to radiation effects, failure of thermocouples at high temperatures or poor gas distribution. The carbon monoxide shift reaction proceeds faster than the reforming reaction does, so the carbon monoxide shift reaction should always be at equilibrium at the exit from the reformer, and this can therefore be used as a check on the reformer exit temperature. The temperature can be found from the carbon monoxide level at the exit by using Figure 2.15. The methane/steam equilibrium temperature can be obtained from Figure 2.14. Poor performance of the primary or secondary reformers can be due to poor gas distribution in the primary reformer, or to poor mixing of the air in the secondary reformer, rather than to poor catalyst activity.

2.3.6.3. HT Shift

The exit concentration of carbon monoxide from the HT shift reactor can be found from analysis or from the temperature rise across the LT shift catalyst. The latter can be used to give the carbon monoxide conversion in the LT shift using Figure 2.22. By adding a rough estimate of the carbon monoxide slip at the exit from the LT shift, the HT shift exit concentration can be found. Another check on the carbon monoxide concentration at the exit can be obtained from its concentration at the exit from the reformer and the temperature rise across the HT shift catalyst. The catalyst activity is calculated from the carbon monoxide level at the exit, using Figure 2.16 and equation (60). It can be checked from the temperature profile using equation (64).

2.3.6.4. LT Shift

The carbon monoxide slip from the LT shift is normally found by exit gas analysis and this can be checked by the temperature rise in the methanator, if the carbon dioxide slip from the CO_2 removal system is known. The active volume of catalyst is calculated from the carbon monoxide concentration at the exit, using Figure 2.17 and equation (62). It can be checked from the temperature profile, as in Section 2.3.5.4.

2.3.6.5. Methanator

The performance of the methanator is measured by the concentration of carbon oxides at the exit. If the temperature in the methanator is held constant, then the carbon oxide concentration at the exit is a sufficient quantitative indication of performance. If the operating temperature is varied, then the performance should be quantified by the catalyst active volume calculated from equation (62).

2.3.6.6. Ammonia and Methanol Synthesis Converter

The concentration of ammonia or methanol at the exit from the converter is obtained by analysis. It can be checked by adding the converter inlet concentration to the conversion, calculated from the adiabatic temperature rise through the converter, or from the product rate and synthesis loop circulation rate. The relationship between conversion and temperature rise for ammonia synthesis is given in Figure 2.23. The performance of the ammonia and methanol synthesis catalyst can be quantified roughly by the approach to equilibrium, but if temperatures and loop composition change, then the reaction rate constant will change, so the performance must be quantified by the catalyst activity. Figure 2.18 can be used to calculate the approach to equilibrium for ammonia synthesis catalyst, but a full gas analysis is needed to calculate the equilibrium temperature for methanol synthesis catalyst. The kinetic equations for ammonia and methanol synthesis are complicated, so the catalyst activity is calculated using computer programs.

2.4. Computer Programs

The preceding part of this chapter has given the methods for designing catalyst volumes, rating the converter for different conditions and calculating the performance of catalyst charges. The basic equations used and the data for typical cases have been presented, in order to obtain results by simple hand-calculations or by graphic methods. In many instances this is not appropriate, and computer programs have to be used.

2.4.1. Reasons for Using Computer Calculations

The reasons for needing to carry out computer simulations, and the sort of calculations where computing is necessary are given below.

2.4.1.1. Accurate Calculation

The simple equations presented in previous sections make a number of assumptions about the reactions (see Section 2.1.6). If these assumptions are not valid, then the calculation through the bed must be done stepwise, changing the temperature, composition, rate constant and equilibrium constant for each step. This process is too time consuming to be done using a calculator.

2.4.1.2. Non-isothermal Reactors

Adiabatic reactors are assumed to be isothermal for the purpose of carrying out simple approximate calculations (see Section 2.1.6), but where there is a large amount of cooling or heating (e.g. primary reformer) the assumption is grossly inaccurate and the reactor must be modelled stepwise.

2.4.1.3. Multiple Reactions

Where there is more than one equilibrium limited reaction, the composition of the gas after it has passed through a certain volume of catalyst has to be found by iteration between the two reactions. This is too laborious to do by hand, and there are usually too many variables for the problem to be solved graphically.

2.4.1.4. Optimization

When a problem is incompletely defined, optimization is needed to find the "best" solution. For design problems this solution is one giving the minimum capital cost plus capitalized operating cost. For performance calculations the "best" solution gives the minimum operating cost. An example is a multibed converter, where it is necessary to find the optimum split of catalyst between beds. Another example is where the conversion in a reactor is optimized to minimize the cost of the reactor plus the cost of the rest of the plant.

2.4.1.5. Simulation

When the operation of a reactor or a section of plant is to be simulated, there are normally extra measurements beyond the minimum set of measurements needed to define the state of the reactor or section of plant. The extra measurements will often conflict with the simulation produced from the minimum set of measurements. In order to use all of the measurement information available to produce a simulation giving the best fit to the measurements, a computerized simulation method is required.

2.4.2. Types of Computer Programs

Computer programs for catalytic reactor calculations can either be "stand-alone" or integrated in a generalized flowsheeting package. "Stand-alone" programs perform one or a few particular calculations. For example, a program may be written to design a cold-shot methanol converter, or it may be written for both design and rating of a cold-shot

2.4. Computer Programs

methanol converter. Once written it cannot be used to find the activity of the catalyst in the converter given the measurements from a simulation. The program would have to be run a number of times with different catalyst activities to see which run gave the best fit to the data.

Computer programs in a generalized flowsheeting package are more flexible. For instance, in the example cited previously, once the equations describing the reactions in a bed of methanol catalyst have been put into the flowsheeting package, then the routines in the package can be used to manipulate the equations as required. Any number of beds can be combined to model a cold-shot converter, and the converter can be connected to models of other equipment in order to model a complete methanol synthesis loop or even a complete plant. The flowsheeting package can be used for design, rating or simulation calculations on the converter. Modern flowsheeting packages also have an optimization routine.

An optimization routine can be used to find the optimum conversion in the reactor and, taking the same example of methanol synthesis as before, the optimum circulation rate in the synthesis loop in terms of the minimum combination of capital and operating costs. The optimizations routine can also be used for plant simulation with "redundant" data — the function usually minimized is the sum of the appropriately weighted squares of the differences between actual plant readings and corresponding values from the simulation.

There is a growing tendency to use reactor models in flowsheeting programs rather than as stand-alone programs, because of the greater flexibility provided. Examples of commercial flowsheeting packages that have been developed include:

FLOWPACK II	ICI
DESIGN 2000	Chemshare Corporation
PROCESS	Simulation Sciences Inc.
CONCEPT	Computer Aided Design Centre
ASPEN	Massachusetts Institute of Technology

Chapter 3

Handling and Using Catalysts in the Plant

3.1. Introduction

In a book of this type it may seem unnecessary to say that the efficient operation of a chemical plant based on catalytic processes depends to a considerable extent on the condition of the catalysts employed. Unfortunately, the practical priorities occupying the minds of the people involved in construction, commissioning and operation of new plants tend to be mainly of a mechanical nature, and proper treatment of the catalysts does not always receive the attention that it deserves, even though the economic success of a plant is closely linked to the way in which the catalysts are treated. Satisfactory gas flow through a catalyst bed depends on correct packing, and poor packing will lead to uneven gas flow and unsatisfactory performance. Similarly, catalyst performance and lifetime are influenced by the way in which it is reduced and operated. Catalyst handling, commissioning and operation should be considered in detail at the design and planning stages so that suitable thermocouples, manholes, access platforms, chutes, sieves, hoppers, and so on, are provided during construction so that the catalysts can be treated with due care.

3.2. Catalyst Storage

Catalysts are generally supplied in mild steel drums with polythene liners and depending on the catalyst concerned, the full drums weigh between 50 and 250 kg. The catalyst drums should be inspected carefully when they arrive on site for damage in transit, so that any insurance claim can be properly supported. Drums must not be stacked on their sides or stacked more than four drums high, even when held on pallets. Taller stacks tend to be unstable, and the lower drums can be crushed. Metal drums are usually suitable for outside storage for a few months, but should be protected against rain and standing water. If prolonged storage is expected or if the drums are not of metal, they should be kept under cover and away from damp walls and floors. The lids should be left on the drums until just before the catalyst is to be charged, and if the lids are accidentally knocked off or removed for inspection, it is important that they should be replaced as soon as possible, so that contamination of the catalyst is avoided. If the drum lid cannot be replaced, then the catalyst should be redrummed without delay. If any contamination occurs it is difficult to assess the extent of the damage, and an analysis of a sample may not give an adequate indication because of the difficulty of getting a typical sample. If there is any doubt about the state of the catalyst it is best not to charge it to the reactor.

3.3. Drum Handling

Catalyst drums should be handled as gently as possible. Suitable space is required for storing the drums between delivery and charging, and double handling can be avoided if this space is close to the reactor. If a mobile crane or fork-lift truck is to be used for drum handling, a smooth paved area is desirable to facilitate movement. When drums are to be lifted to the charging manhole it is generally most satisfactory to use a mobile crane rather than the individual lifting beam on the vessel, because a crane can lift drums off a wide area and this avoids multiple handling. *Drums must not be rolled*, and if manhandling is unavoidable, then suitable drum barrows, upending levers and skids should be provided. More importantly, proper equipment is essential for the safety of the operator, particularly when larger or heavier drums are used. Some typical equipment is illustrated in Figure 3.1.

Catalyst drums are often supplied on pallets, which reduces the likelihood of damage in transit but requires suitable fork-lift trucks and a paved area to handle the pallets. The fork-lift truck to be used for dismantling the pallets should be fitted with rim or body clamps to avoid damage to the drums. The use of containers for either catalyst drums or palleted drums eases shipment and further reduces the likelihood of damage in transit, but again suitable fork-lift trucks must be provided at the reception point. It is important not to use standard forks to lift the drums under the rolling hoops, as damage to the drums and catalyst is almost inevitable.

Chapter 3. Handling and Using Catalysts in the Plant

Figure 3.1. Drum-handling equipment (courtesy of Powell & Co.).

3.4. Intermediate Bulk Containers and Socks

In order to cut down the amount of work to be done during a plant shutdown it can be convenient to transfer the catalyst from drums into intermediate bulk containers (IBCs), or in the case of reforming

3.4. Intermediate Bulk Containers and Socks

catalyst, into the socks from which it will be charged to the tubes. The catalyst can be supplied from the manufacturer in bulk containers in many cases. Sufficient catalyst can be stored in such intermediate containers for each reactor involved in the shutdown, and the associated catalyst drum handling can be done outside the period of the shutdown itself. Various types of intermediate bulk containers are available, and Figure 3.2 illustrates a method of their charging and use. This particular type of intermediate bulk container is made of reinforced plastic and holds about two cubic metres of catalyst. The cheapest types of bulk container are intended for use on one occasion only, but a more expensive type can be obtained which is suitable for re-use. When properly closed, both types can be stored outside for some time, preferably on pallets, as shown in Figure 3.2(b). Pallets are convenient for handling the IBCs and help to keep then out of pools of water. Reformer catalyst is normally packed in canvas or polythene socks before charging, and the catalyst can be stored in the socks indefinately so long as they are kept dry. ICI can deliver the catalyst already packed in socks in suitable boxes. The process of sock filling is illustrated in Figure 3.3.

Figure 3.2. (a) Equipment for charging intermediate bulk containers (IBCs) from drums. From right to left: drum-holding attachment on fork-lift truck; mechanical hoist to lift drum and tip it into hopper; dust extraction gear on left of picture.

Chapter 3. Handling and Using Catalysts in the Plant

Figure 3.2. (b) One-trip IBCs stored ready for use. Note use of pallets. (c) A re-usable IBC being lifted.

3.4. Intermediate Bulk Containers and Socks

Figure 3.2. (d) The underside of a one-trip IBC showing closure. (e) The discharge chute from a one-trip IBC.

(d)

(e)

Chapter 3. Handling and Using Catalysts in the Plant

Figure 3.2. (f) Control of catalyst flow from a one-trip IBC. (g) Control of catalyst flow from a re-usable IBC.

(f)

(g)

3.4. Intermediate Bulk Containers and Socks

Figure 3.3. (a) General view of a sock-charging rig, showing the dust extraction system on the left. (b) Detail of clip holding a sock.

(a)

(b)

Chapter 3. Handling and Using Catalysts in the Plant

Figure 3.3. (c) Sock charging in progress. (d) Catalyst flow shut-off when sock is full.

3.5. Seiving Catalyst

Figure 3.3. (e) Polythene socks stored on pallets ready for lifting to furnace top.

3.5. Sieving Catalyst

Catalysts are sieved by the manufacturer before they are packed into drums for despatch, but in some instances attrition can occur in transit if the drums are roughly handled. Some form of screening is usual before charging, especially if the catalyst appears to contain dust on delivery. A good method of sieving is to pass the catalyst over a simple inclined screen, and this is often the most satisfactory method, since vibrating screens can cause additional unnecessary damage and loss. The mesh spacing should be about half the smallest dimension of the catalyst pellet. A convenient arrangement, which also minimizes handling, is shown in Figure 3.4. This system employs an inclined sieve to convey the catalyst from the drum to the charging hopper or chute. Provided that it is carried out with due care a useful technique is to blow the dust away with a compressed air jet whilst the catalyst is being poured over the screen into the charging bucket or hopper. Some proprietary air-blowing devices are available which can be fitted into the catalyst charging tube to remove dust just as the catalyst is entering the vessel.

Chapter 3. Handling and Using Catalysts in the Plant

Figure 3.4. Charging catalyst over an inclined screen.

Alternatively, simple hand-sieving may be used, but the operators must be instructed to stop sieving when the dust has been removed so that unnecessary attrition is avoided. As always, dust masks must, of course, be provided for those who are engaged in this kind of operation, and some provision should be made to collect the fines for disposal. Reforming catalyst does not usually require sieving, but broken rings should be removed whilst the socks are being charged since relatively few broken rings can affect the tube pressure drop significantly.

3.6. Catalyst Charging

3.6.1. Pre-charging Checks

Before the catalyst is charged it is important that the condition of the catalyst support grid in the vessel (or reformer tube) and any supporting materials such as inert balls be checked, because faults at this point are difficult, if not impossible, to rectify once the catalyst is in place. Some form of light metal shield or "spider" fitted into the discharge manhole prevents this branch from forming a stagnant space full of a catalyst which could give undesirable reactions. The vessel should be clean, dry and free from loose scale and debris. It is important to ensure that the

3.6 Catalyst Charging

charging level is clearly defined, so as to avoid underfilling or overfilling. The desired level can be marked with chalk before charging is commenced. In the case of ammonia converters intermediate level marks can be provided to check bulk density during charging.

It can be valuable to check that the thermocouples are correctly installed before charging is commenced by warming them in turn to ensure that the correct indication is given on the instrument panel.

3.6.2. Charging Vessels

There are two general rules for charging catalysts into vessels:
(a) the catalysts should not have a free fall of more than 50–100 cm and
(b) the catalyst must be disributed evenly as the bed is filled.

The distance that a catalyst can fall without serious damage depends on its strength and shape. A hard spherical granule will withstand dropping better than a soft angular pellet or extrudate. Recent work indicates that a strong catalyst may be charged satisfactorily by a pneumatic conveyor, and this technique can speed the charging process substantially. The catalyst must not all be poured into the vessel in one spot with the resultant heap being raked level. If this is done the particles tend to segregate, as in illustrated in Figure 3.5. Small particles and dust stay mainly in the centre of the heap and restrict the gas flow in this portion

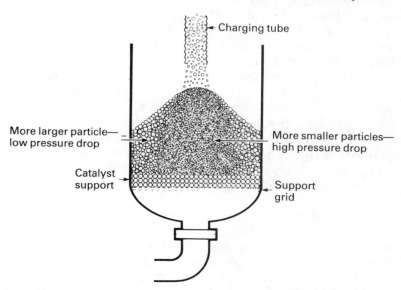

Figure 3.5. Segregation of different sized particles when charging tube is not moved (exaggerated).

of the vessel, whilst larger pieces roll to the edge where the gas velocity will be higher. Uneven packaging and breakage during charging can seriously affect the gas distribution and the effectiveness of the catalyst bed. The degree of packing in the catalyst bed has a marked effect on the voidage, so that even in a bed of regular pellets the voidage can vary by ±10% according to the closeness of packing. If the granules are not all the same size, particularly if fines are present, the voidage variations can be considerably greater. The effect of packing on pressure drop is very marked because the pressure drop is roughly proportional to (voidage)$^{-3}$. Thus, the difference in pressure drops between a loosely packed and closely packed bed of regular pellets is roughly in the ratio of 90^3 to 110^3, or about 1 to 1.8. A loosely packed catalyst bed will settle in use, and so the pressure drop may well increase by about 50% in the process.

The effect of packing variations on the distribution of gas flow through a catalyst bed can be illustrated by considering two identical reactors in parallel which are packed to the same depth but to different voidages, the voidage in the loosely packed bed being ~20% greater than the voidage in the tightly packed bed. In such a system the gas flow will be ~40% higher in the loosely packed reactor than in the reactor with the more dense packing. As a result 42% of the gas will pass through the tightly packed reactor with a space velocity of ~84% of design, whereas ~58% of the gas will pass through the other reactor at a space velocity of ~116% of design. Moreover, the effect of packing variation is more pronounced at low gas flow, as the flow conditions become less turbulent, for instance, during the reduction of a low-temperature shift catalyst, or the initial warm-up of a primary reformer. The effect of this poor gas distribution on catalyst performance will vary with the duty but, for example, in a zinc oxide bed it will mean that breakthrough could occur at about 86% of the expected life when the less densely packed side becomes saturated. In practice this type of maldistribution is most likely in cases where all the catalyst has been allowed to fall into one part of the bed just beneath the charging manhole, or if operators walk on the bed or vibrate it unevenly.

A deep, narrow bed is the easiest shape to charge satisfactorily, and irregularities are more likely to be averaged out along the direction of the gas flow. Extra care is required with wide beds because it is difficult to distribute the catalyst evenly over a large area. Radial-flow beds are particularly susceptible to irregular packing and uneven gas flow. The result of irregular packing and concomitant uneven gas flow can be particularly pronounced in reactors which depend on the gas flow to control the temperature of the bed. Examples include ammonia

3.6 Catalyst Charging

converters, the hydrogenation of acetylene in the presence of ethylene, and also some types of shift converter where the reaction is not expected to reach equilibrium. High local temperatures will occur in regions of low flow if the reaction proceeds further than is desired. Such high-temperature regions cause further resistance to flow, which can aggravate the imbalance within the bed even more.

One of the quickest ways of filling a vessel evenly is with a canvas sock fitted to a hopper which is supported outside the manhole. The sock is kept full of catalyst and raised slowly to allow the catalyst to flow into the vessel in a controlled manner. The sock must be guided in some way, so that it does not always discharge at the same point, and it needs to be shortened periodically to avoid kinking as the vessel is filled. A metal tube, which has the advantage of being more easily steered, can also be used and is better for charging the heavier, and more abrasive, catalysts such as that used for ammonia synthesis. Such a tube can be made in flanged sections which can be removed individually as the catalyst level rises and a shorter tube is needed. Another successful refinement has been a hinged flap at the bottom of the charging tube which can be operated with a cord from the top. The catalyst flow can then be readily controlled.

The use of intermediate bulk containers eases the labour of charging vessels and reduces the amount of crane work involved at the time of the shutdown. This process is illustrated in Figure 3.2. The flow of catalyst into the hopper can be controlled easily by constricting the discharge spout as shown in Figure 3.2 (f) and (g). When operators have to enter the vessel during or after charging, planks should be used so that they do not tread directly on the catalyst. Most catalysts produce some dust in the vessel during the charging process, and respiratory protection must be provided for anybody who has to enter at this time (see Section 3.15). Protection against the dust is particularly important if the material has any toxic properties. If the catalyst bed is wide and access is through a side manhole, then it can be difficult to get the catalyst distributed across the vessel without raking, particularly when the charging is nearly complete. As mentioned above, raking is particularly undesirable as it can lead to concentration of any fines near the charging manhole. A small conveyor, which is narrow enough to fit through the manhole and long enough to reach across the vessel is useful under such conditions. Alternatively the vessel can be charged by the rope-and-bucket method. This may involve having a man standing on a plank inside the vessel to empty the bucket onto the bed. Sometimes a double-rope system may be employed so that the bucket can be tipped from outside the vessel. There are also buckets with opening bottoms, and a successful design is shown in Figure 3.6. Occasionally the catalyst drums themselves can be

Chapter 3. Handling and Using Catalysts in the Plant

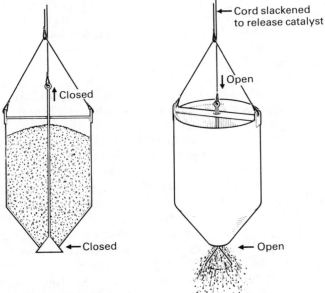

Figure 3.6. Example of a charging bucket.

lowered and emptied into the vessel, and this cuts out one handling stage but denies the user the chance of screening the catalyst before charging. It also involves the risk of damage to the inside of the vessel from heavy drums, a particular risk if the vessel has a refractory lining. The use of plastic drums or rubber buckets lessens the chance of damage to the refractory lining. It is, of course, important that the handles of such buckets be securely fixed.

Metal grids (or a layer of heavy inert material such as fused alumina chips) are sometimes used on the top of the catalyst bed to prevent movement of the pellets by high-velocity gases. Holding down grids however can be a problem during catalyst discharge, as it is sometimes necessary to enter the vessel to remove them whilst the catalyst is in place and blanketed with nitrogen (see Section 3.15). An inert layer of ceramic rings or alumina chips can be useful in intercepting any contaminants in the gas stream before they reach the catalyst itself. In the case of secondary reformers the inert material must be largely silica-free, to avoid silica evolution at the very high operating temperatures, and subsequent deposition down stream in the plant. A fused alumina with up to 1% silica has, however, been found to be satisfactory and it is stronger than the low-silica variety and being non-porous, the silica is not volatilized under operating conditions.

3.6.3. Charging Ammonia Converters

Ammonia synthesis catalyst requires special care in charging because of the irregular shape of the granules, which can lead to a wide variation in bulk density. Most tube-cooled converters can be charged by tipping the catalyst directly into the converter. The tube spacer plates hinder the introduction of catalyst charging socks, but also prevent excessive free fall of the catalyst during the charging. To avoid entry of catalyst granules, the top ends of the cooling tube should be plugged or covered in rows with a small channel section during the charging process. The charging hopper should be moved around, over the top of the converter, to provide an even build-up of the catalyst level. A quench converter can be charged through a hopper and tube in the same way as other vessels can, but ICI Catalyst 35-4 may be allowed a free fall of up to three metres The charged bulk density of the catalyst can vary considerably. Most converters are designed to be charged to a bulk density of about 2.7 kg l^{-1}, but it is possible to get densities between 2.3 and 3 kg l^{-1}. The simplest way of controlling bulk density is to adjust the height from which the catalyst drops onto the bed. If this is not possible, a vibrator can be used to compact the bed. Vibrators must be used with care to ensure the amount of vibration is even over the whole cross section of the bed, because the activity of the catalyst charge and its pressure drop will depend on the degree of compaction. It is therefore most important to keep a close watch on the bulk density during charging. This is usually done by comparing the charged weight with volume marks inside the vessel, or a volume calculated from dip measurements from the top of the vessel. If the density appears to be too low, then additional vibration must be provided or the catalyst must be dropped from a greater height. If the charged density is too high, then the next part of the bed must be charged more gently. Vertical variations in bulk density are not so important, provided the catalyst at that particular level is charged in the same way and vibrated evenly. Continuous checks must be made during the charging process to ensure that the correct weight of the catalyst is charged to the vessel.

The charging of prereduced ammonia synthesis catalyst requries particular care. Charging should be done quickly and, if possible, be controlled from outside the vessel. Excessive vibration or disturbance should be avoided, as this could initiate re-oxidation of the bulk of the catalyst. If it is necessary to have personnel in the vessel, they should be aware of the possible hazards and must be able to get out easily if necessary. During charging, the bed temperatures should be monitored carefully and a nitrogen blanket be established if the temperatures start to rise. Charging should then be completed under nitrogen, and care must be taken to prevent air from being drawn into the vessel. Personnel should be aware of the hazards of working in inert atmospheres, and sound working practice be rigidly followed (see Section 3.15, "Safety precautions").

3.6.4. Charging Reformer Tubes

When catalyst is charged into reformer tubes, it is important that the catalyst is not dropped more than about 50 cm. The catalyst pellets must not be allowed to form bridges across the tubes, as this leads to empty regions in the tubes and overheating during operation. The normal method of charging reformer tubes is with polythene or canvas socks, and the process is illustrated in Figures 3.7(a)–(f). A method for filling socks is shown in Figure 3.3. The catalyst is first put in the sock, which should be about 150–200 cm long and sized so that the filled socks slide easily within the reformer tube. The outside diameter of the charged sock should be about 15 mm less than the internal diameter of the tube. If it is larger than this the sock will not slide easily, and if it is smaller the lower end of the sock may not unfold correctly as the sock is raised. The socks should be filled with catalyst at ground level and be lifted to the top of the furnace in a sling, so as to avoid lifting the drums themselves to the furnace top. It is common practice to precharge the socks before the shutdown is commenced, or to buy the catalyst already charged into suitable socks. A strong cord is tied around the top of the sock or hooked onto it through an eyelet, and about 20 cm at the bottom end is folded up to hold the catalyst in place whilst the sock is lowered into the tube. If the sock is a loose fit in the tube the folded end may not unfold so easily and a shorter portion is required. When the sock has reached the bottom of the tube the cord is withdrawn with a sharp tug, which allows the bottom end of the sock to unfold and the catalyst to fall gently into the tube. It is important that the cord should be longer than the catalyst tube, and it is useful to anchor the free end of the cord so that it cannot drop into the tube, or to fasten it to a ring that is too large to go down the tube.

Before charging begins, the support grids must be checked and both the inlet and exit connections should be tested to ensure that they are clear. When the support grid has been proved to be clear, the tube top should be covered to prevent debris from dropping into it. A useful check for obstructions in the inlet and exit connections can be made with the pressure-drop device shown in Figure 3.8. The expanding bung is clamped into the tube and the air supply opened out to maintain a convenient pressure, say 30 psig at the upper side of the orifice plate. The pressure on the downstream of the orifice, which should be less

3.6 Catalyst Charging

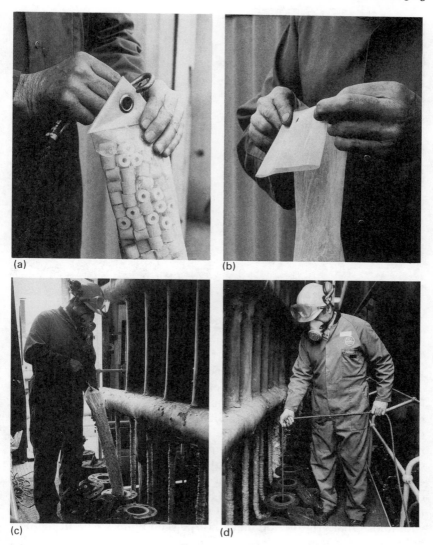

Figure 3.7. Use of socks for filling reformer tubes. (a) Fitting a spring hook to the top of the sock. (b) Folding the open end of the sock before insertion in the reformer tube. (c) Lowering the sock into the tube. (d) Pulling the cord to unfold the sock and release the catalyst.

157

Chapter 3. Handling and Using Catalysts in the Plant

(e)   (f)

Figure 3.7. (e) Topping up tube with catalyst to an exact level. (f) Use of a dipstick to check the catalyst level.

than half the absolute pressure on the upper side, provides an indication of any obstruction in the gas exit connection. The same device is used for checking the pressure drop through the catalyst after charging. The bung may then be raised above the inlet pigtail when the tube has been filled with catalyst, in order to check that the inlet pigtail is clear. A more convenient version of this instrument is available from ICI and is shown in Figure 3.9. In this model the pressure gauges are replaced by digital indicators, which are more accurate and robust. Whichever type of instrument is used it is important to attach it securely to the reformer tube to prevent it from blowing out when in use. This can be achieved with hooks that grip the top flange, or pegs and pins which fit the bolt holes on the top flange, as shown in Figure 3.9.

The reforming catalyst should be weighed before charging to provide a check on the quantity of the catalyst charged to each tube. When the tube has been charged the pressure drop and charged weight should be examined. The pressure drop across each tube should be within ±5% of the mean as detailed in Appendix 19. A higher figure generally indicates broken rings, in which case the whole tube should be discharged and recharged after the catalyst has been sorted. A low pressure drop suggests that there are gaps in the packing and that the tube requires

3.6 Catalyst Charging

further vibration. Pressure-drop variations may be reflected in charged weight, but this is not such a good indication of poor packing as the pressure-drop reading. Some type of extraction gear should be available before charging is begun, as some tubes will generally need to be

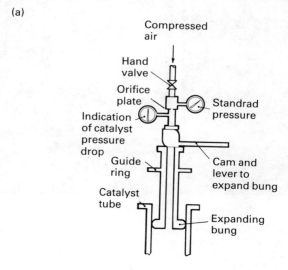

Figure 3.8. Use of pressure-drop equipment. (a) Typical early version of the pressure-drop apparatus. (b) Checks before and after charging.

159

Chapter 3. Handling and Using Catalysts in the Plant

(c) (d)

Figure 3.8. (c) Pressure-drop instrument ready for use. (d) Adjusting the air pressure upstream of the orifice plate to a standard setting. On this instrument the pressure-drop reading is taken from the right-hand gauge, which measures the pressure after the orifice plate.

discharged to correct high pressure drops (see Section 3.12 on discharge). The extraction gear can also be useful for removing debris from the tubes before charging is begun. As reformer tubes are not all exactly the same size and the catalyst is not always exactly the same density, the charged weight will only show major discrepancies in the tube charging, so it is more important to charge to the correct level and pressure drop. The tube should be vibrated periodically during the charging process, either with electric vibrators fastened inside the furnace or by hammering the top flange with soft-faced hammers. The routine for vibration should be defined as charging commences and not altered as charging proceeds, so as to give even packing over the whole furnace. The final catalyst level is measured carefully to ensure that all of the heated length of the tube will contain catalyst when the plant is at operating temperature.

3.7. Catalyst Reduction

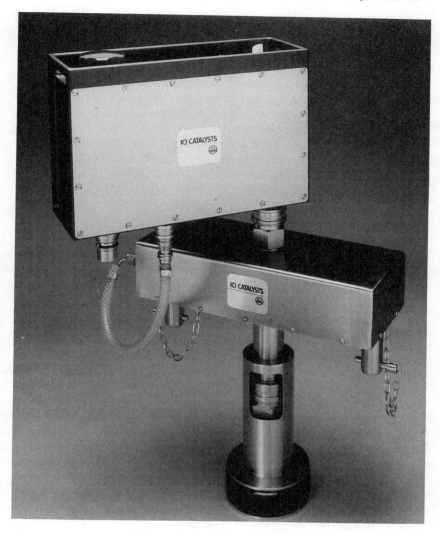

Figure 3.9. ICI pressure-drop instrument with digital readout.

3.7. Catalyst Reduction

Catalysts are designed to be in their active form under process conditions. They are generally supplied in an inert form which is stable under atmospheric conditions, and sometimes they also contain small quantities of impurities such as sulphur which have to be removed in the

Chapter 3. Handling and Using Catalysts in the Plant

course of activation. The activation procedure usually involves the reduction of a metal oxide, and requires careful control if the best catalyst performance is to be achieved. The procedures outlined in the following sections are typical of standard practices for catalyst reduction. It is often possible to speed up the commissioning process but faster methods of reduction can sometimes give rise to problems. For instance, local overheating or uneven reduction can take place so that fast start-up methods are best considered on a case-by-case basis taking into account what instrumentation and other facilities that are available, as well as the operating priorities on individual plants. ICI Catalysts has substantial experience of the considerations involved, and will be pleased to assist in such studies.

3.7.1. Reduction of Reforming Catalyst

Nickel reforming catalysts are supplied as supported nickel oxide and to become active this must be reduced to give metallic nickel. The reaction

$$NiO + H_2 \rightarrow Ni + H_2O \qquad (1)$$

is almost thermally neutral. The reduction is nearly always carried out with a mixture of steam and hydrogen, and a typical schedule is given below. It is important to have sufficient hydrogen present to ensure that reducing conditions are maintained, and this depends to some extent on the temperature of the catalyst. Figure 3.10 shows the boundary between the thermodynamic oxidizing and reducing conditions for the reduction of nickel oxide in a mixture of steam and hydrogen. In practice more hydrogen has to be present than the thermodynamic minimum for kinetic reasons and a steam-hydrogen ratio of eight is typical. Some manuals have suggested that the initial heating of reformers can be done with natural gas in the absence of steam. Although carbon lay-down is unlikely below catalyst temperatures around 300°C, in practice it is difficult to prevent local overheating and carbon formation, so the procedure is not recommended.

The reduction of steam reforming catalyst is relatively straightforward, but the temperature control of the furnace requires care. Regular visual checks must be made on the furnace throughout the start-up. The instrumentation provided to monitor furnace performance is designed for operation under normal running conditions. It is virtually impossible to provide sufficient instrumentation for start-up conditions, with low flow in the tubes and with few burners lit. The appearance of the tubes, the flames and the refractory lining is the best guide to the actual state of affairs. Observations should be made at least after every change in rate of feedstock or fuel. In addition, care should be taken to maintain safe conditions in the heat-recovery section, as this may become overheated because of low flow on the tube side.

The hydrogen for reduction can be provided directly, or from off gases containing hydrogen, or from natural gas, ammonia or methanol which can form hydrogen in the tubes. It is an advantage to ensure there is some hydrogen present in the inlet feed gas, and if there is no external source of hydrogen this is best provided by recycling the reduction gas. If hydrogen is not provided in the inlet gas, the top portion of the catalyst will not be reduced, resulting in a smaller effective catalyst volume and increasing the likelihood of carbon formation. In plants which contain a cobalt/molybdenum catalyst in the desulphurizing

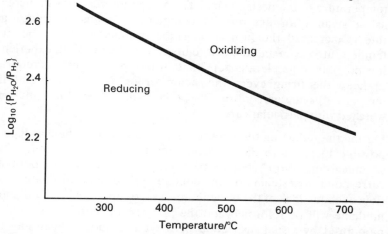

Figure 3.10. Thermodynamic reducing/oxidising conditions for nickel oxide/nickel metal. In practice, considerably more hydrogen is used than indicated.

section, there is a danger of methanation and overheating if the recycle gases contain carbon oxides, and in such cases the cobalt/molybdenum bed must be isolated, or the carbon oxides removed before recycle flow is commenced. The details of start-up procedure will vary from plant to plant, but the sequence set out below illustrates the major principles.

3.7.1.1. Typical reduction with steam and natural gas
1. Purge the plant free of oxygen with nitrogen and heat the reformer above the condensation temperature with nitrogen circulation. The initial heat-up can be done with air in the catalyst tubes, but if there is any possibility of carbon being present this should not be attempted. The rate of heating is determined by the mechanical limitations on the reformer, and typically is 50°C h^{-1}. Particular care is required with new refractory or after repairs have been made to the

refractories in the furnace or flue system. More rapid heating may be possible with an aged furnace.

2. When the temperatures are about 20°C above the condensation temperature steam can be put into the furnace and the steam rate raised to about 40% or 50% of the design rate, so that the furnace can be fired evenly, and the flue gas temperature increased to about 750°C.

3. Introduce natural gas at the lowest controllable flow, preferably at about 5% of the design rate, and increase the flow over 2–3 hours to give a steam/carbon ratio of 7 : 1. At the same time increase the exit temperature to the design value. The firing requirement will increase as the steam reforming reaction commences, and as the natural gas rate is increased the furnace must be trimmed to keep the tube temperatures even and the exit temperature steady. With experience it is possible to accelerate step 3 and also step 4, but care is requried to keep the firing even, the steam/carbon ratio steady and the temperatures under control. The tube skin temperatures must be watched with particular care.

4. During the reduction the inlet temperature should be held as high as possible to promote the reduction reaction at the tube inlet. The steam/carbon ratio should be held at 7 : 1 which normally corresponds to a steam flow of about 50% of design with a natural gas flow of about 25% of the design rate. As the catalyst is reduced, more methane will be reformed and the progress of reduction should be monitored by visual inspection of the tubes and by analysing the methane concentration at the exit. When conditions have stabilized, the steam and gas flows can be adjusted to give the design steam/carbon ratio, and rates can be increased up to the design values.

The steam rate should always be increased before increasing the gas rate to maintain a steam/carbon ratio at or above the design value.

3.7.1.2. Reduction with gas recirculation

With a recirculating system, the initial warm-up is done by circulating nitrogen with hydrogen added at some convenient stage, normally after the steam has been put into the reformer. Ammonia may be used as a hydrogen source, either by cracking it in an ammonia cracker or by injecting it into the process gas stream before the mixed feed preheater. The reformer exit temperature should be at about 800°C to ensure that as much as possible of the ammonia is cracked. Methanol can be used in a similar way, except that the cracked gas must not be recycled through the cobalt/molybdenum section of the desulphurizer because of the danger of exothermic methanation of the carbon oxides which arise from the cracked methanol. The nitrogen circulation is stopped as soon as feed gas is put to the reformer, and recycle hydrogen flow can be commenced as soon as the product gas is sufficiently pure to avoid the possibility of excessive temperatures in the cobalt/molybdenum catalyst, if this is used. The critical level of carbon oxides varies with the plant design. Normally the recycle gas constitutes only about 5% of the feed gas flow, so that any carbon oxides are well-diluted. Difficulty only arises then if the feed gas flow is stopped and the undiluted recycle gas flow maintained.

3.7.2. Reduction of High-temperature Shift Catalyst

High-temperature shift catalysts are supplied in the form of Fe_2O_3/Cr_2O_3, which may contain traces of sulphur in the form of sulphate. During reduction, the Fe_2O_3 (haematite) is converted to Fe_3O_4 (magnetite) and the sulphur is removed as H_2S. In plants where the shift converter follows a reformer, the catalyst is usually reduced in the course of the reformer start-up. Reduction of the shift catalyst can begin at about 150°C, whilst the reforming catalyst is being reduced with hydrogen and steam, and in most cases reduction will be continued after the hydrocarbon feed is added. In most ammonia plants the shift catalyst does not reach full operating temperatures until air is put on the secondary reformer. High-temperature shift catalysts must not be exposed to hydrogen in the absence of steam, as further reduction to iron is possible. When the HT shift converter follows a new charge of primary catalyst containing traces of sulphur, it is preferable that the gas leaving the primary is not put into the HT shift until the sulphur content is less than 1 ppm. No harm will be done to the high-temperature shift catalyst, which will work in a sulphided state, but the sulphur will be retained for a time and will delay the elimination of sulphur from the system. This can extend the start-up period, for instance if the CO_2 removal system is sensitive to sulphur, and the effects on the LT shift catalyst also has to be considered.

The reduction of Fe_2O_3 to Fe_3O_4 with hydrogen shown in equation (2) is only slightly exothermic. No significant temperature rise is likely

$$H_2 + 3Fe_2O_3 \rightarrow 2Fe_3O_4 + H_2O \qquad \triangle H = -9 \text{ kJ mol}^{-1} \qquad (2)$$

unless carbon monoxide is also present and the shift reaction takes place as is shown in equation (3). This reaction will increase the temperature of the gas by about 10°C for each 1% of CO present in the wet gas.

Above 630°C there is a phase change in the Fe_3O_4 which rapidly leads to a permanent loss of activity; some loss of activity can occur if the catalyst is heated above 540°C, and therefore the operating temperatures should be kept below 530°C.

$$CO + H_2O \rightarrow CO_2 + H_2 \qquad \triangle H = -41 \text{ kJ mol}^{-1} \qquad (3)$$

3.7.2.1. Typical reduction of high-temperature shift catalyst

High-temperature shift converters are used in many types of plant, and a typical reduction schedule is as given below for use with ICI Catalysts 15-4 and 15-5, which can withstand condensing steam. However, if possible the catalyst should be heated up above the dew point, before admitting steam. Nitrogen can provide a convenient heating medium.

1. Purge the converter free from air using an inert gas.

2. Start the flow of process steam and gas to give a space velocity of about 200 h^{-1}. The process gas stream should contain less than 15% carbon monoxide in the wet gas, and this can be controlled by adjusting the steam rate. The converter usually requires periodic draining to avoid accumulation of condensate. Heat the converter at about 50–60°C h^{-1} until the inlet temperature reaches 300°C. Reduction usually commences at about 150°C, and is normally almost complete at 300°C.

3. When the first part of the bed reaches 300°C, reduce the amount of heat supplied to the converter, since temperatures will rise when the exothermic shift reaction begins.

4. Temperatures must be kept below 500°C during reduction, by the addition of extra steam if required. The shift converter should be brought to operating conditions slowly to ensure that it is working at the lowest possible temperature consistent with a satisfactory exit composition.

3.7.3. Reduction of Low-temperature Shift Catalyst

Low-temperature shift catalysts are supplied as supported copper oxide, and require careful reduction to give the maximum activity. The reduction is exothermic as shown in equation (*4*), and a temperature front will pass down this bed as the reduction proceeds.

$$CuO + H_2 \rightarrow Cu + H_2O \qquad \triangle H = -81 \text{ kJ mol}^{-1} \qquad (4)$$

Catalyst activity is mainly dependent on the surface area of the copper crystallites which are formed during the reduction. Copper has a low

3.7. Catalyst Reduction

sintering temperature and catalyst temperatures should be kept below 230°C during reduction to ensure maximum activity, although ICI Catalyst 53-1 can be operated at higher temperatures when it has been reduced.

Particular care is needed when reducing catalysts that contain hexavalent chromium, as the reduction is very exothermic, as shown in equation (5). The catalyst is reduced in a stream of heated carrier gas, normally nitrogen or natural gas, to which a controlled low concentration of hydrogen is added.

$$2CrO_3 + 3H_2 \rightarrow Cr_2O_3 + 3H_2O \qquad \triangle H = -649 \text{ kJ mol}^{-1} \qquad (5)$$

It is sometimes difficult to be sure when reduction is complete unless the catalyst bed is very well provided with thermocouples. Particular care should be taken if the lowest thermocouple in the catalyst is some distance above the bottom of the bed, as it can be difficult to ensure that the reduction front has passed completely through the catalyst bed. A thermocouple in the outlet pipe is a useful guide in this situation, but some allowance must be made for loss of heat from the outlet pipework. A similar problem can arise if the reduction front is not flat, which is particularly likely at low space velocities. The catalyst thermocouples cannot be assumed to indicate the temperature across the whole width of the reactor, and the outlet thermocouple should be watched for signs of increased reaction as the hydrogen concentration is increased at the end of a reduction. It is valuable to check the quantity of hydrogen consumed in the course of reduction. For ICI Catalyst 53-1 about 85 m^3 of hydrogen are required per cubic metre of catalyst. If, for any reason, it seems that the expected quantity of hydrogen has not been used, the flow rate or hydrogen concentration measurement may be at fault and the hydrogen level should be gradually increased more cautiously until complete reduction is achieved, still keeping the maximum catalyst temperature below 230°C. The reduction water can be collected, as its measurement provides another check on the progress of reduction. About 100 kg of water should be evolved per cubic metre of catalyst, of which some 60 kg results from reduction of copper oxides and the balance results from "water of crystallization". If there is any oxygen in the feed gas, then this will react with some of the hydrogen and produce an equivalent quantity of water.

The following reduction schedule for ICI Catalyst 53-1 is typical, and is illustrated in Figure 3.11.

Chapter 3. Handling and Using Catalysts in the Plant

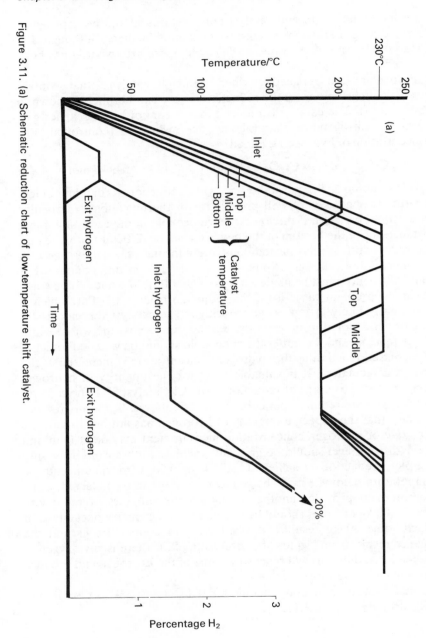

Figure 3.11. (a) Schematic reduction chart of low-temperature shift catalyst.

3.7. Catalyst Reduction

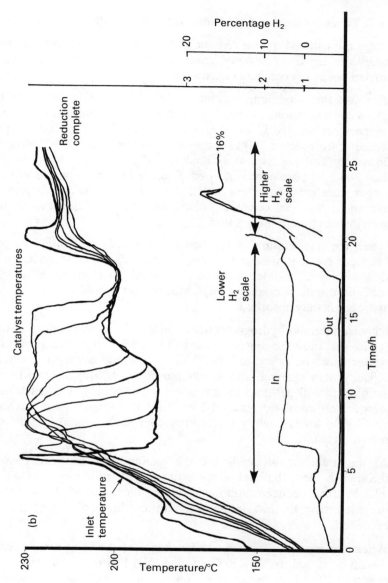

Figure 3.11. (b) Actual reduction chart for a charge of ICI Catalyst 53-1.

Chapter 3. Handling and Using Catalysts in the Plant

3.7.3.1. Typical reduction of low-temperature shift catalyst

1. Purge the catalyst free of oxygen with a dry inert gas, and heat the catalyst to 120°C, at a rate of about 50°C h^{-1}. The reduction may be carried out at any convenient pressure.

2. Establish the maximum possible space velocity. Add hydrogen to give a concentration of 0.5% hydrogen at the bed inlet, and raise the temperature to 165°C at a rate of about 30°C h^{-1}. Before the temperature reaches 140°C check the plant analyser, the hydrogen flowmeter and the laboratory analyser. If the space velocity is less than 300 h^{-1} more care is needed, and the reduction period will be longer since more time will be taken to add the required volume of hydrogen. A low space velocity also makes it more difficult to ensure even distribution of the gas flow.

3. When the temperature of the bed has reached 160°C increase the hydrogen concentration in the carrier gas to 0.5–1.0%. If reduction is slow to begin, the inlet temperature should be raised to 180°C and held there until a definite temperature front is established and the hydrogen is fully absorbed.

4. Increase the inlet hydrogen concentration. If nitrogen is used as the carrier gas, raise the concentration to 1.0–1.5%. With natural gas as a carrier it can be raised to 2.0–2.5%, as methane has a greater heat capacity than nitrogen and allows the higher hydrogen level to be maintained without reaching excessive temperatures. The maximum temperature must not exceed 230°C. At this stage it may be also possible to lower slightly the inlet temperature without loss of the reduction front.

5. As the reduction proceeds and the temperature front has passed through the bed, the inlet temperature may be raised to 200°C. The inlet hydrogen concentration can then be increased to about 3–5%, provided that the maximum temperature limit of 230°C is not exceeded.

6. When the reduction appears to be complete the inlet temperature should be raised to 225–230°C and the inlet hydrogen concentration raised to 20% if this is possible. If a circulating system is employed, the circulator characteristics may limit the maximum hydrogen concentration because of its effect on the gas density. The final increase in hydrogen concentration should take not less than two hours. No temperature rise should occur, and the maximum catalyst temperature should not exceed 230°C. The hydrogen concentration

3.7. Catalyst Reduction

at the inlet and the exit should also be almost equal. In some cases a small difference between inlet and exit hydrogen contents will persist, but if there is no temperature rise and the exit concentration is within 0.5% of the inlet for more than four hours, then the reduction can be considered complete.

3.7.4. Reduction of methanation catalyst

Methanation catalysts are activated with hydrogen, which reduces the nickel oxide to metallic nickel as in equation (6).

$$NiO + H_2 \rightarrow Ni + H_2O \qquad (6)$$

The vessel is first purged free of oxygen with nitrogen or process gas, then heated at about 20–50°C h^{-1} to 200°C. The initial heating is sometimes done with an inert gas such as nitrogen or natural gas, but frequently process gas is employed. When the catalyst is unreduced, process gas can be used without any danger of nickel carbonyl formation (see Section 3.15). Reduction commences at about 200°C. The total carbon oxides content ($CO + CO_2$) of the reducing gas should not exceed about 1%, and the composition of the inlet gas must be controlled at or below this figure by careful attention to the preceding plant. The methanation reaction will commence as soon as some of the catalyst is reduced, and the catalyst temperature will rise by about 6.0°C for each 0.1% of CO_2 and 7.4°C for each 0.1% of CO converted to methane. It is best to raise the inlet temperature at about 25°C h^{-1} to 350°C, so that the catalyst is not overheated. The maximum operating temperature of methanation catalysts is about 425°C, but they are not usually damaged by brief overheating—to temperatures of around 600°C in the case of ICI Catalyst 11-3. However, the vessel design will generally place some limit on the maximum permissible temperature. The best way to cool the catalyst and to even out hot spots is to increase the flow of process gas, provided that the $CO_2 + CO$ content is less than 1%. It is sometimes useful to increase the catalyst temperature to 400°C at the end of reduction to ensure that maximum activity is reached. If it is not possible to reach 400°C by normal adjustment of the inlet gas heater, catalyst temperatures can be increased by by-passing some of the gas flow round the LT shift converter, so as to increase the CO content of the gas entering the methanator, and thus raise the temperature rise across it by additional methanation.

3.7.5. Reduction of Ammonia Synthesis Catalyst

Ammonia synthesis catalysts are usually supplied in the form of Fe_3O_4, which must be reduced to metallic iron before it is catalytically active.

Chapter 3. Handling and Using Catalysts in the Plant

Some catalysts are supplied in a prereduced and stabilized form, with over 90% of the catalyst already reduced. They require a much shorter time for reduction, but the same general considerations apply. Normal catalyst is reduced with hydrogen or synthesis gas as shown in equation (7).

$$Fe_3O_4 + 4H_2 \rightarrow 3Fe + 4H_2O \qquad \triangle H = -150 \text{ kJ mol}^{-1} \qquad (7)$$

The reducing gases must be heated throughout the reduction if hydrogen alone is used, but when reduction is carried out with synthesis gas, ammonia is produced as soon as some of the catalyst is reduced, and the external heat requirements are decreased. The ammonia synthesis reaction is exothermic and provides heat to raise the catalyst temperature and so accelerates the rate of reduction. As soon as the catalyst temperatures show signs of rising of their own accord, the gas rate through the bed can be increased to transfer the heat down the bed and assist in reducing the remainder of the catalyst. As more ammonia is formed the bed becomes self-heating, and the heat input to the gas can be decreased until eventually, when all the available gas is going through the catalyst, a cold by-pass will be required to control the inlet temperature, and to prevent an excessive temperature rise in the catalyst bed. Catalyst activity can deteriorate at temperatures above 550°C, and it is desirable to keep temperatures below 500°C during the reduction. The water which is produced during reduction can deactivate part of the catalyst which has already been reduced, and this effect is more severe at temperatures above 450°C. It is normal to measure the water content of the gas leaving the catalyst and to keep this below 10 000 ppm by adjusting the rate of temperaure rise in the converter. However, the permissible water content is related to the particular catalyst and the converter design, and it is generally advisable to hold the water level below 5000 ppm. At higher rates of reduction there is some sintering of the outside of each catalyst granule, which is reduced first, and then exposed to high levels of water emerging from the centre of the granule.

The circulating gas leaving the ammonia converter must be cooled in order to condense out as much of the water as possible before the gas is recycled over the catalyst. High-pressure operation assists in the condensation of water, and the ammonia produced when using synthesis gas ensures that the condensate is a water/ammonia solution, which condenses more readily than water alone. If the loop contains a refrigeration cooler this can, with advantage, be commissioned as soon as the ammonia content of the condensate is sufficient to avoid freezing. When the refrigeration system is commissioned, the inlet ammonia concentration will fall and the driving force for the synthesis reaction will be increased, thus commissioning the refrigeration system will tend

3.7. Catalyst Reduction

to increase the catalyst temperature. Care should be taken, however, to ensure that the rate of temperature increase is kept under control. The ammonia content of the reduction water usually rises rapidly in the first few hours, and will usually exceed 20% before much reduction has taken place. The freezing point of 20% ammonia in water is $-33°C$, which is the lowest temperature likely to be met in a refrigerated loop. If the reduction schedule is interrupted it is very important to maintain a forward flow of gas through the catalyst bed, so that water cannot diffuse back on to the freshly reduced catalyst. This is normally ensured by closing a valve on the converter inlet line and slowly blowing off gas from the exit line.

NOTE The converter may well come fully into operation before reduction is actually complete, and it is important to keep the operating temperatures steady, and below 500°C if possible, for the first 7 days after start-up.

Typical reduction procedures for ICI Catalyst 35-4 are given below, for two general types of ammonia converter (tube-cooled and multibed quench). These, as well as other types of converter are discussed in more detail in Chapter 8.

3.7.5.1. *Typical reduction of a tube-cooled converter*

1. Bring the loop to the desired reduction pressure and set a low purge rate from the loop. This ensures that fresh synthesis gas is always present and that the circulating gas does become hydrogen-deficient.

2. Start circulation round the loop and adjust the gas flow through the heater and the catalyst bed.

3. Commence heating the catalyst at about $50°C\ h^{-1}$ and raise the inlet temperature to about 340°C. The gas rate should be as high as possible to assist in the heating process. It is generally preferable to start heating up with a high gas rate and then to decrease the gas rate gradually to control the progress of the warm-up. Some reduction may commence at about 300°C but 370–380°C is more typical. The water concentrations at the exit should be checked when temperatures reach 280°C, and measurements should be continued as frequently as possible until the reduction is complete.

4. Decrease the heating rate to about $10°C\ h^{-1}$, by increasing the gas flow rather than by reducing the heater output. Some catalyst temperatures may fall briefly as the gas rate is increased. If they do not recover within a few minutes, decrease the gas rate until

temperatures have risen by 10–20°C and try again. If one side of the converter becomes hotter than the other, wait until the colder side catches up before increasing the gas rate or pressure. If necessary, reduce the pressure to limit the temperature rise in the hot part and increase the inlet temperature to raise the temperature in the cold part. If the temperature becomes very uneven it is possible to reduce one side of the reactor without reducing the other, and this condition can persist even after some months of operation.

5. Measure the water content of the gas leaving the converter and control it to a maximum of 5000 ppm by adjusting the rate of increase in temperature (see Chapter 8). Some cold quench may be needed at the bed inlet if the rate of heating cannot be controlled by increasing the gas rate, but this should not be used until the peak temperature is above 480°C.

6. Keep the peak catalyst temperature below 500°C until reduction is complete but since the converter may come fully into operation before all of the reduction is complete it is important to keep a steady operating temperature, preferably below 500°C for the first week after start-up.

3.7.5.2. Typical reduction of a multibed quench converter

1. Bring the loop to the desired reduction pressure and set a low purge rate from the loop. This ensures that fresh synthesis gas is always present and that the circulating gas does not become hydrogen-deficient.

2. Start circulation round the loop, and adjust the flow to the heater and the catalyst. There are generally two flow paths to the inlet of the catalyst bed: the normal route which cools the converter shell and then passes through the internal exchanger, and that through the heater which meets the gas from the heat exchanger at the top of the catalyst bed. Some flow will be required through the heat exchanger to cool the exit gas and avoid overheating the exit pipe and converter shell, but this flow is normally kept as low as possible to minimize dilution of the hot gas from the heater. It is important to know the design temperature limits of the exit pipework and converter shell, and to take measurements to ensure they are not exceeded.

3. Heat the whole converter up at about 50°C h^{-1} to an inlet temperature of 340°C, and then decrease the rate of heating to about 10°C h^{-1} by increasing the total gas flow to the converter but without reducing the heater flow or the rate of firing.

4. Hold the inlet temperatures of the second and subsequent beds at about 350°C by the addition of quench at the inlet to the second bed.

5. Check the exit water concentration, and control it below 5000 ppm by adjusting the gas rate rather than the heater output (see Chapter 8). Once reduction starts at the top of the first bed, ammonia synthesis will start and the heat of reaction will produce a temperature rise across the bed so that the exit temperature will increase at a greater rate than the inlet. The rate of increase of the exit temperature should be controlled carefully to about 5°C h^{-1} by increasing the total gas flow. At the same time it will be necessary to increase the quench rate to the second bed in order to maintain its temperature at 350°C. If temperatures are allowed to increase too quickly, an uneven reduction can occur as mentioned in Section 3.7.5.1, and the same remedies can be applied.

6. When the water evolution from the top bed begins to decrease, the second bed can be heated-up in the same way as the first, using quench to keep the third and later beds at about 350°C if possible.

7. Reduce the remaining beds in turn in the same way as the first, keeping the unreduced beds at about 350°C.

3.8 Catalyst Shutdown and Restarts

When a catalyst charge is taken offline, care is required to avoid damage to the catalyst particularly when it is in the reduced state. Many reduced catalysts are pyrophoric and under static conditions the plant thermocouples may not give an adequate view of the temperatures in the catalyst bed. It is normal practice to keep an inert atmosphere in the plant when it is not online (see Section 3.10). Nitrogen, natural gas or process gas can be used to exclude air from the plant when it is not in operation. Condensation can occur as the plant cools if the system is not purged free from steam and although most catalysts are not damaged by water, any poisons which are present, such as chloride in the LT shift may be carried further into the catalyst bed if it becomes wet. Ammonia synthesis catalyst, alkalised reforming catalyst and ICI 59-3 chloride removal catalyst must be kept dry to avoid possible loss of alkali. Water and steam will also react with reduced ammonia synthesis catalysts and cause a gradual loss of activity. Unreduced HT shift catalyst must be kept dry to avoid dissolution of and hence loss of the hexavalent chromium.

When the plant is offline and depressured, care is required to avoid the entry of fluids which would normally be kept out by the plant pressure. Purge lines, temporary hoses, drains and vent lines must be reviewed for this possibility and a check kept on the vessel drains for signs of any unexpected ingress of liquid.

The restart of a reduced catalyst can often be done more quickly than the initial reduction. Rates of heating of around 200°C h^{-1} will not damage the catalysts normally used in ammonia, hydrogen and methanol plants. If any of the catalysts have become wet whilst offline, the dry-out process requires more care to avoid fragmentation of the pellets by an over-rapid boiling of water. Primary reformers can be heated at 150-200°C h^{-1}, but care is required not to exceed the mechanical limits of the furnace design and also to avoid local overheating because of uneven firing at low throughputs. Visual inspection is normally required to check for local overheating as it is not usual for furnaces to be sufficiently well instrumented to detect local abnormalities from the control room. It is therefore difficult to lay down general rules for minimising start-up periods, and most catalyst vendors publish somewhat cautionary guidelines. If an operator is in doubt about the advisability of a particular start-up routine it is best considered on a case-by-case basis with the catalyst vendor having due regard to the local situation, instrumentation and the operating priorities.

3.9. Catalyst Regeneration

3.9.1. Regeneration of Reforming Catalyst

The operation of reforming catalysts is sometimes impaired by carbon deposition, usually as a consequence of maloperation. These deposits may be removed by either steaming, in the absence of feedstock, or with a steam/air mixture. The normal procedure is to take off the hydrocarbon feed and adjust the firing to keep the temperatures at about the normal operating level. The steaming can be continued for several days if convenient, although about 6–12 hours is usually sufficient. Some re-reduction may be necessary after a prolonged period of steaming but, after a short regeneration, it is usually sufficient to add the hydrocarbon feed slowly to the steam and to allow the catalyst to be reduced by the process gas. If air is used to remove carbon the air rate must not exceed 2% of the steam flow, and a careful watch must be kept on the reformer tubes for signs of overheating. If too much air is used the tube metal may be melted, and the control system must be designed carefully to eliminate this possibility; for instance in the case of steam failure *the air must be shut off instantly*. Re-reduction is advisable after an air regeneration. In either case the progress of regeneration can be followed by measuring the carbon dioxide level in the exit gas.

3.9.2. Regeneration of High-temperature Shift Catalyst

When HT shift catalysts are used with cyclic reformers they may become sulphided as well as contaminated with gums and carbon formed in the up-stream plant, and they can usually be regenerated by treatment with steam and air. Superheated steam should be passed over the catalyst at 400–500°C and at a space velocity of up to 1000 h^{-1}. The steaming is continued until the concentration of hydrogen sulphide has been reduced to less than 80—100 ppm by volume. The inlet temperature is then reduced to 300–350°C, and air is admitted cautiously to give not more than 1% of oxygen in the steam. It is important that no temperature should exceed 550°C, and if temperatures are rising and appear likely to exceed 500°C the air supply should be cut off, and the catalyst be allowed to cool in the steam flow. Progress during regeneration can be followed by measuring the carbon dioxide content of the exit gas. Some of the catalyst will be oxidized, at the same time as the carbon compounds are being removed, and will require re-reduction before it can be used again, but the catalyst cannot be regarded as completely oxidized when the carbon has been removed.

3.9.3. Regeneration of Low-temperature Shift Catalyst

Attempts to regenerate LT shift catalyst have been only partly successful. The normal procedure is to reoxidize the catalyst (see Section 3.11.3) which will remove some of the sulphur poison, and then to wash with condensate to remove water-soluble poisons such as sulphate and chloride. The catalyst is then re-reduced. The time taken for this procedure is generally much longer than for a normal catalyst change and the regenerated catalyst has a lower activity and shorter life than new catalyst. The procedure is not therefore economical. Moreover, the reoxidation procedure requires particular care to avoid getting excessive temperatures which can reduce the catalyst activity and in extreme cases when the reaction is not properly controlled the catalyst bed may be fused.

3.9.4. Washing Methanation Catalyst

Methanation catalyst can sometimes be contaminated by carry-over from the carbon dioxide removal section. Potassium carbonate does not poison the catalyst, but reduces its effectiveness by physically blocking the surface. ICI Catalysts 11-3 and 11-4 can be reactivated after potassium carbonate contamination by back-washing with clean condensate. Monoethanolamine (MEA) and diethanolamine (DEA) can also be washed off successfully, but experience has shown that catalyst which has been contaminated with Sulphinol or Vetrocoke solutions can rarely be reactivated satisfactorily by washing (see Chapter 7 for additional information).

3.9.5. Regeneration of Ammonia Synthesis Catalyst

Ammonia synthesis catalyst is sometimes deactivated by accidental steaming, washing or reoxidation with impure nitrogen. A carry-over of liquid ammonia can also cause loss of activity by nitriding the catalyst. A satisfactory activity can generally be restored after steaming or reoxidation by a short re-reduction of the catalyst. Full activity may not be regained for some time, and rapid temperature changes should be avoided for the first week after recommissioning. Nitrided synthesis catalyst generally recovers full activity after a few days of operation, during which the catalyst temperatures may have to be kept higher than normal to offset the initial low activity.

3.10. Blanketing of Reduced Catalyst

Many metal-based catalysts are pyrophoric in their reduced state, and must be kept blanketed with an inert atomsphere when not in use. Nitrogen is the most satisfactory blanket gas, as it is usually available with an oxygen content of less than 10 ppm. Higher oxygen levels are tolerable over some catalysts, which can be re-reduced without loss of activity. If the nitrogen contains more than 10 ppm of oxygen care is required, since the catalyst may be damaged by oxidation. In such cases the catalyst should be cooled below 50°C before introducing the oxygen-contaminated nitrogen, and the flow of nitrogen should be kept to a minimum to avoid putting more oxygen than is necessary over the catalyst. If possible, the nitrogen blanket should be static and be maintained at a slight positive pressure. The thermocouples in the catalyst vessel must be left in commission when the catalyst bed is blanketed, to provide indications of unexpected overheating, for instance, following a deterioration in the quality of the nitrogen. Natural gas may be used as a blanketing medium, but its use involves a fire hazard. It can be employed up to temperatures of 300°C without risk of carbon formation, but local overheating can cause cracking to carbon even if the bulk of the catalyst is below 300°C. This danger is only likely in a tubular furnace such as a primary reformer with uneven firing.

When a vessel is being opened up whilst under a nitrogen blanket, careful thought must be given to the blanketing arrangements at each stage. If there is more than one opening, then there is a danger that the vessel will act as a chimney, and draw air inwards at the lower opening, even though nitrogen is still being supplied. This danger is most severe if the catalyst is still warm. A good general rule is to have only one opening to the atmosphere at any one time. Ammonia converters require a detailed plan to protect the catalyst during maintenance work. The ammonia cartridge often has an inlet and exit side which may be disconnected when the vessel is isolated from the rest of the plant, so it is virtually two separate vessels, each of which requires purging and blanketing, and it is important to ensure that a positive purge is maintained through the catalyst at all stages of the operation. It is also desirable to blanket the remainder of the cartridge, which may contain pyrophoric catalyst dust. The top and bottom of the cartridge must not be open at the same time. A set of expanding rubber bungs, or inflatable bladders, can be used to close open pipes. Special sealing discs are provided for many ICI converters. The cartridge can be removed from the pressure vessel after all open ends have been closed, and a long nitrogen hose has been firmly attached. It is essential to ensure that the hose is long enough to remain fixed throughout the lifting operation, and that it will not foul any of the surrounding plant.

When vessels contain nitrogen, or some other blanketing gas, it is most important to ensure that no person enters without proper breathing equipment. Warning notices should be put at all open manholes. Several accidents have occurred, some of them fatal, through ignorance of the danger of entering a vessel containing insufficient oxygen. As the danger is invisible it is easily disregarded. Breathing equipment may be provided, but it will be useless unless everybody involved is aware of the danger, and of the necessity for using the apparatus (see Section 3.15).

3.11. Catalyst Stabilization

Operators sometimes wish to stabilize the catalyst for maintenance work or before the catalyst is extracted from the vessel. Catalyst which is to be re-used must be stabilized carefully to minimize loss of activity. The re-use of stablized catalyst is not generally economical because the loss

Chapter 3. Handling and Using Catalysts in the Plant

of production time during the stabilizing process can well exceed the cost of a new catalyst charge, and the recommissioned catalyst is unlikely to perform as well as a new charge, particularly in the case of LT shift catalyst, whose performance is critical to the economic operation of an ammonia plant. When the catalyst is to be discarded, more rapid stabilization is possible, but care is still required to avoid excessive temperatures. The following methods have been used for stabilizing ICI catalysts.

3.11.1. Stabilization of Reforming Catalyst

The catalysts in primary or secondary reformers are generally stabilized by steaming in the absence of hydrocarbon feed during a standard shutdown. The oxidation of nickel metal in air is strongly exothermic, as shown in equation (8), and if for some reason some catalyst has not been steamed it may require stabilizing as described below for methanation catalyst. Fortunately, the situation is very rare and steaming for a few hours whilst the plant is being cooled down is more than sufficient for safety. In some cases the reoxidation is very slow, and with ICI Catalyst 57-3 no stabilization is required even if the catalyst is to be discharged in the reduced state. If carbon has been deposited in a primary or secondary reformer, very high temperatures can be produced if it is allowed to burn in air. If the presence of carbon is likely, then the steaming must be continued until it has been removed before either reformer is opened to the air. The steaming procedure is described in Section 3.9.1. If steaming is not possible, then the catalyst should be cooled completely and a careful watch kept for signs of overheating when the reformer is opened up.

$$Ni + \tfrac{1}{2}O_2 \rightarrow NiO \qquad \Delta H = -686 \text{ kJ mol}^{-1} \qquad (8)$$

3.11.2. Stabilization of High-temperature Shift Catalyst

Method 1

If the catalyst is to be discarded, and it has been operating in a sulphur-free atmosphere the following procedure should be followed.

1. Cool the catalyst to 350–375°C with process steam at a space velocity of about 1000 h^{-1}.

2. Add 3–5% by volume of air to the steam. To avoid possible local overheating and damage to the converter, the measured catalyst temperature must be kept below 500°C. Increase the air rate slightly

if this can be done without exceeding the catalyst temperature limits.

3. When the catalyst has been oxidized, the steam may be shut off, and the catalysts cooled down by a current of air.

Method 2

If the catalyst has been operating in the sulphided state, for instance in a coal-based plant, FeS will be present and the sulphur must be removed from the catalyst by steaming at 400–500°C and at a space velocity of up to 1000 h^{-1}. The steam temperature should not be less than 350°C, and steaming should be continued until the concentration of H_2S in the steam has been reduced to less than 80–100 ppm by volume. This process may take between two and seven days. The catalyst should then be cooled in steam to 200°C and stabilized with air and steam, either as in Method 1, if the catalyst is to be discarded, or as in Method 3, if it is to be re-used.

Method 3

Although not recommended, if the catalyst has to be re-used the stabilization method is similar to Method 1, but a more stringent control of temperature is required to avoid loss of activity.

1. Cool the catalyst to 200°C with process steam. Add 1% by volume of air to the steam to start the oxidation and observe the temperature rise across the converter which should be approximately 30°C.

2. After about half an hour, if the temperature rise appears to be normal, increase the air rate to give 3% of air in steam. After 3 hours, if the temperature rise is still normal (80–90°C), increase the air rate to a maximum of 5% by volume of air in steam. During the oxidation the catalyst temperature must not exceed 300°C.

3. Analyse the dry exit gas for oxygen to confirm that oxygen is being consumed. When the oxygen content of the dry exit gas has reached 1% the concentration of air in the steam can be increased to 10%.

4. When the oxidation is complete there will be no temperature rise in the catalyst, and the oxygen content of the dry exit gas will approach 20%. The steam flow can then be stopped, and the catalyst can be cooled down with air.

3.11.3. Stabilization of Low-temperature Shift Catalyst

1. Cool the catalyst to 200°C in nitrogen.

Chapter 3. Handling and Using Catalysts in the Plant

2. Add about 0.1% of oxygen to the nitrogen, carefully controlling the temperaure so that it does not exceed 220°C.

3. When the temperatures have steadied out at about 200°C, slowly increase the oxygen concentration until the catalyst is stable at 200°C with 3% oxygen in the inlet gas.

4. Allow the catalyst bed to cool to room temperature with the same gas mixture, and slowly increase the oxygen concentration until only air is present in the catalyst bed.

3.11.4. Stabilization of Methanation Catalyst

The oxidation of reduced nickel catalyst can be very exothermic, as shown in equation (8), so the catalyst must be stabilized carefully if air is to be allowed into the vessel. It is not normal to re-use stabilized methanation catalyst, because of the possible loss of nickel surface area and activity due to overheating. Although a small decrease in activity may not appear very obvious, the extra carbon oxides entering the ammonia loop can cause damage.

1. Cool to 70°C and purge the converter with inert gas (nitrogen, for example).

2. Maintain the flow of inert gas with a space velocity of about 1000 h^{-1}, and add sufficient air to give an initial oxygen concentration of about 0.1% by volume at the inlet.

3. Increase the airflow and maintain the maximum bed temperature below 250°C by adjusting the oxygen concentration, which must be decreased if there is any indication of temperatures in excess of 250°C.

4. A hot spot will travel down the bed as the air is added. When temperatures settle down again, more air should be added in order to continue the oxidation, and the process is repeated until no further temperature rise takes place upon introduction of additional air, or until the oxygen content of the inlet and exit gases is the same.

SAFETY NOTE Gases containing carbon monoxide must be excluded from the methanator at any temperature below 150°C. Nickel carbonyl is formed below 150°C, and this is an extremely toxic gas (see Section 3.15).

3.11.5. Stabilization of Ammonia Synthesis Catalyst

It is not normal to attempt to stabilize ammonia synthesis catalyst in the converter, as the reduced catalyst can be discharged safely with careful attention to the details outlined in Section 3.15.1. Also, an ammonia converter seldom has enough thermocouples to ensure proper control of the stabilization process and some pockets of unstabilized catalyst can remain to trigger-off an unexpected temperature rise. If, however, stabilization is required, a typical procedure is given below.

1. Cool the catalyst below 60°C, normally by circulating synthesis gas, and purge free from hydrogen and ammonia, using nitrogen which should contain less than 100 ppm of oxygen.

2. Start to circulate the nitrogen at 3–5 bar pressure, using the maximum possible circulation rate.

3. Add air carefully to the circulating mixture to give about 500 ppm of oxygen at the converter inlet, and gradually increase this to 0.3%. Ensure that the maximum catalyst temperature remains below 100°C. If the temperature rises above 100°C much more heat will be evolved, as the bulk of the catalyst will be oxidized instead of just the surface.

4. When the catalyst is stable with 0.3% oxygen in the inlet gas, slowly increase the oxygen concentration to about 7%, continuing to keep the catalyst temperature below 95°C. The increase from 7% to 20% can be taken more rapidly, but a careful watch must be kept on the catalyst temperature, and nitrogen should always be kept available to purge the converter in case of an unexpected temperature increase.

3.12. Catalyst Discharge

3.12.1. General

When a catalyst is to be discharged by gravity flow from the bottom of the vessel the operation is a simple process, especially if the catalyst is not to be re-used. Suitable chutes will be required to lead the catalyst away from the vessel into drums, or into a tipper truck, or merely to distribute it over the ground. It is also possible to use a conveyor or vacuum equipment to carry the catalyst directly into the truck or hopper.

The rate of catalyst discharge can be substantially improved by employing large, mobile, air-conveying units such as that illustrated in

Chapter 3. Handling and Using Catalysts in the Plant

Figure 3.12. The air-conveying unit works on the principle of entrainment of particles in a high-velocity air stream, and the power of such units enables them to be used for a variety of ancillary duties, such as the removal of damaged brickwork, cleaning out vessels and discharging packing from an absorption tower. There are several specialist contracting firms who can provide and operate such units.

Figure 3.12. Mobile unit for catalyst extraction. (Courtesy of Wistech).

3.12.2. Discharge of Pyrophoric Catalyst

If the catalyst being discharged is pyrophoric, water hoses should be available to quench the catalyst in case it becomes hot; water sprays can also be added in the conveying equipment. The catalyst can be discharged with a nitrogen blanket in the vessel to prevent oxidation taking place while the catalyst is in the reactor which could cause damage to the vessel. During actual discharge only the discharge port should be open, to prevent air from being drawn into the vessel by the "chimney" effect. Some form of hinged cover can be used to advantage in order to control the flow of catalyst and reduce the chance of air entering the vessel. Temperatures in the catalyst bed should be monitored for signs of overheating. If excessively high temperatures develop in the bed the discharge should be stopped and the inert gas atmosphere re-established. The discharge can be continued when the catalyst is cool enough. Alternatively, the vessel can be filled with water and drained several times before the catalyst is discharged—where the vessel or cartridge is made of stainless steel, demineralized water should be used. Water prevents easy access of air to the active surface, as well as cooling the catalyst and so markedly slows down oxidation of the reduced catalyst. This practice is useful with methanators to avoid the risks associated with possible formation of nickel carbonyl. Water can also be used in some types of ammonia converter, but synthesis catalyst is not stabilized by water and will become pyrophoric again if it is allowed to dry out. Reduced ammonia synthesis catalyst reacts slowly, even with cold water, to produce hydrogen. The small bubbles eventually collect and rise to the surface. **Great care is required to avoid a build-up of hydrogen or hydrogen/air mixtures.** The vessel must be well-ventilated and the wet catalyst roused vigorously with nitrogen from time to time to dislodge the hydrogen bubbles. Once the catalyst is discharged it should be kept away from all inflammable materials. Bins, drums or trucks employed for the removal of the pyrophoric catalyst should be made of steel and should be kept well-ventilated.

3.12.3. Top Discharge

In many primary and secondary reformers, and some ammonia synthesis reactors, bottom-discharge facilities are not provided and the catalyst must be removed from the top; this is usually done by vacuum extraction. There are several possible arrangements for such equipment. For small plants an industrial vacuum cleaner, such as that made by Bivac, is often used. The vacuum gear is usually left on the ground with a fixed riser pipe up to the top of the vessel. In the case of the primary reformer the riser can be connected to a forked manifold, with one branch above each row of tubes. A short flexible pipe then connects the manifold to the extraction tube. The extracted catalyst is separated in a cyclone fitted into a hopper or drum, and is then discharged through a chute into a trailer, or into further drums for retention. Compressed-air ejectors can also be used to provide the vacuum, if a suitable air supply is available. In the case of large plants the type of equipment shown in Figure 3.12 is preferable.

When a vessel has a large top opening, such as a secondary reformer, it is possible for men to enter the vessel and shovel the catalyst into buckets which are pulled out of the vessel on a rope. Before men enter the vessel it is essential that the catalyst be cooled and stabilized, and that the vessel be purged through with air. Appropriate protective clothing should be worn, and a compressed-air hose is useful to assist in clearing the atmosphere in the vessel. To reduce the amount of catalyst breakage and dust formation, boards should be provided in the vessel for the operators to stand on.

3.12.4. Blanketing Pyrophoric Catalyst During Vacuum Extraction

Where a catalyst is to be discharged in a pyrophoric state using vacuum techniques, sufficient inert gas should be available for injection into the bottom of the reactor to compensate for the amount being withdrawn with the catalyst in the vacuum equipment. If the inert gas supply does not match the rate of extraction, then some air will be drawn down into the vessel and into the top of the catalyst bed. This may cause local heating, which could affect the discharge apparatus. In any case, the operators should be provided with insulated gloves for handling the extraction tube. A watch must be kept on the temperature in the bed, and if there are signs of overheating the extraction should be stopped while the inert gas blanket is re-established. It is also important that the inert gas/air mixture is not sucked through the discharged catalyst in the extaction hopper. This can easily be avoided by a purge of inert gas directed through the bottom of the discharged catalyst. A circulating extraction system is sometimes used when there is a shortage of inert gas. This requires a circulating pump which sucks the catalyst from the converter, separates the catalyst from the circulating gas, and returns the inert gas into the top of the catalyst vessel from the delivery side of the pump. A purge of fresh inert gas is maintained up from the bottom of the undischarged catalyst, and this is supplemented at the top of the bed by the gas returned from the pump delivery. Operatives working inside the vessel should be protected by an integrated life support system, when ever catalyst is discharged under nitrogen (see Section 3.15, "Safety precautions").

3.12.5. Discharge of Ammonia Synthesis Catalyst

Ammonia synthesis catalyst may be stabilized before discharge, as described above. The stabilization tends to be a long process, requiring special instruments and equipment. Stabilized synthesis catalyst must still be handled with care, as it may still be slightly pyrophoric, or may contain a proportion of unstabilized catalyst. A cold-shot converter may be discharged on to the ground if it has a suitable discharge port. The undischarged catalyst remaining in the vessel must, of course, be blanketed with nitrogen. The ICI "lozenge" cold-shot converter is particularly convenient for direct discharge, as it has a manhole which allows the catalyst to be discharged without any disturbance to the cartridge itself. The quantity of catalyst to be disposed of can be a problem in the case of large converters. The catalyst cannot simply be poured on to the ground near the converter, and some arrangements are needed to remove it during discharging. The unstabilized catalyst can be handled safely so long as it is kept wet. If a water spray is fitted into the discharge chute, the catalyst can be run directly on to a conveyor, and from there into a metal truck or container. The catalyst is not stabilized by the water, and if it is allowed to dry out it will again become pyrophoric. Some hydrogen is evolved by a reaction between the water and the reduced catalyst. If steam from a wet part of the catalyst passes through another part which has become dry and hot, hydrogen can be formed which can then ignite. In practice, the discharge can be done safely and easily by providing sprays to keep the catalyst wet, but the discharged material should be spread on to open ground in a thin layer as soon as possible. A thin layer will oxidize in a few days, but can be dangerous to walk on until oxidation is complete.

Most tube-cooled converters, and other types, have to be discharged with vacuum extraction gear through the top cover, as described above. Another method of discharging catalyst is to fill the vessel with water and then extract the catalyst together with the water into a mobile air-conveying unit. It must be remembered that even the cold catalyst will react with water to produce hydrogen, and explosive mixtures can build up if the air space is allowed to become static. A continuous purge of the space above the catalyst and water is required to keep the hydrogen concentration at a safe level.

3.13. Re-use of Discharged Catalyst

If a catalyst is performing satisfactorily, then it is generally best to leave it undisturbed in the reactor, and blanketed with an inert gas during shutdown periods. Most catalysts become weakened in use, and discharging and recharging inevitably means loss of material. Reforming catalysts can be discharged and recharged easily and without loss of activity, but some of the material will be lost in the course of extraction and handling. Some check on strength and composition is essential before recharging, and ICI will do this for its customers. Sieving (or even hand-sorting) may be necessary before the recharge, and suitable apparatus must be available. Hand-sorting is made easier by the use of a short conveyer band. The catalyst must be discharged carefully and kept in order according to its position in the tube. If reforming catalyst is to be recharged it must not be put in a higher position than that from which it was removed, and in many cases the bottom layer is discarded, the

Chapter 3. Handling and Using Catalysts in the Plant

remaining catalyst put back further down the tube and fresh catalyst put at the top.

Sulphided zinc oxide can burn on exposure to air if it is hot, so complete cooling is desirable—particularly if it contains carbon, which is often carried into the catalyst from the preheat section where it is formed. Zinc oxide sulphur removal catalyst can also be discharged and re-used if necessary, without loss of activity. High-temperature shift catalyst and methanation catalyst may be stabilized by controlled oxidation as described in the previous section. These catalysts can be recommissioned without loss of activity by the normal reduction procedure. Low-temperature shift catalyst and ammonia synthesis catalyst may be stabilized for inspection or re-use as described above, but very careful temperature control is required to avoid loss of activity, both during stabilization and during the subsequent re-reduction. The practice is not recommended, since the subsequent performance is rarely satisfactory.

3.14. Disposal of Used Catalyst

Catalysts containing nickel, copper, zinc and precious metals may be sold after use to firms dealing in the recovery of such materials. The economics of the sale will depend on the location of the plant, the composition of the catalyst, and the prevailing metal prices. Catalyst which has been discharged in the reduced state must be allowed to become stable in air before despatch. Handling costs are reduced when the catalyst can be discharged directly into a trailer or into old drums. High-temperature shift and ammonia synthesis catalysts are generally scrapped after use and, if discharged in a reduced state, they should be spread out thinly on the ground, away from any flammable material, in order to avoid the danger of fires during the oxidation period. If the material is tipped in a heap the interior can become very hot, so it is important that discharged catalyst should not be tipped together with other waste.

3.15. Safety Precautions

Operators should be aware of the hazards associated with the use of catalysts and draw up the appropriate safety instructions.

3.15. Safety Precautions

1. Discharged Pyrophoric Catalysts
   Catalysts discharged in the pyrophoric state must be kept separate from flammable materials. Dumps of the catalyst should be within reach of water hoses so that any overheating that occurs can be controlled. High temperatures can build up in heaps and it is a prudent precaution to spread the catalyst thinly over the ground until the oxidation is complete and under no circumstances should personnel be allowed to walk over the catalyst until it has been fully stabilized.

2. Nickel Carbonyl Hazard
   Catalysts containing metallic nickel must not be exposed to gases containing carbon monoxide at temperatures below 150°C. Observation of this rule avoids the risk of the formation of nickel tetracarbonyl, $Ni(CO)_4$, an extremely toxic, almost odourless gas which is stable at low temperatures. It is most likely to be formed in methanation reactors when the plant is cooled down, unless the system has been thoroughly purged with nitrogen. There is also a similar dnager with nickel/molybdenum desulphurization catalyst. Formation of nickel tetracarbonyl over reforming catalysts under normal conditions is not very likely because they are usually thoroughly steamed during shut-down. However, nickel tetracarbonyl formation should always be considered particularly when it has not been possible to follow normal shut-down procedures.

3. Entry into Inert Gas Atmospheres
   Extreme care is needed during a shut-down when an entry has to be made into a vessel containing an inert gas. Such atmospheres do not support life and personnel entering must wear a suitable breathing apparatus. Failure to do so will result in a loss of consciousness within seconds of breathing the atmospheres followed within minutes by death. To avoid accidental entry of the vessels openings must be kept closed. When personnel have to work inside the vessel prominent warning notices must be displayed. Everyone working within the area should be made aware of the nature and dangers of asphyxia. They should know how to attempt a rescue and resuscitation of anyone who may be overcome. An integrated life-support system is essential with adequate back-up. Several double fatalities have occurred because operatives have entered the vessel without a breathing apparatus to assist a colleague. If a company has no experience in such activities then the work is often best done by a specialised service firm.

Chapter 3. Handling and Using Catalysts in the Plant

4. Dust Exposure

   Short-term exposure to the metals and metal oxides used in catalysts may give rise to irritation of the skin, eyes and respiratory system. Over exposure can give rise to more serious effects. Product safety data sheets should be consulted for information. Catalysts should be handled as far as possible in well-ventilated areas and in a way which avoids the excessive formation of dust. Operatives who handle catalyst must wear suitable protective body clothing, gloves and goggles. Inhalation of dust should be avoided, and the appropriate occupational exposure limits should be strictly observed. If these limits are likely to be exceeded then respiratory protection should be used. Everyone involved in the handling operation should clean up afterwards and, in particular, must wash before eating. Clothing should be changed at the end of each shift, and more frequently if contamination is heavy.

5. Ergonomics

   Hazards associated with the handling of catalysts are discussed in Section 3.3.

Chapter 4

Feedstock Purification

4.1. Introduction

The catalysts used in modern ammonia, methanol and hydrogen plants are extremely active and have high selectivity, but they are also very sensitive to poisons. In order to achieve the long production runs required for the economic production of ammonia, methanol and hydrogen it is therefore important to ensure that all of the process fluids are free from poisons, and careful control is required to purify the process water, the process air and the hydrocarbon feedstock. Because of the large volumes of gas passing over the catalysts, small levels of poisons (often at the limit of detection) can have a cumulative effect and restrict catalyst life. A modern ammonia plant is, however, a sensitive analytical instrument, and careful monitoring of catalyst performance will soon show the presence of poisons. The commonest poisons found in hydrocarbon feedstocks are sulphur, chloride and organometallic compounds.

Sulphur is a particularly severe poison for the steam reforming catalyst, which contains nickel. It is adsorbed on the nickel as a surface sulphide and interferes with the steam reforming reaction. The loss of activity can lead to carbon deposition and subsequent overheating of the reformer tubes, which may result in tube failure. The reaction with sulphur is reversible and the reformer will often recover once the source of sulphur is removed. Most steam reformers are designed to be able to operate with a sulphur level of up to 0.5 ppm in the feed, but in practice even this level will cause some loss in activity, and efficient operation demands a level of less than 0.1 ppm. Most modern desulphurization systems will reduce the sulphur content in the feed to less than 0.02 ppm. The onset of sulphur poisoning can be detected by the appearance of hot patches at the top of the reformer tubes. If these are ignored, then the hot patches extend down the tubes and the level of methane in the process gas leaving the reformer ultimately rises.

Chloride is also a serious poison. The chloride ion has high mobility and can migrate freely through the plant with the process gas stream, causing damage to equipment and the catalysts. Many alloy steels are sensitive to chloride induced stress corrosion, and chloride attack causes many heat exchangers to fail. Chloride accelerates the sintering of the

Chapter 4. Feedstock Purification

metal crystallites in the catalyst, and this produces an effect similar to thermal ageing but takes place much more rapidly. Chloride will deactivate reforming catalyst, but it is a more serious poison on the copper based low temperature shift and methanol synthesis catalysts. The generally accepted limit for chloride contamination in the feedstock is less than 5 ppb (1ppb = 1 part in 10^9).

Organometallic compounds are common in crude oil (e.g. vanadium porphyrins), but rarely occur in the hydrocarbon feedstocks used on steam reformers. They cause damage by blinding the hydrodesulphurization and reforming catalysts and, once deposited, cannot be removed. The generally accepted limit for heavy metals is also less than 5 ppb.

4.2. Feedstocks for Ammonia, Methanol and Hydrogen Production

4.2.1. Natural Gas

A wide range of hydrocarbon feedstocks is used for the production of synthesis gas by steam reforming. The most common feedstock now is natural gas. This occurs widely throughout the world, and is the easiest feedstock to process. Natural gas consists mainly of methane, with small amounts of low molecular weight hydrocarbons, and often nitrogen and carbon dioxide. Many natural gases have a low sulphur content which is usually present as simple compounds such as hydrogen sulphide, carbonyl sulphide or mercaptans. In many countries compounds containing sulphur are added to natural gas as a stenching agent to aid in the detection of leaks (e.g. diethyl sulphide in the U.K. and tetrahydrothiophene in The Netherlands). Typical gas compositions for some major gas fields are given in Table 4.1.

Table 4.1. Typical composition of natural gas found in some major gas fields

Component	North Sea	Groningen	Ekofisk	Indonesia
CH_4/%	93.81	81.25	85.45	84.88
C_2H_6/%	4.52	2.83	8.36	7.54
C_3H_8/%	0.38	0.41	2.85	1.60
C_4H_{10}/%	0.04	0.14	0.86	0.03
C_5H_{12}/%	0.02	0.09	0.22	0.12
N_2/%	0.73	14.23	0.43	1.82
CO_2/%	0.47	0.96	1.83	4.0
Total sulphur (H_2S)/ ppm	5		30	2

4.2.2. Associated Gas, Natural Gas Condensates and LPG

Associated gas (the gas released in oil production) is often used as feedstock for ammonia and hydrogen plants located in oil-rich regions such as the Middle East. This gas is very similar to natural gas in that it contains a high level of methane, but it also tends to contain small but variable quantities of high molecular weight hydrocarbons. These may condense in the gas pipeline and arrive at the plant as "slugs" of liquids. Also, associated gas tends to contain significant quantities of hydrogen sulphide and organic sulphur compounds. These factors make associated gas harder to process than natural gas. The variable gas density also makes flow measurement and control difficult, and the high and variable sulphur content may necessitate the inclusion of a sulphur recovery unit to reduce the sulphur content to a reasonable level before it is delivered to a chemical plant. Typical gas compositions for a plant receiving associated gas are given in Table 4.2.[88] The analyses are after treatment on the gas-sweetening plant but before final desulphurization. The total sulphur in the raw gas varies from 1.0% to 3.0% (v/v).

Table 4.2. Variation in composition of associated gas supplied to an ammonia plant in the Middle East[89]

Component	A	B	C	D
CH_4/%	70.75	75.02	73.5	85.04
C_2H_6/%	15.62	14.31	12.3	12.37
C_3H_8/%	7.87	6.02	4.7	2.32
C_4H_{10}/%	3.31	2.51	3.42	0.27
C_5H_{12}/%	1.27	0.78	1.47	0.01
C_6H_{14}/%	0.49	0.23	0.66	
C_7H_{16}/%	0.22	0.03	0.34	
C_8H_{18}/%	0.17		1.49	
C_9H_{20}/%			1.50	
CO_2/%	0.18	0.03	0.03	0.02

A, B and C = variations in associated gas over a 3-year period.

D = gas composition after the installation of a NGL station to remove the heavier hydrocarbons.

Natural gas condensates and LPG consisting of mixtures of ethane through to butane are often used as feedstocks for ammonia plants. These are obtained from the oil-processing industry and are usually fairly pure, with only small amounts of hydrogen sulphide and carbonyl sulphide. However, they can be contaminated with chloride during shipping and storage.

4.2.3. Naphtha

A wide range of naphthas is used, varying from simple light distillate to heavy naphthas with a final boiling point of up to 220°C and containing up to 20% of aromatics. Naphthas are the most difficult feedstocks to purify because they often contain significant quantities of organic sulphur compounds such as thiophenes and benzothiophenes, which are difficult to break down. Although naphtha is not now widely used on a large scale, it is often used as a standby feed. Care is needed in this case to ensure that the purification catalysts can cope with the change from a simple gas duty to naphtha. Many small hydrogen plants use naphtha as their only feed, and because of their small size they have difficulty in obtaining sufficient heat for the purification stage and may not be able to desulphurize certain naphtha feedstocks fully.

4.2.4. Refinery Off Gases and Electrolytic Hydrogen

The high price of hydrocarbon feedstocks has led many operators to use waste gases from refineries. These are generally rich in hydrogen (60–80%) and have a low sulphur content. However, they may contain hydrogen chloride if the refinery uses certain precious metal-based catalytic reforming processes which require the addition of organic chlorides (typically propylene dichloride) to control catalyst activity. They may also contain significant quantities of unsaturated hydrocarbons which can cause processing problems, particularly if acetylenes are present. In a few cases hydrogen obtained by the electrolysis of brine is used as a feedstock. In addition to the risk of chloride contamination this feed may contain mercury, which is particularly harmful to copper-based catalysts.

4.2.5. Coal Gasification and Coke Oven Gas

Ammonia and methanol are also manufacturerd using coal as a feedstock. This is currently limited to countries such as South Africa, Poland, Zambia, India and Turkey which have access to cheap supplies of coal but lack sufficient natural gas. Coal gasification is likely to become more common in the future as supplies of oil and gas are depleted. The gases produced by coal gasification contain both gaseous and solid impurities, and require complex purification trains. The nature of the gas produced is governed mainly by the gasification temperature and, to a lesser extent, by the type of coal. In general, high-temperature gasifiers break down all of the hydrocarbons present, but also produce hydrogen cyanide and nitrogen oxides from the nitrogen compounds present. Low-temperature gasifiers do not produce

4.2. Feedstocks for Ammonia, Methanol and Hydrogen Production

appreciable quantities of these compounds, but they do yield a range of hydrocarbons and heterocyclic compounds. Table 4.3 gives the reported gas compositions for the Koppers–Totzek gasifier and the Lurgi gasifier which respectively are examples of high- and low-temperature gasification.[89, 90]

Table 4.3. Composition of synthesis gas produced by the gasification of coal in the Koppers–Totzek and Lurgi gasifiers[89, 90]

Component	Koppers–Totzek	Lurgi
CO/%	58.0	20.0
CO_2/%	12.0	29.0
H_2/%	27.0	39.0
N_2 + A/%	1.5	1.0
CH_4/%	<0.1	10.0
C_nH_m/%	NR	0.6
H_2S/%	0.5	0.4
COS/%	0.04	NR
SO_2/ppm	0.1	NR
HCN/ppm	100	NR
NO_x/ppm	30–70	NR
NH_3/ppm	15	NR
O_2/ppm	100	NR
Ash/mg m^{-3}	0.1	NR

NR = not reported.

Coke oven gas is occasionally used as a feedstock. This has a composition similar to feedstocks derived from low temperature gasifiers, but as coke ovens are operated on a cyclic pattern there can be wide variations in gas composition.

4.2.6. Mixed Feeds

Many ammonia and methanol plants now have to be able to use a mixture of different hydrocarbons as feed and fuel. This is perfectly feasible, provided that the full analyses of the feeds are known and the correct purification stages are installed. The different types of feeds and fuel used on one European 1000 tonnes day^{-1} ammonia plant are given in Table 4.4.[91]

Chapter 4. Feedstock Purification

Table 4.4. Hydrocarbons used as feedstocks and fuels for a 1000 tonnes day^{-1} ammonia plant

Feedstock	Fuel
Natural gas	Refinery off-gas
LPG consisting of a 50 : 50 mixture of butane and butene	Fuel oil from an ethylene cracker
Synthesis gas from an oil gasifier	Low-sulphur fuel oil
Hydrogen from an ethylene cracker	LPG (as standby)
Refined naphtha (as standby)	Natural gas (as standby)

4.3. Desulphurization

Sulphur is a poison for nickel steam reforming catalysts, but the poisoning is reversible and in practice there is a threshold limit for given conditions, below which the poisoning effect is not apparent. The sensitivity of the catalyst to sulphur poisoning increases as the reforming temperature is lowered and Figure 4.1 shows the minimum level of sulphur required to poison nickel catalyst in typical industrial steam reformers at different temperatures.[92] Time effects are also important. With very low sulphur levels in the feed to the reformer significant loss of activity is only apparent after quite a long period on line.

Many plants have been able to reverse the effect of a few days of mild sulphur poisoning by switching to a sulphur-free feed, or by steaming the catalysts for a few hours. However, if carbon laydown has taken place it can be difficult to remove the carbon completely from all of the tubes in a reformer. Also, sulphur tends to migrate with the process gases and accumulates on the HT shift and LT shift catalysts, where it can reduce activity. It is far better to prevent sulphur from entering the plant by using an efficient desulphurization system than to have shortened catalyst lives.

4.3.1. Processes for Single-stage Sulphur Removal

Sulphur removal from natural gas by absorption at ambient temperature on activated charcoal or with molecular sieves has been widely used in North America. The active charcoal may have its capacity enhanced by impregnation with transition metals such as copper. Frequent regeneration of the absorbent with steam is required, and two or more

4.3. Desulphurization

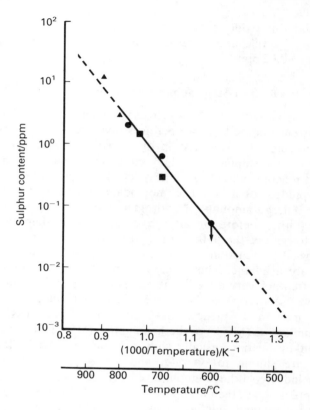

Figure 4.1. The minimum sulphur concentration in process gas to cause practical poisoning of nickel steam reforming catalyst at different temperatures. Hydrocarbon feed: ●, Heptane; ■, naphtha; ▲, methane.

vessels must be available so that one is on-line while the other is being regenerated. The efficiency of these absorbent systems depends both on the type of sulphur compounds and on the amount of high molecular weight hydrocarbons in the natural gas. Low-boiling sulphur compounds such as carbonyl sulphide are not strongly absorbed, and condensable hydrocarbons can rapidly saturate the absorbent. This method of sulphur removal alone has limited application because not all gases can be desulphurized to the level required by a modern high heat flux steam reformer. Also, the regeneration stage requires energy in the form of low-pressure steam, and a malodorous gaseous effluent is produced during regeneration, which can cause environmental problems. If only hydrogen sulphide is present in the natural gas, then it is possible to remove it using zinc oxide. The reaction between hydrogen sulphide and zinc oxide is covered in detail in Section 4.11 and it is worth noting that this method can effectively remove hydrogen sulphide completely (~0.02 ppm).

4.3.2. Processes for Two-stage Sulphur Removal

Because of the variety of feedstocks in use and the wide range of sulphur compounds encountered, most operators now use a two-stage process for sulphur removal. The first involves reaction with hydrogen to produce hydrogen sulphide and the second is the removal of hydrogen sulphide by reaction with porous zinc oxide. The first stage is carried out using a sulphided cobalt or nickel molybdate catalyst with an excess of hydrogen. If large amounts of hydrogen sulphide are present in the feedstock (either before or after hydrogenolysis), then it may be necessary to remove the bulk of the hydrogen sulphide with a washing stage followed by reaction with zinc oxide. A number of proprietary processes are available. These may also remove carbon dioxide, but they are often not very effective at removing carbonyl sulphide. They involve the use of two packed towers, the first being used for the absorbent medium is an aqueous solution of a weak base such as monoethanolamine (MEA) or diethanolamine (DEA). Regeneration is carried out by heating, usually with a steam-heated reboiler. These processes are simple to operate, and are widely used in the oil processing industry. If large amounts of hydrogen sulphide are liberated in the stripping stage, then it may be necessary to use the Claus process to produce elemental sulphur. This, its simplest form, is shown in equation (1) and the reaction is usually carried out over a promoted alumina catalyst.

$$2H_2S + SO_2 \rightarrow 3S + 2H_2O \qquad (1)$$

The final stage of any modern desulphurization process is a bed (or beds) of granules of high-porosity zinc oxide. Reaction of hydrogen sulphide to give zinc sulphide is favoured under almost all process conditions, and it is possible to reduce the level of hydrogen sulphide in the gas leaving the bed to less than 0.02 ppm. The zinc oxide is of course consumed during the process, and is not regenerable. The bed must be discharged and replaced when it is exhausted, so it may be expensive to use zinc oxide alone to remove high levels of sulphur. The economics vary from plant to plant, but for a typical 1000 tonnes day^{-1} ammonia plant it is not usual to use zinc oxide alone for hydrogen sulphide removal when the feedstock contains sulpur levels above 200 ppm.

4.4. Thermal Dissociation of Sulphur Compounds

Primary and secondary mercaptans decompose at approximately 200–250°C, and tertiary mercaptans decompose at slightly lower temperatures. The products are mainly olefins and hydrogen sulphide, but complex products—including polymers—are also formed. Aromatic thiophenols are generally more stable. Aliphatic disulphides decompose at similar temperatures to mercaptans, giving a mixture of products which usually contain mercaptans and hydrogen sulphide. Aromatic disulphides are more stable and tend to form sulphides and elemental sulphur at approximately 300°C. Cyclic and linear straight-chain sulphides are usually stable up to 400°C, then they decompose to form hydrogen sulphide and olefins. Thiophenes are thermally stable at 470–500°C. Table 4.5 shows the temperatures at which a range of sulphur compounds decompose under the same process conditions. Because of the wide range in thermal stabilities it is quite common for some of the sulphur compounds present in the feedstock to dissociate in the preheater or vaporizer sections of the desulphurization unit before coming into contact with the catalyst. This can lead to carbon and polymer formation on the preheater coils and on the top layer of catalyst. The addition of hydrogen tends to suppress such cracking reactions, and it is always beneficial to add recycle hydrogen before the heating stage.

Table 4.5. Thermal decomposition of selected sulphur compounds

Compound	Temperature at which decomposition commences/°C
n-C_4H_9SH	150
i-C_4H_9SH	225–250
$C_6H_{11}SH$	200
C_6H_5SH	200
$(C_6H_5)_2S$	450
$(C_2H_5)_2S$	400
$C_6H_5SC_6H_{11}$	450
Thiophene	Stable at 500
2,5-Dimethylthiophene	475

4.5. Hydrogenolysis of Sulphur Compounds

The term hydrogenolysis in the present context refers to the addition of hydrogen across a sulphur–carbon bond. The reaction is usually carried out over a nickel or cobalt molybdate catalyst and results in the formation of hydrogen sulphide and a saturated hydrocarbon. Some typical hydrogenolysis reactions for a number of different sulphur compounds are shown in equations (2)-(7). All of these reactions are exothermic but, because of the very low levels of sulphur compounds found in most feedstocks, a temperature rise is seldom observed. Enthalpies of hydrogenation of some typically encountered organic sulphur compounds are given in Table 4.6.

$$C_2H_5SH + H_2 \to C_2H_6 + H_2S \tag{2}$$

$$C_6H_5SH + H_2 \to C_6H_6 + H_2S \tag{3}$$

$$CH_3SC_2H_5 + 2H_2 \to CH_4 + C_2H_6 + H_2S \tag{4}$$

$$C_2H_5SSC_2H_5 + 3H_2 \to 2C_2H_6 + 2H_2S \tag{5}$$

$$C_4H_8S \text{ (tetrahydrothiophene)} + 2H_2 \to C_4H_{10} + H_2S \tag{6}$$

$$C_4H_4S \text{ (thiophene)} + 4H_2 \to C_4H_{10} + H_2S \tag{7}$$

Table 4.6. Heats of hydrogenolysis of some organic sulphur compounds

Reaction	ΔH/ kJ mol^{-1}
$C_2H_5SH + H_2 \to C_2H_6 + H_2S$	-70.2
$C_2H_5SC_2H_5 + 2H_2 \to 2C_2H_6 + H_2S$	-117.2
C_4H_8S (tetrahydrothiophene) + $2H_2$ $\to$ n-C_4H_{10} + H_2S	-120.2
C_4H_4S (thiophene) + $4H_2 \to$ n-C_4H_{10} + H_2S	-280.3

In general, the rate of hydrogenolysis is first-order with respect to the sulphur compound. The order of reaction with respect to the partial pressure of hydrogen lies between zero-order and first-order, depending on the nature of the sulphur compounds present. The equilibrium constants for the hydrogenolysis of organic sulphur compounds are large

4.5. Hydrogenolysis of Sulphur Compounds

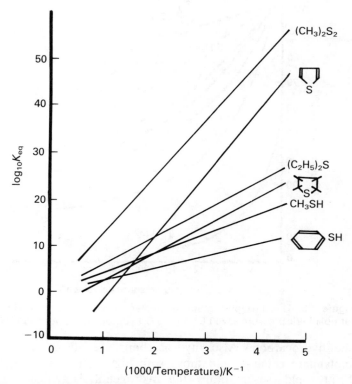

Figure 4.2. The variation of equilibrium constants for the hydrogenolysis of some organic sulphur compounds with temperature.

and positive, even at temperatures as high as 500°C. The rate of hydrogenolysis increases with temperature, and under the usual process conditions between 350°C and 400°C with an excess of hydrogen the hydrogenolysis is normally complete. Figure 4.2 shows the variation of the equilibrium constants for the hydrogenolysis of some sulphur compounds with temperature. There is a marked difference in the rate of hydrogenolysis of thiophene and other commonly found sulphur compounds. This is illustrated in Figure 4.3, which shows the ease of hydrogenolysis of a number of sulphur compounds using a heptane feed at atmospheric pressure and 370°C and 250°C.

Rate constants for hydrogenolysis of a particular sulphur compound are dependent on the types of hydrocarbon present. Diffusion of sulphur compounds is easier through low molecular weight hydrocarbons than through those of high molecular weight, and the rate of hydrogenolysis is therefore faster the lower the molecular weight of

Chapter 4. Feedstock Purification

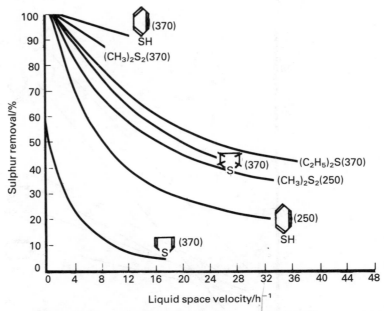

Figure 4.3. The desulphurization of heptane using a cobalt molybdate catalyst (atmospheric pressure at 250°C and 370°C, hydrogen/heptane ratio 1 : 4).

the hydrocarbon. Also, hydrocarbons are absorbed on the cobalt molybdate catalyst, and those with high molecular weight are more firmly held, so inhibiting the hydrogenolysis reaction. Olefins are hydrogenated exothermically over cobalt molybdate catalyst, but if insufficient hydrogen is present for complete hydrogenation it is possible for hydrogen sulphide produced at the top of the bed to add across the double bond to form an organic sulphur compound. Nowadays, because naphtha is rarely used as a feed, complex sulphur compounds seldom have to be treated in the desulphurization unit. With process conditions largely set by the zinc oxide bed and the primary reformer catalyst, it is rare to encounter any problems with the hydrogenolysis stage of the desulphurization process. In practice, if unreacted sulphur compounds are reported in the gas leaving the hydrodesulphurizer the two most common explanations are analytical error or mechanical damage to the catalyst bed which had given rise to channelling, so allowing some process gas to by-pass the catalyst in the reactor.

4.6. Carbonyl Sulphide

Carbonyl sulphide (COS) is usually present in natural gas which contains carbon dioxide and hydrogen sulphide. Carbonyl sulphide has a low boiling point (−50°C) and is therefore not easy to remove with fixed-bed absorbents such as active charcoal or molecular sieves, and it can be a source of sulphur poisoning. Carbonyl sulphide is thermally stable and it may not be fully hydrogenolysed over cobalt molybdate catalyst. This is because of what may be called the "reverse water-gas shift" reaction which is catalysed by cobalt molybdate catalysts, and so limits the extent of conversion. Carbonyl sulphide leaving the cobalt molybdate catalyst will react with zinc oxide, but the rate of reaction is slow and the absorption profile shallow, so a conventional desulphurization system will not fully remove carbonyl sulphide.[93] However, it has been shown that carbonyl sulphide is rapidly hydrolysed by steam to carbon dioxide and hydrogen sulphide as in equation (8), and the level of carbonyl sulphide leaving the cobalt molybdate catalyst can be controlled by steam addition. The steam requirement can be calculated from equation (9) using the data in Table 4.7. The hydrolysis reaction is fast over cobalt molybdate catalyst, and plants with high levels of carbonyl sulphide in the feed gas can use steam hydrolysis with confidence.

$$COS + H_2O \rightleftharpoons CO_2 + H_2S \qquad (8)$$

$$K = \frac{P_{H_2S} P_{CO_2}}{P_{COS} P_{H_2O}} \qquad (9)$$

Table 4.7. Equilibrium constants for the hydrolysis of carbonyl sulphide

Temperature/°C	K_p
25	7.2×10^5
125	2.23×10^4
225	2.86×10^3
325	719
425	268

4.7. Cobalt Molybdate Catalysts

Cobalt molybdate is the most widely used catalyst for the hydrogenolysis of sulphur compounds. It has been the subject of much research work over recent years, but since its discovery many years ago its composition has changed little. The main development has been the change from "premixed" to impregnated catalysts. Modern cobalt molybdate catalysts use a special alumina support which is characterized by a highly porous structure. The internal surface area is typically between 200 m^2 g^{-1} and 400 m^2 g^{-1}. The alumina support is impregnated first with a molybdenum compound and then with a cobalt compound and then processed to give a mixture of complex oxides. The total amount of molybdenum and cobalt present varies according to the formulation, but the amount of molybdenum oxide (MoO_3) is usually three to four times the amount of cobalt oxide (CoO). In the oxide form the catalyst consists of a "thin layer" of MoO_3 bonded to the alumina support, with cobalt held in the upper lattice planes. There is speculation about the active phase or phases. The fresh catalyst has been shown to contain Al_2O_3, $CoAl_2O_4$, CoO, MoO_3, $CoMoO_4$ and a complex cobalt molybdenum oxide. The relative proportions of these compounds depends on the nature of the support and the firing temperature used during manufacture. The higher the firing temperature, the greater the content of $CoAl_2O_4$. The composition of a typical commercially available cobalt molybdate catalyst (ICI Catalyst 41-6) is given in Table 4.8.

Table 4.8. Composition and properties of a typical cobalt molybdate catalyst (ICI Catalyst 41-6)

CoO	4.0%
MoO_3	12.0%
SiO_2	1.0%
Al_2O_3	Balance
Bulk density	0.6 kg l^{-1}
Surface area	220 m^2 g^{-1}
Pore volume	0.6 ml g^{-1}

Cobalt molybdate catalysts have low activity in the oxide form, and to obtain maximum activity they must be sulphided. The composition of the sulphided form is related to the oxide form, but because the sulphiding is carried out in a reducing atmosphere there is some

4.7. Cobalt Molybdate Catalyst

reduction in the oxidation state of the molybdenum ions. It is found in practice that between one and two sulphur atoms are taken up by each molybdenum ion, and water is liberated by condensation of neighbouring hydroxyl groups.

Despite many studies there is still no universal agreement on the nature of the active sites in cobalt molybdate catalyst. It is generally agreed that the active sites are those with sulphur vacancies. It seems likely that these sites contain cobalt atoms present in the MoS_2-like Co–Mo–S phase. The role of the cobalt ions is not clear, but it is assumed that they promote the activity of the molybdenum sulphide. Work on the reactions taking place on cobalt molybdate catalyst is complicated by the highly mobile nature of part of the sulphur. The degree of sulphiding depends on the partial pressure of hydrogen sulphide in the feed, and it is possible for hydrogen sulphide to be released from the catalyst when changing from a feed with a high sulphur content to one with a low sulphur content. Discharged cobalt molybdate catalysts from ammonia or methanol plants usually contain between 0.5% and 3.0%(w/w) of sulphur. The higher value is only found if the catalyst is being used for desulphurizing a feed with a high sulphur content, e.g. a "dirty" naphtha.

4.7.1. Presulphiding Cobalt Molybdate Catalyst

If sulphur is not always present in the feedstock it may be necessary to presulphide the catalyst before bringing it into service. This is carried out using an easily decomposed sulphur compound, such as carbon disulphide, dimethylsulphide (DMS), dimethyldisulphide (DMDS) or even hydrogen sulphide. The reaction is carried out using nitrogen, methane or the normal feed as a carrier. For presulphiding it is usual to use about 1% of the sulphur compound and 5%(v/v) of hydrogen. The reaction is exothermic and these conditions will produce a temperature rise of about 10°C. Care must be taken not to exceed bed temperatures of 300°C until sulphiding is complete. This is to avoid the risk of "over-reducing" the oxides in the catalyst. Once formed, metal does not easily sulphide and the catalyst will not then gain full activity. Sulphiding can be carried out during a normal start-up, but is must be remembered that the feed may not have sufficient sulphur to activate the catalyst fully until after many days of operation.

Hydrogenolysis reactions are carried out over cobalt molybdate catalysts at temperatures between 280°C and 400°C. The catalyst is not normally active enough at lower temperatures, and there is the risk of hydrocarbon cracking at higher temperatures. The reaction pressure is normally set by the inlet pressure to the reformer, and most

Chapter 4. Feedstock Purification

hydrodesulphurization reactors in ammonia and methanol plants work in the range 30–50 bar. The space velocity used depends on the feed and the type of sulphur compounds present, but most plants operate with space velocities of less than 3000 h^{-1}. The reaction needs hydrogen, but the level used is normally set by the need to suppress hydrocarbon cracking and by that required for the primary reformer. Most plants using natural gas work with an inlet hydrogen content of 2–5%, while those using naphtha work with ~25%. The larger excess of hydrogen used with naphtha is mainly to suppress cracking reactions.

4.7.2. Other Reactions over Cobalt Molybdate Catalyst

When sulphided, cobalt molybdate catalysts promote most hydrogenation reactions. This means that a number of side reactions can take place over the catalyst, and operators need to consider possible reactions of all of the components in the feedstock. Some of these reactions are shown in equations *(10)-(13)*.

$$O_2 + 2H_2 \rightarrow 2H_2O \qquad \triangle H = -179 \text{ kJ mol}^{-1} \qquad (10)$$

$$C_nH_{2n} + H_2 \rightarrow C_nH_{2n+2} \qquad \triangle H = -(117\text{--}142) \text{ kJ mol}^{-1} \qquad (11)$$

$$CO + 3H_2 \rightarrow CH_4 + H_2O \qquad \triangle H = -219 \text{ kJ mol}^{-1} \qquad (12)$$

$$CO_2 + H_2 \rightarrow CO + H_2O \qquad \triangle H = +41 \text{ kJ mol}^{-1} \qquad (13)$$

With the exception of the reverse water-gas shift reaction *(13)* all of these reactions are exothermic and can cause a significant rise in temperature across the catalyst bed. A typical natural gas feed containing only 0.8% of unsaturated hydrocarbons would give a temperature rise of 20°C (equation *(11)*). If the feed contains a high level of unsaturated hydrocarbons it will be necessary to hydrogenate in a separate recycle loop with a heat exchanger to deal with the exotherm. One operator has successfully run a 1000 tonnes day^{-1} ammonia plant using a feed with 40–50% butene in this way.[91] Olefins present another problem, since if they are not fully saturated they can react with hydrogen sulphide to form organic sulphides which will not be retained on the zinc oxide bed. Acetylenes pose a particular problem because they tend to polymerize to form high molecular weight compounds over cobalt molybdate catalyst. These block the catalyst pores and can lead to a rise in the pressure drop across the bed. Low concentrations of acetylene (200 ppm) have been found to cause so much polymer deposition that a bed of cobalt molybdate catalyst had to be changed after only a few months of operation.[88]

Oxygen is not normally encountered in natural gas, but can be present if air is injected to reduce the calorific value of a rich gas. It can cause problems with cobalt molybdate catalyst: if insufficient sulphur is present to sulphide the catalyst, sulphur may be converted to sulphur dioxide which is not retained by hot zinc oxide.

4.8. Nickel Molybdate Catalysts

Cobalt molybdate catalysts are the most widely used hydrogenolysis catalysts, but there are certain process conditions under which they cannot be used (in particular, where there are very high levels of hydrogen and carbon oxides). These conditions can be encountered on methanol plants and town-gas plants when process gas is used as a source of hydrogen. It is possible under certain circumstances for cobalt molybdate catalysts to promote the methanation reaction (equation (*12*)) and, because of the large amount of heat released, there can be a very large rise in temperature. For these duties sulphided nickel molybdate catalyst is recommended, because once sulphided there is a much smaller risk of the methanation reaction taking place. Nickel molybdate catalyst behaves in a similar way to cobalt molybdate catalyst, but it is a slightly more powerful hydrogenation catalyst and can cause more hydrocracking. The composition of a typical commercially available nickel molybdate catalyst (ICI Catalyst 61-1) is given in Table 4.9.

Table 4.9. Composition and properties of a typical nickel molybdate catalyst (ICI Catalyst 61-1)

NiO	3.8%
MoO_3	11.8%
SiO_2	1.8%
Al_2O_3	Balance
Bulk density	0.6 kg l^{-1}
Surface area	220 $m^2 g^{-1}$
Pore volume	0.6 ml g^{-1}

4.9. Physical Form of Cobalt and Nickel Molybdate Catalysts

Catalyst manufacturers have developed a number of different shaped catalysts which increase the geometric surface area of the catalyst, and so increase the reactivity. These are used for hydrotreating in refineries where high activity is required. The increased activity is obtained at the cost of greater resistance to flow, and hence a higher pressure drop across the bed. Usually in ammonia and methanol plants the hydrodesulphurization reaction is not kinetically limited. Most ammonia and methanol plant operators use catalyst in the form of simple extrudates with a nominal diameter of ~3 mm. In this form the catalyst has high enough activity for the desulphurization needed and in addition the pressure drop is low (usually less than 3 psig). Refinery operators need higher activity and usually use extrudates with a nominal diameter of ~1.5 mm. They may also use multilobe extrudates, or even extrudates with diameters of ~1 mm. Special catalysts may be used for demetallization operation. These are mechanically stronger and have a greater absorption capacity. Because of the variety of shapes available, operators should be careful when replacing the catalyst to use the same form. Change in physical form may increase the catalyst activity, but it may also significantly increase pressure drop across the bed.

4.10. Replacement and Discharging of Cobalt and Nickel Molybdate Catalysts

Both nickel and cobalt molybdate catalysts can give long lives, and rarely have to be changed because of loss of activity. The most usual reason for changing the catalyst is because of an increase in pressure drop resulting from carbon deposition. The carbon is usually formed as a fine dust in the gas preheater section and is carried forward onto the catalyst. Unsaturated hydrocarbons may polymerize on the catalyst to give tarry deposits. The catalyst can still operate with surprisingly high levels of carbon—up to 10% causes little problem and levels up to ~30% have been observed. It is possible to regenerate the catalyst by burning off the carbon with a steam/air mixture, but this is a very exothermic reaction and few plants have the necessary instrumentation to carry out the operation in a sufficiently well controlled manner. It is therefore usually recommended that the top layer of contaminated catalyst be removed and replaced with fresh material.

Used cobalt and nickel molybdate catalysts are potentially pyrophoric in air, and it is advisable to cool them to below 50°C before they are discharged. It is also advisable to use an inert atmosphere for their vacuum discharge. The pyrophoricity is probably due more to the presence of fine carbon dust, residual hydrocarbons and chemisorbed hydrogen than to the cobalt, nickel and molybdenum sulphides, although in some circumstances high area metal sulphides undergo rapid oxidation to the corresponding sulphate.

4.11. Zinc Oxide

4.11.1. Background to Zinc Oxide Absorbents

For many years hydrogen sulphide was removed by absorption in beds of iron oxide. This was loaded in the form of Fe_2O_3, and then converted in the presence of hydrogen at temperatures above 175°C into Fe_3O_4 as in reaction (14). Absorption of hydrogen sulphide was carried out at 350–400°C according to reaction (15).

$$3Fe_2O_3 + H_2 \rightarrow 2Fe_3O_4 + H_2O \tag{14}$$

$$Fe_3O_4 + H_2 + 3H_2S \rightarrow 3FeS + 4H_2O \tag{15}$$

Iron oxide is a low-cost material and has a high absorption capacity, but it is not a suitable absorbent to use for desulphurizing the feed to a steam reformer. This is because at equilibrium the partial pressure of hydrogen sulphide is too high—see Table 4.10. Also, hydrogen sulphide is relatively easily stripped by hydrogen and steam under conditions which may arise during start-up and shutdown of the plant. The final absorption stage in modern synthesis gas plants is carried out using beds of zinc oxide. Zinc oxide reacts almost completely with hydrogen sulphide to form zinc sulphide as shown in equation (16).

$$ZnO + H_2S \rightarrow ZnS + H_2O \tag{16}$$

4.11.2. Thermodynamics and Reaction Kinetics

The thermodynamic effects of water vapour on the efficiency of iron oxide and zinc oxide for the absorption of hydrogen sulphide are shown in Tables 4.10 and 4.11, and the equilibrium constants for equation (16) over a range of temperatures are given in Table 4.12. The exact values for the equilibrium constants depend on the form of zinc sulphide produced and this is discussed in Appendix 20.

Kinetic studies[95] of the reaction of hydrogen sulphide with powdered zinc oxide have shown that the reaction is first-order with respect to the hydrogen sulphide concentration, with a rate constant given by equation (17).

$$k = 9.46 \times 10^{-2} \exp(-7236/RT) \tag{17}$$

In the reaction between hydrogen sulphide and zinc oxide the hydrogen sulphide molecules must first diffuse to the surface of the zinc oxide. There H_2S reacts with the zinc oxide to form zinc sulphide, and the

Chapter 4. Feedstock Purification

Table 4.10. The effect of water vapour and temperature on the equilibrium H_2S concentration over iron oxide, based on equation (15)

H_2O/%	Equilibrium H_2S concentration/ppm(v/v)				
	200°C	250°C	300°C	370°C	400°C
3.3	1.85	3.1	5.1	8.9	10.8
1.7	7×10^{-1}	1.35	2.02	3.51	4.23
0.33	0.85×10^{-2}	1.57×10^{-1}	2.02×10^{-1}	4.1×10^{-1}	4.9×10^{-1}
0.17	1.57×10^{-3}	0.62×10^{-1}	0.93×10^{-1}	1.64×10^{-1}	1.96×10^{-1}

Table 4.11. The effect of water vapour and temperature on the equilibrium H_2S concentration over zinc oxide, based on equation (16)

H_2O/%	Equilibrium H_2S concentration/ppm(v/v)				
	200°C	250°C	300°C	370°C	400°C
3.3	2.6×10^{-4}	1.7×10^{-3}	0.7×10^{-2}	4.2×10^{-2}	6.5×10^{-2}
1.7	1.3×10^{-4}	0.9×10^{-3}	0.3×10^{-2}	2.1×10^{-2}	3.2×10^{-2}
0.33	2.6×10^{-5}	1.7×10^{-4}	0.7×10^{-3}	4.0×10^{-3}	6.5×10^{-3}
0.17	1.3×10^{-5}	0.9×10^{-4}	0.3×10^{-3}	2.0×10^{-3}	3.3×10^{-3}

Table 4.12. Gas phase equilibrium constants for the reaction of zinc oxide with hydrogen sulphide (see also Appendix 20)

Temperature/°C	$K_p = P_{H_2O}/P_{H_2S}$
200	2.081×10^8
260	2.359×10^7
300	7.121×10^6
360	1.569×10^6
400	6.648×10^5
460	2.185×10^5
500	1.145×10^5

water formed must diffuse away. Finally, the sulphide ion must diffuse into the lattice, and oxide ions diffuse to the surface. The conversion of the hexagonal zinc oxide structure into cubic zinc sulphide involves severe structural changes, and the incorporation of the much larger sulphide ion in place of the oxide ion results in a marked loss in porosity. Under normal conditions the equilibrium is strongly in favour of sulphide formation, but the factors discussed above mean that the overall rate of the reaction is controlled by both pore diffusion and lattice diffusion. As a result of these effects, all of the zinc oxide may not be converted to sulphide in a sensible time.

4.11.3. Formulation of Commercial Zinc Oxide

The factors discussed in the previous section have to be taken into account in the formulation of zinc oxide absorbents. The dependence on lattice diffusion can be reduced by making a highly porous zinc oxide with a large internal surface area. However, such a material would have a low density and little mechanical strength. The low mechanical strength would make handling difficult, and the low density would mean that the theoretical sulphur pick-up on a volume basis would be low. Strength and density can be increased by compaction, but such material would have a low porosity with a low surface area, and the rate of reaction with hydrogen sulphide would be correspondingly low. ICI recognized this problem and does not attempt to make a zinc oxide absorbent using zinc oxide alone. Instead, ICI's zinc oxide absorbent (Catalyst 32-4) consists of zinc oxide with a small amount of cement support. This approach enables a high porosity to be obtained whilst maintaining physical strength, ensuring good utilization of the zinc oxide. A typical composition of ICI Catalyst 32-4 is given in Table 4.13. This material can absorb up to 30%(w/w) of sulphur (measured in discharged catalyst). Zinc oxide is also produced in the form of extrudates. These are more difficult to handle because the extrudates tend to interlock and resist flow during charging. The angle of repose for granular zinc oxide is 32° while that for extrudates is typically 38°. Section 4.11.5 contains a discussion on enhanced low temperature activity zinc oxide.

Table 4.13. Typical analysis and properties of zinc oxide granules (ICI Catalyst 32-4)

ZnO	90.0%
CaO	2.0%
Al_2O_3	Balance
Surface area	25 $m^2 g^{-1}$
Bulk density	1.1 kg l^{-1}
Diameter	3.0–5.0 mm

4.11.4. Use of Test Reactors to Assess Zinc Oxide Absorbents

Large beds of zinc oxide are used on the commercial scale in order to ensure long lives, and hence they are not suitable as test reactors. It is, however, possible to study the efficiency of sulphur removal using small

Chapter 4. Feedstock Purification

laboratory reactors. The most satisfactory procedure is to use the test reactor to compare different zinc oxides rather than to obtain absolute measurements.

For example in a typical experiment a small laboratory reactor was filled with a mixture of equal volumes of two zinc oxides. An operating temperature of 370°C and a gas pressure of 30 bar were used, so the conditions of the test were similar to those encountered in industrial reactors. Natural gas containing hydrogen and hydrogen sulphide was passed through the reactor until hydrogen sulphide was detected in the exit gas. The mixed bed was then discharged in layers, and the sulphur

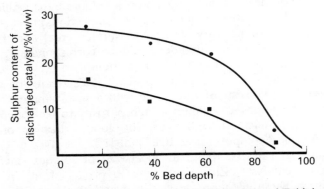

Figure 4.4. Sulphur absorption profiles for ●, ICI Catalyst 32-4 and ■, high-density zinc oxide. Test conditions: temperature, 370°C; pressure, 30 bar; feed, natural gas; space velocity, 400 h^{-1}.

absorption profiles shown in Figure 4.4 were obtained by chemical analysis. The sulphur distribution within individual pellets was determined by different techniques. A small amount of hydrogen sulphide containing the radioactive isotope ^{35}S had been included with the hydrogen sulphide. This allowed the set of autoradiographs shown in Figure 4.5 to be obtained from split pellets of the discharged zinc oxide sampled from different parts of the bed. The material from the inlet is at the top and that from the exit at the bottom. The same split pellets were also examined using an electron microprobe analyser. This gave the photomicrographs shown in Figure 4.6. The electron microprobe is less affected by surface preparation, and gives a more quantitative picture of the absorption pattern across the pellets than is possible with the autoradiographs. The electron microprobe is able to detect sulphur levels greater than ~0.1%(w/w). These two techniques give an interesting indication of the way in which the sulphur is absorbed from the outside of the pellet towards the inner core. They also show that if the zinc oxide is too dense, an impervious layer of zinc sulphide is

4.11. Zinc Oxide

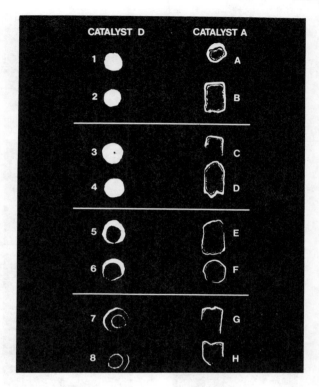

Figure 4.5. Autoradiographs showing the distribution of sulphur within individual pellets of zinc oxide. Pellets 1 and A at bed inlet, 8 and H at bed exit.

formed around the pellets which makes utilization of the inner core difficult.[96, 97]

4.11.5. Effect of Temperature, Pressure and Space Velocity on Efficiency of Zinc Oxide Absorbents

The strong dependence on pore and lattice diffusion leads to the performance of zinc oxide being affected by the process conditions. Because of the high activation energy of the lattice diffusion process, the efficiency of absorption falls rapidly as the temperature is reduced. Figure 4.7 shows the average sulphur content of a bed of zinc oxide in a test reactor at the moment of sulphur slip at different temperatures using the same hydrocarbon feed. It can be seen that for efficient utilization the bed needs to be kept above 350°C. Zinc oxide will react with hydrogen sulphide up to 700°C, but at high temperatures zinc metal

Chapter 4. Feedstock Purification

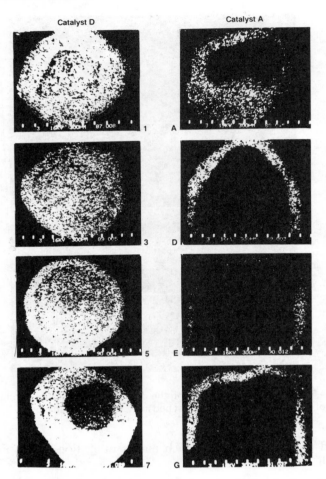

Figure 4.6. Electron microprobe photographs showing the distribution of sulphur within some of the pellets of zinc oxide in Figure 4.5.

starts to be formed. In spite of the low efficiency of sulphur pick-up at low temperatures, zinc oxide is in use in plants down to ambient temperatures. Attempts have been made to improve the low-temperature performance of zinc oxide, and it has been found that the inclusion of certain metals does result in improved absorption. A recently developed commercial zinc oxide (ICI Catalyst 75-1) which has greatly improved low-temperature activity[98] has been achieved by adjusting both the pore structure and the surface area. The enhanced low-temperature activity compared with normal zinc oxide is shown in Figure 4.8.

4.11. Zinc Oxide

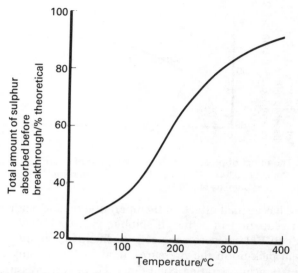

Figure 4.7. The effect of temperature on the absorption of hydrogen sulphide by zinc oxide (ICI Catalyst 32-4). Test conditions: pressure, 30 bar; feed, light hydrocarbon; space velocity, 400 h^{-1}.

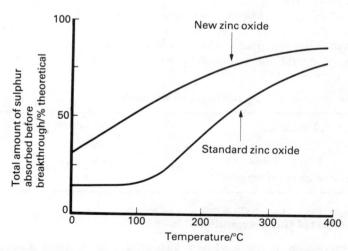

Figure 4.8. A comparison of low-temperature zinc oxide (ICI Catalyst 75-1) with conventional zinc oxide (ICI Catalyst 32-4). Pressure, 1 bar; feed, natural gas; 4% CO_2; 0.2% H_2S.

Chapter 4. Feedstock Purification

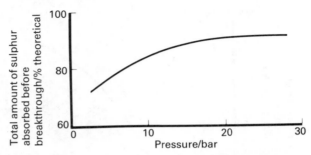

Figure 4.9. The effect of pressure on the absorption of hydrogen sulphide by zinc oxide (ICI Catalyst 32-4). Test conditions: temperature, 370°C; feed, light hydrocarbon; space velocity, 400 h^{-1}.

Pressure has a small effect on the operation of zinc oxide beds. There is a slight reduction in achieved sulphur pick-up below ~10 bar, but from 3 bar to 100 bar performance is unchanged. Figure 4.9 shows the variation in pick-up in the pressure range 1–30 bar using ICI Catalyst 32-4 under typical operating conditions. However, pressure does have an effect if the feedstock contains gases which can condense in the micropores of the zinc oxide. In this case there is then an additional diffusion limitation which slows the reaction of hydrogen sulphide with zinc oxide at the centre of the pellet. Also, at high space velocities (above 4000 h^{-1}) the absorption profile is sharper the greater the pressure. Space velocity influences the degree of absorption because it governs the contact time. Figure 4.10 shows the relationship between space velocity and sulphur absorption capacity using a naphtha feedstock at 20 bar and 370°C. Zinc oxide absorbents were developed to desulphurize the hydrocarbon feedstocks used for steam reformers, and the process conditions on most plants are within the range given in Table 4.14.

Table 4.14. Normal operating conditions for zinc oxide

Temperature	370–400°C
Pressure	20–40 bar
Space velocity	500–1000 h^{-1}
Bed height/diameter ratio	1 : 1

4.11.6. Effect of Gas Composition

The composition of the feedstock also affects the sulphur absorption profile. A very sharp profile is obtained when hydrogen is the process gas. The profile with methane is almost as sharp as with hydrogen, but

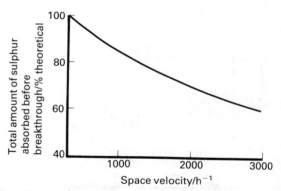

Figure 4.10. The effect of space velocity on the absorption of hydrogen sulphide by zinc oxide (ICI Catalyst 32-4). Temperature, 370°C; pressure, 20 bar; feed, light hydrocarbon; 300 ppm H_2S, 25%(v/v) H_2.

high molecular weight hydrocarbons give a shallower profile. This is because of physical blocking of the capillaries, which causes an additional hindrance to the diffusion process. Carbon dioxide can also interfere with the absorption reaction. Zinc carbonate is stable at temperatures below about 160°C (dependent on the partial pressure of carbon dioxide). Hydrogen sulphide displaces carbon dioxide from zinc carbonate, but at low temperatures these are clearly competing reactions and absorption efficiency may be reduced.

4.11.7. Effect of Reactor Design

The conversion of zinc oxide to zinc sulphide gives an absorption profile which passes down the bed, and a zinc oxide charge is considered to be spent when hydrogen sulphide slip above 0.1 ppm is observed. The shape of the absorption profile is governed by the process conditions and the properties of the zinc oxide, so the efficiency of utilization for any given process conditions and absorbent depends on the height-to-diameter ratio of the zinc oxide bed. Increasing the bed depth does increase the pressure drop, but as the pressure drop is very low for the space velocities normally used, tall thin beds are more practical than is generally realized. Figure 4.11 shows the relationship between pressure drop and height-to-diameter ratio for a 20 m³ charge of ICI Catalyst 32-4 on a typical 1000 tonnes day^{-1} ammonia plant with natural gas feed. The static crush strength of ICI Catalyst 32-4 has been tested, and there is no evidence of breakdown at pressures equivalent to beds 15 m deep. Beds are already in service with depths up to 10 m, and deeper beds are being considered.

Many operators use two beds of zinc oxide in series. This enables the

Chapter 4. Feedstock Purification

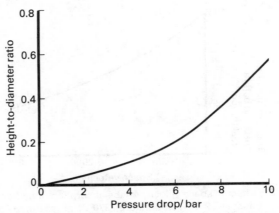

Figure 4.11. The relationship between pressure drop and height-to-diameter ratio for a bed of granular zinc oxide (ICI Catalyst 32-4). Temperature, 370°C; inlet pressure, 40 bar; feed, natural gas; flow rate, 29 000 N m^3 h^{-1}; catalyst volume, 20 m^3.

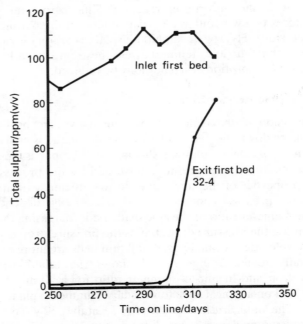

Figure 4.12. Inlet and interbed hydrogen sulphide concentration during the operation of two beds of zinc oxide in series (ICI Catalyst 32-4). Temperature, 370°C; pressure, 40 bar; space velocity, 1100 h^{-1}; catalyst volume, 18.5 m^3.

lead bed to be run to saturation before changing. This approach is particularly attractive to operators with high levels of sulphur in the feed, because by using a by-pass they can change each bed of zinc oxide without having to taking the plant off-line. Figure 4.12 shows the increase in hydrogen sulphide slip between two beds of zinc oxide used by an operator in the Middle East.[99]

4.11.8. Other Desulphurization Uses for Zinc Oxide

Zinc oxide absorbents were developed to remove hydrogen sulphide from hydrocarbons. It has been found that simple organic sulphur compounds are also removed by reaction with zinc oxide to form zinc sulphide, but this is via a cracking mechanism which leads to carbon deposition on the catalyst. This reaction is used by many operators during the start-up of a plant, when no hydrogen is available and the hydrodesulphurization catalyst is not functioning.

Zinc oxide is being used commercially to desulphurize carbon dioxide. For this application the temperature is below 100°C and the pressure up to 40 bar. Under these conditions zinc carbonate is stable and is formed first. The hydrogen sulphide reacts with the zinc carbonate to release carbon dioxide. The absorption profile for this duty are shown in Figure 4.13. Zinc oxide is also used for cyanide removal. Hydrogen cyanide reacts with zinc oxide at temperatures below 150°C to form zinc cyanide. At higher temperatures zinc oxide catalyses the decomposition of HCN.

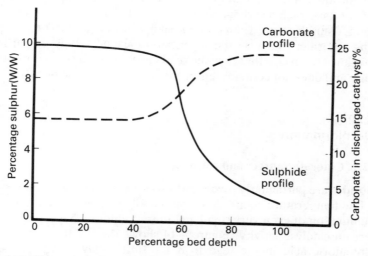

Figure 4.13. The carbonate and sulphide profiles seen in a bed of zinc oxide (ICI Catalyst 32-4) used to desulphurize carbon dioxide at ambient temperature.

Zinc oxide has been widely used in guard beds to protect LT shift catalysts. Typical process conditions for this duty are about 200°C with a steam/gas ratio of 3 : 5 and an operating pressure of ~30 bar. Modern copper zinc LT shift catalysts are self-guarding, and it is better to use LT shift catalyst in the guard bed (see Chapter 6). Zinc oxide is, however, being used to protect methanation catalyst from being poisoned by sulphur compounds which can be carried over from certain carbon dioxide removal processes. Because the carry-over is transient and the total quantity of sulphur small, protection is obtained with a shallow layer of zinc oxide on top of the methanation catalyst. Typical process conditions are pressure ~20 bar, temperature 300°C and space velocity 25 000 h^{-1}.

4.11.9. Impurities in Zinc Oxide

Zinc ores usually contain arsenic, cadmium and lead compounds, so commercial zinc oxide may contain oxides of these metals. It has been found that arsenic and cadmium may be volatalized as sulphides and can be transferred from the zinc oxide. Arsenic is a severe poison to nickel reforming catalyst and the use of zinc oxide with a high level of arsenic (2000 ppm) has resulted in extensive damage to reforming catalyst.[100] Arsenic is a particularly serious poison because it first adheres to the pipework and is then slowly released onto the catalyst. Once arsenic has been deposited on the reformer tubes it can only be removed by mechanical cleaning. Cadmium is less of a problem than arsenic, as it does not have such a serious poisoning effect. It tends to migrate as far as the LT shift catalyst. Cadmium metal has been observed on cold parts of the reformer, such as the reformer tube ends. It is of particular importance to ensure that the zinc oxide installed in a plant is of proven quality and does not contain significant quantities of base metals.

4.12. Dechlorination

4.12.1. Chloride Sources and Absorbents

Chlorides are particularly severe poisons for copper-based catalysts such as low-temperature shift and methanol synthesis catalysts. The accumulation of even very small amounts of chloride has a very marked effect on catalyst activity. Figure 4.14 shows the results of introducing hydrochloric acid in the feed to a LT shift catalyst. An increase of chloride content of the catalyst from 0.01% to 0.03% halved the activity of the catalyst by accelerating the sintering of the active phases (see

4.12. Dechlorination

Chapter 6). Chloride may be present in the process water, the air to the secondary reformer or the hydrocarbon feedstock. It is very difficult to remove chloride from the process air and, as a result, ammonia plants located near the sea or in heavily industrialized environments tend to have short LT shift catalyst lives. Modern water demineralization plants can effectively remove inorganic chlorides but cannot deal so well with organic chlorides, and plants using river water contaminated with pesticides may have problems with chloride poisoning. Reverse osmosis is used to deal with this problem in power stations, but is not really economical for ammonia plants with their high raw water make-up requirement.

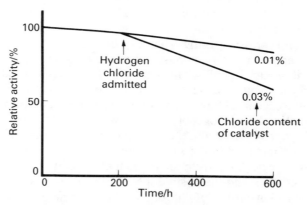

Figure 4.14. The effect of chlorine poisoning on the activity of low temperature shift catalyst (ICI Catalyst 53-1).

The increasing use of seawater injection to enhance oil production is resulting in higher levels of chloride in crude and distillate oils. Up to 15 ppm of aliphatic chlorides have been observed in naphtha fractions. Natural gas only rarely contains chlorides, but the use of rock caverns for storage of NGL and LPG can lead to hydrogen chloride in the gas. The level of hydrogen chloride increases with more dilute salt solution in contact with the liquid hydrocarbon.[101] One operator has found 0.3–0.5 µg g^{-1} in feed gas being drawn from storage in a rock cavern.[102] Certain refinery processes give waste gases containing small amounts of hydrogen chloride. These gases are sometimes used as the feedstock for the related hydrogen reformer, and dechlorination has to be a stage of the refinery process. The arrival of a hydrocarbon feed, containing chloride, on a cobalt or nickel molybdate catalyst at the start of a normal purification train results in similar reactions to desulphurization. The organic chlorides are hydrogenolysed to give hydrogen chloride. If sulphur is present, then both hydrogen sulphide and hydrogen chloride

Chapter 4. Feedstock Purification

are formed, and the nickel or cobalt molybdate catalyst reaches a steady state with a mixture of sulphides and chlorides.

A typical molybdate catalyst can absorb about 1.5% of chloride,[101] which is mobile and eventually will be released. If the cobalt or nickel molybdate catalyst is followed by a bed of zinc oxide, then the hydrogen chloride will displace hydrogen sulphide to form zinc chloride. Zinc chloride melts at 285°C, and as most desulphurizers operate at 350–400°C the formation of zinc chloride would result in extensive sintering and blocking of the porous pellets. This would lead to significant leakage of both hydrogen sulphide and hydrogen chloride. ICI has recognized this problem[102] and developed a chloride absorbent to be installed above the zinc oxide. This (ICI Catalyst 59-3) consists of an activated alumina and is now in use in over 50 plants around the world. In developing the chloride guard, work was carried out at atmospheric pressure using a variety of organic chlorides which had been detected in raw naphthas. All of the compounds identified were simple aliphatic chlorides, and were easily hydrogenolysed over cobalt or nickel molybdate catalyst under the conditions used in plants. No unreactive organic chlorine compounds have been detected, so it is very unlikely that the hydrogenolysis stage would be rate-determining. The chloride guard catalyst and the zinc oxide are not regenerable, and both are usually installed in the same vessel. The volumes of both catalysts charged are calculated from the amount of sulphur and chloride present in the feedstock and the desired interval between catalyst change-outs. The absorption profile of chloride in the chlorine guard is very sharp. The chloride guard reacts with hydrogen chloride, and may be used ahead of the cobalt or nickel molybdate catalyst if hydrogen chloride alone is present in the feed.[101] Also, because of its high reactivity it may even be used to remove hydrogen chloride at close-to-ambient temperatures, and it is being used to purify refinery off-gases at 30°C.

4.12.2. Operating Conditions

Because of the high solubility of chlorides in water it is not practical to use a chloride guard catalyst in conditions where excessive condensation can occur. Chloride is easily washed from any catalyst by water, and is thus easily transported through the plant. This is one reason why operators are advised against using steam as the warm-up medium for beds of catalyst. If only hydrogen chloride is present, then the chloride guard absorbent may be used betweem ambient temperatures and 400°C, without the need for an hydrogenolysis stage. Pressure has no measurable effect on the absorption reaction. Because only small levels of chloride are encountered, space velocity is rarely a significant factor and values up to 10,000 h^{-1} are common.

4.13. Removal of Silica and Fluoride

Because of its high basicity, ICI Catalyst 59-3 (chloride guard) can be used to absorb other acid gases. It has been used to absorb silica released from a high-temperature methanator. In this duty it was operating at 500°C and absorbed 13.3% SiO_2 at the top of the bed.

Hydrogen fluoride is used to catalyse certain alkylation reactors. The process produces by-products containing a few parts per million of hydrogen fluoride and ICI Catalyst 59-3 has been used to remove hydrogen fluoride from such hydrocarbon process streams.

4.14. Demetallization

Many crude oils contain organometallic compounds, some of which are volatile and may pass through the refinery to be present in the derived naphtha and gaseous feedstocks. The most commonly found metals are vanadium, arsenic, lead and nickel. It is particularly important that arsenic is not allowed to enter a steam reformer, as not only does it poison the reforming catalyst but it is absorbed on the reformer tubes from where it is slowly released to poison the next charge of reforming catalyst (see Chapter 5). Demetallization is carried out using cobalt and nickel molybdate catalyst. If organometallic compounds are known to be present, then it is advisable to use a special molybdate catalyst employing a highly absorbent support, usually produced in the form of spheres. The catalyst is operated in the normal way for the hydrogenolysis reactions. On arrival at the catalyst the metal compounds are broken down and hydrogenated to release the metal, which is absorbed on the catalyst. The catalyst has limited absorption capacity and must be discharged when saturated, and it is not possible to regenerate from metal fouling. Typical profiles obtained on a plant using naphtha feedstocks are given in Table 4.15. Metals tends to accumulate on top of the cobalt molybdate catalyst, and it is advisable to carry out a complete analysis of the top layer of catalyst to check for impurities in the feed at shutdowns.

Mercury compounds are occasionally found in natural gas; for instance, levels up to 100 ppb have been found in gas fields in Indonesia. Because of its high volatility mercury cannot be removed with a conventional cobalt molybdate catalyst. Mercury is removed using beds of charcoal which have been impregnated with sulphur. These are operated at ambient temperature, and are not regenerable.

Chapter 4. Feedstock Purification

Table 4.15. Absorption of lead and arsenic on cobalt molybdate catalyst

Distance down bed/cm	Metal/%	
	Pb	As
Top	2.3	0.85
25	0.14	0.12
125	0.05	0.01

4.15. Denitrification

Most crude oils contain organic nitrogen compounds, often in the form of heterocyclic compounds. These are hydrogenolysed using nickel or cobalt molybdate catalysts to give ammonia and the related hydrocarbon. (The reaction is carried out under the same process conditions as are used for desulphurization.) Nickel molybdate is considered to be a more powerful denitrification catalyst than cobalt molybdate, and is used to deal with unreactive nitrogen compounds. Fortunately, organic nitrogen compounds are only very rarely encountered in the feedstocks used for steam reformers. Both cobalt and nickel molybdate catalysts are slightly deactivated by ammonia. This may be present in recycle gas from an ammonia plant. The deactivation is reversible, and the catalyst soon recovers. The general limit set for safe operation in the presence of ammonia is less than 500 ppm.

Chapter 5

Steam Reforming

5.1. History

In 1910 the first commercial attempts to fix nitrogen relied on the electric arc process or the cyanamide process. Both were extremely energy-intensive and this provided the stimulus for the development of the direct synthesis of ammonia from nitrogen and hydrogen. One of the major problems encountered by the early pioneers was to produce hydrogen and nitrogen cheaply in large quantities. Initially, hydrogen was produced in a water-gas generator, and nitrogen by cryogenic separation of air. The low temperature process was quickly replaced by producer-gas sets, and by 1917 the development of the classical Haber Bosch process, which continued in Europe until the 1950s, was complete. This was to react coke with steam to produce hydrogen in the endothermic reaction (1) that was followed by the separate, moderately exothermic, water gas shift reaction (2). Since then, carbon, in one form or another, has been used to extract hydrogen from water relying on simple reactions.

$$C + H_2O \rightarrow CO + H_2 \qquad (1)$$
$$CO + H_2O \rightarrow CO_2 + H_2 \qquad (2)$$

In order to sustain the reaction temperature the steam was intermittently cut off and air was blown through the coke. The gas arising from the cyclic process was rich in hydrogen. Additional nitrogen necessary for stoichiometric ammonia synthesis was generally provided by separate producer-gas generators. The mixed gases were purified, oxides of carbon were removed, and then the mixture was compressed for synthesis into ammonia.

In areas where natural gas was available in large quantities interest centred around steam reforming of methane as the source of hydrogen. The process was more economic than that based on coal. The hydrogen content of the hydrocarbon improved the yield of hydrogen per unit of carbon in the feedstock compared with coal, and there were also fewer unwanted by-products. The methane steam technology was pioneered in the first quarter of this century by BASF who established the essential configuration of the primary steam reformer, and the technology was used in 1931 by Standard Oil of New Jersey to produce hydrogen from off-gases at its Baton Rouge and Bayway refineries. The steam reforming reaction took place over catalyst in vertical tubes which were supported in parallel rows in a radiant furnace. The endothermic heat of reaction was supplied by burning fuel in the furnace (see Section 5.8.1). The process was considerably improved by ICI, who developed the fundamental engineering data for the design of the furnace, improved the catalyst formulation and introduced the desulphurization step using zinc oxide. The process was used to produce hydrogen from off gases for coal hydrogenation plants which ICI built in 1936 and 1940. The ICI technology was subsequently used in the development of the North American ammonia industry when plants were constructed at El Dorado, Baxter Springs, Etters, Sterlington and Calgary. All used natural gas which contained mainly methane with low concentrations of higher hydrocarbons, and nearly all used catalyst developed by ICI. Natural gas was not a readily available feedstock in the UK before the discoveries in the North Sea, but as more refineries were built other hydrocarbons, such as naphthas, became increasingly available. It was apparent in the 1950s that if naphthas could be steam reformed economically they would provide a cheap source of hydrogen for the manufacture of ammonia. Work by ICI at this time led to the development of a catalyst which would reform naphthas at economic steam ratios without carbon formation. The catalyst was stable, resistant to poisons and had an economical life.

In 1959 ICI started up the first large-scale pressure steam reformer using naphtha as a feedstock, and this became the forerunner of over 400 plants subsequently licensed around the world in areas where natural gas was not available. From 1959 to date development of the catalyst continued in order to allow plants to be run at higher pressure and temperature, and with feedstocks containing different hydrogen/carbon ratios. It also allowed feedstocks containing quantities of unsaturated and aromatic compounds to be reformed. In more-recent years the increasing availability of natural gas has resulted in its use as a major source of reformer feedstock, and this is likely to remain so for some time. Development of catalysts for natural gas reforming has concentrated on extending catalyst life, improving activity, inhibiting carbon forming reactions, and by improving the physical properties.

5.2. Feedstock and Feedstock Pretreatment

Feedstocks for a catalytic steam reforming process can vary widely in composition. The simplest, methane, occurs as natural gas and will probably contain other saturated hydrocarbons of low molecular weight, as well as nitrogen, carbon dioxide, and organic and inorganic sulphur compounds. Chloride can also be present, arising from storage in brine well cavities. Liquid feedstocks can vary from pure distillates such as propane and other compounds with a general formula C_nH_{2n+2} to more-complex compounds having empirical formulae which lie between $CH_{1.9}$ and $CH_{2.4}$. They may contain sulphur or chlorine compounds (see Chapter 4), which arise as a result of contamination in shared storage tanks or transport vessels.

5.2.1. Natural Gas

Typical analyses of natural gas from four locations are given in Table 5.1. Before being used as a feedstock for steam reforming, liquid condensates must be removed in order to ensure a constant gas composition, and any solid suspension must be removed by filtration. As was discussed in Chapter 4, sulphur compounds, both inorganic and organic, must be removed by hydrogenolysis and absorption in a bed of zinc oxide. Chlorine, if present in concentrations greater than 1 ppm, must be removed by using, for example, an alkalized absorbent such as ICI 59-3, usually placed in the same vessel but as a separate bed above the desulphurization catalyst. Gas is normally supplied to a site at pressure and at ambient temperature. Expansion across valves results in a reduction in temperature and, as there is often a small quantity of moisture present in the gas, there is the possibility of forming hydrates[103] on pressure let-down. It is therefore advisable to heat the gas by 50–100°C before passing it into the process.

Carbon dioxide present in the gas supplied to an ammonia or hydrogen plant is a diluent and decreases the efficiency of the reforming

Table 5.1. Typical analyses of natural gas from four locations [a]

Component	North Sea	Qatar	Netherlands	Pakistan
CH_4	94.85	76.6	81.4	93.48
C_2H_6	3.90	12.59	2.9	0.93
C_3H_8	0.15	2.38	0.4	0.24
i-C_4H_{10}		0.11		0.04
n-C_4H_{10}	0.08	0.21	0.1	0.06
C_5+		0.02		0.41
CO_2	0.20	6.18	1.0	0.23
N_2	0.79	0.24	14.2	4.02
S	4 ppm	1.02	1 ppm	N/A

[a] Unless otherwise stated, composition is expressed as %.
N/A = not available

process. It is heated in the reforming process, participates in the shift reaction in the reformer, but subsequently must be removed from the gas stream. Also, there is a possibility that it will methanate in the presence of hydrogen in the desulphurization process. This could cause an unacceptably high temperature rise in that vessel, and also reduce the quantity of hydrogen available for sulphide hydrogenolysis and reduce the rate of desulphurization. To avoid methanation the concentration of carbon dioxide present should be less than 5% (see Chapter 4). Nitrogen present in natural gas is an inactive diluent in the reforming process and remains with the hydrogen produced. If the hydrogen is required for the production of ammonia it can contribute towards establishing the required nitrogen/hydrogen ratio. However, if the hydrogen is required for methanol synthesis nitrogen will eventually have to be purged from the synthesis loop.

5.2.2. Naphthas

Naphtha fractions with a final boiling point of less than 220°C are generally considered suitable for catalytic steam reforming, and typical analyses are shown in Table 5.2. Naphthenes must not constitute more than 40% of the feedstock to avoid the possibility of aromatization in the hydrodesulphurization process. This would increase the aromatic content of the feed to the reformer, which should normally be less than 30% at the reformer inlet. The amount of olefinic compounds in the feed gas to the reformer must be less than 1%, unless special precautions are taken to cater for the associated temperature increase when they hydrogenate in the hydrodesulphurization process. As with methanation, hydrogen consumed by a hydrogenation process will not be available for hydrodesulphurization, and the overall quantity of hydrogen used in the process must take this into account. The sulphur content must be reduced by hydrofining or hydrogenation followed by desulphurization to reduce it to less than 0.5 ppm at the exit of the hydrodesulphurization process. Chlorine should be reduced to *less than* 1 ppm by a chlorine guard installed before the zinc oxide, and there must be no arsenic contamination in the naphtha. There must also be less than 1 ppm lead present (often as a contamination from tankers previously used for transporting leaded petrol). The purification processes are considered in more detail in Chapter 4.

Table 5.2. Typical analyses of naphtha

	I	II	III	IV
Specific gravity at 15.5°C	0.713	0.713	0.729	0.741
IPT distillation				
Initial boiling point/°C	33.0	34.0	32.5	43.0
2% vol distilled at °C	45.0	45.0	45.5	57.0
5% vol distilled at °C	53.0	55.0	55.0	68.0
10% vol distilled at °C	64.0	62.0	66.0	82.0
20% vol distilled at °C	76.0	76.5	84.5	94.0
30% vol distilled at °C	89.0	89.5	99.5	105.5
40% vol distilled at °C	102.5	100.5	111.0	119.0
50% vol distilled at °C	113.0	112.0	121.5	131.0
60% vol distilled at °C	123.0	122.0	131.5	143.5
70% vol distilled at °C	132.0	132.5	143.0	155.5
80% vol distilled at °C	143.5	146.0	154.5	168.5
90% vol distilled at °C	155.5	154.5	167.0	187.5
95% vol distilled at °C	165.0	163.0	178.0	200.0
Final boiling point/°C	172.0	174.0	183.0	214.0
Total distillate/%	98.5	98.0	96.5	99.0
Distillation residue/%	1.0	1.2	1.0	1.0
Distillation loss/%	0.5	0.8	2.5	0
Bromine number/g (100 g)$^{-1}$	1.4	not determined	1.2	1.4
FIA[a]				
Unsaturates/%(v/v)	1	1	1	1
Aromatics/%(v/v)	6	6	9	10
Saturates/%(v/v)	92	93	91	89
Total sulphur/ppm(w/w)	231	258	1510	1430
[b] H_2S/ppm(w/w)	nd	nd	22	nd
RSH/ppm(w/w)	70	84	557	252
R_2S_2/ppm(w/w)	21	7	48	156
R_2S/ppm(w/w)	105	126	831	743
Elemental sulphur/ppm(w/w)	6	4	17	1
Unreactive sulphur/ppm(w/w)	42	56	83	272
%C	84.5	84.9	85.7	85.4
%H	15.4	15.2	14.6	14.9
Gross calorific value/kJ g^{-1}	46.05	45.06	45.73	45.87
Flash point/°C	−17.8	−17.8	−17.8	−17.8
Average molecular weight	100–2	103–2	106–2	110–3
Reid vapour pressure	9.0	9.7	not determined	4.5

[a] Fluorescent indicator analysis.

[b] Owing to certain systematic errors, the sulphur-types analysis totals slightly more than the total sulphur as determined by combustion.

Chapter 5. Steam Reforming

5.3. Chemistry of Steam Reforming

5.3.1. Thermodynamics

The objective of the catalytic steam reforming process is to extract the maximum quantity of hydrogen held in water and the hydrocarbon feedstock. Thereafter the subsequent manipulation of the gas stream depends on the purpose for which the gas is intended. Common uses are: synthesis gas for ammonia and methanol; hydrogen and carbon monoxide for oxo-alcohols and Fischer–Tropsch synthesis; hydrogen for refineries, hydrogenation reactions and fuel cells; and reducing gas for direct reduction of iron ore. In the case of gas for ammonia synthesis the reforming is done in two stages, using a primary and a secondary reformer. This allows nitrogen to be added in the form of air between the stages, and permits reforming to continue, to achieve a higher final equilibrium temperature at the exit of the secondary reformer.

The reforming of natural gas utilizes two simple reversible reactions: the reforming reaction (*3*) and the water-gas shift reaction (*4*).

$$CH_4 + H_2O \rightleftharpoons CO + 3H_2 \qquad \triangle H = +206 \text{ kJ mol}^{-1} \qquad (3)$$

$$CO + H_2O \rightleftharpoons CO_2 + H_2 \qquad \triangle H = -41 \text{ kJ mol}^{-1} \qquad (4)$$

The reforming reaction is strongly endothermic, so the forward reaction is favoured by high temperature as well as by low pressure, while the shift reaction is exothermic and is favoured by low temperature but is largely unaffected by changes in pressure. To maximize the overall efficiency (and hence economics) of the conversion of carbon to carbon dioxide and the production of hydrogen, reformers are operated at high temperature and pressure. This is followed by the shift process which, by using two different catalysts, permits the shift reaction to be brought to equilibrium at as low a temperature as possible (see Chapter 6). It can be seen that with methane the stoichiometric requirement for steam per carbon atom is 1.0. However, it has been demonstrated that this is not practicable because all catalysts so far developed tend to promote carbon forming reactions under steam reforming conditions. These reactions can only be suppressed by using an excess of steam, with the result that the minimum ratio is in the region of 1.7. However, the reforming reaction itself is also promoted by an excess of steam (see Section 5.3.2) and hence some advantage is derived from this necessity, and in practice ratios of 3.0–3.5 are commonly used, but there can be economic attractions in using lower steam ratios and there is now a trend in this direction. Knowledge of the thermodynamic data associated with the reforming reaction allows graphs to be constructed

5.3. Chemistry of Steam Reforming

from which equilibrium concentrations of reactant gases leaving a reformer can be determined for specified feed-gas and reforming conditions. This is demonstrated in Figures 5.1–5.3. Equilibrium constants for the methane reforming and shift reaction are tabulated in Appendices 6 and 7.

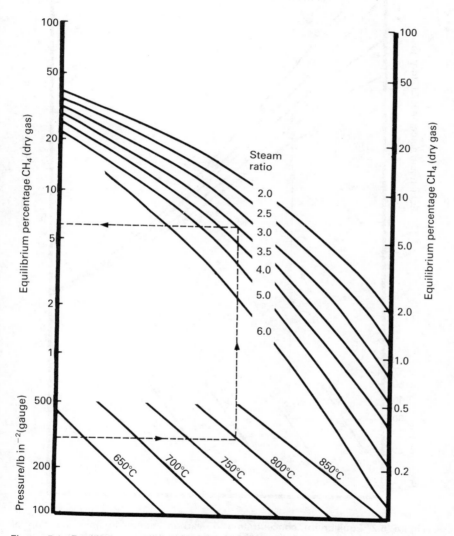

Figure 5.1. Equilibrium concentration of methane as a function of temperature, pressure and steam/carbon ratio for methane steam reforming.

Chapter 5. Steam Reforming

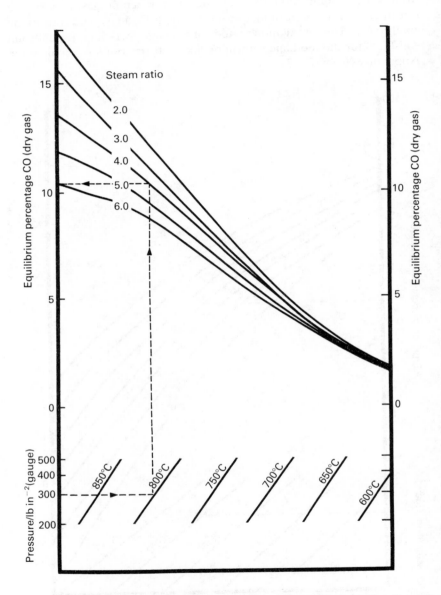

Figure 5.2. Equilibrium concentration of carbon monoxide as a function of temperature, pressure and steam/carbon ratio for methane steam reforming.

5.3. Chemistry of Steam Reforming

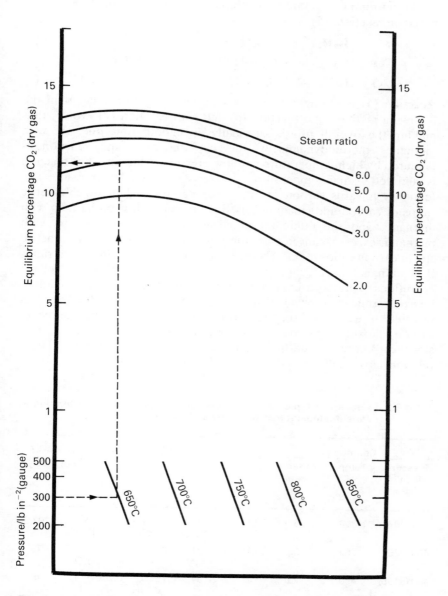

Figure 5.3. Equilibrium concentration of carbon dioxide as a function of temperature, pressure and steam/carbon ratio for methane steam reforming.

Chapter 5. Steam Reforming

The reforming of saturated naphthas of general formula C_nH_{2n+2} is based on reactions (5)-(7).

$$C_nH_{2n+2} + nH_2O \rightarrow nCO + (2n+1)H_2 \qquad (5)$$
$$CO + 3H_2 \rightleftharpoons CH_4 + H_2O \qquad \Delta H = +206 \text{ kJ mol}^{-1} \qquad (6)$$
$$CO + H_2O \rightleftharpoons CO_2 + H_2 \qquad \Delta H = -41 \text{ kJ mol}^{-1} \qquad (7)$$

Reaction (5) is strongly endothermic, absorbing more heat than the following methanation reaction (6) and shift reaction (7) evolve, thus making the overall process normally an endothermic one. As with methane, thermodynamically the naphtha reforming reaction is favoured by high temperature and low pressure, while the shift reaction is inhibited by high temperature but largely unaffected by pressure. The tendency towards carbon formation on catalysts when stoichiometric ratios of carbon and steam are used is greater with naphtha than with methane, and the minimum practical ratio is about 2.2. As with methane, excess steam favours the reforming reaction, and in practice steam/carbon ratios of 3.5–4.5 are common. At low steam/carbon ratios the methanation reaction begins to dominate, and under certain conditions of pressure and temperature can cause the overall reaction to be exothermic, as shown in Table 5.3. As with methane reforming, it is possible to calculate the equilibrium methane, carbon monoxide, carbon dioxide, hydrogen and water concentrations from inlet steam/carbon ratios and the operating conditions of a reformer. These relationships are illustrated in Figures 5.4–5.7, respectively.

Table 5.3. Examples of typical heats of reaction in naphtha steam reforming at different temperatures, pressures and steam/carbon ratios

Conditions			Reaction	ΔH (25°C)/
Pressure/ lb in^{-2} (kPa)	Temp./ °C	Steam ratio		kJ mol^{-1} CH$_{2.2}$
300 (2070)	800	3.0	$CH_{2.2} + 3H_2O \rightarrow 0.2CH_4 + 0.4CO_2 + 0.4CO + 1.94H_2 + 1.81H_2O$	+102.5
400 (2760)	750	3.0	$CH_{2.2} + 3H_2O \rightarrow 0.35CH_4 + 0.25CO + 0.4CO_2 + 1.5H_2 + 1.95H_2O$	+75
450 (3105)	450	2.0	$CH_{2.2} + 2H_2O \rightarrow 0.75CH_4 + 0.25CO_2 + 0.14H_2 + 1.5H_2O$	−48

5.3. Chemistry of Steam Reforming

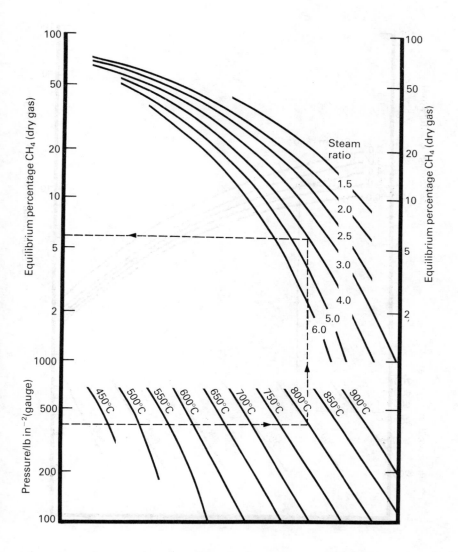

Figure 5.4. Equilibrium concentration of methane as a function of temperature, pressure and steam/carbon ratio for naphtha steam reforming.

Chapter 5. Steam Reforming

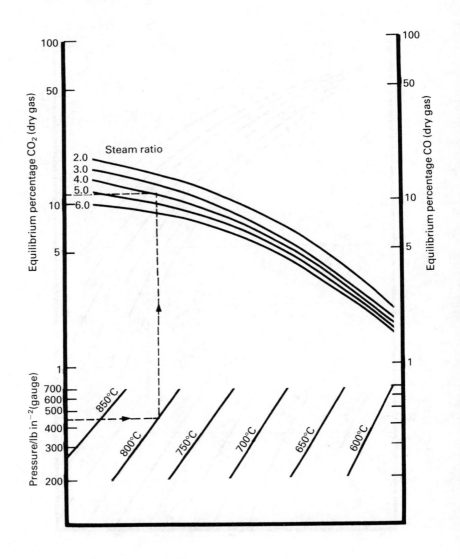

Figure 5.5. Equilibrium concentration of carbon monoxide as a function of temperature, pressure and steam/carbon ratio for naphtha steam reforming.

5.3. Chemistry of Steam Reforming

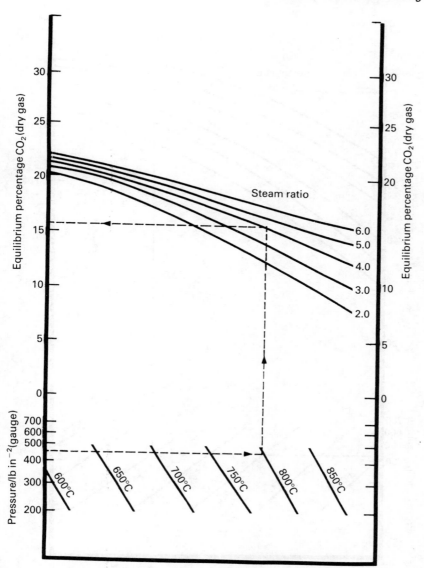

Figure 5.6. Equilibrium concentration of carbon dioxide as a function of temperature, pressure and steam/carbon ratio for naphtha steam reforming.

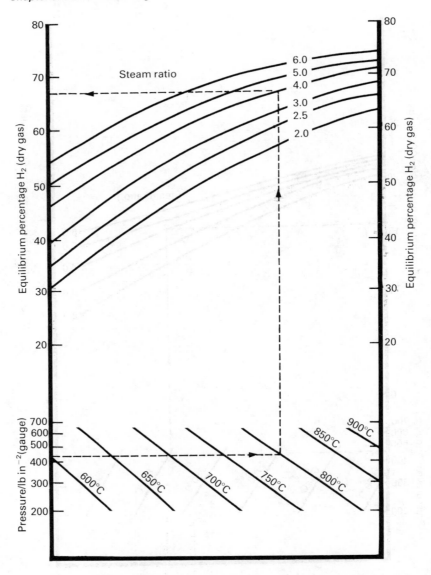

Figure 5.7. Equilibrium concentration of hydrogen as a function of temperature, pressure and steam/carbon ratio for naphtha steam reforming.

5.3.2. Kinetics

The results of a number of studies on the kinetics of the methane steam reforming reaction have been published.[104] There is general agreement that the reaction is first order in methane, but there is less agreement with other kinetic parameters. In part this is due to the use of different catalysts and experimental conditions, but often it has resulted from a lack of appreciation of diffusion and heat transfer limitations. Thus, reported activation energies span a wide range of values caused by varying degrees of diffusion limitation, and these can also cause misleading total pressure effects. Indeed, with the relatively large catalyst particle sizes used in industrial steam reformers these effects result in very low effectiveness of the catalyst. Effectiveness factors (η) may, depending on conditions, only be as high as ~0.3 at the inlet region, and perhaps as low as 0.01 at the exit. Because of this, apparent activity increases as the particle size is made smaller, but the increased pressure drop which arises across the reformer restricts the size of the catalyst that can be used in practice (see Section 5.6).

In a now classic study, Russian workers[105] used a recirculation laboratory reactor to reform methane over nickel foil in order to obviate any pore diffusion limitation. From their results obtained at 800–900°C and atmospheric pressure they concluded that the water-gas shift equilibrium was always established, and that their data could be satisfactorily described by equation (8).

$$-dCH_4/dt = kP_{CH_4}/[1 + a(P_{H_2O}/P_{H_2}) + bP_{CO}] \quad (8)$$

At 800°C the constants a and b were 0.5 and 2.0 atm^{-1} (1 atm $\simeq$ 1.01 bar), and at 900°C they were 0.2 and 0 atm^{-1}, respectively. The derived activation energy over this temperature range was 130 kJ mol^{-1}. They also concluded that the decomposition of methane to carbon on the nickel surface was very much slower than the reaction of methane with steam, and as a result mechanisms involving formation of carbon followed by its reaction with water were unlikely. The same workers[106] also examined the kinetics using a conventional catalyst comprising nickel on alumina, and obtained the simple rate equation (9)

$$-dCH_4/dt = kP_{CH_4} \quad (9)$$

with activation energies of 77 kJ mol^{-1} at 800–900°C, and 100 kJ mol^{-1} at 700–800°C. These results clearly demonstrate the effect of diffusion on the kinetic parameters, and the thickness of the "working layers" was calculated to be ~0.04 mm for 1.2 mm particles. It is of interest to compare the results of methane steam reforming,[105] methane cracking[107] and its exchange with deuterium[108] over nickel films. In each case the activation energy (in the absence of diffusion effects) is ~130 kJ mol^{-1}.

A selection of more-recently reported rate equations for hydrocarbon steam reforming are given in Table 5.4. Several of those relating to methane reforming include the term $(1 - K'/K)$, where K is the conventional equilibrium constant for the reaction (to give CO or CO_2) and K' is the corresponding ratio for the reactants and products present under the actual operating conditions. This term takes into account the reverse reaction, which becomes important when operation is close to equilibrium, and its inclusion in some form is essential in the design of full-scale reformers.[109, 110]

There has been some debate about the first formed products of the steam reforming reactions,[109] and it appears that the relative concentrations of carbon monoxide and carbon dioxide leaving the catalyst surface depend on the efficiency of the catalyst in the water-gas shift reaction. With rhodium-based catalysts the CO/CO_2 ratio of the initially formed carbon oxides is relatively high (in keeping with poor shift activity), whereas with nickel catalysts the amount of carbon monoxide is lower.

The mechanism of steam reforming of higher hydrocarbons is more complex than that for methane, since fission of carbon–carbon bonds is necessary to produce single carbon surface species. The order with respect to higher hydrocarbons in most published work is zero over nickel or rhodium catalysts, reflecting the ease of fragmentation of higher hydrocarbons and a consequent high surface coverage. The order with respect to water ranges from zero to one. Both single- and two-site mechanisms have been proposed,[109, 110] and work on steam dealkylation as typified by the reaction of steam with toluene to give benzene, hydrogen and carbon oxides, has shed light on the mechanisms of reforming.[111] Rhodium catalysts have been used in most of these studies because nickel has poor selectivity. The reaction proceeds through intermediate formation of ethane or directly to products. It appears that the catalysts are bifunctional, with dealkylation taking place on the metal to give carbene intermediates that react with hydroxyl groups derived from water adsorbed on the support.

5.3. Chemistry of Steam Reforming

Table 5.4. Selected kinetic equations for hydrocarbon steam reforming

Hydro-carbon	Catalyst	Temperature/°C	Pressure/bar	Rate equation	Remarks	Reference
CH_4	Ni	500–900	1–15	$[hc](1 - K_5'/K_5)$	Plant design model	112
CH_4	Ni	500–900	1–15	$[hc](1 - K_6'/K_6)$	Plant design model	113
CH_4	Ni	500–900	1–15	$[hc][H_2O]^2(1 - K_6'/K_6)$	CO_2 from CH_4, then reverse shift to give CO	114
CH_4	Industrial Ni catalyst	500–900	21–41	$\dfrac{[hc]}{(1 - K_5'/K_5)[H_2]}$	Rate constant is pressure-dependent because of diffusion	115
CH_4	Ni foil	470–800	1–41	$\dfrac{[hc][H_2O]}{[H_2O] + a[H_2]^2 + b[H_2]^3}(1 - K_5'/K_5)$	Rate constant is pressure-dependent. Adsorption of hc to form $=CH_2$, then reaction with gas-phase H_2O	116,117
CH_4	Ni/α-Al_2O_3	350–450	1–2	$\dfrac{[hc][H_2O]^2}{1 - a[hc]}(1 - K_6'/K_6)$	Rate-determining step reaction of adsorbed hc and gas-phase H_2O. H_2 inhibits reaction	118
CH_4	Commercial Ni catalyst	638	1–18	$\dfrac{-K_{CO}([CO] - K[CH_4][H_2O]/[H_2O]^3)}{1 + a[H_2O] + b[CH_4][H_2O]/[H_2]^3 + c[CH_4][H_2O]^2/[CH_2]^4}$	Equation is for CO formation, similar one for CO_2 formation. Rate-determining step is desorption of CO and CO_2 after reaction CH_4 gas with adsorbed H_2O	119

241

Chapter 5. Steam Reforming

Hydro-carbon	Catalyst	Temperature/°C	Pressure/bar	Rate equation	Remarks	Reference
CH_4	Ni/Al_2O_3 or SiO_2	670–770	16–26	$[hc][H_2O]^2(1 - K_6'/K_6)$	Rate constant is pressure-dependent because of diffusion: $E = 38.5$–62 kJ mol^{-1}	114
CH_4	Rh, etc./SiO_2	350–600	1	$[hc]^0[H_2O]^{0.5}$	Gaseous products at equilibrium for shift reactions and $CH_4 + H_2O \rightarrow CO + 3H_2$	120
C_3H_6	Ni/SiO_2 or C	500–750	1	$[hc]^{0.75}[H_2O]^{0.6}$	Two-site mechanism; non-dissociative hc adsorption: $E = 64$ kJ mol^{-1}	121
C_2H_6	Ni/Cr_2O_3	300–360	1	$\dfrac{[hc]^0}{1 + a[H_2O]/[H_2]}$	Rate-determining step hc adsorption; reaction via $=CH_2$	122
n-C_6H_{14}, etc.	Ni	500–800	1–30	$[hc]^0[H_2O]^0$	$E \simeq 46$ kJ mol^{-1}	123
n-C_7H_{16}	Rh/γ-Al_2O_3	550–800	1		CH_4 and CO greater than thermodynamics involving $CH_4 + H_2O = CO + 3H_2$ and shift reaction. CH_4 direct from hc and from methanation	124
n-C_7H_{16}	Rh/$MgWO_4$	500	1	$\dfrac{[hc][H_2O]}{1 + a[hc]}$	As above. Zero order for hc at high [hc]. CO_2 primary product?: $E = 78$ kJ mol^{-1}	125

5.3. Chemistry of Steam Reforming

Hydrocarbon	Catalyst	Temperature/°C	Pressure/bar	Rate equation	Remarks	Reference
n-C_6H_{14}	Ni/K poly-aluminate	500–800	1–20	—	CH_4 less than thermodynamics involving $CH_4 + H_2O = CO + 3H_2$ and CO_2 less than shift equilibrium at high H_2O/hc ratio. CO primary product?	126
C_3H_8, etc.	Ni/Al_2O_3 or Mg silicate	450–500	1	—	As above, but CO_2 less than shift equilibrium. CO from reverse shift, CH_4 from CO/CO_2 + H_2	127
Toluene	Rh/γ-Al_2O_3	520	1	$[hc]^0[H_2O]^0$	Two-site adsorption. Some gasification via dealkylation, π-bonded ring and $=CH_2$	128
Toluene	Rh/γ-Al_2O_3	500–600	1	—	Gasification via adsorbed 6-membered aromatic and CH_x species	129
Toluene	Rh/γ-Al_2O_3	400–500	1	$[hc]^n[H_2O]^{(1-n/2)}$	$n \simeq 0.1$. Two-site adsorption followed by surface reaction: $E \approx 138$ kJ mol^{-1}	130
Toluene	Rh/α-Cr_2O_3, etc.	625	1–20	$\dfrac{[hc][H_2O]}{(1+a[hc])+b[H_2O]^2}$	Single-site adsorption. Rate-determining step is surface hc and H_2O reaction: $E = 115$ kJ mol^{-1}	131

hc = hydrocarbon; [x] = partial pressure of species x; a, b and c are constants.
E = activation energy; K = equilibrium constant; K' = K calculated from non-equilibrium concentrations.

5.4. Design of Steam Reforming Catalysts

In designing a practical catalyst for a steam reforming duty it is necessary to keep a number of objectives in view, and these are discussed in this section.

5.4.1. Selectivity

The catalyst must promote the desired reaction and be as inactive as possible towards unwanted side-reactions, particularly to the formation of carbon. The catalyst should also be as resistant as possible to poisons.

5.4.2. Thermal Stability

The catalyst must be able to maintain its activity under the demanding process conditions necessary to promote the desired reaction (e.g. ~800°C, $P_{H_2O} \simeq 20$ bar). With impregnated catalysts (see below) an important parameter in maintaining activity over prolonged periods is the nature of the support material and its pore structure.

5.4.3. Physical Properties

The catalyst must be strong enough to withstand the handling it receives, from manufacture to charging into the reformer, as well as the stresses generated by the process conditions and the thermal cycles arising from plant start-up and shutdown. The catalyst also must be of a suitable physical shape to provide an appropriate geometric surface area to give an acceptable activity per unit volume of packed bed whilst possessing acceptably low pressure-drop characteristics. The support must not be affected by water condensing on it, nor must it produce an unacceptable quantity of dust and material carryover, which could foul heat exchangers and other catalysts downstream.

5.4.4. Nickel as a Steam Reforming Catalyst

For many years nickel has been recognized as the most suitable metal for steam reforming of hydrocarbons. Other metals can be used; for example cobalt, platinum, palladium, iridium, ruthenium and rhodium. Although some precious metals are considerably more active per unit weight than nickel, nickel is much cheaper and sufficiently active to enable suitable catalysts to be produced economically. The reforming reaction takes place on the nickel surface, so the catalyst must be manufactured in a form which produces the maximum stable nickel surface area available to the reactants. This is done by dispersing the

5.4. Design of Steam Reforming Catalysts

nickel as small crystallites on a refractory support which must be sufficiently porous to allow access by the gas to the nickel surface. This is usually achieved by precipitating nickel as an insoluble compound, from a soluble salt, in the presence of a refractory support such as mixtures of aluminium oxide, magnesium oxide, calcium oxide and calcium aluminate cement. Alternatively, the nickel can be incorporated by impregnating a preformed catalyst support, such as alumina or an aluminate, with a solution of a nickel salt which is subsequently decomposed by heating to the oxide. In either case the nickel oxide is reduced to the metal by hydrogen supplied from another plant, or by cracking a suitable reactant gas (e.g. ammonia) over the catalyst as the reformer is being started up (see Section 5.8.7.1). In some instances process gas itself is used to reduce the nickel oxide to metal as the reformer is gradually brought on-line.

Catalytic performance and strength are determined by the catalyst formulation. Impregnated catalysts are generally stronger than precipitated catalysts, and this is one of the reasons for their widespread use. When comparing the activity of different catalyst types it is necessary to take into account their nickel contents. Table 5.5 demonstrates the variation of strength and nickel surface area for typical natural gas reforming catalysts with varying nickel content. The activity of a steam reforming catalyst in service is closely related to the available surface area of the nickel metal and the access the reactants have to it. Most commercial natural gas catalysts are now of the impregnated type and give a relatively high nickel surface area when first reduced, but under normal reforming conditions the surface area falls as sintering of nickel crystallites occurs. The higher the temperature is, the more

Table 5.5. Nickel content, in service strength and nickel surface area of some experimental steam reforming catalysts

Catalyst	Type	NiO/%	In-service strength (side)		Nickel surface area/
			kg	lb	m^2g^{-1}
A	Precipitated	33.0	12–20	25–45	0.05
B	Precipitated	30.0	14–23	30–50	0.04
C	Precipitated	25.0	23–32	50–70	0.03
D	Impregnated	10.0	36–45	50–100	0.03

Strength is the mean of 20 pellet-crushing tests. "In-service" corresponds to 1000 hours of operation in a full-size reformer tube.

Chapter 5. Steam Reforming

rapidly the sintering proceeds. This effect can be seen from Table 5.6, which gives the nickel surface area (per gram of catalyst) for a precipitated and an impregnated catalyst before and after use. It is not surprising that activity is a function of the overall nickel content. However, it has been demonstrated that with both impregnated and precipitated catalyst there is an optimum beyond which an increase in nickel content does not produce any further significant increase in activity. Typically these optima are approximately 20% for precipitated and up to about 15% for impregnated catalyst, but this does depend on the nature and physical properties of the actual support. Table 5.7

Table 5.6. Nickel surface area of reduced reforming catalyst before and after use

Catalyst type	Nickel surface area/$m^2 g^{-1}$	
	Before use	After use
Precipitated (B)	8.7	0.04
Impregnated (D)	0.7	0.03

"Before use" is after reduction at 750°C.
"After use" corresponds to 1000 hours of operation in a full-size reformer tube.

Table 5.7. Relationship between reforming activity and nickel content of precipitated catalyst in laboratory tests

Catalyst 1		Catalyst 2	
Nickel content/%	Methane conversion/%	Nickel content/%	Methane conversion/%
10.6	10.3	19.3	15.5
13.9	13.4	21.0	18.2
17.9	19.8	22.1	20.4
20.8	20.1	23.8	19.6
25.8	20.6		

Feed: methane and steam (steam ratio 3.0).
Temperature: inlet 450°C, exit 600°C.
Pressure: 26 bar.
Space velocity (methane + steam): 35 000 h^{-1}.

5.4. Design of Steam Reforming Catalysts

shows the results of two series of experiments with precipitated catalyst where this effect is clearly demonstrated. A graphic representation of the relationships between nickel content, surface area, crystallite size and activity for one particular precipitated catalyst is given in Figure 5.8.

Steam reforming catalysts differ in the ease with which they reduce, and the extent of reduction is influenced by the chemical nature of the catalyst support, the reduction temperature and time, and the composition of the reducing gas. The highest initial nickel surface area is obtained when the reduction is done using pure hydrogen (rather than steam and hydrogen) and when the reduction temperature is ~600°C. Below this temperature reduction can be slow and incomplete. Above ~600°C some sintering may take place, which lowers the nickel surface area. When steam is present lower surface areas can result, because sintering is enhanced, and this process proceeds further to give even lower surface areas if excessive reduction periods are employed. Nevertheless, for some catalysts, particularly precipitated catalysts for naphtha reforming, reduction periods up to twenty four hours are recommended. Some of these effects are illustrated in Figures 5.9 and 5.10 for particular precipitated and impregnated catalysts.

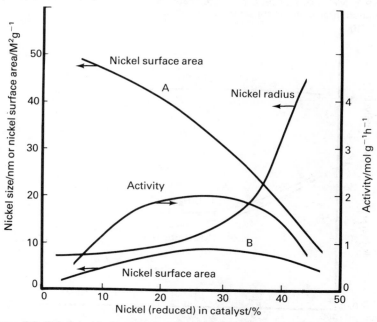

Figure 5.8. Relationship between nickel content, surface area, crystallite size and activity of an experimental precipitated steam reforming catalyst. Nickel surface area is expressed as A: m^2 per gram of nickel present, and B: m^2 per gram of catalyst.

Chapter 5. Steam Reforming

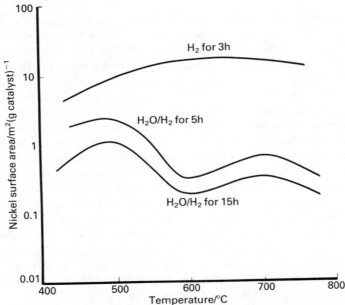

Figure 5.9. Nickel surface area in reduced catalyst related to temperature and time of reduction; precipitated catalyst ($H_2O/H_2 = 8/1$).

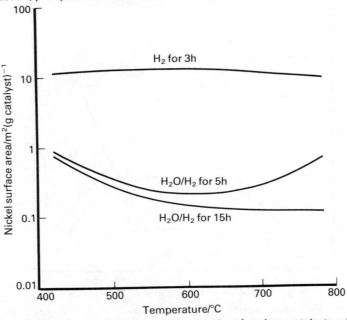

Figure 5.10. Nickel surface area in reduced reforming catalyst related to temperature and time of reduction; impregnated catalyst ($H_2O/H_2 = 8/1$).

5.4. Design of Steam Reforming Catalysts

If a reduced nickel steam reforming catalyst is subjected to normal reforming conditions of temperature and pressure in the presence of steam but no hydrogen, the nickel may become re-oxidized—a situation that probably exists at the top of tubes in reformers which operate with no hydrogen recycle. Full activity is not restored until the catalyst is re-reduced. Reducing conditions are maintained if a H_2O/H_2 ratio of 8 : 1 or less is sustained. Full reduction procedures are detailed in Section 5.8.7.

5.4.5. Supports for Nickel Steam Reforming Catalysts

The design of a steam reforming catalyst support must reflect the need for it to be robust at high temperature and pressure. It must also be suitable for the dispersion of nickel crystallites and allow access of the reacting species, but it must not interfere with their activity. If possible it should promote or at least sustain the activity of the nickel, but it must not catalyse side-reactions, particularly those which produce carbon deposits. Good physical properties can be obtained by using a simple support of α-alumina that is calcined at a temperature around 1500°C. A particularly useful range of catalysts are produced from a support which is derived from a mixture of alumina and an hydraulic cement that is processed to a porous calcium aluminate. This support is less acidic than α-alumina, and is therefore less susceptible to carbon formation initiated by hydrocarbon-cracking reactions (see Section 5.11). The presence of silica would further increase the already adequate strength of the support, but as silica is volatile in the presence of steam at high temperatures only very small quantities can be used, except in naphtha steam reforming catalysts when it is combined with potassium which significantly reduces its partial pressure under reforming conditions. Magnesia can also be included in some formulations, but has to be used with caution and understanding, however, as under certain conditions it can hydrate and markedly weaken the catalyst. Table 5.8 shows the relationship between temperature and partial pressure of water in equilibrium with $Mg(OH)_2$. Thus, at about 425°C the partial pressure of water in equilibrium with $Mg(OH)_2$ {$Mg(OH)_2 \rightleftharpoons MgO + H_2O$} is about that commonly used in reforming. Therefore during start-up or shutdown in the presence of reaction steam only, hydration of the magnesia can take place below ~425°C and, since the molar volume of $Mg(OH)_2$ is almost twice that of MgO, a dramatic weakening of the support may result unless the MgO is formulated in such a way that it is chemically associated with other refractory oxides. If this is not done reformer start-ups and shutdowns must ensure that the catalyst is in a dry atmosphere whilst temperatures are below the critical hydration temperatures.

Table 5.8. Relationship between temperature and partial pressure of water in equilibrium with Mg(OH)$_2$

Temperature/°C	250	300	350	380	400	450	500	600
Partial pressure of water/bar	0.32	2	8	14	22	50	140	630

5.4.6. Carbon Formation on Reforming Catalyst

All hydrocarbons will spontaneously decompose into carbon and hydrogen as illustrated in equation (*10*), at the elevated temperatures experienced in a steam reformer if steam is not present.

$$CH_4 \rightarrow C + 2H_2 \quad \text{(thermal cracking)} \quad (10)$$

For naphtha the decomposition is more complex, being represented in Figure 5.11, where carbon can be formed by direct thermal cracking and also from various intermediates, particularly unsaturated species. In the presence of steam, and in particular with less than the stoichiometric quantity of steam, other carbon forming reactions become possible. These are the disproportionation and reduction of carbon monoxide, reactions (*11*) and (*12*) respectively.

$$2CO \rightarrow C + CO_2 \quad \text{(disproportionation)} \quad (11)$$

$$CO + H_2 \rightarrow C + H_2O \quad \text{(CO reduction)} \quad (12)$$

When methane or naphthas are reformed the formation of carbon within the nickel catalyst can be prevented by ensuring the steam/hydrocarbon ratio exceeds a certain minimum ratio. This minimum varies with pressure and temperature, and thermodynamic data derived by Dent et al.[132] can be used to calculate the minimum steam ratio under different conditions, as illustrated in Figure 5.12. The exact values calculated depend on the thermodynamic parameters assumed for "carbon", and a considerable amount of research has shown that different forms of carbon can be produced, depending on the prevailing conditions. Accordingly, graphs of the kind shown in Figure 5.12 should only be taken as being indicative, rather than definitive.

There is a greater tendency for higher hydrocarbons than for methane to form carbon, because on pyrolysis the readily formed initial intermediates polymerize and deposit carbon. The concentration of the intermediates is an important factor, and is critical in influencing the

5.4. Design of Steam Reforming Catalysts

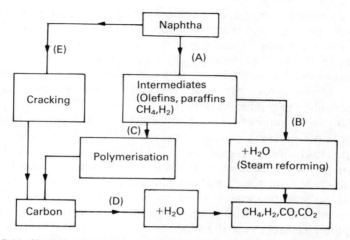

Figure 5.11. Naphtha reforming: simplified scheme for production of carbon.

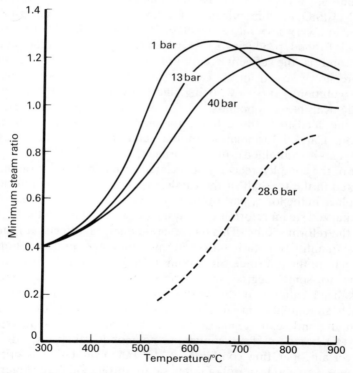

Figure 5.12. Thermodynamic minimum steam/carbon ratios to prevent carbon formation in naphtha (———) and methane (– –) steam reforming.

Chapter 5. Steam Reforming

delicate balance between carbon-forming and carbon-removing reactions. Both the nickel and the support play dual roles, contributing to the reforming process and also to the formation of carbon. As shown in Figure 5.11, the nickel catalyst, depending on its activity, will catalyse the decomposition of feedstock (A) and reaction of steam with the olefin intermediates (B), but will also promote cracking and polymerization to produce carbon (C).

Acidity in the support is known to facilitate the reaction A, but it will also promote cracking (E) and polymerization (C), again producing carbon. This problem was solved by ICI by introducing an alkali metal component into the catalyst. This accelerates the carbon–steam reaction (D) and at the same time the alkali neutralizes the acidity in the catalyst support, so retarding cracking and polymerization. The most effective alkali was found to be K_2O (potash), and most reformers today running on naphtha feedstocks use the alkalized catalyst system. This is the basis of ICI's 46-1 series of catalysts. The potassium is effective by being mobile on the catalyst surface. Accurate formulation combines the potassium as a complex potassium alumina-silicate (e.g. Kalsilite, $K_2O.Al_2O_3.SiO_2$) and monticellite ($CaO.MgO.SiO_2$). The potassium is liberated at a very slow rate as involatile K_2CO_3 which is hydrolysed as fast as it is formed, producing KOH, which is very mobile on the catalyst surface and is the effective carbon-removing agent. Potassium is therefore slowly lost from the catalyst into the product gases, but the rate of evolution is very slow, being kinetically controlled by its release from the Kalsilite compounds. The higher the temperature and the higher the feedstock throughput, the more rapid is the potassium depletion. Careful formulation of the catalyst ensures that lives of several years are obtained in most reformers.

Most of the complex reactions associated with naphtha reforming are completed in the top half of the catalyst bed, with methane reforming taking place in the lower part of the tubes. It is therefore possible to use an unalkalized steam reforming catalyst, such as ICI 46-4, in the bottom half of the reformer tube. This has several beneficial effects. It reduces the total quantity of potash in the reformer, and takes potash out of the hottest part of the reformer. Since potash depresses the activity of nickel catalysts to some degree with respect to methane reforming, a non-alkalized catalyst at the bottom of the reformer improves the approach to equilibrium at the exit of the reformer for a given throughput and exit temperature. Non-acidic magnesium spinel ($MgAl_2O_4$) based catalysts containing no mobile alkali are effective under certain operating conditions,[133–135] but formulation is critical. However, given good control of temperature throughout a reformer it is possible to achieve the delicate balance between carbon formation and removal. Systems of this sort have lower tolerance to variation in operating conditions, changes in reformer firing patterns or temperature profiles which can disturb the balance and give carbon laydown.

Compared with naphthas, feedstocks such as ethane, propane and butane have a lower tendency to form carbon on a methane reforming catalyst, which at high steam/carbon ratios and lower throughputs can effectively reform these materials for two or more years. However, the catalyst will be susceptible to a slow accumulation of carbon, particularly in the top of the tubes, if the plant is subject to frequent changes in operating conditions. To overcome this ICI has developed a catalyst that allows normal throughputs and steam ratios to be sustained and yet avoid the problem. The catalyst, ICI 46-9, was developed from a methane reforming catalyst, and incorporates potassium promotors to resist carbon deposition. It is normally used in the top half of the reformer tubes. Additional applications of this and related catalysts are discussed in Section 5.11.

5.5. Secondary Reforming

The reforming process in an ammonia plant is divided into two stages to permit the addition of the necessary nitrogen, as air, to the process and to raise the temperature of the reactant gases as high as construction materials will economically allow. In this way as much methane as possible is converted to carbon oxides and hydrogen. Introduction of air into the process gas produces the highly exothermic reaction between the oxygen and hydrogen, and this is partially balanced by a continuation of the endothermic steam reforming reaction. Nevertheless, temperatures of about 1000°C are achieved, and the major requirements of a secondary reforming catalyst are that physically it should withstand high temperatures without excessive sintering or shrinkage occurring, and it should be able to retain an acceptable activity at these temperatures. The support is of refractory oxides similar to primary methane reforming catalyst. The air and reactant gases are mixed in a specially designed burner above the secondary catalyst bed, which typically on a 1000 tonnes day^{-1} ammonia plant contains about 30 m^3 of catalyst. Poor mixing of the gases from a badly designed or faulty burner can result in local temperatures much higher than 1000°C. Such temperatures can fuse the catalyst, causing maldistribution of gas and a high pressure drop. A layer of refractory fused alumina chips is often put on top of the bed to protect the catalyst and to promote good mixing before the gas enters the catalyst bed.

5.6. Catalyst Dimensions

Reforming catalysts can be produced in different shapes and sizes by pelleting or extrusion techniques. Since it is used in tubular reactors, it must be of a shape and size which packs easily and homogeneously into the tubes to give an active bed which does not have an unacceptably high pressure drop. As much as possible of the active nickel surface must be accessible to the reactant gas, while the catalyst must be strong enough to resist abrasion and breaking during handling or during any thermal cycling that might occur. It must also generate enough turbulence in the gas to give good heat transfer between the tube wall and the body of the catalyst. Obvious possible simple shapes are pellets, spheres, rings and various extrusions in the shape of tubes, cylinders, rods and bars of different sizes and cross sections. For reforming it has been found that a thick-walled ring meets all of the above criteria, and the dimensions most commonly used are a diameter of ~17 mm with lengths of ~17 mm, ~10 mm and even ~6 mm. If the outside dimensions of the ring are reduced, then an equivalent packed volume will have an increased pressure drop and slightly higher activity resulting from the larger geometric surface area. By using different ratios of the various sized catalyst it is possible to balance gas flows through individual reformer tubes and to compensate for asymmetric heat fluxes in the furnace. The relationship between the catalyst size and pressure drop is discussed in Chapter 2. In the past breakage of reforming catalyst during use (caused by thermal cycles) was a common problem. Modern catalysts do not suffer from this to any significant extent, and the strength of a 17mm long steam reforming catalyst ring should be in excess of 50 lb (23 kg) (i.e. a 50 lb weight placed on the curved surface of the pellet should not break it).

Modern commercial ring-shaped steam reforming catalysts meet these requirements, and they have been in use for many years. Although reliable service is obtained with ring catalysts there has been a move in recent years to enhance steam reformer performance by catalyst improvements. This has been done by modifying catalyst shape rather than changing the fundamental chemistry of the support or the catalytically active phase. The two main objectives are to reduce pressure-drop across the reformer, and/or lower tube wall temperatures (particularly in the region of maximum heat flux) via increased activity through higher geometric surface area and heat transfer properties. The major benefits that can be obtained are either longer life (see Section 5.8.1) or the possibility of increased throughput, or a combination of both effects.

5.6. Catalyst Dimensions

In designing shaped steam reforming catalysts a number of factors have to be taken into account. These include the packing characteristics of the catalyst particles in relatively narrow tubes, pressure drop, geometric surface area, heat transfer properties and physical strength. Whilst some of these properties may be enhanced with a particular shape, others might be adversely affected. For example, some high geometric surface area shaped particles tend to bridge across the tube walls when they are being charged, and this makes uniform packing difficult. Subsequently, when the reformer is running, this problem can lead to severe 'hot-spots' at regions where there is little catalyst in the tube. Other high geometric surface area shapes may have low strength, and a further consideration is the pattern of breakage of the catalyst particles should this occur. If a ring breaks, it usually forms two fairly large pieces, which do not cause a detrimental increase in pressure drop. However, with shapes having open structures there is the danger of them shattering into a number of small pieces when they break and these can give rise to a high pressure drop.

A range of shaped steam reforming catalysts is commercially available, and most of these have been designed on the basis of increasing geometric surface area per unit tube volume, and/or reducing pressure-drop across the reformer. Although the heat transfer properties associated with different shapes have often been neglected during the design stages of shaped catalyst, getting heat into the process gas stream is critical for good performance and importantly the internal heat transfer coefficient usually varies with the inverse of the pressure drop. The relative importance of catalyst geometric surface area and heat transfer can be demonstrated using the tube wall temperature profiles shown in Figure 5.13. These were obtained from a computer model of a natural gas top fired, ammonia plant steam reformer. The effect of arbitrarily doubling catalyst geometric surface area is shown in curve B, while the base case is curve A. The main consequence of doing this is to lower the tube wall temperature in the top portion of the tube, where the thermodynamic driving force for the reforming reaction is greatest, and the reaction rate is limited to a significant extent by available catalyst geometric surface area. On the other hand, the effect of doubling the heat transfer coefficient, shown in curve C, has the profound effect of lowering the tube temperature over the whole tube length. This is because heat transfer is governed mainly by the fluid mechanics of the system, and is independent of the chemical reaction. The effect of doubling both catalyst surface area and heat transfer coefficient is shown in curve D—the effects are additive, clearly showing heat transfer effects need to be considered as much as geometric surface area when designing new catalyst shapes for steam reforming duties.

Chapter 5. Steam Reforming

The benefits of improved heat transfer properties are of particular importance in high heat flux applications, Moreover, as the tube diameter increases, the catalyst volume increases with respect to the tube wall heat transfer area, and as result the catalyst geometric surface area becomes relatively less important as heat transfer properties become more dominant.

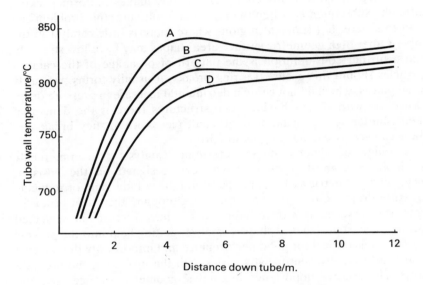

Figure 5.13. Effect of catalyst variables on tube wall temperature profile in a typical top fired steam reformer. Curve A: base case; curve B: catalyst with twice the geometric surface area of that in A; curve C: catalyst with twice the heat transfer coefficient of that in A; curve D: catalyst that has twice the geometric surface area and twice the heat transfer coefficient.

5.7. Uses of Catalytic Steam Reforming

The following are common uses of the catalytic steam reforming process employing a hydrocarbon feedstock. The examples refer to natural gas as a feedstock unless otherwise stated.

5.7.1. Ammonia Synthesis

The steam reforming process is used to convert methane and steam to hydrogen which, with the addition of nitrogen, produces a mixture of about three parts of hydrogen to one part of nitrogen. It is performed in two stages. In the primary stage the endothermic reactions take place at pressures around 30 bar and temperatures of 800°C or higher. This is followed by an exothermic secondary reformer, where air is added to the partially reformed gas stream. In this case the catalyst is charged as a single bed in a vessel capable of withstanding temperatures of up to 1000°C. Temperature and pressure in both reformers are limited by the materials available for construction of the tubes, vessels and pipework. Typical inlet and exit conditions and analyses for primary and secondary reformers are given in Table 5.9. A small amount of hydrogen (3–5 mol%) contained in the inlet gas stream to the primary reformer ensures that the nickel in the catalyst in the top of the tubes is maintained in the reduced form. As discussed in the next chapter, the carbon monoxide in the gas leaving the secondary reformer is converted to carbon dioxide in shift reactors, and then removed by scrubbing from the gas stream. Any residual carbon oxides are then converted back to methane by methanation before compression of the hydrogen and nitrogen to ammonia synthesis pressure.

Table 5.9. Typical operating conditions for primary and secondary reformers on an ammonia plant (dry gas analysis)

	Inlet primary reformer	Exit primary reformer	Exit secondary reformer
Pressure/bar	35	30	29
Temperature/°C	525	790	971
CH_4/%	91.9	9.4	0.2
C_2H_6/%	2.9		
C_3H_8/%	0.6		
C_4H_{10}/%	0.2		
C_5H_{12}/%	0.1		
C_6H_{14}/%	0.1		
CO_2/%	0.3	11.6	8.8
N_2/%	1.0	0.5	22.1
CO/%		8.3	11.5
H_2/%	2.9	70.2	57.1
A/%			0.3

Chapter 5. Steam Reforming

5.7.2. Methanol Synthesis

Since only hydrogen and carbon oxides are used in the synthesis of methanol, no secondary reforming or carbon dioxide removal is needed on a methanol plant. The reformer is required to convert as much of the feedstock as possible to hydrogen and carbon oxides, and to leave as little methane as possible. This requires a design which provides as low an operating pressure and as high an exit temperature as the economics of the process will allow. Typical inlet and exit conditions for a naphtha feed are shown in Table 5.10. With a naphtha feedstock the carbon/hydrogen ratio is close to stoichiometry for the methanol synthesis reaction. With a methane feedstock the carbon/hydrogen ratio is further from the stoichiometric requirement, and if necessary additional carbon in the form of carbon dioxide can be added either at the reformer inlet or at the inlet to the synthesis gas compressor. If it is added to the reformer inlet gas it will have a significant effect on the thermodynamics of the steam reforming reaction. The composition achieved at the reformer exit will change, as will the equilibrium which effectively prevents carbon formation.

Figure 5.14 shows the relationship between the carbon dioxide content in a naphtha reformer feed gas, the reformer temperature and the minimum steam/carbon ratio based on thermodynamic data. This indicates where carbon cannot form. It does not, however, indicate when carbon will form, as this depends on kinetic considerations and can only be confirmed by experiment.

Table 5.10. Typical operating conditions for a methanol plant reformer (dry gas analyses)

	Inlet reformer	Exit reformer
Pressure/bar	23.0	20.4
Temperature/°C	514	850
CH_4/%	89.1	4.5
C_2H_6/%	3.3	
C_3H_8/%	1.2	
CO_2/%	0.1	8.1
N_2/%	2.3	0.6
H_2/%	4.0	73.1
CO/%		13.7

5.7. Uses of Catalytic Steam Reforming

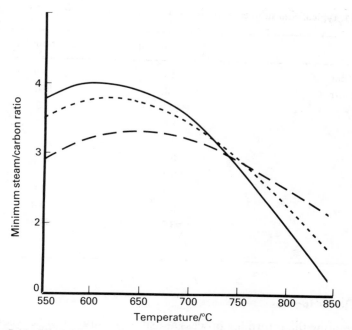

Figure 5.14. Effect of CO$_2$ content on minimum steam/carbon ratio in naphtha steam reforming. Pressure, 14.8 bar, 25°C approach to equilibrium, CO$_2$/naphtha ratio: ———, 3 : 1; – – –, 2 : 1; — —, 1 : 1.

5.7.3. Oxo Synthesis Gas

The oxo process provides a means of converting an olefin to an aldehyde containing one more carbon atom by catalytic carbonylation with an approximately equimolar mixture of hydrogen and carbon monoxide. This reaction is called hydroformylation. Butyraldehyde obtained in this way from propylene is usually subjected to aldol condensation, and the resulting C$_8$ aldehyde is hydrogenated to 2-ethylhexanol. Long-chain aldehydes derived from the appropriate straight-chain terminal olefin when hydrogenated to the primary alcohol are used in biodegradable detergents. A soluble cobalt catalyst is used in the conventional oxo process, while the more recently introduced low-pressure process uses a rhodium catalyst. Both methane and naphtha reforming can be used to produce oxo synthesis gas. To achieve the desired hydrogen/carbon oxides ratio carbon dioxide is added to the reformer inlet gas, and a reformer pressure of around 15 kg cm^{-2} with exit temperature around 850–900°C are common. A typical reformer for an oxo duty using natural gas as feedstock has inlet and exit conditions shown in Table 5.11.

Table 5.11. Typical operating conditions for an oxo gas reformer (dry gas analysis)

	Inlet reformer	Exit reformer
Pressure/bar	16.5	13.5
Temperature/°C	540	870
CH_4/%	62.2	2.9
C_2H_6/%	1.8	
C_3H_8/%	0.3	
C_4H_{10}/%	0.3	
C_5H_{12}/%	0.1	
CO/%	2.1	21.9
CO_2/%	29.6	10.8
H_2/%	3.0	64.2
N_2/%	0.6	0.2

5.7.4. Reducing Gas

By operating the reforming process at exit temperatures of between 800 and 1000°C, at low pressures of between 3 and 10 bar and with low steam/carbon ratios of ~2.2, it is possible to produce a gas with a high hydrogen concentration suitable for use in refinery hydrogenation processes. These gases may also be used for the direct reduction of iron ore to iron, although the actual reformer designs are different. Typical reformer inlet and exit conditions are shown in Table 5.12. In the iron ore reduction case large-diameter tubes are often used and the inlet gas to the reformer usually contains significant amounts of "top gas", containing carbon oxides from an iron-ore shaft furnace as well as fresh natural gas, so that the operating conditions and catalyst have to be carefully designed to prevent carbon being produced from the carbon oxides in the cooler regions of the tubes. The catalyst also has to withstand high temperatures and any sulphur which is present in the feed gas. Catalysts for this duty usually contain nickel oxide supported on α-alumina or a related refractory material. Because the process is conducted at a low pressure it is advantageous to use a large catalyst ring size (e.g. ICI Catalyst 24) to give a small pressure drop across the reformer.

5.7. Uses of Catalytic Steam Reforming

Table 5.12. Typical conditions for a reducing gas reformer (dry gas analysis) steam/carbon ratio 2.3

	Inlet reformer	Exit reformer
Pressure/bar	10.7	7.0
Temperature/°C	510	830
CH_4/%	85.5	4.0
C_2H_6/%	9.1	
C_3H_8/%	0.6	
CO_2/%	3.7	7.0
N_2/%	1.1	1.0
CO/%		16.0
H_2/%		72.0

5.7.5. Town Gas

During the 1960s many processes were developed for the production of town gas from naphtha in countries which at the time had little natural gas available—UK, Germany and Japan, for example. Of particular interest were those processes which operated at relatively high pressures (20-30 bar) and the most significant of these were those developed by ICI and British Gas in the UK; by BASF/Lurgi in FRG; by Haldor Topsøe in Denmark and by Mitsubishi in Japan. ICI developed two processes and they dominated the market. The first, the ICI Lean Gas Process made use of ICI Catalyst 46-1 naphtha steam reforming catalyst, and was used in around 200 plants with a total capacity of 120 million m^3d^{-1} of gas. In this process naphtha was reformed to give a hydrogen rich gas ('Lean Gas') which was enriched with a gaseous hydrocarbon to raise its caloric value to 500 Btu ft^{-3}. Typical operating conditions were 400 psig, exit temperature 750°C and steam to carbon ratio 3.0 with exit gas composition being H_2 59.9%, CO 9.9%, CO_2 16.2% and CH_4 14.9%. The second process which produced a gas containing 500 Btu ft^{-3} without further addition of gaseous hydrocarbons was the ICI 500 Process. Twenty-four of these plants were built in the UK having a total capacity of 17 million m^3d^{-1}. The first plants came on stream just as North Sea Gas was introduced into the UK and as a consequence no further plants were built. In this process naphtha was reformed in two stages. The first was a conventional naphtha steam reformer operating under conditions corresponding roughly to those in the Lean Gas reformer process and this was followed by an adiabatic reactor into which extra naphtha and steam were added to the gas from the primary

Chapter 5. Steam Reforming

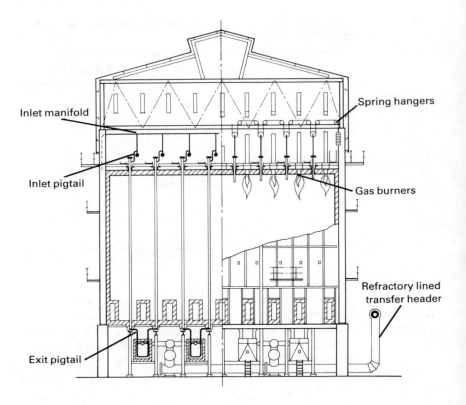

Figure 5.15. Schematic arrangement of a top-fired furnace: based on the primary reformer on ICI's No. 4 ammonia plant at Billingham.

reformer. Effectively the arrangement permitted the naphtha to be reformed at a low steam ratio and allowed the overall operating conditions chosen to give high methane concentration in the final product gas. Typical operating conditions were 400 psig, exit temperature 690°C and overall steam to carbon ratio 1.8 with exit gas composition after carbon monoxide conversion and partial carbon dioxide removal being H_2 48.5%, CO 3.0%, CO_2 14.5% and CH_4 34%. The exit temperature of the second stage although lower than at the outlet of the Lean Gas reformer was high enough to permit the use of stable high temperature nickel catalysts similar to those which had been used for many years in the tubular reforming of naphtha and natural gas.

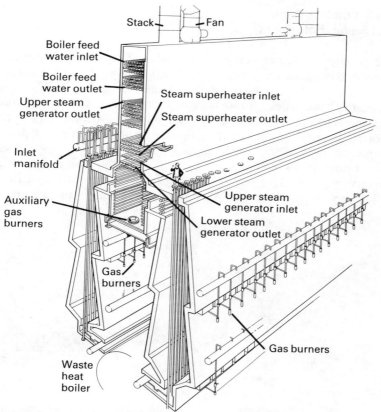

Figure 5.16. Schematic arrangement of a terrace wall fired furnace: based on the primary reformer on ICI's No. 2 methanol plant at Billingham.

5.7.6. Substitute Natural Gas (SNG)

Although the discussion in this chapter on steam reforming has been related solely to those processes which use externally fired tubular reactors, it is possible to react some hydrocarbons and steam in adiabatic catalyst beds. One such process, the catalytic rich gas (CRG) process, is overall exothermic and produces a methane-rich gas. To maximize the proportion of methane formed in the process it is operated at a low steam/feedstock ratio and the carbon oxides are methanated in one or two methanation stages while excess carbon dioxide is removed in a conventional process. In this way a suitable natural gas substitute containing up to about 95% methane can be produced. A small amount of propane may be added to enrich the gas to give it the desired quality. Other routes to SNG are discussed in Chapter 7.

5.8. Practical Aspects of Steam Reformers

The primary reformer consists essentially of two main sections: the furnace, containing the tubes charged with catalyst, and the convection section, where heat is recovered from the flue gas by such duties as preheating feedstock, process air and/or combustion air, boiler feedwater heating and steam raising, and superheating. The steam reforming reactions are usually carried out at pressures up to 35 bar and temperatures of 800°C or higher, while the flue gas may reach a temperature in excess of 1000°C. Consequently, the design of the primary reformer is complex and depends on the duty and on the philosophy of the chemical contractor engineering the plant. The

Figure 5.17. The top-fired reformer on ICI's No. 4 ammonia plant at Billingham.

5.8. Practical Aspects of Steam Reformers

Figure 5.18. Arrangement of burners and tubes in a typical top-fired furnace.

furnace can be top-fired, terrace wall-fired or side-fired, and in the case of small hydrogen plants it can be of a bottom-fired design. Typical throughput, which is usually expressed as the amount of steam plus feedstock per hour per litre of catalyst, is in the range 2–7 kg h^{-1}l^{-1}, with a heat flux of of 170–430 MJ m^{-2} (15 000–38 000 Btu ft^{-2}). The overall length of reformer tubes is usually in the range 7.5–12.0 m (25–40 ft) although the heated charged length may be up to ∼9.0 m (30 ft); tube diameter usually lies between 7cm and 13cm (2.8 and 5 in). The number of tubes depends on output, and for a large reformer there may be as many as 650 tubes. Figures 5.15 and 5.16 show schematic arrangements

Chapter 5. Steam Reforming

Figure 5.19. Terrace wall-fired reformers on ICI's No. 2 methanol plant at Billingham. This photograph was taken during the commissioning of the methanol plant.

of a top-fired and a terrace wall-fired furnace, with associated heat recovery sections and transfer ducts. Figure 5.17 shows the top-fired furnace used in No. 4 ammonia plant at ICI's plant at Billingham, in the UK, while Figure 5.18 shows the juxtaposition of burners and tubes in a typical top-fired furnace. Figure 5.19 shows the side-fired furnace used in No. 2 methanol plant at Billingham. There are a number of common problems associated with the design of a reformer which are of interest to an operator, and these are discussed in the following sections.

5.8.1. Containing the Catalyst

It is necessary to contain the steam reforming catalyst at high pressure and high temperature in a way which permits the transfer of sufficient heat to satisfy the endothermic reaction taking place on the catalyst. Conventional steel tubes do not possess the material characteristics to withstand the pressure and temperature at which a modern reformer operates.[132a] A suitable cost-effective material is a chromium/nickel alloy with the following composition: Cr 24–28%; Ni 18–22%; C, 0.35–0.45%; Mn, 2%; Si 2%; P and S, 0.05%. The melting point of the alloy is close to 1370°C, and it is suitable for use at temperatures up to 1150°C. The alloy is spun-cast in approximately 3-m (10-ft) sections of the appropriate diameter, and these are welded together to produce a tube of the length required. Other materials such as Pyrotherm G24/24 Nb and Manaurite 36X can also be used, since they allow operation of the reformer at higher temperatures and pressures. However, they are more expensive. Nevertheless, they are now used by a growing number of operators. Spun-cast tubes, even at normal and particularly at occasional abnormal transient operating temperatures, generate very high stresses, and these are minimized by using as thin a tube wall as possible. All weld protrusions should be machined off to minimize tube stresses at welds and also to prevent catalyst hold-up. If the diameter of the reformer tubes is too great, heat transfer to the catalyst in the centre of the tube will be restricted and the reaction rate limited. On the other hand, if the tube diameter is too small the pressure drop will be high. Moreover, with small-diameter tubes a smaller catalyst will also be required, and this will increase the pressure drop further. Most designers consider the optimum tube size to lie between 7 and 13 cm (2.8 and 5 in.) internal diameter, depending on the feedstock, the catalyst and the required composition of the reformed gas.

In operation there is a gradation of temperature longitudinally from the inlet to the outlet of the tube, as well as radially across the wall of the tube. Creep occurs with time at normal operating pressures and temperatures. The temperature which the tube wall experiences depends on the distribution of heat input, and the heat absorbed by the reaction taking place on the catalyst in the tubes. Uneven heat input and uneven catalyst activity caused by uneven packing or catalyst poisoning will cause local overheating, resulting in excessive creep in that locality, which will hasten tube rupture. Normally reformers are designed with a tube life of about 10 years (100,000 h) using creep strength data based on creep-rupture tests of varying duration available from a number of sources. A correlation derived by ICI and reported in the form of a

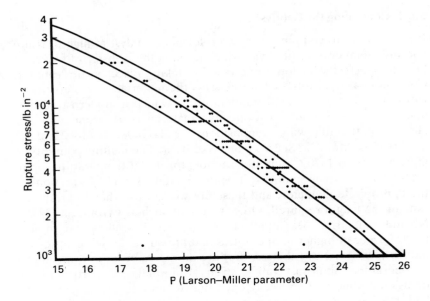

Figure 5.20. Larson–Miller diagram summarizing the results of 170 rupture tests on cast 25/20 chrome nickel alloy. $P = 10^{-3} \times T(15 + \log t)$, T = temperature/K, t = time/h.

Larson–Miller diagram is shown in Figure 5.20, which summarizes the results of 170 rupture tests on 25/20 chrome nickel alloy at different temperatures and pressures. The higher the temperature and pressure are, the greater the creep and the shorter the tube life. This applies to all parts of each tube, and if part of any tube is subjected consistently to higher-than-average temperatures it will fail prematurely. It is therefore important that hot spots, due to catalyst poisoning or carbon deposition, are removed as soon as possible. A simple relationship between the mean tube wall temperature, the pressure in a tube of 25/20 chrome nickel alloy and the life of the tube is given in Figure 5.21.

5.8.2. Reactant Gas Distribution

It is necessary to distribute the inlet gas and collect the reacted gases in a way which produces a predetermined flow through each tube—usually,

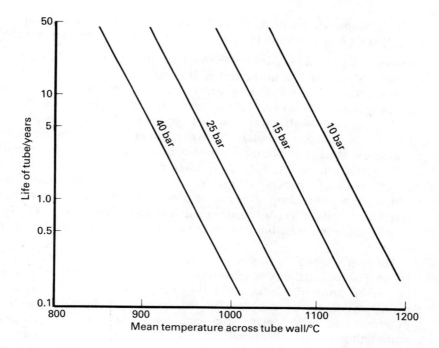

Figure 5.21. Diagram showing influence of internal pressure and wall temperature on tube life for typical reformer tube made of cast 25/20 chrome nickel alloy.

but not always the same flow through each tube. Any design which introduces multiple parallel flow paths requires careful attention to ensure that the desired flow is achieved through each tube. Even gas distribution will depend on the following:

(a) all tubes being of the same length and internal diameter;
(b) the gas being evenly distributed to the tube inlets to give uniform tube inlet pressures;
(c) the catalyst being uniformly packed into each tube to give a uniform pressure drop—this means that the catalyst must have uniform pressure drop characteristics, and that it must be charged in a manner which guarantees sufficient uniformity when first put on-line and also after numerous shutdown and start-up cycles;

(d) the reformed gas being collected from the tubes in such a manner that there is a uniform pressure distribution in the collecting header.

The most significant variable in the hands of the operator is the packing density of the catalyst in the tubes. If distribution of pressure drop is uneven, the flow of reactant gas will vary among tubes and the reaction heat required will also vary. If the combustion heat applied is uniform, then some tubes with lower flow will appear hotter while others with higher flow will appear cooler. This will result in a decrease in the life of the tubes which are hottest, and also in a reduction in the overall reformer efficiency. Pressure drop differences across the tubes caused by differences in packing density can be reduced by introducing an additional pressure drop. This can be done by the installation of orifice plates at the tube inlets or by using relatively narrow inlet and exit pipes, usually called pigtails, but of course the overall pressure drop is then greater.

The combustion heat flux in the furnace may not be even, for example because of the proximity of some tubes to a hot furnace wall, or because of uneven firing of the fuel. The flue gas collection system may also cause asymmetry of gas flow in the furnace. Deliberate selective differences in orifice plate sizes can adjust the gas flows to give more-uniform tube temperatures. Another way of producing a correction is to introduce varying quantities of two sizes of catalyst into the tubes. The ratio of the two sizes can be calculated to modify the gas flow to correct the tube skin temperature.

5.8.3. Firing the Reformer

It is necessary to fire the furnace to give an even heat flux across the furnace, and a vertical heat flux gradient close to the heat demanded by the reactions taking place within the tubes. Uneven firing creates an uneven heat flux, which in turn gives uneven tube heating and this produces an inefficient reformer. Designers of both top- and side-fired reformers aim to distribute heat as evenly as possible across the reformer, and to collect the combusted gas in a way which allows an even flow of hot gas through the furnace. Ideally, a high heat flux should be available where a lot of endothermic reaction is taking place, and a lower heat flux where there is little taking place. Once the reactants

5.8. Practical Aspects of Steam Reformers

have been heated to reaction temperatures (typically ~600°C) most of the reforming reaction takes place in the upper parts of the reformer tube and a high heat flux is required. As the process gas moves down through the tubes it requires a diminishing heat flux. The heat flux profile varies with different feedstocks. Naphtha reacts at a lower temperature and demands more heat higher in the furnace than methane does. The point of maximum available heat flux can be adjusted in top-fired furnaces by using burners giving different flame lengths. For example, a naphtha reformer may use a flame length of about 1.2 m (4 ft) whilst the same reformer for a methane feedstock may require a flame length of 2.4–3.0 m (8–10 ft). With side-fired reformers the number of burners or the firing pressures at different levels on the furnace walls can be adjusted to achieve the desired flux profile. Typical tube skin (tube outside wall) temperatures and process gas temperature profiles for a top-fired methane reformer are shown in Figure 5.22. The measurement of tube skin temperature is discussed in Appendix 12.

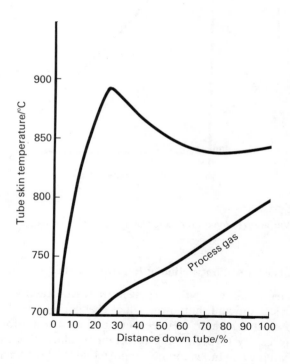

Figure 5.22. Illustrative example of a steam reformer tube skin temperature profile and the corresponding process gas temperature profile for a top-fired reformer.

271

Chapter 5. Steam Reforming

Changes in catalyst activity, for example due to poisoning, cause a change in the tube skin temperature achieved, and this effect can be seen from the example given in Figure 5.23. In some instances such changes, when they are sufficiently large, will have a significant adverse effect on tube life, and it may be necessary to limit reformer throughput so that the heat flux can be reduced and tube skin temperatures kept within acceptable limits.

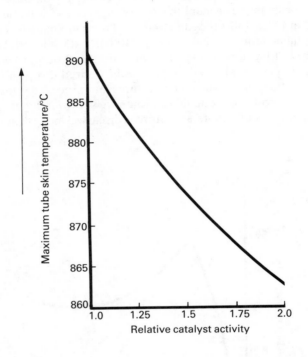

Figure 5.23. Variation of maximum tube skin temperature with catalyst activity.

5.8.4. Expansion and Contraction of Reformer Tubes

It is necessary to allow for the expansion and contraction of reformer tubes during start-up and shutdown. Tubes change in length by 10–13 mm (4–5 in.) in most reformers between hot and cold conditions, and it is therefore necessary to design the reformer to allow this expansion to take place safely. The tubes can be supported at the bottom and allowed to expand upwards, or supported at the top and allowed to expand downwards. It is important that the system is designed so that as little stress as possible is imposed on the tubes when they are full of catalyst. Free movement of the tubes at all times during the lifetime of the reformer is most important if major damage to the reformer is to be avoided. Restrictions to this movement will cause rupture of the tube, pigtail or header.

The catalyst expands and contracts with the tubes, though not to the same extent, and some settling will occur with repeated thermal cycles. As a consequence, sufficient catalyst must be added to each tube to ensure that the heated length of the tube always contains catalyst. The catalyst must not break down when subjected to the stresses of thermal cycles which sometimes might be rapid. If breakdown occurs the reformer pressure drop will rise, and uneven flow between tubes will result.

5.8.5. Facilities to Charge and Discharge Catalyst

A reformer must be designed to make charging and discharging of catalyst as easy as possible, and details of how this should be done are described in Chapter 3 and Appendix 19. It is important that good quality dust-free catalyst is evenly loaded into each tube to give as even a pressure drop as possible across all tubes in the furnace. It should be loaded in a manner which avoids any breakage of the catalyst or formation of voids, and the heated part of the tubes must always contain catalyst. A badly charged furnace is apparent immediately the plant starts up, since hot spots appear where there are voids and where the catalyst has not been charged to a high enough level. Tubes with loosely filled catalyst appear mottled, while a poor pressure-drop distribution results in a broad range of tube skin temperatures, often visible to the eye. Hot spots will reduce the life of the reformer tubes, but sometimes they can be removed by further vibrating the tube to settle the catalyst. Generally, however, a badly charged furnace remains so for the life of the catalyst charge, and the catalyst performance will deteriorate more rapidly than a well-charged furnace.

Reformers can be designed for top or bottom discharge, although most new plants are designed for top discharge. This requires a flange and blank which can be removed to expose the catalyst. The catalyst will normally be free-flowing, and can be drawn out of the tube by a suction device. It is important that all of the old catalyst is removed and that there is no blockage in the catalyst support grid at the bottom of the tube. This is usually checked visually using a light and binoculars, although television equipment that provides a clear view of inside the tube is often used by ICI. With a bottom discharge the support grid is removed with a flange on the bottom of the tubes. On occasions, if the catalyst is being removed following massive deposition of carbon arising from maloperation which, for example, gave a low steam/carbon ratio, it may actually be necessary to drill the catalyst out of the tubes with a conventional mining bit.

5.8.6. Designing a Reformer for Efficient Operation

There are two interdependent aspects of reformer efficiency. The first is the efficiency of conversion of feedstock to hydrogen and carbon oxides. This is enhanced by using as active a catalyst as possible, and designing the heat flux so that as much heat is transferred to the catalyst as is possible within the tube material limits. The second aspect is the thermal efficiency of the combustion process. The reformer should be designed with sufficient efficient burners to give an even heat flux across the furnace with the desired vertical heat flux profile. There should be a minimum ingress of tramp air through holes in the furnace roof or ductwork, as well as effective insulation to minimize conduction heat losses. In addition the design must incorporate an efficient flue gas heat recovery system. A facility for measuring oxygen in the flue gas is also needed, to allow good reformer combustion control and identification of air leakage into the flue gas system.

5.8.7. Catalyst Reduction

When the reformer is started up the operator should observe the furnace designer's instructions, especially those concerning the rate of increase of temperature and the way in which it is increased. Instructions concerning the burner pattern and the rate at which the firing is increased should be followed. As a general principle, whenever throughput is increased the increase in steam should always precede the change in hydrocarbon, to ensure that the steam/carbon ratio is always held at or above the design value. Throughout the start-up regular visual checks must be made on the furnace, because the instrumentation provided to monitor furnace performance is designed primarily for operation under normal running conditions. As a result, during reduction the indicated exit tube temperatures are invariably lower than the true value, because of heat losses, and allowance must be made for this discrepancy. Depending on the location of thermocouples, the indicated temperature may be 25–100°C lower than the actual value. It is virtually impossible to provide sufficient instrumentation for start-up conditions, and the appearance of the tubes, flames and refractory lining is the best guide to the actual furnace conditions. Care should be taken to maintain safe conditions in the heat-recovery section, since mechanical failures here can be a major cause of delay during start-up. Although details will vary from plant to plant, the procedures set out

5.8. Practical Aspects of Steam Reformers

below are sufficiently general for most modern plants. The establishment of the normal operating pressure of the plant is not expressly mentioned. This may be carried out when it is convenient during the start-up, providing condensation does not occur.

The actual reduction of the catalyst in a natural gas reformer may be carried out with hydrogen, off-gases containing hydrogen, natural gas, ammonia or methanol. For most reformers the catalyst should be reduced in the presence of hydrogen before feed is introduced. In this way a catalyst with high activity, particularly at the lower temperatures at the tube inlet, will be obtained. The plant should first be purged of oxygen, with nitrogen, and the reformer should be heated above the condensation temperature with nitrogen circulation. Steam is introduced as soon as is convenient when the exit header temperature (and secondary reformer in an ammonia plant) is at least 20°C above the condensation temperature, preferably at 40–50% of the design rate to allow the furnace to be fired evenly.

5.8.7.1. Reduction with Hydrogen

Hydrogen, or hydrogen-rich gas free from catalyst poisons, may be used with the steam/hydrogen ratio in the range 6–8 : 1. Throughout the reduction the tube exit temperature should be at the design value, while the tube inlet temperature should be as high as possible to promote reduction of the catalyst at the tube inlet. Once the exit temperature has been reached, reduction conditions should be held for six to eight hours. The reformer is then brought on-line by introducing natural gas and at the same time stopping nitrogen. Hydrogen addition is usually continued, if sufficient is available, until the hydrogen recycle loop is established. The natural gas rate is then increased to 25% of the design rate as quickly as possible, with the steam/carbon ratio at least 5 : 1. The firing of the furnace is adjusted, as natural gas is reformed, to maintain the exit temperature at the design level. The natural gas flow rate is increased in stages over two to three hours until the design steam/carbon ratio is reached. The methane content in the exit gas has to be checked after each change, to ensure that it remains low and steady. If the exit methane level increases or the tubes show hot zones, the catalyst is re-reduced in hydrogen for two to four hours or at a steam/carbon ratio of 7 : 1. Throughput is then increased in stages to the design value.

5.8.7.2. Reduction with Ammonia

Ammonia may be used as a hydrogen source, either by cracking it in an ammonia cracker or in the reformer itself. It may be injected as a liquid upstream of the primary reformer, often at a point before the process steam superheater. If it is introduced before the feedstock hydrocarbon heater, the desulphurizer should be bypassed if possible. The procedure is then similar to that with hydrogen, except that the reformer exit temperature should be as high as possible, preferably 800°C or higher, to maximize the amount of ammonia which is cracked.

5.8.7.3. Reduction with Methanol
Methanol can be cracked in the primary reformer and used as a source of hydrogen for reduction. The flow of methanol should be adjusted to give a steam/hydrogen ratio of between 6 : 1 and 8 : 1 at the reformer exit. The cracked gas should not be recycled, because of the possibility of methanation of the carbon oxides that are produced.

5.8.7.4. Reduction with Natural Gas
Once the plant has been purged of oxygen and steam has been introduced, the exit reformer temperature is increased to at least 750°C. Natural gas is introduced at about 5% of the design rate, and is then increased over two to three hours to give a steam/carbon ratio of 7 : 1. At the same time the exit temperature is raised to the design value, while the tube inlet temperature is held as high as possible to promote the reduction of the catalyst at the tube inlet. As the heat requirement increases, due to the endothermic steam reforming reaction, the furnace firing is trimmed to keep the exit temperature steady. Over the next few hours the methane concentration in the outlet gas falls to a low steady value as the catalyst is reduced. Once this stage has been reached reduction conditions are maintained for four to eight hours. The natural gas flow rate is then increased over two to three hours, with the exit temperature steady, until the design steam ratio is reached. If the methane concentration remains steady the throughput may be increased in stages to the design value. If the exit methane level increases or if the tube shows hot bands it is necessary to continue reduction conditions at a steam/carbon ratio of 7 : 1 for two to four hours before proceding.

5.8.7.5. Reduction with Other Hydrocarbons
The use of hydrocarbons other than methane or light refinery gases as a source of hydrogen during reduction is not recommended because of the greater likelihood of carbon formation. Ethane, propane and butane may be used if it is absolutely necessary, but great care is needed. More time should be taken over the reduction, and more precision is required in all of the measurements made. Any hydrogen present in the feed gas to the reformer will greatly assist reduction.

5.8.7.6. Reduction after Shutdown

After a shutdown in which the catalyst has been steamed, it is usually recommended that the catalyst should be re-reduced to restore its full activity. This is done in the same way as with new catalyst, except that a rather shorter reduction period of two to four hours is normally sufficient. Reduction does not have to be made at temperatures above the operating temperature unless the catalyst has been contaminated with sulphur.

5.9. Factors Affecting the Life of Reforming Catalyst

The life of a catalyst charge can be affected by the following factors:

(a) catalyst breakdown;

(b) tube blockage;

(c) overheating of the catalyst;

(d) poisoning of the catalyst;

(e) thermal ageing.

Catalyst breakage and blockage of the tubes causes an increased pressure drop across the reformer, and if the effect is random it shows as an uneven appearance of the tubes in the furnace. More fundamentally it can lead to overheating of the catalyst, loss of activity and a reduction of throughput. All of these effects may be caused by the deposition of carbon. Overheating of the catalyst can also be caused by maloperation of the reformer. Loss of activity through poisoning by contaminants in the process gas is important, since this can cause carbon deposition and result in overheating, catalyst breakage and, in extreme cases, even partial blocking of the tubes. Gradual loss of activity or thermal ageing caused by progressive loss of nickel surface area through sintering places a limit on the life of a catalyst charge, and for a particular catalyst this depends on the actual operating conditions. In practice the most important effects are catalyst poisons and carbon formation, and these are considered in detail in the following two sections.

Chapter 5. Steam Reforming

5.10. Catalyst Poisons

5.10.1. Sulphur

Sulphur is invariably present as inorganic and/or organic sulphides in most naturally occurring feedstocks, and is normally a powerful poison for nickel catalysts. It has to be reduced to less than 0.5 ppm in the feed gas to the reformer, and this is usually done using hydrodesulphurization catalysts in combination with a bed of zinc oxide, as discussed in Chapter 4. As the zinc oxide becomes saturated with sulphur, hydrogen sulphide will start to slip, and as the level rises in the feed gas the activity of the primary reforming catalyst decreases. As a result the tubes become hotter and the methane concentration in the gas leaving the primary reformer rises, and additionally with a naphtha feedstock the aromatic "slip" also increases. Such a progression requires the throughput to be reduced, since if it is allowed to continue it will lead to carbon deposition, a large pressure drop and irreversible damage to the catalyst. Carbon deposition results when relatively large amounts of hydrocarbon are exposed to high temperatures in the presence of insufficient hydrogen to prevent cracking of the hydrocarbon, and this situation can result from having very low activity catalyst in the upper portions of the tubes. Sulphur will also eventually affect the secondary catalyst performance, and will irreversibly destroy the activity of the low-temperature shift catalyst. If the source of sulphur is removed before damage occurs, then primary and secondary catalysts will recover their original activity.

The sensitivity of the reforming catalyst to poisoning increases at lower operating temperatures. This is demonstrated in Figure 5.24, so while poisoning of the catalyst occurs with about 5 ppm of sulphur in the feedstock at a reformer exit temperature of 800°C, concentrations of the order of 0.01 ppm poison the catalyst at 500°C, because the poisoning process may be represented as a simple exothermic adsorption process. Some sulphur is often unavoidably introduced into the catalyst during manufacture, and this is liberated when the catalyst is being reduced for the first time. Sulphur will be envolved as hydrogen sulphide, and the catalyst should not be used at its design throughput until this is removed, nor should the product gas be allowed to enter the low-temperature shift catalyst bed until it is essentially free of sulphur.

5.10.2. Arsenic

Arsenic in very small quantities can irreversibly destroy the activity of primary reforming catalysts. Such poisoning can occur on plants which

5.10. Catalyst Poisons

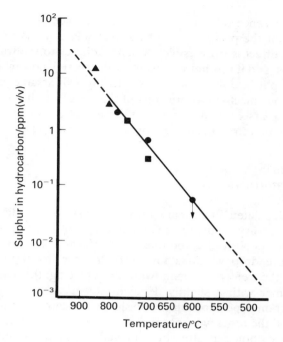

Figure 5.24. Minimum sulphur concentration to poison nickel at different temperatures. ●, Heptane; ■, naphtha; ▼, methane.

have an arsenic-containing carbon dioxide removal system (e.g. Vetrocoke). Leakage of arsenic-containing solutions across boiler feedwater heaters or into condensates from steam reboilers can put arsenic compounds into the boiler drum, from where it is carried through the process steam into the primary reformer. Here the arsenic is deposited on the top of the catalyst and onto the reformer tube walls, moving progressively down the tubes. The early symptom is a high methane slip (or aromatic slip on a naphtha-based plant) from the reformer tubes, which become hot at the top. As the arsenic content rises, so the catalyst activity decreases and carbon deposition occurs, causing a rise in the pressure drop.

Analysis of the reaction steam will detect arsenic when it is present. Arsenic will also be revealed by analysis of the catalyst at the top of the tubes when the reformer is shut down. A concentration of as little as 50 ppm of As_2O_3 on the catalyst will seriously affect the reformer performance, and with 150 ppm there is a serious risk of carbon

Chapter 5. Steam Reforming

deposition. The presence of 1 ppm of As_2O_3 in the steam entering a reformer will impair the performance of the reformer in a matter of a few days, and the effect is irreversible. Arsenic adheres to the walls of the reformer tubes, and if it is not removed it will severely poison a new charge of catalyst put into the same tubes. The layer of arsenic can be removed by vigorous mechanical brushing of the inside of all tubes. In severe cases, where arsenic is detected on the walls of the steam mains leading to the primary reformer, these should be acid-cleaned.

5.11. Hot Bands in Natural Gas Reformers

Carbon can be deposited in primary reforming catalyst by different mechanisms and to varying degrees. Complete loss of reaction steam results in a massive deposit of carbon, and the reformer will develop a very large pressure drop within a few seconds. It will not then be possible to run the reformer again without replacing the catalyst. Running a reformer with a slightly deficient steam/carbon ratio will result in slight carbon deposition which will slowly increase the reformer pressure drop, and the tubes will appear hotter than normal. If detected soon enough this carbon can often be removed satisfactorily, but this depends on the type of catalyst being employed. Removing the feed flow and sustaining normal reformer temperatures with only steam and hydrogen will convert the carbon to carbon dioxide, which can be detected by an analyser at the reformer exit. The hydrogen will keep the catalyst in a reduced form. When the carbon has been deposited within the catalyst pores by carbon monoxide disproportionation, steaming will increase the reformer pressure drop. In this case the carbon expands within the catalyst pores and cracks the pellet and, although the carbon *in situ* retains the pellet strength, when the carbon is removed the pellet collapses. If this happens it is necessary to change the catalyst.

Slow deposition of carbon can occur for a number of reasons. Careful catalyst formulation is essential to maximize selectivity, and to eliminate acid sites which can promote carbon formation. When a predominantly methane feedstock is reformed, low catalyst activity in the inlet portion of the tube can lead to carbon deposition, which restricts heat transfer and gives rise to the phenomenon known as "hot bands". Both the carbon monoxide disproportionation reaction, equation (*11*), and carbon monoxide reduction reaction, equation (*12*), are always in the carbon-free side of the equilibrium throughout the reformer tube, regardless of catalyst activity. However, the methane-cracking reaction, equation (*10*), is on the carbon-forming side of equilibrium for a

5.11. Hot Bands in Natural Gas Reformers

significant portion of the tube. Carbon is not, however, produced at the lower temperatures near the inlet, because both reactions which remove carbon (reverse-CO disproportionation and reverse-CO reduction) are faster at these temperatures than the rate of carbon formation by methane-cracking. However, as the temperature increases, so does the rate of carbon formation and, at a temperature of about 650°C, the carbon-forming reaction becomes faster than the carbon-removing ones. If the rates of these reactions are fixed, then it is essential that the catalyst has enough activity to produce sufficient hydrogen via steam reforming below this temperature, so that the gas composition lies on the carbon-removal side of the methane-cracking equilibrium. In the example shown in Figure 5.25 a fresh high activity catalyst is able to do this, but by the time the gas temperature reaches 650°C with a lower activity catalyst the gas composition is still well on the side of the carbon formation. Thus, "hot bands" always form at about the same position on all tubes in the furnace and approximately the same position in all reforming furnaces.

As expected, heavily loaded top-fired furnaces are the most susceptible to forming "hot bands". The low catalyst activity can arise from a number of causes—catalyst may be old and at the end of its useful life, it may be poisoned or inadequately reduced. If no hydrogen is recycled with the feedstock, then the catalyst in the inlet portion remains in the oxidized state until reforming or cracking of the feedstock occurs, and produces some hydrogen. This increases the load on the catalyst further down the tube, since the the inlet portion is then functioning simply as a heat exchanger. Further, if reformer conditions change, then unreduced catalyst may be called upon to do some reforming. It will be unable to do so, carbon will be deposited and hot areas will appear at the top of the tubes. At the other extreme of feedstock composition, such as with naphthas, the tendency to form carbon is much higher, because of the nature of the hydrocarbons and the high level of unsaturated intermediates which can polymerize and dehydrogenate and form carbon. In this situation support neutrality and high activity are insufficient to prevent gradual carbon build-up and to allow operation at economic low steam/carbon ratios. ICI Catalyst 46-1 was successfully developed to do this in naphtha steam reformers (see Section 5.4.6). ICI has also recently extended the approach of using potash to facilitate reaction of carbon with steam in catalysts for handling feedstocks such as LPG and lighter hydrocarbons, where there have been recurring problems with hot bands, or poor performance attributable to carbon formation from the heavier hydrocarbons or methane-cracking, particularly in the more highly stressed furnaces.

Chapter 5. Steam Reforming

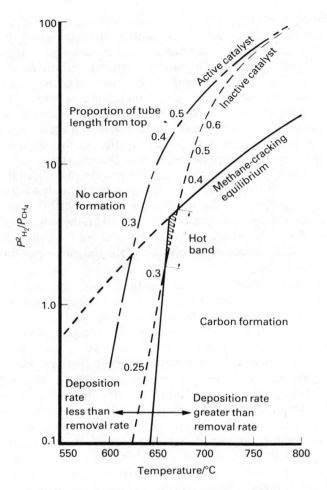

Figure 5.25. Formation of a hot band in a methane steam reformer. In this example initial catalyst activity is adequate to provide sufficient hydrogen to prevent carbon forming from methane cracking but if activity is impaired by poison or the catalyst is old, carbon can form and a hot band results. The text describes how this can be overcome by incorporating alkali into the catalyst formulation.

Here ICI Catalysts 25-3 and 46-9 contain the potash in the form of potassium ß-alumina (a complex potassium aluminate), which hydrolyses slowly, releasing volatile alkali at the required concentration to ensure continuous carbon-free operation. This approach has been demonstrated to be very successful, even with feedstocks at 99% methane purity where hot-band formation has been eliminated and catalyst lives extended from 1½–2 to 4½–5 years.

Chapter 6

The Water-gas Shift Reaction

6.1. Introduction

Water-gas is an equilmolar mixture of hydrogen and carbon monoxide. It is formed when steam is passed through incandescent coke, and the original process for its manufacture was developed by the town-gas industry during the latter part of the nineteenth century to supplement coal gas. The calorific value of water-gas is less than that of coal-gas and had to be raised by addition of oil, which could range from light distillate to heavy fuel oil, to give "carburetted" water-gas before the two gases could be mixed. This led to an improvement in the efficiency of gas making, and also provided a convenient means of meeting the temporary and sudden demands for gas that occurred at peak periods.[136] At temperatures above 1000°C steam reacted with coke according to equation (*1*) but at lower temperatures carbon dioxide, which was produced according to equation (*2*), became increasingly important. Both reactions are endothermic, and consequently the temperature of the incandescent coke fell as the reaction proceeded.

$$C + H_2O \rightarrow CO + H_2 \qquad \triangle H = 131.2 \text{ kJ mol}^{-1} \qquad (1)$$

$$C + 2H_2O \rightarrow CO_2 + 2H_2 \qquad \triangle H = 90.0 \text{ kJ mol}^{-1} \qquad (2)$$

Heat to sustain water-gas production was supplied at regular intervals by cutting off the steam and then blowing air through the bed to oxidize coke to carbon monoxide according to the equation

$$2C + O_2 \rightarrow 2CO \qquad \triangle H = -220.8 \text{ kJ mol}^{-1} \qquad (3)$$

As soon as the temperature of the bed of coke was sufficiently high the steam cycle was restarted. In practice these reactions took place over a temperature range so that a small amount of carbon dioxide was invariably present in the water-gas. In addition, the gas also contained a small amount of nitrogen because of contamination from the "air blow" cycle.

Except in a few plants using electrolytic hydrogen, the water-gas process provided the only economic means of producing hydrogen in the quantity required by the Haber ammonia synthesis process.[137,138] In the first synthetic ammonia plant carbon monoxide was removed by liquefaction and scrubbing with hot caustic-soda solution. Nitrogen was supplied from a cryogenic air separation unit. However, it was soon realized that the carbon monoxide liquefaction process was unsuitable for use on such a large scale, and a catalytic process was developed which converted carbon monoxide to carbon dioxide (and additional desired hydrogen) by reaction with steam according to the equation

$$CO + H_2O \rightarrow CO_2 + H_2 \qquad \triangle H = -41.1 \text{ kJ mol}^{-1} \qquad (4)$$

This reaction was termed the "water-gas shift reaction" (WGSR). The reaction was first reported in the literature[139] as early as 1888, but its technical importance was not recognized until the development of the Haber process.

The rate of the homogeneous water-gas shift reaction in the gas phase is very low at practical temperatures, but in 1912 Bosch and Wild[140] discovered a catalyst consisting of oxides of iron and chromium that could be used at 400–500°C to reduce the carbon monoxide to around 2%. The catalytic water-gas shift reaction was incorporated into the first coal-based ammonia process flowsheet in 1915, and since then it has played a vital role in the ammonia process flowsheet. Its use meant that not only could carbon dioxide be removed more easily than carbon monoxide by dissolving it in water, but there was the added advantage that the yield of hydrogen from a given amount of coke was significantly increased.

Shortly after the introduction of the ammonia process the cryogenic nitrogen unit was replaced by separate producer-gas generators in which coke was reacted continuously with air according to equation (3). The producer-gas so formed, consisting of nitrogen and carbon monoxide, was mixed with water-gas in the appropriate ratio and passed over the shift catalyst. This established the process route for the synthesis of ammonia from coal, although in later plants the two gas streams were frequently "shifted" before they were mixed. Typical gas analyses obtained on one of ICI's old coal-based plants are shown in Table 6.1. Carbon dioxide and residual carbon monoxide were then removed by absorption in water and copper liquor respectively, and the resulting hydrogen/nitrogen mixture was passed to a gas compression stage and then added to the synthesis loop.

Similar iron-based shift catalysts are used in modern ammonia and hydrogen plants, and in order to achieve a very high conversion of carbon monoxide to carbon dioxide a second catalyst, based on copper, is used. The iron-based catalyst is operated at a relatively high temperature, and the more recently introduced higher activity copper-based catalyst is operated at a much lower temperature. In this way carbon monoxide concentrations are reduced to less than 0.3% before most of the carbon dioxide is removed and the process gas enters the methanator, as shown in Table 6.2.

Table 6.1. Typical gas analyses in two early coal-based plants (% dry basis)

Description of process stream	Prudhoe (UK) Mixed gas stream		Billingham (UK) Hydrogen-rich stream		Nitrogen-rich stream	
Component	A	B	A	B	A	B
N_2	23.2	16.5	2.0	1.5	35.0	27.8
H_2	35.1	49.5	53.1	61.2	28.8	41.5
CO	34.6	4.5	37.9	7.1	25.3	3.9
CO_2	6.6	29.1	6.6	30.0	10.1	26.1
Inerts	0.5	0.3	0.3	0.3	0.8	0.7

A, before shift conversion; B, after shift conversion.

Table 6.2. Typical gas analyses before and after shift converters in a modern natural gas-based ammonia plant (% dry basis)

Component	Inlet HT shift	Exit HT shift/ inlet HT shift	Exit LT shift
N_2	22.5	20.5	20.0
H_2	56.5	60.3	61.2
CO	12.9	3.0	0.2
CO_2	7.5	15.6	18.0
Inerts	0.6	0.6	0.6

6.2. Thermodynamics

The water-gas shift reaction is moderately exothermic ($\triangle H = -41.1$ kJ mol^{-1}) and hence its equilibrium constant decreases with temperature, and high conversions are favoured by low temperatures, as shown in Figure 6.1. The position of equilibrium is virtually unaffected by pressure. As expected, additions of greater than stoichiometric quantities of steam improve conversion; equilibrium constants for the reaction over a wide range of temperatures are tabulated in Appendix 7. Under adiabatic conditions conversion in a single bed of catalyst is thermodynamically limited—as the reaction proceeds the heat of reaction increases the operating temperature, and

Chapter 6. The Water-gas Shift Reaction

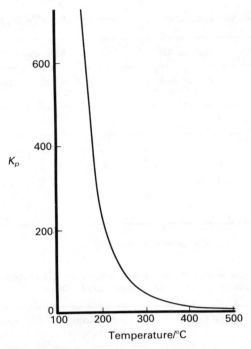

Figure 6.1. Variation of equilibrium constant (K_p) for the water-gas shift reaction with temperature (see Appendix 7 for tabulated values).

so restricts the conversion possible. Typical carbon monoxide levels that are achieved[141] at the exit with a single adiabatic bed of chromia-supported iron-based shift catalyst in an ammonia plant are in the range 2–4%, and because it is necessary to operate these catalysts at high inlet temperatures (typically now in the range 370–400°C) they are known as *high temperature (HT) shift* catalysts.

The thermodynamic equilibrium limitation on the reaction can be reduced by using two or more beds of HT shift catalyst with inter-bed cooling and, perhaps, removal of carbon dioxide between the stages. In this way it was possible in the late-1950s to decrease the thermodynamic limitation, and achieve[142] carbon monoxide levels of less than 1% at the exit. The limitation on conversion was then catalyst activity. When a single stage, or two stages, of HT shift were used, the final carbon monoxide removal stage was generally absorption in copper liquor (Chapter 7), although some plants used a methanator, because of the simplicity of the process, and accepted the attendant hydrogen loss. At this time attempts were made to improve the carbon monoxide conversion so that methanation could be used more economically as a

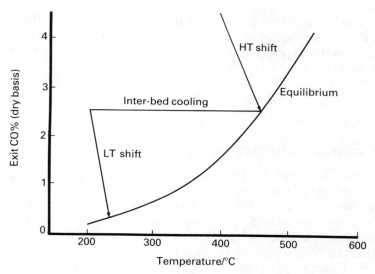

Figure 6.2. Typical variations of carbon monoxide levels in HT and LT shift catalyst beds.

means of removing the remaining traces of carbon oxides. It was this impetus that led to the development and introduction of copper-based shift catalysts in the early 1960s.

A significant improvement in the conversion of carbon monoxide could then be obtained with two-bed operation, with the second bed operating at the lowest possible inlet temperature, which in practice was dictated by the dew-point (about 200°C). The conventional iron-based catalysts are not sufficiently active for such low-temperature operation, but copper-based water-gas shift catalysts are, and when they are used in the second bed, carbon monoxide exit concentrations as low as 0.1–0.3% can be achieved, which is an acceptable economic level for subsequent methanation. These copper-based catalysts are known as *low temperature (LT) shift catalysts*, and typical operating conditions showing carbon monoxide concentrations and temperatures for a two-bed system using HT and LT shift catalysts are shown in Figure 6.2. In this arrangement it is necessary to lower the temperature of the process gas at the exit of the HT shift converter, where it generally exceeds 400°C, to a temperature of around 200°C which is suitable for the inlet of the LT shift catalyst. Inter-bed cooling is usually achieved by heat exchange which, depending on the steam pressure levels in the plant, may be used to heat boiler feedwater or to raise steam. In some cases the temperature may also be trimmed by injecting either steam or condensate into the process gas. In such plants the life of the LT shift catalyst may be shortened because of physical damage from entrained water droplets or because of the presence of catalyst poisons in the water itself.

6.3. Kinetics and Mechanism

6.3.1. Kinetics over HT Shift Catalyst

Numerous studies of the kinetics of the water-gas shift reaction over iron oxide/chromia catalysts have been reported during the past 30 years, and more than 20 different kinetic equations have been proposed (see selected examples in Table 6.3). Differences among authors have been especially marked in their opinion of the reaction mechanism, and the effect of pressure on the reaction rate. The reasons for this conflict have been attributed to the presence of impurities[143] in the gases used, to varying degrees of mass-transfer limitation[144] and to the fact that kinetic measurements have been mostly obtained with integral rather than differential reactors which were often only operating at or near atmospheric pressure. Fott et al.[145] asserted that the effect of pressure on the reaction rate is the main criterion which enables the suitability of a given kinetic equation to be evaluated. Their data, obtained using a differential reactor operating up to 10 bar pressure, showed an increase in rate with pressure up to about 5 bar, after which further increases in pressure had little further effect on the reaction rate.

The five main classes of reaction models, together with their associated kinetic expressions that have been proposed for the water-gas shift reaction over iron/chromia catalyst, are shown in Table 6.3. Podolski and Kim[146] carried out experiments in a recycle reactor and used statistical techniques to discriminate between rival models. They also re-examined some of the previously published data, and concluded that only Langmuir–Hinshelwood and the power-law models could adequately accommodate all of the experimental results. The importance of mass-transfer effects in the reaction over a commercial catalyst was assessed in a more recent publication,[147–149] and the conclusions obtained were:

1. The activation energy in the absence of diffusion effects for the water-gas shift reaction over iron-oxide/chromia catalyst is 121.8 kJ mol^{-1}.

2. The reaction over 5.4 × 3.6 mm pellets is not pore-diffusion limited

at temperatures below 370°C and at pressures up to 31 bar. Over larger, 8.5 × 10.5 mm, pellets the reaction becomes diffusion limited at temperatures above 350°C at 31 bar pressure.

As a result of this work a method of optimizing HT shift converters was the subject of a patent application.[150]

Table 6.3. Rate equations for the water-gas shift reaction over iron oxide/chromia catalysts

Model type	Kinetic equation	References
Langmuir–Hinshelwood	$r = \dfrac{k K_{CO} K_{H_2O} \{[CO][H_2O] - [CO_2][H_2]/K\}}{\{1 + K_{CO}[CO] + K_{H_2O}[H_2O] + K_{CO_2}[CO_2] + K_{H_2}[H_2]\}^2}$	145-149
Hulburt–Vasan	$r = \dfrac{k[H_2O]}{1 + K[H_2O]/[H_2]}$	172
Kodama	$r = \dfrac{k\{[CO][H_2O] - [CO_2][H_2]/K\}}{1 + K_1[CO] + K_2[H_2O] + K_3[CO_2] + K_4[H_2]}$	173
Oxidation–reduction	$r = \dfrac{k_1 k_2 \{[CO][H_2O] - [CO_2][H_2]/K\}}{k_1[CO] + k_2[H_2O] + k_{-1}[CO_2] + k_{-2}[H_2]}$	155, 156, 174
	$r = \dfrac{k_1 K' \{[CO][H_2O] - [CO_2][H_2]/K\}}{K'[H_2O] + [CO_2]}$	
Power law	$r = ak[CO]^m[H_2O]^n[CO_2]^p[H_2]^q$	143, 146, 155

6.3.2. Kinetics over LT Shift Catalyst

Fewer publications have appeared on the mechanism and kinetics of the water-gas shift reaction over copper-based catalysts than over high-temperature iron-based catalysts, but similar types of kinetic expressions have been proposed. Indeed, the pore-diffusion limited version of the Langmuir–Hinshelwood equation (5) for copper-based shift catalysts is consistent with plant data and semi-technical scale results.[151] There is little doubt that the copper LT shift catalysts generally operate in a pore-diffusion limited regime, although there is debate over the extent of this limitation on typical industrial catalyst pellets. In addition they are often self-guarding with respect to poisons (sulphur and chlorine compounds), the pick-up of which is evidently a very rapid process and may itself also be pore-diffusion limited (see Section 6.6.6).

$$\text{rate} = \frac{k[CO][H_2O]^{1/2}(1 - [CO_2][H_2]/K[CO][H_2O])}{P^{1/2}(1 + C_1[CO] + C_2[H_2O] + \ldots)} \quad (5)$$

6.3.3. Mechanism of the Catalytic Water-gas Shift Reaction

The mechanism of the catalysed shift reaction remains in dispute for both copper- and iron-based catalysts. Briefly, two types of mechanism have been proposed—adsorptive and regenerative. In the former mechanism reactants adsorb on the catalyst surface, where they react to form surface intermediates such as formates, followed by decomposition to products and desorption from the surface. In the regenerative mechanism the surface undergoes successive oxidation and reduction cycles by water and carbon monoxide, respectively, to form the corresponding hydrogen and carbon dioxide products of the water-gas shift reaction. Dealing first with iron oxide/chromia catalysts, the adsorption mechanism has been supported by tracer studies and apparent stoichiometric number analyses.[152-154] Unfortunately, however, the kinetics of the water-gas shift reaction can equally well be described by either the adsorptive mechanism[146] or by the regenerative mechanism[155] which was first proposed[156] for these catalysts by Kul'kova and Temken in 1949. Support for the regenerative mechanism came from the work of Boreskov et al.[157] who showed the rate at which water oxidizes the magnetite surface and carbon monoxide reduces it, corresponds to the rate of the water-gas shift reaction. Recently an *in situ* gravimetric study[158] of oxygen removal from and incorporation into magnetite/chromia catalysts in CO_2/CO and H_2O/H_2 gas mixtures at about 350°C further supported the regenerative mechanism.

For the copper-based LT shift catalysts the regenerative mechanism was rejected by van Herwijnen and co-workers[159, 160] on the basis that neither cupric nor cuprous oxide could be formed under reaction conditions from copper metal and steam. They proposed an adsorptive mechanism with a surface formate intermediate. However, an analysis of transient kinetics of the shift reaction over a Cu/ZnO catalyst by Fiolitakis and Hofmann[161] supports the regenerative mechanism. More-recent work[162] with both $Cu/ZnO/Al_2O_3$ catalysts and unsupported polycrystalline copper catalysts has also produced results consistent with the regenerative mechanism. The occurrence of the four separate steps needed for both forward and reverse reaction shown in equations (6) and (7) (in which S is a vacant site) has been demon-

$$CO + O_{(a)} \rightleftharpoons CO_2 + S \qquad (6)$$

$$H_2O + S \rightleftharpoons H_2 + O_{(a)} \qquad (7)$$

strated on copper. For oxygen coverage of half-monolayer or less there is no equilibrium limitation[163] to reaction (7), in contrast with the equivalent reaction forming cuprous oxide, so that van Herwijnen's objection to the regenerative mechanism is invalid for adsorbed oxygen formation. It therefore seems likely that the water-gas shift reaction proceeds by the same regenerative mechanism on both Fe_3O_4/Cr_2O_3 and $Cu/ZnO/Al_2O_3$ catalysts.

6.4. Converter Design

A wide range of reactor designs and flowsheet configurations are possible for the water-gas shift stage in ammonia and hydrogen plants. In practice, however, the process is invariably carried out using two simple adiabatic beds containing different catalysts, operating at different temperatures. This section outlines some of the practical considerations associated with converters for these duties.

As with all fixed-bed converters, manholes should be provided to allow access for inspection and to permit safe charging and discharging of the catalyst. There should be a well-designed gas distributor at the converter inlet to reduce the space needed above the catalyst to a minimum and to enable the minimum size of vessel to be used. It must also ensure that the incoming gas does not disturb the catalyst bed, which would result in attrition of the pellets and cause a high pressure drop across the bed. The gas collection system should also be designed carefully to minimize pressure drop, and to avoid poor gas distribution through the bed. A perforated plate or mesh box is usually secured over the outlet pipe and the space in the bottom of the converter is charged with a bed of graded alumina balls. Often a mesh consisting of several bolted segments may then be placed on top of the balls. This arrangement avoids the use of steelwork in the base of the converter, and reduces the number of internal attachments which, in time, could give rise to cracks at points of high stress. Once the catalyst has been charged it is common practice to add a further layer of fused alumina in the form of lumps, which in some cases is covered by a mesh or grid. This not only avoids attrition of the catalyst if there are problems with gas distribution, but it also catches solids carried over from upstream operations. Figure 6.3 shows the arrangement of the HT shift converter on one of ICI's ammonia plants in the UK.

Both the HT and LT shift catalyst beds should have thermocouples in the bed which will allow monitoring of temperature gradients during catalyst reduction and normal operation. These can be inside a thermosheath inserted through the top of the reactor, as illustrated in Figure 6.3. One profile is normally sufficient for the HT shift bed, but it is advisable to have two thermosheaths in the LT shift vessel. This

Chapter 6. The Water-gas Shift Reaction

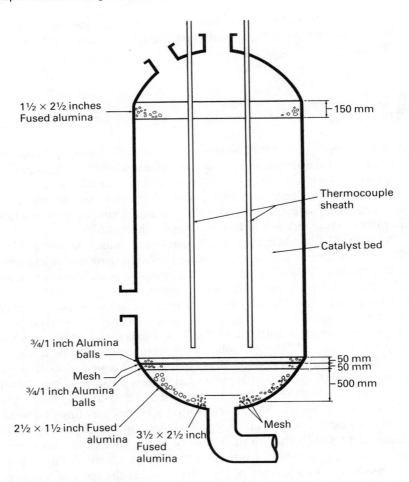

Figure 6.3. Typical arrangement of shift converters.

permits two temperature profiles in different parts of the bed to be recorded, which can provide useful additional information. For instance, a difference in the two profiles provides evidence of maldistribution of gas through the bed. The LT shift vessel must also be provided with inlet and outlet isolation valves, and a bypass to permit isolation from the process gas stream at start-up and shutdown. A nitrogen supply and vessel vent must also be provided to allow the vessel to be purged free from steam before cooling down in order to prevent condensation in the catalyst. A positive pressure of nitrogen is also used to keep air out of the catalyst bed when the vessel is opened. However, care must be taken never to allow free entry of personnel into the nitrogen-filled LT shift vessel, as discussed in Chapter 3.

6.5. High-temperature Shift

The catalytic water-gas shift reaction operating at relatively high temperatures has now been used for more than 70 years to maximize hydrogen efficiency in synthesis-gas production at pressures ranging from atmospheric pressure up to 80 bar, and with feed gases containing 3–75% of carbon monoxide. Under operating conditions the catalyst used for this duty consists of magnetite crystallites stabilized with the addition of a small amount of chromia, although fresh unreduced catalyst contains haematite (Fe_2O_3) 90–95% with 5–10% Cr_2O_3 depending on the manufacturer (see Section 6.5.1). Catalysts have also been operated with a wide range of steam/dry gas ratios, and with impurity levels which have often made reliable operation almost impossible. Fortunately, iron-based catalysts are not very sensitive to poisoning by sulphur which is commonly present in feed gas derived from coal or fuel oil, and if the catalyst must be operated under highly sulphiding conditions the deactivating effect can be compensated for by increasing the volume of catalyst used.

6.5.1. High Temperature Shift Catalyst Formulation

There has been little change in the apparent composition of catalysts used for the high temperature carbon monoxide conversion reaction since the process was first used commercially by BASF in 1915; however, there have been marked formulation improvements to provide the stability needed for the more severe conditions of high pressures and temperatures encountered in modern plants. Methods of manufacture have changed to give lower sulphur concentrations in the product, and the particle shape has been optimized to give improved activity. The analysis and physical properties of typical HT shift catalysts are given in Table 6.4. Magnetite (Fe_3O_4) has good activity for the reaction at moderately high temperatures and, provided the magnetite crystallites are stabilized by the addition of a refractory oxide component such as chromia (Cr_2O_3), satisfactorily long lives are obtained. During operation of the early coal-based ammonia plants at low pressures catalyst operating lives of up to 15 years were common! Today these catalysts are usually produced by precipitation methods using a solution of iron sulphate (which is readily available from the aluminium, iron and steel industries) and sodium carbonate. It is important that the precipitate so formed is washed very carefully to remove traces of residual sulphate which may subsequently be converted to hydrogen sulphide during reduction. In the early

Table 6.4. Typical analyses and physical properties of HT shift catalysts.

Property	Catalyst A	Catalyst B	Catalyst C
Diameter/mm	9.5	5.4	9.5
Height/mm	4.8	3.6	4.9
Bulk density/kg l^{-1}	1.25	1.08	1.12
BET surface area/m^{2}g^{-1}	120	61	80
Pore volume/cm^{3}g^{-1}	0.22	0.35	0.26
Fe$_2$O$_3$/%	89	90	87
Cr$_2$O$_3$/%	9	9	10

coal-based plants this washing step was not very critical since the process gas itself could contain concentrations of sulphur compounds in excess of 100 ppm of sulphur and commercial catalysts would commonly contain more than 1% by weight of residual sulphate. In modern plants, where operation depends on several sulphur-sensitive catalysts, it is desirable to have almost sulphur-free HT shift catalyst, which avoids the formation of hydrogen sulphide. This is not difficult to achieve, and modern catalysts made by carefully controlled procedures contain less than 0.1% by weight of residual sulphur, generally as a simple sulphate, that is rapidly converted to hydrogen sulphide during the reductive activation of the catalyst. The process gas containing the relatively small amount of hydrogen sulphide is usually vented until it reaches an acceptably low concentration. However, these modified production techniques do lead to low-density product that may lose a significant amount of strength during normal operation. Low-density catalyst can also be susceptible to damage by water, which may result in some disintegration of the particles, causing gradual increase of pressure drop across the converter. These difficulties have been overcome by the introduction of new manufacturing procedures that involve improved pelleting technology, and the production of optimum catalyst sizes for particular duties (see Section 6.6.3).

Following precipitation of the iron and chromium components, and after the washing and drying steps, the resulting product must be calcined at a temperature which will convert the basic carbonates to haematite (Fe$_2$O$_3$). This step is important because it must be carefully controlled to avoid the formation of large quantities of chromic oxide (CrO$_3$) from the air-oxidation of chromia. This apparently simple reaction is shown in equation (8).

$$2Cr_2O_3 + 3O_2 \rightarrow 4CrO_3 \qquad (8)$$

6.5. High-temperature Shift

The presence of Cr(VI) compounds in HT catalysts is undesirable not only because of the potential health hazard to operators who must handle the catalyst, but also because of possible operating problems as the catalyst is reduced. Hexavalent chromium is soluble and, if condensation takes place in the reactor on unreduced catalyst, then leaching of chromium may occur. In addition the exothermic conversion of hexavalent chromium to trivalent chromium during the reduction procedure during plant commissioning leads to considerable evolution of heat and (see Section 6.5.3) large increases in bed temperatures.

6.5.2. Diffusion Effects and Pellet Size

Good pellet strength is essential to achieve long operating life from charges of HT shift catalyst, and there is generally a strong correlation between strength and pellet density. In the absence of any mass-transfer limitations on reaction rates, catalyst activity per unit volume of catalyst bed is also directly related to pellet density although, in the case of HT shift catalyst, for commercially sized pellets operating at typical plant pressures a pore-diffusion limitation becomes increasingly significant at temperatures of 350°C and above.[147] In the pore-diffusion limited regime the effect of pellet density on the activity per unit volume of catalyst bed is reduced and, over the practical range of pellet densities, apparent activity is almost constant. Selecting an appropriate pellet density then becomes a balance between strength and cost.

In any mass-transfer limited regime the pellet dimensions influence catalyst activity, as well as gas distribution and pressure drop. Catalyst strength is also affected by the pellet size, and this can in turn influence the life of a catalyst charge. It is thus theoretically possible to optimize pellet sizes for any given catalytic duty. In the following two sections the effect of pellet size on activity and pressure drop are considered in relation to HT shift catalyst.

6.5.2.1. Effect of pellet size on activity

When a reaction is highly pore-diffusion limited, only the outer surface of the catalyst pellet is used by the reactants. In fact, the effectiveness of the catalyst becomes inversely proportional to the pellet radius (for a fixed pellet height/radius ratio). Activity measurements on different sized ICI HT shift catalysts (15-4, 8.5 × 10.5 mm; 15-5, 5.4 × 3.6 mm) at 31 bar pressure over the temperature range 310–450°C as shown in Table 6.5, established typical extents of pore-diffusion limitations.[147] At 350°C or less there is little diffusion limitation with either pellet size, but at 400°C or more there is an increasingly severe diffusion limitation with both. From the foregoing it is clear that there is an advantage from the

Table 6.5. Effectiveness and intrinsic activity of HT shift catalyst operating at 31 bar over the temperature range 310–450°C

Temperature/°C	Catalyst[a]	Observed rate constant/ $cm^3\,g^{-1}\,s^{-1}$	Effectiveness	Relative activities of ICI Catalysts 15-4/15-5	Intrinsic rate constant/ $cm^3\,g^{-1}\,s^{-1}$
310	15-5	0.013	0.99		0.013
				0.98	
	15-4	0.013	0.97		
330	15-5	0.033	0.98		0.035
				0.96	
	15-4	0.034	0.94		
350	15-5	0.071	0.96		0.080
				0.93	
	15-4	0.076	0.89		
370	15-5	0.146	0.90		0.166
				0.72	
	15-4	0.108	0.65		
400	15-5	0.311	0.70		0.444
				0.63	
	15-4	0.202	0.44		0.459
425	15-5	0.469	0.53		0.885
				0.59	
	15-4	0.282	0.31		0.910
450	15-5	0.827	0.45		1.86
				0.53	
	15-4	0.445	0.24		1.85

[a] Dimensions of ICI Catalyst 15-4: 8.5 × 10.5 mm; ICI Catalyst 15-5: 5.4 × 3.6 mm.

point of view of activity to use the smaller-sized pellets, since this would give rise to a smaller catalyst bed and hence a smaller converter. However, before the catalyst volume can be optimized the pressure drop must also be considered.

6.5.2.2. Effect of pellet size on pressure drop

The pressure drop through a bed of catalyst pellets is determined by the bed geometry and voidage. The size and shape of pellets thus affects the pressure drop and, generally, long thin pellets or short fat pellets are preferable to nearly square ones in terms of pressure drop. Design catalyst volumes decrease as the pressure is increased, being approximately inversely proportional to the square root of the pressure for a pore-diffusion limited reaction.

At low pressures larger volumes of catalyst are therefore required and the pressure drop per unit volume is fairly high, so that pressure drops

6.5. High-temperature Shift

are high for low-pressure operation but become less important as the pressure is raised above about 10 bar. Of much greater importance, however, is the manner in which the pressure drop builds up as a result of fragmentation of the catalyst during use. In most modern ammonia plants the HT shift converter is downstream of several heat exchangers, and whenever there is mechanical failure in these units condensate or steam leaks into the process stream and impinges on the catalyst. This can lead to break-up of the pellets as well as to carry-over of scale which is deposited on the top of the HT shift bed.

Both phenomena lead to increased pressure drop. This is illustrated in Figure 6.4, which shows an extreme example of the effect of a heat-exchanger problem on HT shift catalyst over an extended time. The leak occurred from a waste heat boiler and on several occasions was as high as 60 tonnes h^{-1} for protracted periods. Efforts to plug leaking tubes were only partially successful and did not eliminate the leak completely, and for eight months the catalyst was subject to water impingement. The data show that for the first two months, when the leak was massive (60 tonnes h^{-1}), the pressure drop began to increase, presumably due to the presence of liquid water and perhaps some

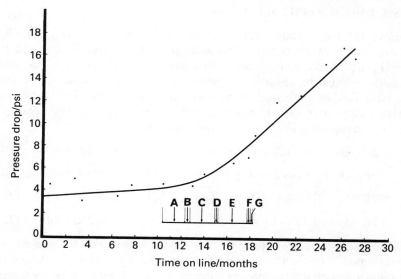

Figure 6.4. Plant operating data: increase in pressure drop over the HT shift converter as a result of a steam leak. During time A after 10 months on-line there was a massive and persistent boiler leak. Attempts to repair the boiler during time B were unsuccessful and the leak persisted during time C at a smaller rate. During time D there was another massive leak and the boiler had to be repaired again. A small leak persisted during time E and was followed by yet another massive fracture during time F. The boiler was replaced during time G.

catalyst fragmentation. Attempts to repair the boiler were unsuccesful and eventually it was replaced. By this time the catalyst had been so badly maltreated that the pressure drop continued to increase. Eventually output was affected, and the catalyst had to be changed after 27 months on-line.

High pellet strength is therefore essential in achieving a useful HT shift catalyst life. Experience has shown that large pellets can usually withstand maloperation better than smaller ones can. However, provided adequate strength and resistance to attrition can be maintained there are benefits in using smaller pellets in pore-diffusion limited reactions. While smaller pellets attract a higher pressure drop per given volume of bed, the converter can be designed to achieve the same pressure drop as would occur with the necessary larger volume of larger pellets. Because of these considerations HT shift catalyst is supplied by ICI in two sizes that are chemically and structually identical. The smaller pellet size combines a high level of activity with conventional pressure-drop characteristics, whereas the larger size possesses the same activity as other typical HT shift catalysts, but gives an exceptionally low pressure drop.

6.5.3. Reduction of HT Shift Catalyst

Before HT shift catalyst can be used the haematite (Fe_2O_3) must be converted to magnetite (Fe_3O_4) and any CrO_3 present converted to Cr_2O_3. The reduction is invariably carried out with process gas, and conditions must be chosen that permit the required reactions to occur without further reduction to metallic iron, since this would promote methanation and carbon monoxide disproportionation. The required reduction reactions are shown in equation (9)-(12).

$$3Fe_2O_3 + H_2 \rightarrow 2Fe_3O_4 + H_2O \quad \triangle H = -16.3 \text{ kJ mol}^{-1} \quad (9)$$

$$3Fe_2O_3 + CO \rightarrow 2Fe_3O_4 + CO_2 \quad \triangle H = +24.8 \text{ kJ mol}^{-1} \quad (10)$$

$$2CrO_3 + 3H_2 \rightarrow Cr_2O_3 + 3H_2O \quad \triangle H = -684.7 \text{ kJ mol}^{-1} \quad (11)$$

$$2CrO_3 + 3CO \rightarrow Cr_2O_3 + 3CO_2 \quad \triangle H = -808.2 \text{ kJ mol}^{-1} \quad (12)$$

The equilibrium between the Fe_2O_3 and Fe_3O_4 phases is determined[164] by the ratios of H_2O/H_2 and CO_2/CO. At 450°C the gas in equilibrium with the two phases contains 96% H_2O and 4% H_2 (reaction (9)) and 99.5% CO_2 and 0.5% CO (reaction (10)). With the usual compositions encountered with process gas the environment is more reducing than this, and consequently the stable phase is Fe_3O_4. If steam *alone* is passed over the reduced catalyst at normal operating temperatures, then Fe_3O_4 is slowly oxidized to Fe_2O_3 (see Section

6.5. High-temperature Shift

6.5.6). Under some conditions further reactions shown in equations (13)–(16) are possible. Although FeO is only stable above 565°C, while in the range 300–565°C the only stable phases are metallic iron and magnetite.[165, 166] At 400°C Fe_3O_4 is the stable phase when the H_2O/H_2 ratio exceeds 0.09 or the CO_2/CO ratio exceeds 1.16, while at 550°C the ratios are 0.28 and 1.0, respectively. Since these ratios are exceeded with normal plant conditions, the stable phase is Fe_3O_4, and neither FeO nor metallic iron are formed.

$$Fe_3O_4 + H_2 \rightarrow 3FeO + H_2O \qquad \triangle H = -63.8 \text{ kJ mol}^{-1} \qquad (13)$$

$$Fe_3O_4 + CO \rightarrow 3FeO + CO_2 \qquad \triangle H = -22.6 \text{ kJ mol}^{-1} \qquad (14)$$

$$FeO + H_2 \rightarrow Fe + H_2O \qquad \triangle H = -24.5 \text{ kJ mol}^{-1} \qquad (15)$$

$$FeO + CO \rightarrow Fe + CO_2 \qquad \triangle H = +12.6 \text{ kJ mol}^{-1} \qquad (16)$$

It is clear that neither pure hydrogen nor a hydrogen/nitrogen mixture should be used to reduce HT shift catalyst, or exothermic reduction to metallic iron as in equation (17) will occur. Although not normally encountered, the presence of metallic iron in the operating catalyst would promote the strongly exothermic formation of methane as in equation (18), together with other hydrocarbons, and probably also carbon monoxide disproportionation as shown in equation (19). Related problems that may be encountered when operating at low steam ratios are discussed in Section 6.7.2.

$$Fe_3O_4 + 4H_2 \rightarrow 3Fe + 4H_2O \qquad \triangle H = -149.4 \text{ kJ mol}^{-1} \qquad (17)$$

$$CO + 3H_2 \rightarrow CH_4 + H_2O \qquad \triangle H = -206.2 \text{ kJ mol}^{-1} \qquad (18)$$

$$2CO \rightarrow C + CO_2 \qquad \triangle H = -172.5 \text{ kJ mol}^{-1} \qquad (19)$$

Therefore steam must always be present during reduction of HT shift catalyst, and the thermodynamically minimum steam/hydrogen ratio at different temperatures is displayed in Figure 6.5. Two curves are shown, one giving the conditions necessary for the reduction of Fe_3O_4 to FeO, and the other giving conditions necessary for the reduction of FeO to metallic iron. These conditions are to be avoided.

Reduction of HT shift catalyst is most conveniently done during the reformer start-up, as discussed in Chapter 3. Whenever possible the initial temperature of the HT shift catalyst bed should be high enough to avoid condensation of steam, since liquid water could wash out any chromate and other soluble impurities that may be present. Although most commercial catalysts are strong enough not to disintegrate when wetted, problems may be experienced if dust in the bed is agglomerated by water to form a cake that can then interfere with the flow distribution of process gas. The converter should first be purged free from air with an inert gas such as nitrogen, and if possible heated above the condensation temperature. Provided prudent precautions are observed,

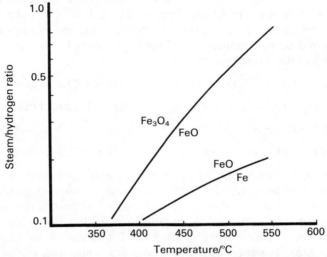

Figure 6.5. Minimum ratio of steam to hydrogen for reduction of conventional HT shift catalysts. Fe_2O_3 becomes stable with respect to Fe_3O_4 at $H_2O/H_2 > 2 \times 10^3$ at ~550°C and about half this value at ~400°C.

it is possible that with care preliminary heating can be carried out with air, saturated hydrocarbon, superheated steam and even carbon dioxide. Regardless of the method used to heat the catalyst, it is important to observe the following:

(a) purge the converter free from air before introducing reduced gas;

(b) maintain adequate gas flow to ensure good distribution;

(c) avoid sudden changes in pressure and flow rate;

(d) observe limits on rate of catalyst warm-up;

(e) avoid the presence of particulate matter in the reducing gas.

Once the catalyst has been heated, process gas or hydrogen along with steam should be introduced, and the temperature raised to the operating level. Although it is preferable to avoid condensation on the catalyst, this may not always be possible when hot process gas is used to heat the catalyst. Some condensation occurs at first, and the condensate must be drained from the reactor until the catalyst is above the temperature at which evaporation takes place. At this point a plateau will be seen on the catalyst-bed temperature profile. Rapid evaporation of water within the charge is to be avoided because the pellets could disintegrate, and a high pressure drop ensue. The catalyst reduction starts at about 150°C, although complete reduction will probably not be achieved until the normal operating inlet temperature to the HT shift converter of about 400°C is achieved. Particular care is needed to ensure that the process gas used to carry out the reduction contains an adequate amount of steam because, as discussed above, in the absence of water conditions favour reaction (17) at the converter inlet, which leads to the formation of metallic iron.

The actual heat of reduction of new catalyst varies with the hydrogen/carbon monoxide ratio in the reducing gas. For reduction of ferric oxide (reactions (9) and (10)) the heat evolved is small or possibly even negative. The heat of reduction of CrO_3 (shown in equations (11) and (12)) is far more significant. The heat release for 1000 kg of catalyst is 33.5–37.5 MJ (based on 1% CrO_3 present), and corresponds to a temperature rise of about 40°C, and for different CrO_3 levels the temperature rise varies in proportion. However, this is modified by the heat being partly absorbed by the colder catalyst. The heat of reduction of reoxidized catalyst is very small, since CrO_3 is not formed under the conditions used to stabilize reduced catalyst. When the reduction of HT shift catalyst is carried out in the presence of carbon monoxide a much larger increase in temperature can be expected as the shift reaction begins to take place. This will cause a temperature rise of approximately 13.5°C per 1% of carbon monoxide in the process gas. Thus, the carbon monoxide content should not exceed that which will cause the catalyst temperature to rise above 500°C as the reduction proceeds.

Residual sulphate in HT shift catalyst is converted to hydrogen sulphide during reduction when catalyst-bed temperatures are in the range 350–400°C. It is therefore necessary to maintain the catalyst at these temperatures for a period long enough for the sulphate to be completely reduced. The sulphate content of most modern catalysts is usually less than 0.1% sulphur, to limit the time taken to commission the catalyst, and also to limit the effect of sulphur poisoning on downstream catalysts. As a general rule, the desulphurization period required is less than 15 hours from the time when the HT shift catalyst is first exposed to hot process gas, as shown in Figure 6.6. As soon as the hydrogen sulphide concentration in the gas at the exit is at a level sufficient to avoid poisoning the LT shift catalyst, which for some catalyst may be as high as 5 ppm, the LT shift converter may also be commissioned.

It was not normally necessary to desulphurize HT shift catalysts to this extent when coal or fuel oil were used as feedstocks, because there were no sulphur-sensitive catalysts immediately downstream of the shift

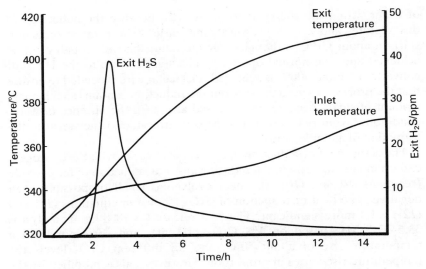

Figure 6.6. Evolution of hydrogen sulphide during commissioning of a typical charge of conventional HT shift catalyst.

vessel, and sulphur was normally removed from the process gas in the scrubbing processes used to remove carbon oxides. As a result HT shift catalysts containing more than 1% of sulphur (some of which was in the form of complex sulphates) were used, and it could take several weeks to convert all of the sulphate they contained to hydrogen sulphide. These catalysts needed a much higher reduction temperature than modern types, but despite the disadvantages of high sulphur impurity levels, these old catalysts were extremely strong and could be operated over prolonged periods.

A recent development has been the introduction of an ultra-low sulphur content HT shift catalyst (Katalco 77-1) which is produced from essentially sulphur-free raw materials by a quite different manufacturing route from conventional "sulphate-based" products. With this catalyst the start-up procedure is restricted to a reduction period which normally takes only 4–8 hours.

6.5.4. Operation of HT Shift Catalyst

In the early coal-based ammonia plant steam/dry gas ratios were often as high as 1.2 with operation at just above atmospheric pressure. Because of the relatively high concentration of carbon monoxide (around 30%) the amount of heat liberated was large, and it was necessary to use two beds of catalyst with inter-bed cooling. Inlet temperatures to these beds were typically 400 and 450°C, with

6.5. High-temperature Shift

temperature rises around 50–100°C. In spite of the high temperatures the duty was light compared with modern operating conditions employing pressures up to 35 bar.

In the design of a modern HT shift converter common practice is to use the lowest practicable inlet temperature consistent with economic operation at the steam ratio demanded by overall plant design. Although catalyst volume could be decreased if higher temperatures were used, this would require more expensive construction materials to be used. Consequently, in practice the usual inlet temperature is in the range 340–360°C. The catalyst should be operated at the minimum practical inlet temperature that affords the desired conversion of carbon monoxide, since this results in an increased life of the catalyst by limiting sintering. Magnetite crystallites are not completely stabilized by the chromium oxide and, as a result, during its life the catalyst gradually loses activity as the surface area decreases due to thermal sintering. This is exacerbated at high temperatures, particularly above 400°C. A consequence of this is that new catalyst will operate satisfactorily at a lower inlet temperatures than older, slightly deactivated catalyst. As the catalyst ages it is necessary to increase the converter inlet temperature in order to maintain a satisfactory exit carbon monoxide level, and the temperature profile through the bed typically changes with time, as shown in Figure 6.7. Eventually the overall carbon monoxide conversion decreases, and the carbon monoxide at the exit runs to a level at which the catalyst has to be changed. During operation,

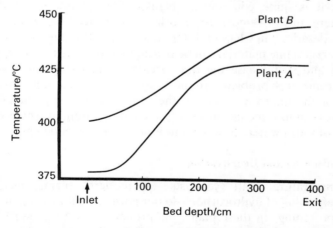

Figure 6.7. Temperature profiles through the HT shift catalyst in a modern Western European ammonia plant: A was obtained after 45 days of operation on-line, and B after 842 days. Operating exit percentage of CO: A, 2%; B, 3%.

303

therefore, care should be taken not to raise inlet temperatures unnecessarily, since this may result in some loss of low-temperature activity and a somewhat shortened economic life of the charge.

The usual lifetime of a charge of HT shift catalyst in an ammonia plant is from 3 to 5 years, depending upon how it has been treated, although there are examples of lives of up to 10 years. End of life can be indicated by a rise in pressure drop caused by catalyst breakage or an accumulation of deposits (see Section 6.5.2.2). More often it is as a result of decreased activity which, during the life of the charge, is compensated for by increasing the temperature of the gas entering the converter, as shown in Figure 6.7. Recent experimental work[148] suggests that there are two stages in the decay of activity: an initial fast fall to a more stable point from which the rate of decay is significantly less. In this stage the loss in activity correlates with loss of surface area and fits conventional sintering kinetic equations.

High-temperature shift catalysts are generally robust and can withstand a good deal of maltreatment. However, during use the catalyst may be affected by a number of factors other than thermal sintering. Because upstream processes are operating at high temperatures there is usually a gradual deposition of steam-volatile components on the top of the bed. In severe cases these deposits can eventually prevent the passage of gas into the catalyst particles and interfere with the overall gas distribution through the bed. This may be associated with an increase in pressure drop through the bed. When this happens it is quite common to blanket the catalyst in a nitrogen atmosphere and to remove the top layer of catalyst using one of the methods described in Chapter 3. Often a layer of Raschig rings made of some inert ceramic material can be placed on top of the bed to trap solid material and so increase the time a charge can be operated without encountering this problem. If the vessel design does not allow easy access for the removal of part of the charge in this way, then soluble impurities can in some instances be removed by careful washing with an up-flow of warm water, although this is not a recommended procedure.

6.5.5. Poisoning and Deactivation

High temperature shift catalysts in modern plants based on steam reforming of hydrocarbons do not normally suffer from problems due to poisoning. In these plants the primary reforming, and LT shift catalysts are much more susceptible to poisons than the iron/chrome HT catalyst, and so care is taken to ensure that feedstocks are virtually poison-free. However, if for some reason hydrogen sulphide is present in the process gas in large amounts this will not seriously affect activity of the HT shift catalyst which, depending on the concentration of hydrogen sulphide, may begin to sulphide. Since the sulphiding reaction is reversible according to equation (20) when the hydrogen sulphide level in the process gas is decreased any sulphiding will be reversed and hydrogen sulphide will slowly evolve from the catalyst. Approximate values of the equilibrium constant for reaction (20) over a range of temperatures are given in Table 6.6.

$$Fe_3O_4 + 3H_2S + H_2 \rightarrow 3FeS + 4H_2O \qquad \Delta H = -75.0 \text{ kJ mol}^{-1} \qquad (20)$$

Table 6.6. Variation of equilibrium constant with temperature for the reaction $Fe_3O_4 + 3H_2S + H_2 \rightleftharpoons 3FeS + 4H_2O$

Temperature/°C	K_p[a]
300	3.22×10^{-10}
350	8.70×10^{-10}
400	2.11×10^{-9}
450	4.58×10^{-9}
500	9.32×10^{-9}
550	1.82×10^{-8}

[a] $K_p = P^4_{H_2O}/P_{H_2}P^3_{H_2S}$

As with all ammonia plant catalysts, the HT shift catalyst can be affected by halides, but they are rarely encountered at sufficiently high levels in feedstocks to cause major problems, unless they have been introduced by use of chlorinated solvents for degreasing during a maintenance shutdown. Arsenic compounds in small concentrations could poison the HT shift catalyst, but normally they are unlikely to be present, unless they could arise inadvertently from a carbon dioxide removal unit, for example through a common compressed air or nitrogen system. Phosphorous compounds or silica, which will cause deactivation by fouling, may be introduced into the reactor from upstream equipment, or as the result of a boiler failure.

High temperature shift reactors in coal-based plants, or those which follow cyclic reformers and operate with feed gases containing traces of unsaturated hydrocarbons and nitric oxide, can become contaminated with gum containing a high proportion of carbon. This deposits on the catalyst and prevents access of the gas into the catalyst pellets. Carbon formed in partial oxidation processes that is carried into the HT shift converter can also block the catalyst pores. These effects can be mitigated by installing a separate guard bed of HT shift catalyst which can be regenerated. Regeneration can be effective provided the catalyst is not physically damaged by the deposit, and can be done by treating the catalyst with steam at 450°C containing between 1 and 2% of oxygen. The actual oxygen content must be carefully adjusted to ensure that the catalyst is not overheated, and must never be increased beyond 2%. The progress of regeneration can be monitored by observing bed temperatures and analysing the gas at the exit of the converter for carbon dioxide, and the process is complete when this falls to zero.

In coal-based plants the total amount of sulphur compounds, generally H_2S and COS, can be significant. Consequently, the HT shift catalyst becomes sulphided, the Fe_3O_4 being converted to FeS according to equation (20). Any carbonyl sulphide present is converted to hydrogen sulphide according to equation (21) and equilibrium constants for this reaction at several temperatures are given in Table 6.7. Sulphided catalyst does have activity in the water-gas shift reaction, although its activity is only about half that of catalyst containing iron as magnetite. Therefore, in circumstances where the catalyst will be sulphided the volume used has to be doubled compared with that normally employed to obtain satisfactory performance.

$$COS + H_2O \rightarrow CO_2 + H_2S \qquad \triangle H = -34.6 \text{ kJ mol}^{-1} \qquad (21)$$

Table 6.7. Variation of the equilibrium constant with temperature for the reaction $H_2S + CO_2 \rightleftharpoons COS + H_2O$

Temperature/°C	K_p[a]
300	8.35×10^{-4}
350	1.52×10^{-3}
400	2.53×10^{-3}
450	3.55×10^{-3}
500	6.04×10^{-3}
550	8.00×10^{-3}

[a] $K_p = P_{COS}P_{H_2O}/P_{H_2S}P_{CO_2}$

6.5.6. Reoxidation and Discharge

In the reduced magnetite form HT shift catalyst is pyrophoric. The heat released during the oxidation process shown in equation (22) is equivalent to some 420 MJ per 1000 kg of catalyst, and corresponds to an adiabatic temperature rise of about 450°C. As a consequence,

6.5. High-temperature Shift

reduced HT shift catalyst must not be exposed to oxygen except under carefully controlled conditions. Therefore, during plant shutdowns it is usually kept in the reduced state under a reducing or inert atmosphere. As discussed in Chapter 3, when old catalyst is being discharged it may be removed from the converter in the reduced state, provided appropriate methods that prevent rapid oxidation are employed. Cold catalyst can be removed by suction from the vessel under nitrogen, and then deposited in a safe place where it can slowly oxidize. Alternatively, the converter may be filled with water, and the wet catalyst removed from a bottom discharge point, or by suction from the top. In either case it should be remembered that some hydrogen gas may be formed according to reaction (23).

$$2Fe_3O_4 + \tfrac{1}{2}O_2 \rightarrow 3Fe_2O_3 \qquad \triangle H = -464.6 \text{ kJ mol}^{-1} \qquad (22)$$

$$2Fe_3O_4 + H_2O \rightarrow 3Fe_2O_3 + H_2 \qquad (23)$$

There are sometimes situations in which it is desirable to re-use reduced catalyst; for example, after partial discharge. Under these circumstances it is possible to reoxidize the catalyst *in situ* under controlled conditions. Typically these involve cooling the catalyst to 200°C and passing steam containing 1% of air through the bed at a space velocity of about 1000 h^{-1}. The actual amount of air used is adjusted to give a temperature rise in the bed of about 30°C, and reaction is complete when there is no further temperature rise, and oxygen is no longer consumed. If the catalyst has been used with sulphur-containing process gas, then it may be possible to remove much of the sulphur contamination by steaming, as shown in equation (24) until a satisfactory low level of hydrogen sulphide is present in the exit gas. If this is not done, subsequent oxidation of the sulphide according to equation (25) results in damage to the catalyst due to the extremely exothermic nature of this reaction.

$$3FeS + 4H_2O \rightarrow Fe_3O_4 + 3H_2S + H_2 \qquad (24)$$

$$6FeS + 13\tfrac{1}{2}O_2 \rightarrow 2Fe_2(SO_4)_3 + Fe_2O_3 \qquad \triangle H = -5370 \text{ kJ mol}^{-1} \quad (25)$$

Recommissioning reoxidized HT shift catalyst involves the same procedures as that for fresh catalyst, but since no hexavalent chromium is formed during reoxidation the process can usually be carried out rather more quickly. However, the performance in terms of activity and life of such catalyst is rarely comparable with that of fresh catalyst, so overall economics seldom justify reoxidation and re-use of reduced HT shift catalyst. A summary of the reactions that could occur during the life of the HT shift catalyst is given in Table 6.8.

Table 6.8. Summary of reactions that may take place during operation of conventional HT shift catalyst

Reduction in the presence of excess steam

$3Fe_2O_3 + H_2 \rightarrow 2Fe_3O_4 + H_2$ $\quad \Delta H = -16.3 \text{ kJ mol}^{-1}$
$3Fe_2O_3 + CO \rightarrow 2Fe_3O_4 + CO_2$ $\quad \Delta H = 24.8 \text{ kJ mol}^{-1}$
$2CrO_3{}^a + 3H_2 \rightarrow Cr_2O_3 + 3H_2O$ $\quad \Delta H = -648.7 \text{ kJ mol}^{-1}$
$2CrO_3{}^a + 3CO \rightarrow Cr_2O_3 + 3CO_2$ $\quad \Delta H = -808.2 \text{ kJ mol}^{-1}$

Reduction in a deficiency of steam

$Fe_3O_4 + H_2 \rightarrow 3FeO + H_2O$ $\quad \Delta H = -63.8 \text{ kJ mol}^{-1}$
$Fe_3O_4 + CO \rightarrow 3FeO + CO_2$ $\quad \Delta H = -22.6 \text{ kJ mol}^{-1}$

$FeO + H_2 \rightarrow Fe + H_2O$ $\quad \Delta H = -28.5 \text{ kJ mol}^{-1}$
$FeO + CO \rightarrow Fe + CO_2$ $\quad \Delta H = 12.6 \text{ kJ mol}^{-1}$

$Fe_3O_4 + 4H_2 \rightarrow 3Fe + 4H_2O$ $\quad \Delta H = -149.4 \text{ kJ mol}^{-1}$
$Fe_3O_4 + 4CO \rightarrow 3Fe + 4CO_2$ $\quad \Delta H = 15.2 \text{ kJ mol}^{-1}$

Reoxidation

$4Fe_3O_4 + O_2 \rightarrow 6Fe_2O_3$

Sulphiding

$Fe_3O_4 + 3H_2S + H_2 \rightarrow 3FeS + 4H_2O$

Other reactions

$3FeSO_4 + 11H_2 \rightarrow Fe_3O_4 + 8H_2O + 3H_2S$
$FeSO_4 + 5H_2 \rightarrow Fe + 4H_2O + H_2S$

[a] Trace amount formed during the manufacturing process.

6.6. Low-temperature Shift

6.6.1. General

During the 1960s there was a fundamental change in the ammonia industry as operators began to appreciate the economies associated with large production units. It was this situation that gave rise to the large single-stream ammonia plants of about 1000 tonnes day^{-1} or more. Developments in catalyst technology played a vital role in this change, and by using efficient feedstock purification catalysts it was possible to use a range of sulphur-sensitive catalysts for the production of synthesis gas. Among these were the active copper-based water-gas shift catalysts that could operate at relatively low temperatures and so give equilibrium carbon monoxide concentrations of less than 0.3% in the process gas entering the final stage of carbon oxides removal. As a direct consequence of having such low levels of carbon monoxide, it was economic to incorporate a methanation stage in the process in place of the very much more complicated copper liquor scrubbing system that was formerly used. Therefore the operating efficiency in modern large single-stream plants depends heavily on the good performance of LT shift catalyst. They are used on virtually all ammonia and refinery hydrogen plants, and the total volume of catalyst installed is now in excess of 20 000 tonnes.

In situations where a very pure process gas is available, and providing a thermally stable catalyst is used, the carbon monoxide level at the exit from a LT shift converter will remain close to the thermodynamic equilibrium value for many years. However, if poison levels are relatively high the catalyst life can be appreciably shortened. In such a case the high frequency with which the catalyst must be changed, together with the associated loss of production when carbon monoxide conversion is low, will often make operation unacceptably expensive. If operation cannot be made more economic by use of a better LT shift catalyst, then the use of a separate guard vessel as discussed in Section 6.6.8 can be particularly beneficial. This effectively concentrates poisons in a small bed and prevents them from reaching the main catalyst bed. The use of such guard beds provides flexibility in planning the intervals between plant overhauls, and as a result they have become quite widely used.

6.6.2. Low-temperature Shift Catalyst Formulation

The formulation of LT shift catalyst is important in terms of selectivity and resistance to poisoning, as well as activity. Selectivity is important because under LT shift conditions methanation of both carbon monoxide and carbon dioxide is thermodynamically very favourable as shown in equations (26) and (27).

$$CO + 3H_2 \rightarrow CH_4 + H_2O \qquad \Delta H = -206.2 \text{ kJ mol}^{-1} \quad (26)$$

$$CO_2 + 4H_2 \rightarrow CH_4 + 2H_2O \qquad \Delta H = -164.9 \text{ kJ mol}^{-1} \quad (27)$$

If these reactions took place to any extent, then valuable hydrogen would be consumed; in addition, since they are very exothermic they could give rise to dramatic and dangerously destructive rises in temperature. Copper-based catalysts have good activity for the water-gas shift reaction, and have no methanation activity so they are well suited for use in water-gas shift duties. However, because copper catalysts are particularly prone to easy sintering they can only be used in

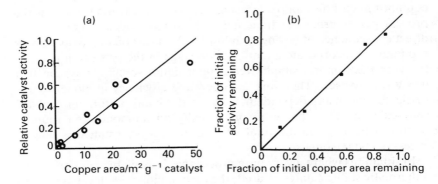

Figure 6.8. Laboratory results showing the dependence of water gas shift activity on copper surface area. a) Atmospheric pressure data for the conventional shift reaction; b) data obtained for the reverse shift reaction at 30 bar.

low-temperature duties.

The ability of supported copper to catalyse the water-gas shift reaction has long been known. Catalysts prepared from mixed oxides of copper and zinc were being demonstrated in the late-1920s, but the problem of producing catalyst with a sufficiently long life to become a commercial proposition was not resolved, due to the high level of poisons in process streams at that time. Later, when much purer synthesis gas was widely available from the steam reforming of hydrocarbons it became possible to use copper-based catalysts, and the first charge[167] of LT shift catalyst was used in the USA during 1963. This was a standard copper oxide/zinc oxide hydrogenation catalyst, based on a 1 : 2 mole ratio of copper to zinc, which when reduced afforded copper crystallites supported on zinc oxide. Experience with early charges of this catalyst showed that the operating life was as short as six months, which was mainly the result of non-optimum formulation and manufacturing procedures, together with the anticipated sensitivity of the catalyst to operating conditions and residual traces of poisons.

Activity results from having a high copper surface area (Figure 6.8 a and b), and this is obtained from the copper being distributed throughout the support material in the form of small crystallites. Sustained activity depends on inhibiting the natural sintering tendency of these crystallites to form progressively larger crystallites having smaller surface area. Experience with HT shift catalyst initially suggested chromia (Cr_2O_3) might stabilize activity by inhibiting sintering.

Unfortunately, this was found to be a poor support for catalysts of this type, and formulations containing chromia did not have significantly improved operating lives. Further experimental work showed alumina

6.6. Low-temperature Shift

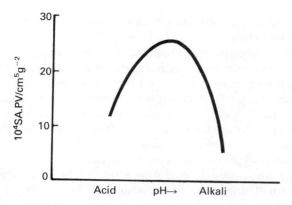

Figure 6.9. Effect of pH during precipitation on the product of copper surface area (SA) and pore volume (PV) for a particular LT shift catalyst.

together with zinc oxide was a much more desirable support that not only could significantly stabilize the copper crystallites against thermally induced sintering, but also enhanced the strength of the catalyst and minimized shrinkage during reduction. The actual ratio of the oxide components for optimum activity and stability, as well as poison resistance (Section 6.6.6) depends on the relative size of the crystallites of the different components, their degree of dispersion and, hence, the method of manufacture. The way in which the catalyst is made also influences the final physical properties such as surface area and pore volume, that in turn affects activity since normal operation usually involves some pore-diffusion limitation. Accordingly, and as with almost all catalysts of this type, simple elemental composition is not a good guide to the likely performance. For instance, catalyst containing high copper levels may display very high initial activity that quite rapidly decays, whereas a formulation with less copper can maintain relatively high activity over prolonged periods.

Up to the time when LT shift catalysts were first developed, manufacture of precipitated catalysts was invariably by batch process in which precipitant was added to a solution of metal salt. During this operation pH does not remain constant, and it was quickly established that the properties of particles precipitated under acid conditions were quite different with respect to both particle size and composition from those formed when the solution was basic. For example, precipitate formed at pH 5 is rich in aluminium and deficient in zinc, while at pH 9 the catalyst is deficient in copper. Catalyst precipitated at pH 7 had the required composition as well as the smallest particle size, and by refining production techniques such precipitate could be produced using a continuous procedure. Catalyst made from this optimum precipitate based on formulations containing copper oxide, zinc oxide and also alumina were introduced during 1965. In catalysts such as ICI Catalyst 52-1 that contain precipitated alumina, both the zinc oxide and the active copper component could be made thermally stable. The resulting catalyst had a relatively low bulk density, reduced impurity levels and a high resistance to sulphur poisoning. Further improvements have subsequently been made so that modern LT shift catalysts such as ICI Catalysts 52-8 and 53-1 are not only more efficient in resisting the effects of sulphur, but are also more resistant to chlorides. (Table 6.9 gives typical chemical and physical properties for these catalysts.) Thus, the success of the continuous manufacturing process is to produce a uniform dispersion of small stable crystallites which, when the copper oxide is reduced to copper metal, significantly resist the combined effects of temperature and poisons. Operating experience has confirmed that for these catalysts the optimum copper oxide content in the unreduced catalyst consistent with these requirements that provides long economic operating life is in the range 30–40%.

Table 6.9. Typical physical and chemical properties of LT shift catalysts

Properties	ICI Catalyst 52-1	ICI Catalyst 52-8	ICI Catalyst 53-1
Diameter/mm	5.4	5.4	5.4
Height/mm	3.6	3.6	3.6
BET surface area/m^2 g^{-1}	90	75	87
Pore volume/cm^3 g^{-1}	0.33	0.25	0.34
Mean pore radius/nm	8.5	6.7	7.7
CuO/%	32	33	33
ZnO/%	53	34	34
Al$_2$O$_3$/%	15	33	33

6.6.3. Diffusion Effects and Pellet Size

The rate of the water-gas shift reaction over commercial copper-based LT shift catalyst is at the borderline of being pore diffusion limited under most operating conditions using typical industrial sized pellets. In these circumstances increasing the pellet geometric surface area per unit reactor volume somewhat enhances performance in terms of activity. This can be achieved by the use of pellets which are smaller or which have a more elaborate shape, but practical limits are set by the balance between activity gained and physical properties such as strength and

6.6. Low-temperature Shift

pressure drop, and also by the manufacturing cost. As a result the size and shape of LT shift catalyst pellets (and indeed most of the catalysts used in ammonia and hydrogen plants) offered by different catalyst vendors are of a similar size.

It is clear from detailed characterization of discharged spent LT shift catalyst, and the results of laboratory experiments, that the poisoning reactions with hydrogen chloride and hydrogen sulphide are strongly diffusion limited. Haynes[163] considered in detail the situation of a diffusion limited reaction suffering diffusion limited poisoning, and explored the effects of parameters such as the Thiele modulus for both reactions on the poison profiles in the catalyst bed and the consequent activity of the total charge. Theoretically, the simplest way to improve the relative poison resistance (by retaining a higher weight percentage of poison at the top of the bed) is also to increase the pellet geometrical surface area per unit reactor volume. By using smaller or differenly shaped pellets the only major kinetic parameters which are changed are the effectivenesses (Thiele moduli) for the shift and poisoning reactions, and since the poisoning reactions are more strongly diffusion limited than the shift reaction itself, they are more sensitive to pellet size.[168] Accordingly, as the pellet size is reduced the shift reaction pore-diffusion limitation will become less severe, with the result that a higher amount of poison will be required to produce a given decrease in activity. The poisoning diffusion limitations will also become less marked, so the poison profile through the bed will change such that it will take longer to poison the same depth of catalyst bed, and the life of the charge will be extended. In practice the resulting sharper poisons profile means that a higher proportion of a charge will be guarded more efficiently. These effects could be important in plants where levels of poisons (especially chloride) are particularly severe, and where the use of small catalyst pellets may be advantageous in guard beds. However, since pressure drop is usually an important consideration in practice, normal-sized pellets which perform well in guard-bed duties are invariably preferred.

In plants experiencing LT shift catalyst poisoning problems, where any increased pressure drop is of major concern, some of the benefits associated with the use of small pellets can be obtained by using a partial charge of small-sized catalyst in the upper part of the bed. Although a number of plants have improved the life of their LT shift catalyst charges by using smaller pellets, this need not always be the case. For example, this approach is not viable in situations where liquid water may wash soluble chlorides from the upper part into the bulk of the charge.

Chapter 6. The Water-gas Shift Reaction

6.6.4. Reduction of LT Shift Catalyst

6.6.4.1. General considerations

Low-temperature shift catalysts are supplied in the oxide form and they must be reduced to metallic copper, which is the active species, before they are used. The reduction reaction (28) is highly exothermic and the heat evolved for 1000 kg of ICI Catalyst 53-1 which contains 30–35% cupric oxide is about 340 MJ, which is sufficient to raise the temperature of the catalyst bed by about 500°C. This is unacceptable because if conventional copper catalysts are heated much above 260°C they begin to sinter and hence lose activity.

$$CuO + H_2 \to Cu + H_2O \qquad \triangle H = -80.8 \text{ kJ mol}^{-1} \qquad (28)$$

In order to obtain maximum catalyst activity it is therefore of paramount importance to control the reduction carefully to limit the temperature to which the catalyst is exposed. This is done by conducting the reduction in the presence of an inert gas using a low concentration of hydrogen. The inert gas should be free from catalyst poisons, and if it contains traces of oxygen allowance must be made for the exothermic reaction that will occur with hydrogen over the LT shift catalyst. Nitrogen or natural gas is usually used as the carrier. Steam has also been used, but since it will cause considerable sintering of the copper crystallites, and lead to shortened operating lives, its use is *not* recommended. If steam has to be used as the carrier gas, then considerable care must be taken to avoid condensation on the catalyst.

If natural gas is used as the carrier gas, which is often the case with gas-based plants, it should be remembered that some hydrocarbons can undergo thermal cracking in the reformer or preheater at temperatures below 300°C to give additional hydrogen and, perhaps, carbon. If this occurs the hydrogen concentrations at the LT shift converter inlet will increase, and unless the reduction is being carefully monitored it could lead to a temperature runaway with consequent damage to the catalyst. Hydrogen itself (from factory main or bottles) and almost any gas containing hydrogen is suitable as the reducing agent—it may be purge gas from the methanator, from the carbon dioxide removal unit or from the outlet of the HT shift converter. The reducing gas must, however, be free from catalyst poisons, and care should be taken to allow for any extra temperature rise that will occur as the water-gas shift reaction takes place if carbon monoxide is present.

Typical reduction procedures for LT shift catalyst are outlined in Chapter 3 and more fully in the following sections. The reduction can be carried out separately, or during plant warm-up with nitrogen or natural gas using a once through system or a recycle system. There are many

6.6. Low-temperature Shift

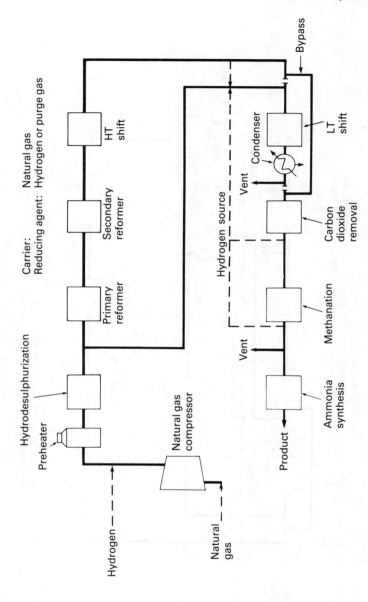

Figure 6.10. Schematic arrangement for reduction of LT shift catalyst using a typical once-through system.

Chapter 6. The Water-gas Shift Reaction

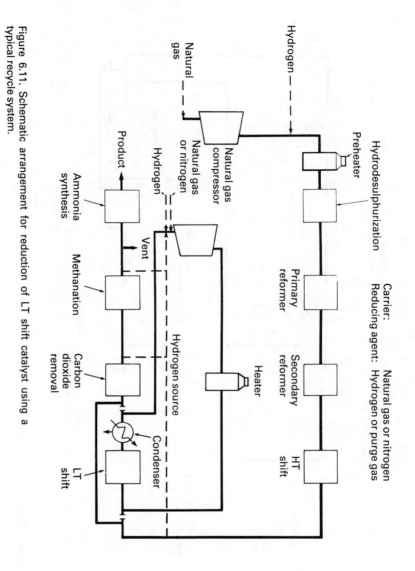

Figure 6.11. Schematic arrangement for reduction of LT shift catalyst using a typical recycle system.

6.6. Low-temperature Shift

ways of engineering this, and Figures 6.10 and 6.11 illustrate two of them. If a recycle system is used the water formed (equation (28)) should be removed from the gas at the exit before it is recycled. Whichever system is used, the progress of the reduction should be carefully controlled and the following points noted.

1. Bed temperatures should be monitored throughout the reduction.

2. The inlet and exit hydrogen concentrations should be determined at regular intervals. Usually this is done with a portable analyser, and it enables the amount of hydrogen consumed to be estimated from a knowledge of the gas rate. As a guide, about 85 m^3 of hydrogen are required for every cubic metre of ICI Catalyst 53-1 that is to be reduced.

3. The amount of water formed also gives an indication of the progress of the reduction although this is usually not so reliable as monitoring bed temperatures and hydrogen consumption. As a general guide each cubic metre of ICI Catalyst 53-1 will produce 40 kg of water of crystallization and 60 kg of water of reduction. Additional water will also be produced if oxygen is present in the reduction gas.

The time needed to complete reduction of a charge of LT shift catalyst depends on the procedure used; as the space velocity of the carrier gas increases, so the overall time decreases because the heat of reaction is removed more quickly. Recycle systems are usually designed to operate at a relatively low space velocity of about 300 h^{-1}, whereas once-through systems can usually operate at much higher flow rates and consequently reduction can be completed more quickly.

Unfortunately, the price of natural gas normally makes once-through reduction more expensive, so that despite the higher initial capital cost of the recycle system it is frequently the preferred method.

6.6.4.2. Once-through reductions

With a typical charge of LT shift catalyst in an ammonia plant the once-through reduction technique takes about 12–24 hours to complete. The plant may be taken to operating pressure at any convenient time during the reduction, and typical guidelines for the procedure are as follows.

1. Purge the converter free of oxygen with inert gas and heat the catalyst to 120°C, at a rate of about 50°C h^{-1}. On most plants the reduction may be carried out at any convenient pressure.

2. Establish the maximum possible flow rate of carrier gas, and add hydrogen to give a concentration of 0.5% at the bed inlet. Raise the catalyst bed temperature to 165°C at a rate of about 30°C h^{-1}.

Before the temperature reaches 140°C check the plant gas analyser, the hydrogen flowmeter and the portable analyser. If the space velocity is less than 300 h^{-1} more care is needed, and the reduction period will be protracted since more time will be taken to add the required volume of hydrogen.

3. When the temperature of the bed has reached 160°C increase the hydrogen concentration in the carrier gas to 0.5–1.0%. If reduction is slow the inlet temperature should be raised cautiously to 180°C and held there.

4. Increase the inlet hydrogen concentration. If nitrogen is used as the carrier gas, raise the concentration to 1.0–1.5%, with natural gas raise it to 2.0–2.5% because of its higher specific heat. With either carrier gas the maximum temperature must not exceed 230°C. At this stage it may also be possible to reduce the inlet temperature providing the reduction of the inlet part of the bed is complete.

5. As the reduction proceeds and the temperature rise begins to diminish, the inlet temperature may be raised to 200°C. The inlet hydrogen concentration can then be increased to about 3–5%, provided that the maximum temperature limit of 230°C is not exceeded.

6. When the reduction appears to be complete the inlet temperature should be raised to 225–230°C and the inlet hydrogen concentration raised to approximately 20% if this is possible. This procedure should not take less than 2 hours (but see Section 6.6.4.3 on recycle systems). No temperature rise should occur, and the maximum catalyst temperature should not exceed 230°C. The hydrogen concentration at the inlet and exit should be equal on the completion of reduction.

6.6.4.3. Recycle reduction systems

Procedures with recycle systems are very similar to those with once-through reduction systems. However, because of the low space velocity used, special attention to detail is necessary, and care must be given to the following points.

1. If the reformer is being used as the start-up heater care must be taken to ensure that the carbon dioxide does not react with hydrogen to give some methane which could then undergo thermal cracking and lay down carbon on the reforming catalyst.

2. Care must be taken to ensure that the concentration of hydrogen

6.6. Low-temperature Shift

entering the LT shift reactor does not exceed 0.5% in the early stages of the reduction.

3. The effect of high concentrations of hydrogen on the recycle compressor should be checked. In some plants the minimum gas density limit of the compressor will govern the hydrogen concentration which may be used in the final stages of reduction.

4. Water envolved during reduction should be drained from the system.

6.6.4.4. Commissioning reduced catalyst

Whichever procedure is used catalyst reduction is complete when the inlet and exit hydrogen concentrations are the same, and the entire catalyst bed is at a temperature of 225–230°C. The quantity of water collected and the amount of hydrogen consumed can be used to confirm this, but in some cases the hydrogen concentrations do not become exactly equal, and a small difference may persist. This difference may indicate a small amount of continued slow reduction, but the reduction procedure can be considered complete if the difference between the exit concentration and the inlet concentration has been less than 0.5% for more than four hours. The reduced catalyst may be brought on-line immediately, or maintained under an inert atmosphere until the rest of the plant is ready for it to be commissioned. If the LT shift vessel is isolated some care is necessary if carbonate in the catalyst has not been fully decomposed, since a gradual release of carbon dioxide could cause a significant pressure increase. It is therefore essential that the pressure and temperature within the vessel be monitored during this period, and any gas evolved from the catalyst vented. Similar steps should be taken if reduction is interrupted and a partially reduced catalyst is isolated in the reactor.

When cold reduced LT shift catalyst is commissioned it should first be warmed to a temperature above the dew-point before process gas is introduced. When process gas is first introduced, the temperatures usually increase rapidly as the catalyst comes to equilibrium with process conditions. The peak temperature may reach 260°C at this stage, but this will not result in catalyst damage because the period at this temperature is short. The temperature rise should be minimized by increasing the flow of process gas as quickly as possible. The inlet temperature should be kept as low as possible provided it is 20°C or more above the dew-point. At low pressure this will be achieved with an inlet temperature of 150°C, but on other plants 200°C will be required. Delays in removing the heat can be avoided by using an inert gas to bring the reactor up to pressure before process gas is introduced.

Alternatively, process gas can be introduced while the reactor is kept at low pressure by venting the gas at the exit. This technique can be particularly useful after a once-through reduction with natural gas, then process gas can be gradually added to the natural gas stream until it eventually replaces it. The pressure is then raised by closing the vent and opening the inlet and exit valves fully to commission the converter.

6.6.5. Operation and Monitoring Performance

The amount of LT shift catalyst used in a plant is larger than the design volume needed to achieve the required carbon monoxide level at the exit, because extra catalyst is included to compensate for the deactivation that inevitably takes place during operation. Typically, for a large plant to operate continuously for at least two or three years this additional catalyst volume amounts to about 70% of the total actually charged. The main causes of deactivation are thermal sintering and poisoning by relatively small amounts of halide and sulphide present in the process gas, and deactivation of LT shift catalyst is discussed in detail in Section 6.6.6.

In most plants the concentration of carbon oxides in the exit gas is usually analysed continuously so that the operating efficiency of the LT shift stage can be readily assessed. A simple method often used to check the carbon monoxide analyses is to measure the temperature rise across the methanator. There will be a temperature rise of about 7.5°C for every 0.1% of carbon monoxide converted to methane in the methanator, and about 6.0°C for every 0.1% of carbon dioxide converted. A new charge of LT shift catalyst should be operated with as low an inlet temperature as is practical while still achieving an acceptable close approach to the equilibrium carbon monoxide concentration at the exit. In this way thermal sintering of the catalyst is minimized and its overall life maximized. In time, however, activity decreases, and the operating temperature can then be progressively increased to compensate for this. The actual rate at which this has to be done depends on both the thermal stability of the catalyst and poisons present in the process gas. In practice the lowest inlet temperature is about 200°C. This limit is determined in part by activity, though more often by the need to avoid condensation of steam on the catalyst, and usually the inlet temperature is set about 20°C higher than the dew-point for the process conditions.

Initially almost all of the reaction takes place in the top part of the catalyst bed, and the temperature profile through this region is steep, reflecting the catalyst's high activity. Over a relatively short period this may decrease as the catalyst reaches its "stabilized" activity. With a

6.6. Low-temperature Shift

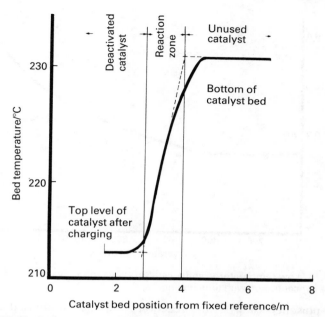

Figure 6.12. Typical temperature profile through a bed of ICI Catalyst 53-1 LT shift catalyst.

well-formulated catalyst with good thermal stability these effects are not large.[169] With less-stable catalysts the activity (and hence the slope of the temperature profile) continues to decrease quite markedly, and it becomes necessary to increase the inlet temperature more rapidly than would otherwise be the case to compensate for loss of activity. The process gas in all plants contains sufficient poisons to cause gradual deactivation of LT shift catalyst, and as poisons accumulate in catalyst at the top of the bed it is deactivated and the temperature profile in the bed gradually moves downwards. The exact behaviour of a charge of LT shift catalyst towards poisons depends on how well the catalyst retains poisons at the top part of the bed. Under normal circumstances, with catalysts such as ICI's 52 and 53 series that are thermally stable and capable of retaining a high level of poisons, the rate of movement of the temperature profile through the bed is constant because the concentration of poisons in the process gas does not normally change. Any differences in the rate of movement which may be observed are therefore important. They can indicate changes in the rate of poisoning, the effects of condensation during shutdown periods or the effect of water released into the process gas following boiler failures. A typical temperature profile is shown in Figure 6.12, that was taken from an operating plant after 250 days on-line. This shows the extent of the

Chapter 6. The Water-gas Shift Reaction

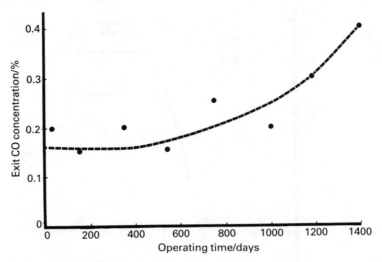

Figure 6.13. Low-temperature shift catalyst operation: variation of carbon monoxide concentration with time on-line.

inactive poisoned catalyst, the reaction zone and the unused catalyst; additional plant data are included in Chapter 3.

As the reaction zone approaches the lower part of the catalyst bed the amount of carbon monoxide in the gas at the exit gradually increases as shown in Figure 6.13. This indicates that the catalyst must be changed as quickly as possible to avoid loss of overall plant production efficiency. This is necessary because, as the reaction takes place in a gradually decreasing volume of catalyst, maldistribution of gas through the small depth of active catalyst is quite common, and rapid unpredicted increases in the carbon monoxide levels at the exit often result. Since such bypassing of active catalyst is not easy to anticipate, monitoring catalyst performance towards the end of the life of a charge becomes much more critical than during the early stages of operation. It is therefore obviously worthwhile to change the LT shift catalyst as soon as the temperature profile approaches the bottom of the bed before the carbon monoxide concentration at the exit increases significantly.

The importance of properly monitoring the performance of LT shift catalyst as a routine procedure to establish the exact timing of catalyst replacement in relation to plant overhauls cannot be overemphasized. In addition to the obvious loss of several days' production during an unexpected plant shutdown, the loss of potential production if carbon monoxide slip increases from say 0.2 to 0.4% while waiting for a planned shutdown can be very significant, and over a year amounts to considerably more than the cost of a charge of catalyst. The normally

6.6. Low-temperature Shift

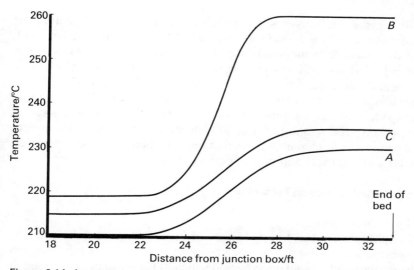

Figure 6.14. Low temperature shift catalyst plant operating data from one of ICI's ammonia plants when there was the steam leak upstream of the HT shift converter. Curve A represents normal operation and curve C after the major leaks. Curve B was measured during the leak when the inlet CO was 9.5% and the steam/dry gas ratio 1.35. The exit CO was then only 0.2% (Quoted analyses are on a dry basis).

predictable rate at which a good LT shift catalyst deactivates makes routine monitoring of operation easy, and accurate forecasts of catalyst life can usually be made. Variations from the normal die-off rate can be associated with plant problems, changes in feedstock quality, or the efficiency of purification catalyst. Interpretation of LT shift catalyst operating data therefore provides an easy diagnostic technique which contributes to maintaining plant efficiency.

Excessive condensation of water on LT shift catalyst is almost invariably detrimental, and generally it must be avoided. After the top part of the bed has been deactivated by poisons the catalyst there continues to absorb poisons and, if condesation occurs, soluble poisons can be washed to the lower parts of the bed. This can lead to premature deactivation of the charge. In addition, if the catalyst has already been weakened as a result of plant maloperation, condensing water may lead to its fragmentation. However, should condensation occur the condensate must immediately be drained from the converter. High steam levels in the process gas can be tolerated, provided condensation does not take place. This is illustrated by the incident referred to in Section 6.5.2.2, when a boiler leak upstream of the HT shift catalyst was so high and the process gas temperature so reduced that there was effectively no reaction occurring over the HT shift catalyst. Figure 6.14 shows the effect on the LT shift catalyst which at the time had been on-line for about 3½ years. Curve A, obtained immediately before the final massive leak, is typical for this plant with the LT shift catalyst used, and the exit carbon monoxide was 0.16%. Curve B was obtained after the HT shift catalyst stopped working due to the large quantity of accidental water-quench, and the carbon monoxide level at the LT shift inlet was about 9.5% (dry basis). Notwithstanding this, the exit carbon monoxide from the LT shift catalyst was only 0.2%, and the associated temperature rise was only about 40°C because of the very high amount of steam present. Curve C shows that once the boiler had been replaced the LT shift catalyst continued to perform well.

6.6.6. Deactivation and Poisoning

6.6.6.1. Deactivation

The literature records a long history of investigation into the catalytic properties of copper in the water-gas shift reaction. Early workers always observed a rapid deactivation which, it is now known, was due to both poisoning and a rapid loss of copper surface area brought about by sintering. When a serious effort was made to develop industrial LT shift catalysts it was quickly confirmed that copper was the most appropriate catalytic species. However, the first copper/zinc oxide LT shift catalysts lacked sufficient thermal stability because the operating conditions were considerably more forcing than those encountered in their original roles as catalysts for mild hydrogenation/dehydrogenation reactions of organic compounds. Much research effort provided formulations which were significantly more thermally stable,[169] and these (before reduction) consisted of alumina in addition to oxides of copper and zinc. Susceptibility to thermal sintering (in the absence of other effects) is, however, an intrinsic property of all dispersed heterogeneous catalysts. In general copper-based catalysts are more susceptible than the other metallic catalysts used in ammonia and hydrogen plants, and this arises because of copper's low Hüttig temperature,[170] which reflects its relatively low melting point of 1063°C, compared with that of iron (1535°C) and of nickel (1435°C). Because of this copper-based catalysts have to be operated at relatively low temperatures. However, it is apparent from data shown in Figure 6.15, as well as extensive other results, that thermal sintering can almost be neglected with well-formulated catalysts, provided they are not subjected to temperatures significantly higher than that at which they were designed to operate. Figure 6.15 also shows that the thermal stability of copper-based LT shift catalysts depends on the manufacturing procedures used to produce them, and not only on their composition. Only recently have modern techniques such as EXAFS been used to

6.6. Low-temperature Shift

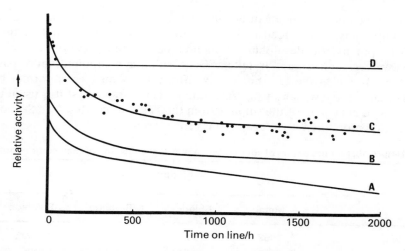

Figure 6.15. Relative activities of typical copper/zinc oxide LT shift catalysts under poison-free conditions. A, Cr_2O_3-based catalyst; B, conventional Al_2O_3-based catalyst; C, high copper-content Al_2O_3-based catalyst; D optimized Al_2O_3-based catalyst. For clarity experimental points are shown on only one of the curves.

probe the intimate mechanism of the thermal sintering of copper catalysts.[171]

Operating a catalyst at low temperature favours exothemic adsorption of poisons and results in high surface coverages, and as a consequence LT shift catalysts are particularly sensitive to even very low levels of poison such as sulphide. This situation is further exacerbated by the low melting point and high surface mobility of copper(I) and zinc halides (see Table 6.10), which if present even in extremely small amounts provide sufficiently mobile species to initiate sintering. As a result, copper LT shift catalysts are extremely sensitive towards conventional site-blocking by poisons such as sulphur species, as well as excessive sintering induced by very low levels of halide. It was for these reasons that the introduction of copper catalysts was delayed until sufficiently pure process gas became available with the introduction of high-pressure steam reforming.

Both sulphur- and chlorine-containing compounds can originate in the feedstock, steam or process air, so that even the location of a plant is relevant to the behaviour of its LT shift catalyst. Steam can be eliminated as a source of poisons by ensuring that the boiler design prevents carry-over of boiler solids into the process steam system. Process air can be filtered. In particularly contaminated environments that may result from adjacent plants and be very specific to that location, consideration may be given to washing the process air to ensure that it is clean. In locations close to the sea, salt-water spray must

Chapter 6. The Water-gas Shift Reaction

be prevented from being drawn into the air compressor suction. Sulphur compounds in the feedstock will normally be reduced to less than 0.1 ppm by the desulphurization unit before the primary reformer. However, unless care is taken, some sulphur compounds may pass downstream to the LT shift catalyst from newly installed reformer or HT shift catalysts, and with conventional HT shift catalyst it is usual to remove almost all of the sulphur from the HT shift catalyst (see Section 6.5.3) before the LT shift catalyst is brought on-line.

Table 6.10. Melting points of copper, iron and nickel and some of their compounds

	Melting point/°C			
	Metal	Chloride	Bromide	Sulphide
Copper	1063	430[a]	492[a]	1100[a]
Iron	1535	674	684	1195[b]
Nickel	1455	1001	963	790[c]

[a]Copper(I) compounds; [b]FeS; [c]Ni_3S_2.

6.6.6.2. Sulphur poisoning

Under LT shift operating conditions sulphur is a powerful poison for copper and, as indicated by the bulk thermodynamics in equation (29) sulphiding is very favoured with the equilibrium constant at typical operating temperatures being about 10^5. It is therefore important to prevent even very low levels of sulphur contacting the catalyst. With LT shift catalysts containing reactive zinc oxide this can be achieved by trapping the sulphur at the top of the bed as zinc sulphide, which is more stable than sulphided copper as indicated by equation (30).

$$2Cu + H_2S \rightarrow Cu_2S + H_2 \qquad \triangle H = -59.4 \text{ kJ mol}^{-1} \qquad (29)$$

$$ZnO + H_2S \rightarrow ZnS + H_2O \qquad \triangle H = -76.6 \text{ kJ mol}^{-1} \qquad (30)$$

The equilibrium constant for formation of bulk zinc sulphide is almost three orders of magnitude larger than that for formation of copper(I) sulphide from copper metal.

There is evidence to suggest that initially sulphur is preferentially adsorbed from the feed gas onto the more active small copper crystallites (~5 nm), and then transferred to the zinc oxide where it reacts to form the thermodynamically more stable zinc sulphide. For this to happen rapidly, the catalyst pellets need to be porous to minimize diffusion effects, and to have small copper crystals as well as the maximum surface area of reactive "free" zinc oxide. If a LT shift

6.6. Low-temperature Shift

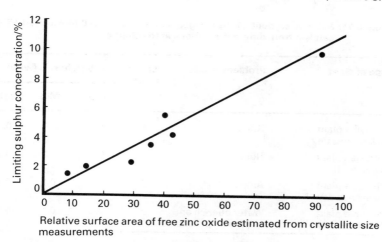

Figure 6.16. Influence of surface area of zinc oxide on the maximum sulphur pick-up of LT shift catalysts under laboratory conditions.

catalyst has only a limited amount of free zinc oxide, or if the zinc oxide is combined with another component as a thermodynamically more stable compound, sulphur will not be retained by the catalyst so readily, and sulphur will be held in a broad poisoned region of the bed. On the other hand, a catalyst with a significant content of reactive "free" zinc oxide will retain sulphur in a well-defined concentrated layer of the catalyst bed, so that the bulk of the charge remains completely unaffected by sulphur and will continue to operate satisfactorily. However, the overall kinetics of this process demand that the zinc oxide surface area be as high as possible, and indeed in controlled laboratory experiments the limiting amount of sulphur combined was a linear function of the estimated surface area of free zinc oxide. This is illustrated in Figure 6.16.

In full-scale plants sulphur levels in the gas are very much lower than those used in relatively short-term laboratory experiments, with the result that the sulphur content of even well-formulated catalyst is rarely more than about 2% when discharged from the top of a bed. Some representative analytical results obtained on discharged catalyst are given in Table 6.11, but in "problem" plants, such as those located in unfavourable locations with high atmospheric pollution levels, the sulphur content of discharged catalyst from the top of the bed can as much as 7%.

Table 6.11. Sulphur content of discharged samples of a well-formulated LT shift catalyst from different positions in the charge

Type of plant	Problem	Life/years	Sulphur/% (w/w)		
			Top	Mid	Bottom
ammonia plant 1350 tonnes day^{-1}	None	4	0.8	NA	0.05
hydrogen plant 80 000 m^3 h^{-1}	None	3	0.5	0.2	0.05
ammonia plant 850 tonnes day^{-1}	Air pollution	1.75	0.9	0.3	0.03
ammonia plant 900 tonnes day^{-1}	Water quench and sulphur air pollution	2	1.6	0.3	0.06

NA, not available.

6.6.6.3. Chloride and other poisons

In practice the only halide encountered in plants is chloride. Both the copper metal and zinc oxide components of LT shift catalysts react with hydrogen chloride. Reaction of bulk copper metal with hydrogen chloride to form cuprous chloride, as in equation (31) is thermodynamically much less favourable than the reaction with hydrogen sulphide. It might therefore be expected that surface adsorption of chloride is correspondingly less important, with the result that chloride is not such a severe poison for LT shift catalysts than sulphide. However, the low melting point of cuprous chloride (430°C) and by inference other copper(I) chloride species, results in them having high mobility under LT shift operating conditions. Consequently, the presence of even extremely small amounts of chloride provides the "catalytic" species necessary for a surface-migration sintering mechanism of the copper crystallites. Because of this, chloride acts as a powerful poison that decreases the activity of LT shift catalysts irreversibly. Similar processes involving the zinc oxide take place when chloride is present because zinc chloride species also have high surface mobility, and this too contributes to a destructive destabilization of catalytic activity via structural changes that decrease thermal stability. In keeping with the thermodynamics for the simple bulk chlorides indicated in equations (31) and (32) under normal operating conditions almost all of the chloride present in LT shift catalyst is associated with zinc rather than copper species.

$$Cu + HCl \rightarrow CuCl + \tfrac{1}{2}H_2 \qquad \Delta H = -43.5 \text{ kJ mol}^{-1} \quad (31)$$

$$ZnO + 2HCl \rightarrow ZnCl_2 + H_2O \qquad \Delta H = -125.0 \text{ kJ mol}^{-1} \quad (32)$$

6.6. Low-temperature Shift

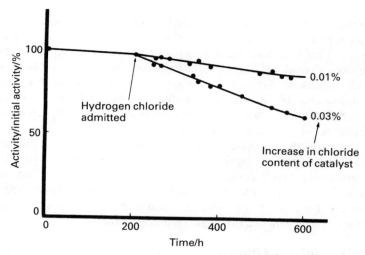

Figure 6.17. Dramatic effect of very small amount of hydrogen chloride on the activity of LT shift catalyst.

The results of laboratory experiments shown in Figure 6.17 illustrate the decrease of activity following the introduction of a small amount of hydrogen chloride into the feed gas. Chloride poisoning of LT shift catalyst therefore presents a particularly difficult problem for plant operators because chloride is usually present in process gas in amounts below the normal levels of detection. However, with a well-formulated catalyst the chloride poison is strongly retained at the top (inlet region) of the LT shift bed, usually in the form of zinc hydroxychlorides, with the result the charge is partially self-guarding. These species are, however, soluble in hot water, and it is therefore important to avoid conditions that give condensation, otherwise condensate will wash chlorides further into the bed. It is clear that admission of chloride to the process streams of ammonia and hydrogen plants should be kept to an absolute minimum. Strict attention must be paid to feedstock purity and traces of chlorides should be removed with an alkaline absorbent during feedstock purification (see Chapter 4). Similarly, chlorinated solvents should not be used during maintenance operations, and on sites where gases such as hydrogen are imported care is necessary to confirm their quality.

Silica is frequently found in discharged LT shift catalyst, and it can be a serious problem because it deposits on the surface and finally into the pores of the pellets, where it reacts to form zinc silicate. In severe cases this will result in a decrease in carbon monoxide conversion and more-rapid movement of the temperature profile down the bed. The presence of silica in a LT shift catalyst also has the more serious effect of exaggerating problems associated with both sulphur and chloride poisons. This is because the reaction of silica with zinc oxide reduces the quantity of available "free" zinc oxide at the top of the bed, which interferes with the normal absorption of poisons. The resulting transfer of sulphur and chlorine further into the bed shortens the life of a charge by decreasing the average concentration of poisons (particularly sulphur) required to cause deactivation.

Other poisons for copper catalysts include arsenic, which might come from some types of carbon dioxide removal systems, and trivalent phosphorous, which could originate from boiler water feed, but in practice these are seldom encountered in process gas. Similarly, relatively high levels of ammonia can temporarily inhibit the shift reaction, but in practice this is also not important.

6.6.7. Oxidation and Discharge

During a plant shutdown the LT shift catalyst should not normally be left in contact with process gas even when the temperature remains substantially above the dew-point. The pressure should be lowered, and if at all possible the converter should be purged with dry nitrogen. Reduced LT shift catalysts are potentially pyrophoric as equation (33) shows, and uncontrolled oxidation in air could result in exotherms of between 800 and 900°C, and care must therefore always be taken when LT shift catalyst is discharged.

$$Cu + \tfrac{1}{2}O_2 \rightarrow CuO \qquad \triangle H = -157.2 \text{ kJ mol}^{-1} \qquad (33)$$

The usual procedure employed when discharging LT shift catalyst is to let down the pressure, purge the vessel with nitrogen, cool to less than 50°C and discharge the catalyst from the bottom of the converter under a flow of nitrogen. As the catalyst falls from the manhole it is sprayed with water, collected and dumped on a suitable site where it is allowed to oxidize slowly. In plants where there is insufficient available nitrogen for it to be used during catalyst discharge, air must not be allowed to enter the converter, or flow through it (for example by the chimney effect) when it contains reduced catalyst, otherwise local overheating will take place. In these situations it may be appropriate to fill the converter with water and discharge the catalyst wet. With this technique catalyst should not be allowed to sit in water for any length of time, otherwise breakdown of catalyst pellets can occur, and it is therefore advisable to drain the vessel as soon as possible after filling it. In recent years it has been possible, in many areas, to employ contractors who use specialized techniques to withdraw the catalyst from the vessel. This may be done with top or bottom discharge, and these techniques are discussed in detail in Chapter 3.

It is also possible to use low levels of air in nitrogen to stabilize the catalyst before it is discharged, but this should be done with extreme caution and only if absolutely necessary. The main danger is that stabilization may be incomplete, due to maldistribution of gas through the bed, and subsequent local oxidation may result in overheating of the vessel. The catalyst may be stabilized according to the following general procedure.

1. Circulate nitrogen through the charge.
2. Maintain an initial bed temperature of 200°C.
3. Add air to the inlet gas such that the oxygen level is about 0.1% and bed temperatures remain below 220°C.
4. When there is no further temperature rise, gradually increase the air rate to establish 3% oxygen in the inlet gas, checking for any sign of temperature rise.
5. Allow the bed to cool close to ambient temperature.
6. Increase the air flow while gradually decreasing that of the nitrogen, so a flow of undiluted air is established.

Low-temperature shift catalyst that has been successfully stabilized in this way may be discharged, and even re-used although subsequent overall performance is invariably inferior to that of new catalyst. In practice this procedure is hardly ever carried out and LT shift catalyst is seldom re-used. However, when the stabilization procedure is carried out the resulting material must always be stored and handled with care, and consideration must be given to its potential pyrophoric nature.

6.6.8. Guard Beds

The importance of guard beds as a means of removing catalyst poisons from the process gas stream has been appreciated for many years. Indeed, one of the earliest examples was the use in ammonia plants of spent synthesis catalyst to protect fresh active synthesis catalyst. The life of a well-formulated LT shift catalyst that has good thermal stability is usually only limited by the presence of poisons. When LT shift catalyst was first introduced it was believed that poisoning was caused predominantly by sulphur compounds, and some operators used a small amount of zinc oxide as a separate bed in the LT shift converter to remove hydrogen sulphide. Although it was reasonably effective as a sulphur guard, and may still be used by a few operators, experience showed that the LT shift catalyst itself was much better—even though it has a lower theoretical total sulphur capacity than zinc oxide alone. Since the sulphide/oxide solid-state diffusion rates are extremely slow, at the relatively low temperatures encountered in the LT shift converter, the sulphur absorption efficiency is determined by surface area. In the case of the LT shift catalyst the zinc oxide is present as very small crystallites, so its surface area is significantly higher than that of conventional zinc oxide normally used for desulphurization duties. Consequently, LT shift catalyst is the more effective sulphur guard.

When it was realized that chloride was a virulent poison some operators considered the introduction of a small bed of alkalized alumina into the top of the LT shift converter to remove hydrogen chloride from the process gas before it entered the bed of LT shift catalyst. Under dry conditions this system operated quite satisfactorily. More usually, however, when condensation occurred the soluble chloride salts were washed from the bed of absorbent into the main catalyst bed and the possibility of caking in the bed of absorbent was also a potential cause of increased pressure drop. Experience has shown that the LT shift catalyst was also a much better absorbent for chloride and, although it also suffers from leaching under condensing conditions, it has not been as prone to caking as alkalized alumina.

The use of a separate guard vessel containing LT shift catalyst is the best method of preventing all known poisons entering the main catalyst bed, and this arrangement has now been adopted by many operators. With a separate guard vessel poisoned catalyst can be changed

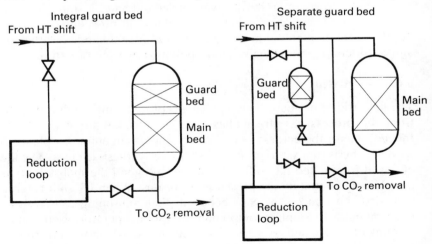

Figure 6.18. Possible arrangement of guard and main beds of LT shift catalyst.

6.6. Low-temperature Shift

regularly, and this may often be done while the main bed is still operating. Typical arrangements for the use of LT shift guard beds are shown in Figures 6.18 and 6.19. By using a guard bed all but very small amounts of poisons are prevented from entering the bulk of the charge, and several plants so equipped have operated for a number of years at peak performance in terms of carbon monoxide conversion.

The example shown in Figure 6.20 is taken from an ammonia plant based on partial oxidation of fuel oil with inefficient feedstock purification, which originally did not include a LT shift stage. When a LT shift stage was added it was guarded with a bed of zinc oxide, but LT shift catalyst lives were very short and uneconomic. By replacing conventional zinc oxide with LT shift catalysts in the guard bed and changing it annually, the overall catalyst performance improved dramatically. In plants such as this that have quite serious poisoning problems, the use of a guard bed not only eliminates a considerable loss of potential ammonia production (see Section 6.6.9) by avoiding high carbon monoxide levels at the exit, they also reduce the total amount of

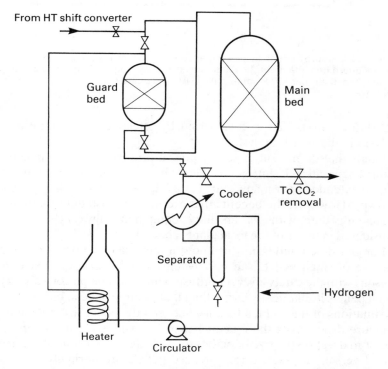

Figure 6.19. Typical layout for separate reduction of guard bed and main bed.

Chapter 6. The Water-gas Shift Reaction

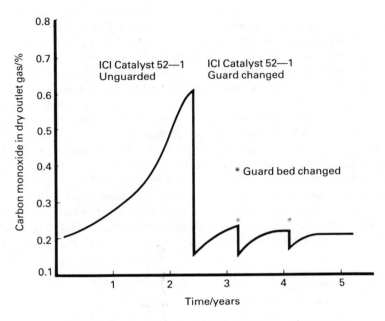

Figure 6.20. Improvement in performance obtained in a partial oxidation plant using a guard bed. Plant output 120 tonnes day^{-1} with equivalent catalyst volumes in main bed 83 m^3 per 1000 tonnes day^{-1} NH$_3$ and guard bed 20 m^3 per 1000 tonnes day^{-1} NH$_3$.

LT shift catalyst used over a period by effectively concentrating the poisons in the guard catalyst.

Plants that do not suffer severe poison problems can also benefit by using a separate LT shift guard. Here the life of the main catalyst bed can be extended over many years (examples in excess of eight years are known). However, the benefits that accrue from installing a new guard vessel together with its ancillary equipment always have to be considered in terms of the local plant economics.

For guard-bed duties in which poison removal is the prime objective, the use of small pellets can be advantageous because the reaction of poisons with the catalyst is very diffusion limited. The poison capacity of small pellets is marginally larger than that of large pellets, but as always, in situations of this kind, a balance between the pellet size used and the pressure drop across the reactor has to be made so that an important factor in deciding the size of pellets to use is the dimensions of the actual guard vessel. In practice regular-sized pellets are normally used, and they provide satisfactory performance. An additional small, but often real, advantage of using a LT shift guard bed is that there is some carbon monoxide conversion within the guard bed, so that the potential reduction of carbon monoxide concentration before the guard catalyst is deactivated by poisons improves the overall performance of the plant.

6.6.9. Economics of Operation

The LT shift stage has an important role in the overall plant operating economics of hydrogen and ammonia plants that use a methanator to remove residual carbon oxides. This is because every mole of carbon monoxide which undergoes reaction over the LT shift catalyst produces one mole of hydrogen as in equation (*34*) and avoids the subsequent loss of three further moles of hydrogen that would have taken place in the methanator according to equation (*35*). Economically this may not be of major importance in hydrogen plants, but the situation is more complex with ammonia plants, where the additional methane formed in the methanator has to be removed from the synthesis loop with a significantly higher purge rate. This is important on most plants because the purge gas has only a fuel value. It therefore follows that the achievement of a low concentration of carbon monoxide at the exit from the LT shift converter is of considerable financial benefit. The detailed calculation of this benefit is complicated, as illustrated in Figure 6.21. In a typical 1000 tonnes day^{-1} plant the net result of decreasing the carbon monoxide level by 0.1% by using an active, well-formulated catalyst and trimming operating conditions is to increase the daily production of ammonia by some 11.6 tonnes of ammonia for the consumption of only an additional 6.0 tonnes of methane. The sensitivity of plant efficiency to the performance of the LT shift catalyst highlights the need for continuous monitoring of the LT shift catalyst and carefully adjusting the operating conditions to optimize its performance.

$$CO + H_2O \rightarrow CO_2 + H_2 \qquad (34)$$

$$CO + 3H_2 \rightarrow CH_4 + H_2O \qquad (35)$$

6.7. Recent Developments

6.7.1. Sulphur-tolerant Shift Catalysts

Partial oxidation plants using heavy fuel oil or coal play only a minor part in the manufacture of synthesis gas for production of ammonia. However, the situation may change in future years as supplies of natural gas and naphtha for steam reforming decline and more-efficient partial oxidation units operating at high pressure are developed. The synthesis

Chapter 6. The Water-gas Shift Reaction

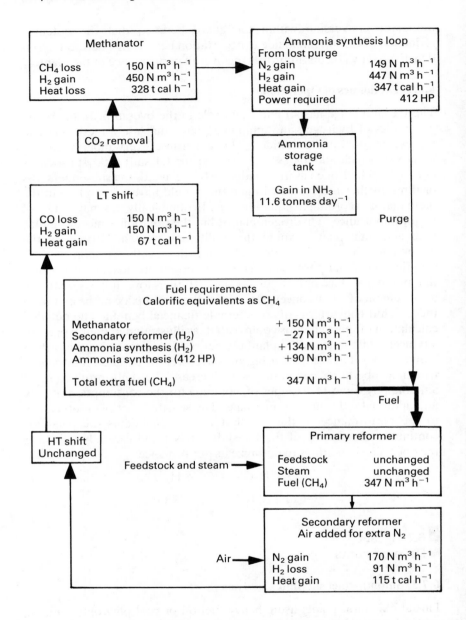

Figure 6.21. Effect on plant economics of reducing carbon monoxide content from the LT shift converter by 0.1% in a typical 1000 tonnes day^{-1} ammonia plant (t cal = 1×10^6 cal).

gas produced by these processes contains high concentrations of carbon monoxide (typically around 50%), with steam/dry gas ratios usually between 0.8 and 1.5. In addition the gas contains relatively large concentrations of sulphur compounds, and this usually precludes the use of copper-based LT shift catalysts because of their sensitivity to sulphur. Traditionally, in these situations the shift reaction has been carried out using conventional chromia supported iron HT shift catalysts. As previously noted in Section 6.5.1, sulphiding this catalyst reduces its activity by about 50%. As a result it was necessary either to increase the steam/gas ratio significantly to achieve a desirably low level of carbon monoxide, or to remove sulphur compounds from the process gas before it entered the shift stage. The former would adversely affect the economics of the process, and usually the latter approach has been adopted. Operating data obtained from a large coal-based ammonia plant, where sulphur compounds were removed by methanol wash, are given in Table 6.12.

Table 6.12. Operation of conventional HT shift catalyst on a coal-based plant in which sulphur compounds are removed by a methanol wash

		Bed 1 (7.7 m^3)		Bed 2 (15.4 m^3)		Bed 3 (15.4 m^3)	
		Inlet	Exit	Inlet	Exit	Inlet	Exit
% dry basis	CO	54	24	24	7	7	4
	CO$_2$	12	28	28	37	37	39
	H$_2$	32	45	45	53	53	55
Temperature/°C		335	470	380	470	400	420
Steam/gas ratio		1.6	1.3	1.3	1.1	1.1	1.0
Wet space velocity/h^{-1}		19 000		9 900		10 000	

In some processes a sulphur-tolerant shift catalyst has been used, and the sulphur compounds removed after the shift converter. These catalysts are based on cobalt/molybdate formulations and are less active than copper LT shift catalysts operating with pure feed gas. They are similar in composition to hydrodesulphurization catalysts and they only reach full activity when they are properly sulphided. They have been used to a limited extent for more than 25 years to convert relatively high concentrations of carbon monoxide in gas streams that contain hydrogen sulphide or other sulphur compounds at levels that would cause significant problems with conventional HT shift catalyst. No loss of activity or major deterioration of physical properties takes place during normal operation, and lives of up to ten years can be expected. More recently[174a,174b] it has been shown that addition of alkali promotes the water-gas shift activity of cobalt/molybdate catalysts, and this further increases the potential uses for this type of catalyst. Perhaps the ideal use for such catalyst is in plants using feed gas from the gasification of oil or coal containing up to 50% carbon monoxide and from 10 ppm up to 3% hydrogen sulphide. The amount of carbon monoxide could be reduced to less than 1% in the gas at the exit, provided the catalyst activity is sufficiently high to permit inlet temperatures as low as about 230°C. As with normal LT shift catalyst, the minimum inlet temperature (consistent with the operational dew-point) should be used to achieve the minimum equilibrium carbon monoxide level at the exit. For such a potentially versatile catalyst it is disappointing that there are so few immediate applications.

6.7.2. Operation at Very Low Steam Ratios

As the cost of hydrocarbon feedstocks increases there will be significant advantages to be gained by operating ammonia plants at yet lower steam/gas ratios. This is because many plants have been designed to use significantly larger amounts of process steam than is actually necessary for the reforming and shift reactions. This leads to thermal inefficiencies associated with raising steam, even though most of the excess steam is used as a heat source. The high steam ratios originally used reflects conservative primary reformer design associated with the early steam reforming catalysts. Modern steam reforming catalysts are capable of operating at lower steam ratios, and there will be efforts to move further in this direction in the future.

Use of very low steam ratios has consequences on the operation of the water-gas shift reaction. With HT shift catalyst, in the absence of steam formation of metallic iron can occur and this would lead to methanation and perhaps carbon formation via carbon monoxide disproportionation. This is unlikely to occur in a severe form under foreseeable operating conditions. However, as the steam ratio is lowered it is possible to form intermediate reduced-iron species. Lowering the steam ratio in a conventional reformer increases the amount of carbon monoxide at the inlet to the HT shift converter relative to the amount of carbon dioxide present, and under some conditons this could lead to the formation of iron carbides. These are known catalysts for the synthesis of hyrocarbons from carbon oxides and hydrogen (Fischer–Tropsch reactions), and this could become a matter of some concern. Formation of hydrocarbons consumes valuable hydrogen, and so reduces the overall plant economics. Moreover, their formation is considerably exothermic and if large amounts of hydrocarbon are formed thermal runaways might result. There could also be downstream problems resulting from the presence of large quantities of hydrocarbons in the process gas. For efficient operation at very low steam ratios it is possible that iron-containing HT shift catalyst will require to be modified to inhibit the formation of hydrocarbons, or alternative iron-free catalyst will have to be used.

Chapter 7

Methanation

7.1. Introduction

Carbon monoxide and carbon dioxide are pronounced catalyst poisons in many hydrogenation reactions, including ammonia synthesis. It is therefore necessary in ammonia plants and most hydrogen plants to reduce the final amount of carbon oxides remaining in process gas after the carbon dioxide removal stage to extremely low levels. In the early ammonia plants the most common method used in the final purification stage employed "copper liquor" to remove carbon monoxide and residual carbon dioxide. This comprised an aqueous solution containing copper(II) and copper(I) compounds of an organic acid salt, such as formate or acetate, in the presence of an excess of ammonia. Under these conditions a moderately stable copper(I) carbon monoxide complex is formed.[175] Carbon dioxide is dissolved as "ammonium carbonate" and both oxides of carbon are recovered when the solution is heated to reverse the absorption reactions.[176] However, the process suffers from problems of waste disposal and corrosion, and as a consequence has now been largely superseded.

Nitrogen wash is a second method of purifying synthesis gas. It uses liquid nitrogen in a packed column to absorb residual oxides of carbon (as well as inert gases which may be present). This method is employed in ammonia plants using partial oxidation to produce synthesis gas from coal, or hydrocarbon feedstock, where liquid nitrogen is available at low cost.

In almost all ammonia and hydrogen plants constructed since the early 1960s the simple and relatively inexpensive catalytic conversion of traces of carbon oxides to methane and water has been used. This process is called *methanation*, and if necessary it can be operated over a wide range of conditions. The methanation reaction was studied in the laboratory during the classic work of Sabatier,[177] who first produced methane catalytically by passing carbon oxides and hydrogen over finely divided metals. Relatively crude methanation catalysts (nickel on porcelain) had been developed before the First World War, but they were very susceptible to poisoning by sulphur compounds present in the gas streams available at that time.

Methanation as a method of carbon monoxide removal was actually used as early as 1920 by George Claude, in France, and Casales, in Italy, who conducted the reaction at very high pressures. It was also used in isolated cases in ammonia and hydrogen plants in the USA in the 1930s. Interestingly, some old ICI coal-based plants had methanators, but these contained iron catalyst which, while having disadvantages compared with nickel catalyst, had the ability to operate in the presence of quite high amounts of poisons. It was not until the 1950s that the process step using a nickel-based catalyst, operating at low pressures, began to be incorporated more widely into ammonia process flowsheets.[178, 179] This arose because of the use of multi-stage high-temperature shift conversion units that enabled the carbon monoxide to be reduced to acceptably low levels, so avoiding large temperature rises and undue hydrogen consumption during methanation. Its almost universal application in the 1960s is largely attributable to the introduction of the then newly developed low-temperature shift catalyst.

The methanation reactions are the reverse of those for methane steam reforming, and they are strongly exothermic. Because they are operated at relatively low temperatures and in the absence of large amounts of water, thermodynamics never controls conversion in ammonia and hydrogen plant duties.

The major part of this chapter is concerned with these applications, but methanation catalysts designed for use in ammonia and hydrogen plants can be used for the purification of hydrogen in refineries and ethylene plants where recycle streams free from carbon oxides are required, and examples of this type of duty are also mentioned. Methanation reactions are also used in the production of various substitute natural gases (SNG) that contain large amounts of methane. These processes may well be of importance in the future. They require special catalysts and make use of a different engineering approach from that used in ammonia and hydrogen plants due to the large amount of heat released. For completeness some of these processes, and the catalysts developed for them, are also discussed in this chapter.

7.2. Methanation in Ammonia and Hydrogen Plants

In an *ammonia* plant methanation is the final stage in the purification of the synthesis gas in which small concentrations of carbon monoxide and carbon dioxide (0.1–0.5%) are removed catalytically by reaction with hydrogen. The synthesis catalyst is extremely susceptible to the presence of carbon oxides, so much so that its long-term activity is

Chapter 7. Methanation

adversely affected unless they are reduced to very low levels (see Section 8.5.2.). In some plants the presence of carbon dioxide in the synthesis gas can also lead to the formation of ammonium carbamate (NH_4COONH_2) where the make-up gas meets ammonia-containing circulating gas, forming a solid deposit which can restrict the gas flow and/or foul the compressor. Ammonium carbamate is also known to cause stress corrosion cracking at critical points in the synthesis-gas compressor.

In a *hydrogen* plant a methanation stage is normally required because a concentration of 0.1–0.5% of carbon oxides is often unacceptable in the product hydrogen, since they may poison a downstream hydrogenation catalyst. The additional methane formed in this way, typically up to 0.5%, is relatively unimportant compared with the 2–3% of unconverted methane which is already present in the gas from the reformer. The concentration of oxides of carbon remaining in the process gas leaving the methanator is typically less than 5 ppm.

The simplified diagram in Figure 7.1 shows how the methanator fits into the process scheme of a typical ammonia plant, while Table 7.1 illustrates gas compositions and plant data for a typical shift/methanation section of an ammonia plant.

Table 7.1. Process gas composition (dry basis) and conditions for shift/methanation sections from an operating ammonia plant

	HT shift		LT shift		Methanation	
	Inlet	Exit	Inlet	Exit	Inlet	Exit
CO/%	12.76	3.33	3.33	0.40	0.49	<5 ppm
CO_2/%	8.18	15.85	15.85	18.24	0.20	
H_2/%	56.34	60.02	60.02	61.15	74.68	74.06
CH_4/%	0.22	0.21	0.21	0.20	0.24	0.95
N_2/%	22.20	20.16	20.16	19.77	24.10	24.69
A/%	0.30	0.25	0.25	0.24	0.29	0.30
Temperature/°C	370	432	220	242	318	365
Steam ratio	0.51	0.48	0.48	0.44	0.011	0.020
Wet gas space velocity/h^{-1}	4100		4475		6409	

7.2. Methanation in Ammonia and Hydrogen Plants

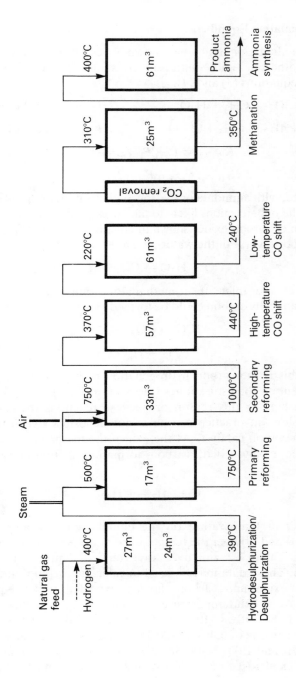

Figure 7.1. Simplified block diagram of typical 1000 tonnes day^{-1} ammonia plant, showing types and approximate volumes of catalyst used in the major stages together with typical converter temperatures.

7.2.1. Methanation Equilibria

The two methanation reactions shown in equations (1) and (2) are strongly exothermic, and equilibrium constants for these reactions are defined by equations (3) and (4).

$$CO + 3H_2 \rightarrow CH_4 + H_2O \qquad \triangle H_{298} = -206.2 \text{ kJ mol}^{-1} \quad (1)$$

$$CO_2 + 4H_2 \rightarrow CH_4 + 2H_2O \qquad \triangle H_{298} = -165.0 \text{ kJ mol}^{-1} \quad (2)$$

$$K_{P(CO)} = P_{CH_4}P_{H_2O}/P_{CO}P_{H_2}^3 \qquad (3)$$

$$K_{P(CO_2)} = P_{CH_4}P_{H_2O}^2/P_{CO_2}P_{H_2}^4 \qquad (4)$$

In practice, with ammonia or hydrogen plants under normal operating conditions the only reactions likely to take place in the methanator are the methanation of carbon oxides shown in equations (1) and (2), and either the forward or reverse of the water-gas shift reaction (5).

$$CO + H_2O \rightarrow CO_2 + H_2 \qquad (5)$$

However, at the inlet of a methanator conditions are such that thermodynamically carbon could be formed via reactions (6) or (7).

$$2CO \rightarrow CO_2 + C \qquad (6)$$

$$CO + H_2 \rightarrow H_2O + C \qquad (7)$$

Opposing this is the carbon hydrogenation reaction (8) which, in the light of recent mechanistic studies (see Section 7.2.2), is expected to be rapid. Whether for this reason or because reactions (6) and (7) are extremely slow, in practice carbon is not deposited over commercial nickel methanation catalyst. At the exit of the methanator, and within most of the catalyst bed, carbon formation is thermodynamically impossible.

$$C + 2H_2 \rightarrow CH_4 \qquad (8)$$

The effect of temperature on the equilibrium constants given by equations (3) and (4) over the temperature range 200–600°C is shown by the data in Table 7.2 and over an extended temperature range in Appendix 6. These data may be used to calculate the concentrations of carbon monoxide and carbon dioxide expected under equilibrium conditions. At an exit temperature of 320°C, $K_{P(CO)} = 3.26 \times 10^6$ bar^{-2} and $K_{P(CO_2)} = 1.10 \times 10^5$ bar^{-2}. Typically, the inlet gas to the methanator consists of 0.4% CO, 0.1% CO_2, 73.7% H_2, 24.4% N_2 and A, 0.4% CH_4 and 1.0% H_2O. Since conversion of the carbon oxides to methane goes almost to completion, the exit gas contains about 1.0%

7.2.1. Methanation Equilibria

CH_4 and 1.5% H_2O, with hydrogen and nitrogen in the 3 : 1 proportion required for ammonia synthesis. At atmospheric pressure and 320°C the equilibrium partial pressures are $P_{CO} = 1.1 \times 10^{-10}$ bar and $P_{CO_2} = 6.8 \times 10^{-11}$ bar, and correspond to equilibrium concentrations of about 1×10^{-4} ppm. Obviously, the performance of a methanator is unlikely to be limited by thermodynamic approach to equilibrium.

Table 7.2. Equilibrium constants for methanation of carbon monoxide and carbon dioxide over a range of temperatures (equations (1) and (2) respectively)

T/°C	$K_{P(CO)} = P_{CH_4}P_{H_2O}/P_{CO}P_{H_2}^3$	$K_{P(CO_2)} = P_{CH_4}P_{H_2O}^2/P_{CO_2}P_{H_2}^4$
200	0.215×10^{12}	0.947×10^9
220	0.235×10^{11}	0.156×10^9
240	0.304×10^{10}	0.294×10^8
260	0.456×10^9	0.627×10^7
280	0.784×10^8	0.149×10^7
300	0.152×10^8	0.387×10^6
320	0.326×10^7	0.110×10^6
340	0.773×10^6	0.337×10^5
360	0.200×10^6	0.111×10^5
380	0.560×10^5	0.389×10^4
400	0.169×10^5	0.144×10^4
420	0.542×10^4	0.566×10^3
440	0.185×10^4	0.233×10^3
460	0.671×10^3	0.100×10^3
480	0.256×10^3	0.450×10^2
500	0.102×10^3	0.210×10^2
520	0.427×10^2	0.102×10^2
540	0.186×10^2	0.508×10^1
560	0.842×10^1	0.262×10^1
580	0.395×10^1	0.139×10^1
600	0.192×10^1	0.761×10^1

From a knowledge of the heat of reaction the temperature rise corresponding to a given conversion of carbon oxides can be calculated using the relevant thermodynamic data given in Table 7.3. Normally operation is effectively adiabatic, because the heat loss from a

Chapter 7. Methanation

well-lagged plant converter is negligible compared with the heat input. If it is assumed that specific heats are constant over the usual range of operating conditions, then the temperature rise for a typical methanator gas composition in an ammonia or hydrogen plant is 74°C per 1% of carbon monoxide converted and 60°C for 1% of carbon dioxide converted.

7.2.2. Kinetics and Mechanisms

The mechanism of catalytic methanation of carbon monoxide over nickel has been the subject of controversy since the reaction was discovered[180] by Sabatier and Senderens in 1902. The early work, such as that reviewed by Vlasenko and Yuzefovich[181] and Vannice,[182]

Table 7.3. Thermodynamic data relevant to methanation

Specific heats/Jmol^{-1}K^{-1}

Catalyst

Nickel oxide	$c_p = 57.22 + 3.47T \times 10^{-3} - 12.19T^{-2} \times 10^5$
Nickel metal	$c_p = 16.97 + 29.43T \times 10^{-3}$
Alumina	$c_p = 109.2 + 18.34T \times 10^{-3} - 30.38T^{-2} \times 10^5$

At 300°C specific heats per gram are:

Nickel oxide	$c_p = 0.743 \text{Jg}^{-1}\text{K}^{-1}$
Nickel metal	$c_p = 0.577 \text{Jg}^{-1}\text{K}^{-1}$
Alumina	$c_p = 1.083 \text{Jg}^{-1}\text{K}^{-1}$

For unreduced catalyst containing 30% nickel oxide and 70% alumina, at 300°, $c_p = 0.98 \text{Jg}^{-1}\text{K}^{-1}$.

For a similar catalyst in the reduced state $c_p = 0.96 \text{Jg}^{-1}\text{K}^{-1}$.

Gas stream

Hydrogen	$c_p = 27.67 + 3.39T \times 10^{-3}$
Nitrogen	$c_p = 28.26 + 2.53T \times 10^{-3} - 0.54T^{-2} \times 10^5$
Carbon monoxide	$c_p = 27.59 + 5.02T \times 10^{-3}$
Carbon dioxide	$c_p = 32.19 + 22.15T \times 10^{-3} - 3.47T^2 \times 10^{-6}$
Methane	$c_p = 22.32 + 48.07T \times 10^{-3}$
Steam	$c_p = 34.36 + 6.30T \times 10^{-3} - 0.56T^2 \times 10^{-5}$

Specific heats of gas mixtures may be calculated assuming that the contribution of each component to the total specific heat is the product of the mole fraction of that component and its specific heat.

Table 7.3 continued

Heats of formation/kJmol^{-1} (kcal mol^{-1}) at 298 K

Nickel oxide	$\triangle Hf = -239.51\ (-57.30)$
Water	$\triangle Hf = -241.60\ (-57.80)$
Carbon monoxide	$\triangle Hf = -110.42\ (-26.42)$
Carbon dioxide	$\triangle Hf = -393.13\ (-94.05)$
Methane	$\triangle Hf = -74.74\ (-17.88)$

Heats of reaction/kJmol^{-1}

	298K	573K
$CO + 3H_2 \rightarrow CH_4 + H_2O$	-205.92	-216.69
$CO_2 + 4H_2 \rightarrow CH_4 + 2H_2O$	-164.81	-177.61
$CO + H_2O \rightarrow CO_2 + H_2$	-41.11	-39.08
$NiO + H_2 \rightarrow Ni + H_2O$	-2.09	-7.89
$NiO + CO \rightarrow Ni + CO_2$	-43.22	-47.11
$Ni + \tfrac{1}{2}O_2 \rightarrow NiO$	-239.51	-236.23

postulated a sequence of steps involving reaction of *adsorbed* carbon monoxide and hydrogen. Vannice[183] derived a theoretical rate equation on the assumption that the rate-determining step was the interaction of adsorbed hydrogen atoms with a surface CHOH species as in the equation (9).

$$CHOH(ads) + (y/2)H_2(ads) \rightarrow CH_y(ads) + H_2O(g) \qquad (9)$$

However, it was established in 1974 that carbon monoxide can dissociate into surface carbon and oxygen on nickel,[184] and the previously accepted concept that CHOH species were intermediates in the reaction lost credibility. There is now considerable evidence that the reaction proceeds via an active surface carbon intermediate formed by the dissociation of carbon monoxide as shown in equations (10) and (11).

$$CO(g) \rightarrow CO(ads) \qquad (10)$$

$$CO(ads) \rightarrow C(ads) + O(ads) \qquad (11)$$

Several studies have provided support for this mechanism, and this work has been reviewed by Ponec[185] and Bell.[186] Wise and co-workers[187, 188] observed that two types of carbon can be formed by disproportionation of carbon monoxide, which they designated alpha and beta carbon. It was postulated that the reactivity of the dispersed alpha carbon made it a feasible reaction intermediate in methanation. Winslow and Bell[189] found that the surface coverage of alpha carbon rapidly came to a steady-state value, and the rate of methanation was found to be a linear function of the coverage of alpha carbon. Using transient response experiments Happel's group[190–192] deduced that the most abundant reacting intermediate on the nickel surface is partially hydrogenated carbon (CH), with the carbidic carbon consisting of a relatively small pool of "active" carbon that exchanges with a larger pool of "inactive" carbon.

The rate-determining step for the methanation of carbon monoxide is still far from resolved. Happel et al.[190] suggested that the rates of hydrogenation of surface carbon and partially hydrogenated carbon (CH) are important factors in determining the reaction rate, Klose and Baerns[193] postulated that the hydrogenation of surface carbon to a CH_2 species (involving two adsorbed hydrogen atoms) is the rate-determining step. However, Underwood and Bennett[194] proposed that both adsorbed C and adsorbed CO occupy important sites on the surface, and it appears that the rates of both CO dissociation and carbon hydrogenation are important in determining the global reaction rate. It therefore appears there is no unequivocal rate-determining step.

Several rate equations have been published for the methanation of carbon monoxide, but few of these relate to actual ammonia/hydrogen plant conditions.[195–197] Power-law rate equations of the form in equation (12) are frequently used to describe the formation of methane. The

$$\text{rate} = kP_{CO}{}^n P_{H_2}{}^m \qquad (12)$$

value of m for most conditions is positive and close or equal to unity. However, negative as well as positive values of n ranging between -1 and 0.5 have been mentioned[193] in the literature. Only at low CO partial pressures combined with high hydrogen pressure can hydrogen compete sufficiently well with carbon monoxide for adsorption sites; n then becomes greater than zero. Under process conditions the methanation of carbon monoxide is first order, but the reaction may not show first-order dependence on total pressure because of diffusional and retardation effects. Table 7.4 contains selected rate equations for carbon monoxide methanation, while those for carbon dioxide are given in Table 7.5.

Although there is less published work on the methanation of carbon dioxide than of carbon monoxide, in connection with hydrogen and ammonia plants interest centres around the methanation of mixtures of both carbon oxides at a total concentration of ususally less than one per cent (Table 7.1 contains some typical values). One of the intriguing aspects of the methanation of mixtures of carbon monoxide and carbon

Table 7.4. Kinetics of carbon monoxide methanation over nickel catalyst

Temperature range/°C	Pressure range/bar			Rate equation	E_a/kJ mol^{-1}	Ref.
	H_2	CO	Total			
135–175	1	$4\text{–}30 \times 10^{-4}$	1	$r_{CH_4} = kP_{CO}^{0}$	67.7	201
170–210	1	$2\text{–}24 \times 10^{-3}$		$r_{CH_4} = kP_{CO}^{1}/(1 + KP_{CO})^2$	42.2	196
200–294	1–15	0.1–0.75	1–15	$r_{CH_4} = kP_{H_2}^{0.15}/(1 + KP_{CO}/P_{H_2})^{1/2}$	75–117	195
25–300	$0\text{–}8 \times 10^{-2}$	$0\text{–}8 \times 10^{-2}$	$7\text{–}19 \times 10^{-2}$	$r_{CH_4} = kP_{H_2}^{1.4}P_{CO}^{-0.9}$	84	202
243	1–100	0.01–1	1–100	$r_{CH_4} = kP_{Total}^{0.5}$	100–109	203
300–350			1	$r_{CH_4} = kP_{CO}P_{H_2}^{3}/(A + BP_{CO} + CP_{CO_2} + DP_{CH_4})^4$		204
250–300		0.1–1		$r_{CH_4} = kP_{H_2}^{0.9}P_{CO}^{-0.2}$	150	205
261–281			1–21.4	$r_{CH_4} = kP_{Total}^{0.3}$	121–134	206
240–280			1	$r_{CH_4} = Ae^{-Em}P_{H_2}^{0.77} \times P_{CO}^{-0.31}$	105	183
180–284	1–25	0.001–0.6		$r_{CH_4} = kK_{CO}K_{H_2}^{2}P_{CO}^{0.5}P_{H_2}/(1 + K_{CO}P_{CO}^{0.5} + K_{H_2}P_{H_2}^{0.5})^3$		193

Chapter 7. Methanation

Table 7.5. Kinetics of carbon dioxide methanation over nickel catalysts

Rate equations	Reference
$r_{CH_4} = k \{P_{CO_2}P_{H_2}^2 - (P_{CH_4}P_{H_2O}/KP_{H_2}^2)\}/(P_{H_2}^{0.5} + aP_{CO_2} + b)^5$	207, 208
$r_{CH_4} = kP_{CO_2}P_{H_2}^4/(1 + K_{H_2}P_{H_2} + K_{CO_2}P_{CO_2})^5$	209
$r_{CH_4} = kP_{CO_2}^{0.5}$	210, 211
$r_{CH_4} = kP_{CO_2}$	181
$r_{CH_4} = kP_{CO_2}/(1 + aP_{CO_2})$	196
$r_{CH_4} = kP_{CO_2}^{0.5}P_{H_2}^{0.5}/(1 + aP_{CO_2}^{0.5}/P_{H_2}^{0.5} + bP_{CO_2}^{0.5}P_{H_2}^{0.5} + cP_{CO})^2$	212

dioxide under laboratory conditions is that reaction of carbon dioxide is inhibited until the concentration of carbon monoxide has been reduced to about 200–300 ppm. Thus, a mixture of the two carbon oxides is much more difficult to methanate completely than carbon monoxide alone is. A number of mechanisms have been proposed for the methanation of carbon dioxide, and these can be summarized under two categories. The first involves the conversion of carbon dioxide to carbon monoxide via the reverse water-gas shift reaction (13) followed by carbon monoxide methanation.

$$CO_2 + H_2 \rightarrow CO + H_2O \qquad (13)$$

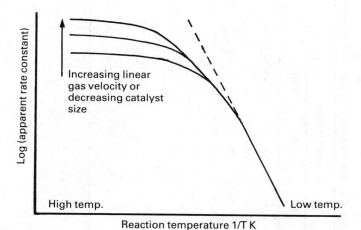

Figure 7.2. Illustrative graph of log (apparent rate constant) against 1/T K for a reaction suffering progressive diffusion limitation. At high temperature with relatively low linear gas velocity the reaction becomes film diffusion limited. This can be countered to some extent by decreasing the catalyst size and/or increasing the linear gas velocity.

7.2.2. Kinetics and Mechanisms

Figure 7.3. Photograph of a typical methanator in an ammonia plant. This particular vessel contains about twenty-five tonnes of catalyst.

The second category involves the direct hydrogenation of carbon dioxide to methane. Recent studies[198-200] have provided evidence which supports the first mechanism and which accounts for the high selectivity to methane as well as the inhibition by carbon monoxide.

The intrinsic rates of the methanation of both carbon monoxide and carbon dioxide, reactions (1) and (2), can become sufficiently fast under some conditions at higher temperatures for diffusion effects to become very important, and not fully appreciating these effects has probably contributed to the differences between some of the published rate expressions for these processes. The graph of log (apparent rate constant) against 1/T K shown in Figure 7.2 illustrates how diffusion limitations can control the rate of methanation at high temperatures. This kind of behaviour can be observed in laboratory experiments with commercial catalyst when methanating low levels of carbon monoxide and carbon dioxide at relatively low linear gas velocities. Under these conditions film diffusion controls the overall reaction rate at high temperatures. Increasing the temperature then has little influence on the reaction rate. However, as shown in Figure 7.2 this limitation can be overcome to some extent by operating with a higher linear gas velocity (while keeping constant space velocity), and/or using a smaller catalyst particle size to provide a higher geometric surface area for mass transfer to take place across. These factors have to be considered when designing methanators for operation under unusual conditions.

7.2.3. Catalyst Formulation

Early experimental work on the methanation of carbon oxides mainly involved the used of nickel, although some work was done on other Group VIII metals. Subsequently, iron catalysts were extensively studied, but were found to be subject to excessive carbon deposition, leading to blockage of the catalyst pores and consequent deactivation. Moreover, compared with nickel catalysts they were less active and less specific—showing a tendency to form higher hydrocarbons, which appeared as a liquid product.

This is the basis of the Fischer–Tropsch synthesis process which has found large-scale commercial application in the production of hydrocarbons from coal-derived synthesis gas in the Sasol plants in South Africa.[213] Precious metals, particularly ruthenium, possess considerable methanation activity, and catalysts are available containing about 0.5% metal supported on alumina. These catalysts can operate at low temperatures, but under normal plant conditions they are no more effective than conventional nickel catalysts. Ruthenium catalysts have found limited commercial application, for example in the "ultramethanation" duty, in caprolactam plants, where the carbon oxides concentration can be reduced from 50–500 ppm to less than 1 ppm at 150–175°C. Reaction rate is inhibited by carbon monoxide, so that in a typical ammonia plant where the feed gas contains 0.2–0.3% carbon monoxide it would be necessary to operate with a reactor inlet temperature of 230°C to achieve 2–5 ppm outlet of carbon oxides in the outlet gas.

Commercial methanation catalysts are therefore mainly in the form of nickel metal dispersed on a support consisting of various oxide mixtures, such as alumina, silica, lime and magnesia, together with compounds such as calcium aluminate cements. In practical terms, a good methanation catalyst is one which is physically strong, reducible at about 300°C (the typical operating temperature) and has high activity. In order to provide long life, it must retain these properties in use, and this can be obtained by careful attention to the formulation and manufacture of the catalyst. For example, lives of 8–10 years are commonly obtained from charges of ICI Catalyst 11-3, depending on the temperature of operation and on the poisons in the synthesis gas.

Under conditions where the rate is limited by chemical control, methanation activity is directly related to the surface area of the nickel metal obtained when the catalyst is reduced. The highest surface area of metal and the highest activity are obtained when the nickel is produced as very small crystallites, usually less than 10 nm in diameter. One of the functions of the support is to provide a matrix on which the nickel crystallites can be finely dispersed, and another is to retard their sintering which would give rise to large crystallites having a lower overall surface area and hence lower activity. The nickel can be dispersed on the support in one of two general ways—the catalyst may be prepared by impregnation methods or by precipitation.

In the former case a preformed support, usually comprising refractory oxides, is calcined to give it the necessary strength and internal structure. It is then impregnated with a solution of a suitable nickel compound, which is then decomposed to the oxide by heating. One problem with methanation catalysts prepared in this way is they may be difficult to reduce at the temperatures available in plant, but this can be overcome by prereducing the catalyst.

Precipitated catalysts are usually prepared by precipitating the basic carbonate in a finely divided, high surface area form which is washed free of unwanted soluble salts, dried and decomposed to give oxides. It is then formed into the appropriate shape by methods such as granulation or (better) by pelleting.

In practice the formulation of a methanation catalyst is a compromise of opposing requirements to give optimum reducibility, activity and stability, together with appropriate physical properties. Intimate mixing of the components can lead to formation of compounds or of solid solutions that are relatively difficult to reduce at normal operating temperatures, but which when reduced contain well-dispersed and well-stabilized nickel crystallites.

As an example of compound formation, alumina and nickel oxide readily form spinel compounds of the type $NiO.Al_2O_3$. The reaction takes place at around 1000°C with nickel oxide and α-alumina, but with finely divided nickel oxide and the more reactive γ-alumina it takes place at much lower temperatures. As a result, with some formulations when the oxides are precipitated "spinel precursors" can be detected in the dried precipitate, and such catalysts may have to be reduced at temperatures as high as 500°C. Catalysts of this type are therefore unsuitable for use in conventional ammonia plant methanation units, since they cannot achieve the conditions necessary to reduce them.

Both magnesia and nickel oxide have face-centred cubic lattices, and because of the similarity of their ionic radii (Ni^{2+} = 6.9 nm, Mg^{2+} = 6.5 nm) the formation of solid solutions with any proportion of the two oxides is possible. Because such solid solutions are thermodynamically more stable than the separate oxides, they are more difficult to reduce than pure nickel oxide, and this may be exacerbated by kinetic effects. Takemura et al.[214] demonstrated that nickel oxide reduced completely at 230–400°C, whereas a 10% NiO/90% MgO solid solution reduces in two

Chapter 7. Methanation

stages, one at 230–400°C and the other at 500–600°C. Therefore, for ease and completeness of reduction excessive solid solution formation of this type should also be avoided.

However, in other respects solid solution formation is beneficial, because it retards crystal growth of nickel oxide. During manufacture a precipitated nickel compound such as the carbonate has to be converted into nickel oxide, and in order to obtain small nickel oxide crystallites it is desirable that the calcination temperature is the minimum compatible with efficient conversion of carbonate to oxide. Differential thermal analysis (DTA)(Figure 7.4) reveals that this endothermic process occurs in two main stages, with maxima around 150°C and 340°C, and that the presence of magnesia raises the required temperature by only about 15°C. However, the presence of magnesia does retard the growth of nickel oxide during calcination (see Figures 7.5–7.7). For example, Figure 7.5 shows that calcination at 500°C for four hours increases the nickel oxide crystallite size to 30–40 nm, whereas the crystallite size of

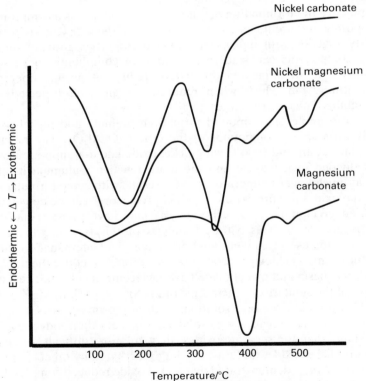

Figure 7.4. Differential thermal analysis of basic carbonates.

7.2.3. Catalyst Formulation

NiO/MgO solid solution (60 : 40w/w) is only about 8 nm after the same treatment. Figures 7.6 and 7.7 illustrate the effect of calcination duration on crystal growth at different temperatures.

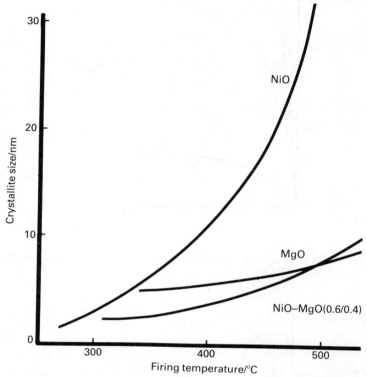

Figure 7.5. Effect of magnesia on thermal stability of nickel oxide—variation of crystallite size with calcination temperature.

Crystallite size is proportional to $T^{0.25}$ for nickel oxide alone but only to $T^{0.12}$ for nickel oxide/magnesia and or magnesia alone (where the temperature T is expressed in Kelvin). Figure 7.8 depicts the effect of calcination temperature on subsequent catalyst activity following reduction at 300°C in laboratory tubular reactors operating at 1 bar with an inlet gas composition of 0.4% CO, 25% N_2 and 74.6% H_2, and an inlet temperature of 300°C. In these tests conversion of carbon monoxide to methane was measured and catalyst activity expressed as the rate coefficient k in the first-order equation (14).

$$\text{rate} = kp_{CO}P^{0.3}(1 - K/Kp)e^{-E/RT} \qquad (14)$$

The reducibility of this catalyst is illustrated in Figure 7.9, which

Chapter 7. Methanation

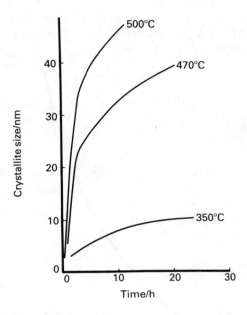

Figure 7.6. Variation of NiO crystallite size with time at calcination temperatures over the range 350–500°C.

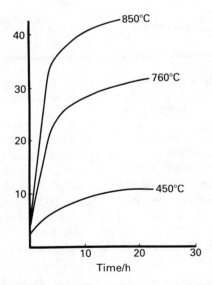

Figure 7.7. Variation of crystallite size (0.6/0.4 NiO : MgO) with time at calcination temperatures over the range of 450–850°C.

7.2.3. Catalyst Formulation

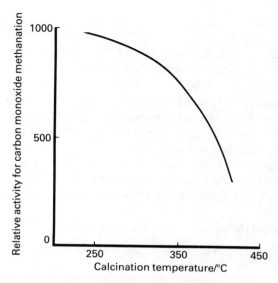

Figure 7.8. Effect of calcination temperature on methanation activity.

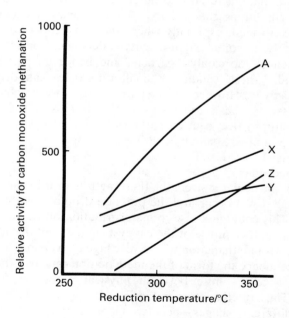

Figure 7.9. Effect of reduction temperatures on methanation activity for catalysts containing different amounts of reactive alumina. The optimized commercial catalyst A is easily reduced and also has superior long-term activity.

shows the relative activity after reduction to constant activity at temperatures in the range 280–350°C. It will be seen that an optimized commerical catalyst A is superior to other catalysts containing larger amounts of alumina, because they are more difficult to reduce at acceptable temperatures.

In summary, ICI found it beneficial to include a small amount of magnesia in methanation Catalyst 11-3 because this provides the ideal compromise between ease of reduciblity and sintering resistance. By this means, a catalyst is produced which is readily reduced at 300–350°C; and, in operation, loss of activity caused by sintering is not a problem during several years of normal operation at temperatures up to 350°C—even with occasional temperature excursions up to 650°C!

7.2.4. Physical Properties of Methanation Catalysts

Commercial methanation catalysts contain between 15% and 35% nickel oxide and are made either by a precipitation method or by impregnation of nickel solution onto a preformed support. There is generally no correlation between nickel content and catalyst activity if catalysts are manufactured in different ways. Moreover, under normal operating conditions the activity of a methanation catalyst may be diffusion controlled, especially where the bed temperature is highest. Consequently, to some extent activity is dependent on total geometric surface area of the catalyst particle, and hence on particle size and shape—under these conditions smaller particles can display higher apparent activity. However, if the particle size is too small the pressure drop across the bed is too high and the process power requirement excessive. In practice, catalyst particles such as pellets with a diameter of about 5 mm are suitable, although in special situations smaller particles may be used.

The mechanical properties of methanation catalyst are also important, particularly strength. If strength is insufficient, breakdown of the catalyst will occur and the pressure drop across the reactor will increase. This could give rise to maldistribution of the gas through the bed, giving inefficient use of the catalyst charge, and this is particularly important with methanation where very high conversions are necessary. As a concequence the form of the methanation catalyst is invariably the result of a compromise between physical properties and effective activity. Thus, methanation catalysts are (or have been) offered in several different shapes—pellets, spheres, irregular shapes, solid extrudates and hollow, tubular extrudate. Packing densities are normally around 1.0 kg l^{-1}. Because of their usual long life in operation, catalyst stability is usually more important than initial activity, which makes the physical strength particularly important. The ICI methanation catalysts are usually supplied in the form of strong cylindrical pellets having a diameter of 5.4 mm and height of 3.6 mm, that normally charges into a reactor to a packing density of 1.10–1.25 kg l^{-1}. This form has, over a number of years, been shown to have good handling characteristics and resistance to mechanical attrition.

Methanators are normally vertically-mounted with downward gas flow, and in order to ensure good gas distribution in the catalyst bed a bed height : diameter ratio greater than 1.0 is recommended. Thus, the methanator in a typical modern 1000 tonnes day^{-1} ammonia plant has a catalyst volume of 20–25 m^3 with a reactor diameter of about 2.5 m and catalyst bed depth of 3.5–4.0 m.

7.2.5. Catalyst Reduction

Methanation catalysts are almost always manufactured and transported in the oxidized form, and therefore they must be reduced in the reactor to give nickel metal in order to make them active. The reduction is usually carried out in process gas and occurs by the two reactions (15) and (16).

$$NiO + H_2 \rightarrow Ni + H_2O \qquad \triangle H_{25°C} = +2.6 \text{ kJ mol}^{-1} \qquad (15)$$

$$NiO + CO \rightarrow Ni + CO_2 \qquad \triangle H_{25°C} = -30.3 \text{ kJ mol}^{-1} \qquad (16)$$

Since neither reaction is strongly exothermic the reduction process itself does not cause a large temperature rise in the catalyst bed. However, once some metallic nickel has been formed by reduction with process gas methanation will start. The exothermic heat of reaction will augment the temperature and accelerate reduction of the catalyst. The temperature rise must not be excessive, and for this reason the gas used for reduction should contain as little carbon monoxide and carbon dioxide as possible, and preferably not more than 1% in total. It is worth making checks to ensure that the concentration of carbon oxides does not increase during the reduction; for example as a result of malfunction of the carbon dioxide removal unit. These precautions protect not only the catalyst, but also the methanation vessel.

Although most methanation catalysts can withstand temperature excursions, the maximum design temperature for the converter itself is frequently about 450°C. In the later stages of reduction it is advantageous to raise the temperature to about 400°C, because this will increase the proportion of reduced nickel. To achieve this temperature it may be necessary to increase the carbon monoxide or carbon dioxide

Chapter 7. Methanation

content of the inlet gas by a controlled bypassing of the low-temperature shift catalyst or of the carbon dioxide removal system. These techniques are often the only means for providing extra heat to the catalyst when the gas entering the methanator is heated by exchange with the exit gas.

Progress of the reduction can be monitored by following the rapid fall in the exit carbon monoxide and carbon dioxide to the design level, which is normally less than 5 ppm $CO + CO_2$. The complete procedure from beginning the heat-up to passing methanator exit gas to the ammonia synthesis loop can be completed in less than 12 hours, and less than 24 hours is normal, but this does depend on the catalyst used. Figure 7.10 illustrates how temperature and exit carbon monoxide and carbon dioxide concentrations varied during a typical reduction of ICI Catalyst 11-3 in a 980 tonnes day^{-1} ammonia plant in Western Europe, and illustrates the effect of partial bypassing of the low-temperature shift converter.

CAUTION: Great care must be taken during the heating up or cooling down of methanation catalyst in the reduced state to avoid formation of highly toxic nickel carbonyl (see Chapter 3, Section 3.15.2).

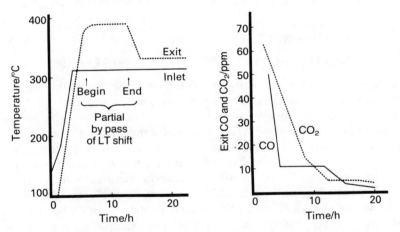

Figure 7.10. By-passing low-temperature shift unit during reduction of methanation catalyst in an ammonia plant to increase methanator temperature to complete reduction.

7.2.6. Catalyst Poisons

With a well-formulated methanation catalyst, sintering is not an important cause of activity loss at normal operating temperatures (250–350°C), even if the catalyst is occasionally overheated to 650°C. In

7.2.6. Catalyst Poisons

practice the principal cause of loss of activity is usually poisoning. Sulphur compounds are virulent poisons for nickel catalysts, but the methanation catalyst is protected by the low-temperature shift catalyst in the preceding stage which removes sulphur from the process gas very effectively. Therefore, in normal operation the methanation catalyst is unlikely to be exposed to sulphur which arises from the feedstock. The exception to this would be if the low-temperature shift converter were partially bypassed. Serious deactivation of the methanation catalyst can then occur, for example one catalyst that contained about 30% nickel oxide (before reduction) lost a significant amount of activity when the sulphur content of the catalyst at the top of the charge exceeded ~0.1%.

The poisons most likely to be encountered under normal operating conditions in an ammonia plant are those originating from the carbon dioxide removal system which precedes the methanator. Carry-over of a small amount of liquid into the methanator, which is almost inevitable, is not normally serious. However, plant malfunction can sometimes result in large quantities of carbon dioxide removal liquor being pumped over the catalyst. This can be very deleterious, particularly if the liquid contains sulphur or arsenic, since irreversible loss of activity results. Of

Table 7.6. Poisoning effect of liquids used in carbon dioxide removal systems on methanation catalysts

Process	Chemical	Effect
Benfield	Aqueous potassium carbonate	Blocks pores of catalyst by evaporation of K_2CO_3 solution
Vetrocoke	Aqueous potassium carbonate plus arsenious oxide	Blocks pores of catalyst by evaporation of K_2CO_3 solution. As_2O_3 is also a poison—0.5% of As on the catalyst will reduce its activity by more than 50%
Benfield DEA	Aqueous potassium carbonate with 3% diethanolamine	Blocks pores of catalyst by evaporation of K_2CO_3 solution. (DEA is harmless)
Sulphinol	Sulpholane, water, di-2-propanolamine	Sulpholane will decompose and cause sulphur poisoning
MEA, DEA	Mono- or diethanolamine in aqueous solution	None
Rectisol	Methanol	None
Catacarb	Aqueous potassium carbonate with borate additive	Blocks pores of catalyst by evaporation of K_2CO_3 solution
Selexol	Dimethyl ether of polyethylene glycol	None

the commonly used carbon dioxide removal processes, shown in Table 7.6, only Vetrocoke liquor (containing an arsenic) and Sulphinol liquor (containing sulphur) will adversely affect the catalyst permanently. However, if only potassium carbonate solution or organic solvents are involved the effects are less important.

In the former case physical blocking of the catalyst pores can occur, but the potassium carbonate can be removed by washing the catalyst with water. Indeed, plant operators in several countries have shown that after a solution carry-over incident it is possible to remove the contaminant by cooling the catalyst (under nitrogen) and back-washing with demineralized water. When the concentration of potassium in the wash water has fallen to a low level the catalyst is dried by heating in an inert gas stream and commissioned in the normal way. The procedure is not applicable to all methanation catalysts, but the physical strength of ICI Catalyst 11-3 is unaffected by this procedure, and subsequent lives of more than 5 years have been achieved.

7.2.7. Prediction of Catalyst Life

It is most important to be able to estimate the remaining useful life of a charge of methanation catalyst at any time. The question to be answered is: "should the catalyst be changed during the next shutdown, or is it good enough to last for several more years?". With most catalysts in this situation the requirement is for a yes/no answer rather than a precise prediction, because changing a catalyst should never be the sole reason for a shutdown. This is reasonable because the cost of plant downtime compared with the cost of a catalyst charge shows that it is clearly economical to change catalyst early during a planned shutdown rather than to run the risk of a catalyst failure, causing an additional unexpected shutdown.

Methanation converters on most ammonia and hydrogen plants are conservatively rated because of original conservative design and as a result of increases in catalyst activity since plants were built. Consequently, at the beginning of the life of a catalyst charge most of the methanation reaction is virtually completed in the first quarter of the bed, and as a result monitoring the exit gas composition gives no information about catalyst die-off. Catalyst deactivation occurs normally by a poisoning mechanism, with a poisoned zone moving progressively through the bed. As poisoning continues the volume of active catalyst remaining will eventually be insufficient to meet the required duty, and the carbon oxide levels at the exit will rapidly increase.

The best way of routinely monitoring the movement of the poisoned

7.2.7. Prediction of Catalyst Life

zone is by accurate measurement of the temperature profile within the catalyst charge. This can be done by using a series of fixed thermocouples, or by a movable thermocouple that can be moved in a sheath that passes through the bed. The latter method is the preferred one, and a typical temperature profile so obtained is shown in Figure 7.11. Data of this kind can be treated in the way outlined in Chapter 6 (Section 6.6.5), and approximate predictions about the remaining useful life of the catalyst charge can be made by plotting the depth of active catalyst against time on line. Considerable experience is needed when deciding the minimum volume of active catalyst needed for a particular duty, since this depends to varying extents on the operating conditions, the actual plant and the catalyst used. Advice on matters of this kind should always be obtained from the catalyst vendor.

A more sophisticated graphical technique for analysing methanator temperature profiles to predict the remaining useful life of the catalyst charge has been devised.[215] This involves selecting a point on an accurately measured temperature profile at which conversion is nearly complete. For example, this can be taken as the point at which a 2.8°C temperature rise remains. The total temperature rise across the methanator is typically 30°C, so the 2.8°C point is sufficiently near the top of the temperature profile that pressures and temperatures can be regarded as constant for the remainder of the bed. The method assumes that carbon monoxide methanates before carbon dioxide, which is commonly accepted[199] (see Section 7.2.2), so that over the last part of the catalyst bed only the completeness of carbon dioxide methanation need be considered. First-order reaction kinetics with respect to carbon

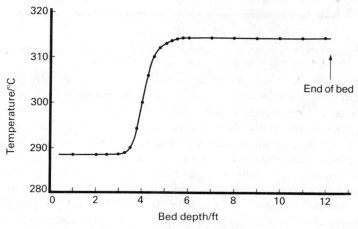

Figure 7.11. Typical ammonia plant methanator temperature profile obtained using a movable thermocouple in a fixed sheath.

Chapter 7. Methanation

dioxide[199] are assumed, and the reaction rate is estimated by drawing the tangent on the temperature profile at the 2.8°C point. This value can then be used to estimate the amount of catalyst required to achieve any selected carbon dioxide exit level. The last 2.8°C temperature rise corresponds to methanation of about 465 ppm of carbon dioxide, and it can be shown that a level of 2 ppm carbon dioxide at the exit corresponds to the point at which the tangent intersects a horizontal line drawn 16°C above the 2.8°C point, as shown in Figure 7.12. In this way the beginning of the active zone can be identified, and so the amount of active catalyst remaining can be calculated.

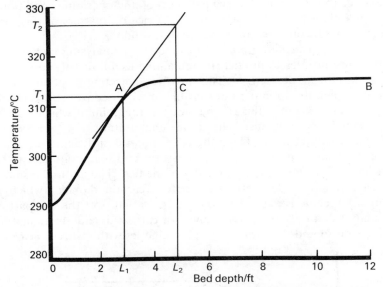

Figure 7.12. Prediction of remaining useful life of a methanation catalyst charge by plotting position of the beginning of the active zone against time.

In practice the effective beginning of the bed is usually located 30–50 cm (1–1.5 ft) below the point at which the maximum temperature is measured. Some typical results obtained in this way are presented in Figure 7.13. In this particular example, given the same die-off rate the end of useful catalyst life would be reached after about six years on-line. Irrespective of how the remaining volume of active catalyst is estimated, it is necessary to have a knowledge of the history of the charge and the plant in order to predict the probable future life of the charge. If catalyst die-off has been caused by a reversible process, such as potassium carbonate deposition, it may be possible to remove the contaminant and to move the temperature profile up the catalyst bed by *in situ* washing, as described in Section 7.2.6.

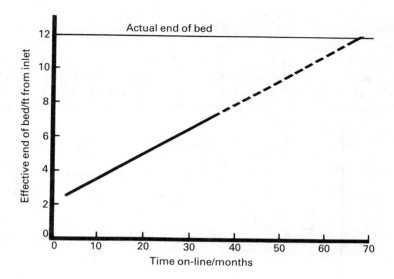

Figure 7.13. Estimation of remaining useful life of methanation catalyst.

7.2.8. Operating Experience

A methanation stage has been included in virtually all ammonia plants designed during the past 20 years, and operating experience in ICI Agricultural Division's four ammonia plants at Billingham, England, is typical of the performance obtained with ICI Catalyst 11-3. For example, No. 1 and No. 2 Kellogg plants (970 tonnes day^{-1}) were charged with 30 m^3 of catalyst in 1973 and 1972, respectively. Both charges gave trouble-free service for more than 10 years until they were discharged in 1983. Each unit produced more than 3 million tonnes of ammonia during the life of one methanation catalyst charge. The catalyst in No. 3 plant was changed in 1977 to allow a statutory vessel inspection to be performed, and this charge is expected to run until the next vessel inspection is required.

Similarly, the No. 4 plant (1100 tonnes day^{-1}) was charged with 36.6 m^3 of catalyst in January 1977 when it was commissioned, and this charge has operated successfully since then. Figure 7.14 displays temperature profiles in these four charges at various times. There are many examples of charges of methanation catalyst giving design CO + CO$_2$ levels at the exit of less than 5 ppm after more than five years of operation, and Table 7.7 reviews the performance data in a selection of West European plants.

Chapter 7. Methanation

The key to satisfactory operation of methanation catalyst is that it must operate well under normal and under abnormal conditions, which may be protracted, or result from a short-term plant upset. Plant A in Table 7.7 is the ICI unit at Severnside near Bristol, England, and illustrates operation at a particularly low inlet temperature, in this instance caused by limitations in the methanator inlet/exit heat exchanger. Resistance to high-temperature excursions is also important,

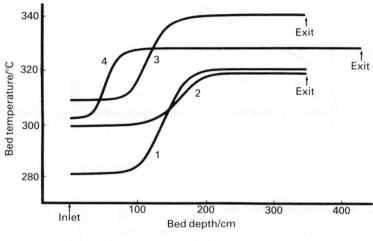

	Operating conditions			
Plant no.	1	2	3	4
Inlet CO/%	0.25	0.2	0.3	0.25
Inlet CO_2/%	0.35	0.15	0.7	0.15
Exit CO + CO_2/ppm	2	2	<4	4
Years on-line	7.5	8.0	7.0	7.0

Figure 7.14. Methanator temperature profiles from ICI's ammonia plants at Billingham.

because upsets in the carbon dioxide removal unit can lead to high carbon dioxide levels at the methanator inlet, and hence to a large temperature rise. Some methanation catalysts can be heated several times to around 550°C without detrimental effect. Indeed, after such an incident the operator may be more concerned with the condition of the vessel rather than that of the catalyst! In one extreme case a charge of ICI Catalyst 11-3 was heated to almost 800°C. Although the catalyst retained a significant amount of its methanation activity it had to be discharged in order to inspect the vessel.

High levels of water can inhibit methanation reactions, so the

7.3. Methanation in Hydrogen Streams for Olefin Plants

Table 7.7. Western European ammonia plant performance data for methanation catalyst (ICI Catalyst 11-3)

	Plant								
	A	B	C	D	E	F	G	H	I
Capacity/tonne day^{-1}	600	1350	800	600	800	900	1000	750	1200
Catalyst volume/m^3	13.2	37	22	13	17	27	24	17	25
Space velocity/h^{-1}	6000	4500	4800	6000	5900	4500	5600	5300	6800
Pressure/bar	16	28	24	21	25	27	24	26	28
Inlet % CO	0.35	0.23	0.20	0.25	0.25	0.35	0.25	0.20	0.32
Inlet % CO$_2$	0.15	0.10	0.05	0.25	0.50	0.15	0.10	0.10	0.05
Inlet temperature/°C	226–259	280–300	300	300	320	295	225–230	310	285
Exit CO + CO$_2$/ppm	2	2	2	2	7	2	8	5	5
Years on line	6.5	8	6.5	7	3.5	3.0	10	4	4

presence of higher than the normal 1–2% of water in the feed is a potential problem. However, with catalyst having good activity that does not suffer physical harm by the presence of higher water levels this need not be important. Thus, examples are known of methanation catalyst operating well with up to 7% H$_2$O in the inlet gas, and still achieving carbon oxides slippage of only 3–4 ppm, even when the plant was run at design inlet temperature and above design load.

7.3. Methanation in Hydrogen Streams for Olefin Plants

Hydrogen recovered from olefin plant tail gas contains methane with traces of ethylene and carbon monoxide. In most cases the impurities do not have a significant effect on other processes on site. However, if the hydrogen is required in the olefins plant itself to hydrogenate acetylenes and dienes in separate C$_2$ and C$_3$ streams (which is usually done over a palladium catalyst), then the concentration of the carbon monoxide must be reduced, and this may be done by methanation using the kind of catalyst discussed in the preceding section. The purity of the hydrogen is not as critical as in an ammonia plant, since levels of up to 150 ppm of carbon monoxide are usually permissible. However, it is important that the methanation catalyst used for this purification duty does not promote the cracking of ethylene to carbon, so the catalyst used needs to be selected carefully. Some typical operating data are given in Table 7.8.

Chapter 7. Methanation

Table 7.8. Typical methanation operating conditions for hydrogen purification in an olefin plant tail-gas duty

	Plant		
	A	B	C
Feed Gas/% H_2	72.0	91.3	84.1
CH_4	27.5	8.6	15.4
C_2H_4	0.2 (max 1.0)	0.01	0.05
CO	0.3	0.08	0.3
Space velocity/h^{-1}	6200	6000	2500
Inlet temperature/°C	274	270	260
Pressure/bar	30	29	29
Exit CO concentration/ppm	<1	<1	<10

7.4. Substitute Natural Gas (SNG)

7.4.1. Oil-based Routes to SNG

Recognition during the early 1970's of a future shortfall of natural gas supplies led to the development and introduction, principally in the USA, of oil-based processes for the manufacture of synthetic natural gas (SNG). The first of these to be commercialized was the British Gas "Catalytic Rich Gas" (CRG) process,[216] and this was followed by two alternative processes, the "Gasynthane" (Lurgi) route and MRG (Japanese Gasoline Company) process. Plants began starting up in 1972–73 and many more were planned, but the oil price increases in late 1973 had major adverse effects on process economics, which caused shelving or cancellation of projects that were far from completion. Only one plant operated the Lurgi and one the MRG route, while the CRG process, because it was first in the field, was used in the remainder of plants.

In the CRG process, which was developed and used in the UK for production of town gas before exploitation of North Sea gas, catalytic gasification of a naphtha or liquefied petroleum gas (LPG) feedstock with steam is used to produce a medium-energy gas which is rich in methane. This is uprated by methanation to the required heating value of 37.25 MJ m^{-3} (1000 Btu ft^{-3}). In the first version of the process, the double methanation route illustrated in Figure 7.15, desulphurised naphtha feedstock was mixed with steam (steam/naphtha weight ratio

7.4 Substitute Natural Gas (SNG)

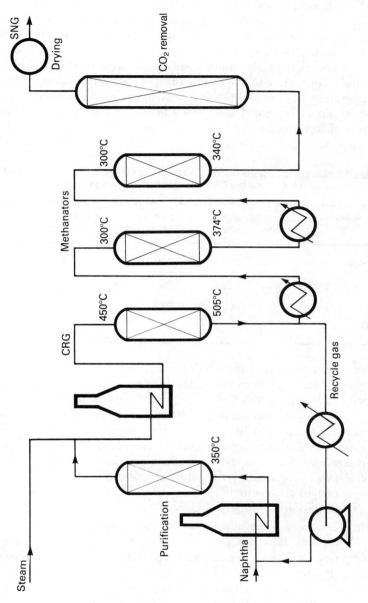

Figure 7.15. The CRG double methanation route with typical inlet and exit temperatures.

Chapter 7. Methanation

2.0), preheated to 450°C and gasified in a CRG reactor. The "rich" gas leaving this reactor at about 505°C was cooled to 300°C before entering the first methanator, where part of the hydrogen reacted with the carbon oxides and gave a temperature rise of 74°C. The gas was then cooled and part of the unused steam was removed to allow further methane formation to take place in a second methanator, where the temperature rise was 40°C. The final stages of the process comprised carbon dioxide removal and drying of the product gas. An important aspect of this process is the recycle loop shown in Figure 7.15. The gas compositions and temperatures in the various stages are given in Table 7.9.

Table 7.9. Gas compositions (as vol.%) at various stages in the production of SNG from naphtha by the CRG double methanation process

	Outlet CRG reactor wet	dry	Outlet first methanator, wet	Outlet second methanator, wet	SNG
CO_2	10.7	20.8	9.5	17.55	0.5
CO	0.5	1.0	0.1		
H_2	8.7	17.0	2.15	1.0	1.45
CH_4	31.4	61.2	34.8	69.75	98.05
H_2O	48.7		53.45	11.7	

In a later version of the CRG process, the "hydrogasification route" (Figure 7.16), gas from the CRG reactor was mixed with additional desulphurised feedstock and passed to the second, or hydrogasification, reactor. The presence of hydrogen in the gas from the first stage was beneficial to catalyst life in the hydrogasifier. A methanation stage followed by carbon dioxide removal completed the process and gas compositions and temperatures are given in Table 7.10. A special precipitated nickel catalyst (containing about 75% nickel oxide before reduction) in the form of pellets, is used in the CRG reactor, the hydrogasifiers and methanators.

In 1977 there were more than 65 CRG plants in operation (Table 7.11), and in 1984 the process was still widely used for town gas production in Brazil, France, Greece, Italy, Japan, South Korea and Spain. By this time in the UK—except for one plant at Granton, Scotland, producing SNG—the CRG plants had been shut down while only four North American plants were running continuously, six were operating seasonally while the remainder were closed down. There is little sign that American demand for SNG will increase significantly in the near future.

7.4 Substitute Natural Gas (SNG)

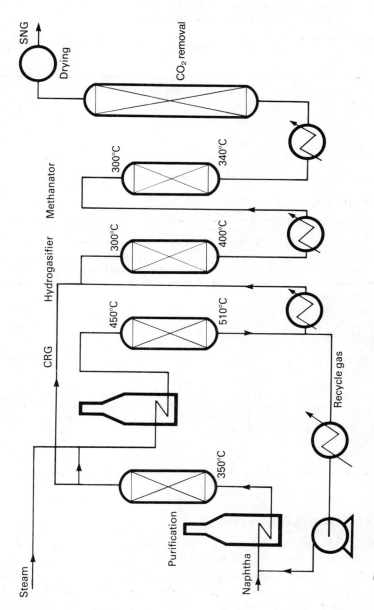

Figure 7.16. The CRG hydrogasification route with typical inlet and exit temperatures.

Chapter 7. Methanation

Table 7.10. Gas compositions (as vol.%) at various stages in the production of SNG from naphtha by the CRG hydrogasification process

	CRG reactor exit wet	CRG reactor exit dry	Hydrogasifier exit wet	Methanator exit wet	SNG
CO_2	12.0	22.0	16.7	19.9	0.5
CO	0.5	0.9	0.55	0.1	0.15
H_2	8.5	15.6	4.75	1.7	2.4
CH_4	33.5	61.5	56.0	68.6	96.95
H_2O	45.5		22.0	9.7	

Table 7.11. CRG plants operating in 1977

	Town gas production	SNG production	
		Double methanation route	Hydrogasification route
Calorific value			
Btu ft^{-3}	650	1000	1000
MJ m^{-3}	24.6	38	38
Number of plants	53	5	10
Total output/ m^3 day^{-1}	27.7 × 10^6	10.4 × 10^6	25 × 10^6
Comments	Plants installed in Japan, South Korea, Brazil, Spain, France, Greece and Italy	Process used in 5 plants in USA	Process used in 9 plants in USA, 1 plant in Japan and in peak load plants in UK converted from town gas units

7.4.2. Coal-based Routes to SNG

The oil-based processes for SNG manufacture were seen to be important in the USA during the 1970s. However, they were never likely to provide a *long-term* solution to the forecast shortfall of natural gas supplies. It was recognized that this must ultimately be provided by coal, and during 1974 and 1975 plans were made for the construction of huge (7 million m^3 day^{-1} ~250 million ft^3 day^{-1}) coal-based SNG plants. Several coal gasification processes were avialable, but the Lurgi route was selected for every plant for which firm plans were announced.

7.4 Substitute Natural Gas (SNG)

However, because of the economic changes which have occurred since the projects were conceived in the late-1960s, only one of the projects reached mechanical completion. This was the Great Plains Gasification Associates project at Beulah, North Dakota, which first produced SNG commercially in 1984. However, the role of this plant is that of a demonstration[217] unit rather than of a major gas supply venture.

7.4.2.1. Lurgi coal/SNG process

This process relies on proven coal gasification technology and methanation to uprate the product synthesis gas to pipeline quality. Feasibility of the methanation section was demonstrated[218] by operation of a sidestream pilot plant on the Fischer–Tropsch plant at SASOL in South Africa and by an American consortium on a British Gas plant at Westfield, Scotland. A flowsheet for the Lurgi coal-to-SNG process is shown in Figure 7.17, and gas compositions and temperatures in the methanation section are given in Table 7.12. The key feature of the methanation section is *recycle of product gas* to moderate the temperature rise in the methanators. The maximum temperature reached is 450°C, and the duty is suitable for "CRG type" catalysts.

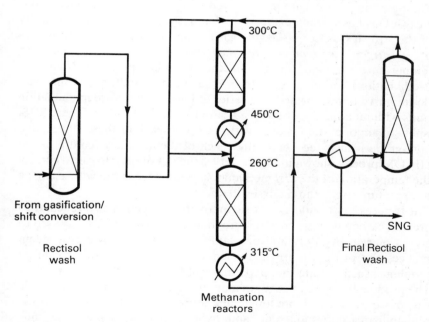

Figure 7.17. The simplified Lurgi coal-to-SNG process, featuring recycle of product gas to moderate the temperature rise in the first methanator. In this schematic flow diagram no recycle compressor is shown.

Chapter 7. Methanation

Table 7.12. Methanation conditions in Sasol SNG plant

		Main methanation		Final methanation	
	Syngas	Feed	Effluent	Feed	Effluent
Pressure/kg cm^{-2}		18.0			
Temperature/°C	270	300	450	260	315
Gas composition					
CO_2/vol.%	13.0	19.3	21.5	21.5	21.3
CO/vol.%	15.5	4.3	0.4	0.4	0.1
H_2/vol.%	60.1	41.3	7.7	7.7	0.7
CH_4/vol.%	10.3	53.3	68.4	68.4	75.9
C_{2+}/vol.%	0.2	0.2	0.1	0.1	0.1
N_2/vol.%	0.9	1.7	2.0	2.0	2.0
H_2O/Nm^3 Nm^{-3}		0.37	0.50	0.04	0.08

7.4.2.2. HICOM coal/SNG process

Taking a long view that coal wll eventually be used to produce SNG, British Gas has worked extensively demonstrating[219, 220] the versatility of the British Gas/Lurgi slagging gasifier and the economic advantages of a *direct high-temperature methanation* route with operating temperatures around 700°C. In the HICOM process (*High CO Methanation*) product gas from the gasifier, after cooling and desulphurization, is passed to a series of high-temperature methanation units without adjusting the H_2/CO ratio to 3.0, as shown in Figure 7.18. Split-stream operation is used with recycle to each of these stages, and high-grade heat is recovered after each of them. A key feature of the HICOM process shown in Figure 7.19 is the use of recycle gas to control the temperature of the *first* methanator. Each stage is run with as high a temperature rise as the catalyst will tolerate.

It is necessary to add excess steam over that needed as a reactant to prevent carbon deposition and more than the minimum excess is used to give a margin of safety and operational flexibility. This excess steam reduces the process thermal efficiency and may cause other catalyst problems, such as sintering, particularly at the high outlet temperatures. The product from the main methanation stages, after removal of recycle gas, passes to one or more low-temperature methanators to convert the remaining carbon monoxide and hydrogen to methane. The final product contains mainly methane and some carbon dioxide.

7.4 Substitute Natural Gas (SNG)

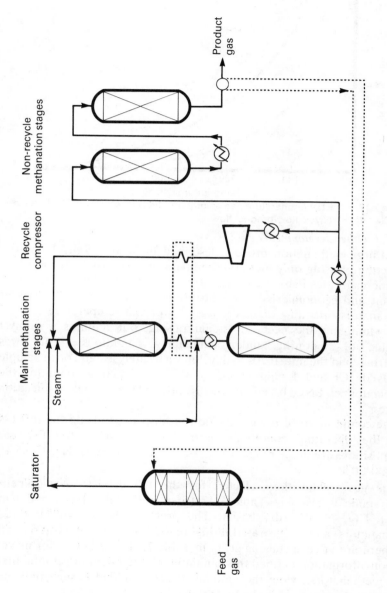

Figure 7.18. Simplified process flow diagram for the HICOM process, featuring recycle of gas to control the temperature of the first methanator.

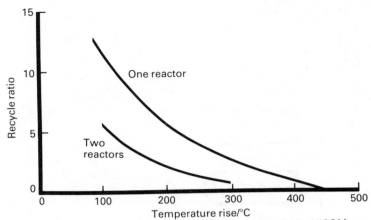

Figure 7.19. Effect of temperature rise on required recycle ratio for HICOM process. Recycle ratio = recycle flow/product flow.

7.4.2.3. Other developments

The Lurgi methanation process described in Section 7.4.2.1 has the attraction of using only moderate temperatures (250–450°C) for which commercially available catalysts are satisfactory. However, more elegant and economic is a straight-through process using a series of methanators operating at successively lower exit temperatures.[221, 222] This eliminates the need for steam raising, but requires a catalyst capable of operating under demanding conditions. Work was undertaken in the mid-1970s to develop such catalysts, and one example involved ICI and Krupp-Koppers, who developed a route for the production of SNG based on the Koppers–Totzek coal gasificiation process.

The composition of the gasification product depends on the process and the operating conditions used, but in general the dry gas compositions are in the range: CO_2, 0–30%; CO, 10–60%; H_2, 25–75%; CH_4, 0–20%.

The gas composition selected for the ICI catalyst development programme is given in Table 7.13 (first methanator inlet) and had $H_2 + CO = 74\%$ (dry basis). The process design involved three methanators, and is illustrated in Figure 7.20, the gas composition and temperature at each stage is given in Table 7.13. Conditions for nickel carbonyl formation defined the minimum inlet temperatures to the first two methanators, while the amount of steam added is such that the maximum temperature in the first methanator was below 750°C.

The outcome of the first phase of the catalyst evaluation and development programme was the production of a new high-nickel

7.4 Substitute Natural Gas (SNG)

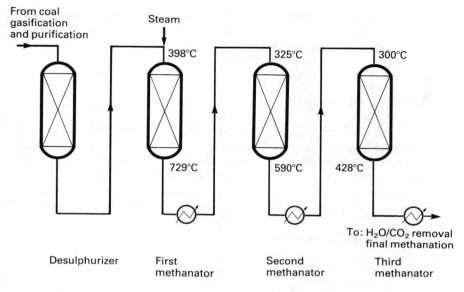

Figure 7.20. ICI high-temperature once-through methanation process.

Table 7.13. Gas compositions (as vol.%) and temperatures in ICI high-temperature methanation process

Composition*	1st Methanator		2nd Methanator		3rd Methanator	
	Inlet	Exit	Inlet	Exit	Inlet	Exit
CO/%	31.14	14.47	14.47	4.29	4.29	0.34
CO_2/%	24.66	40.15	40.15	53.93	53.93	62.70
H_2/%	42.91	35.50	35.50	20.26	20.26	5.83
CH_4/%	0.08	8.52	8.52	19.84	19.84	29.13
N_2 + A/%	1.21	1.36	1.36	1.68	1.68	2.00
H_2O[†]	67.3	72.3	72.3	94.4	94.4	118.2
T/°C	398	729	325	590	300	428

*Dry gas basis, [†]steam relative to 100 volumes of dry gas.

formulation (nickel oxide ~60%) which appeared to have the necessary activity, stability and physical strength. To demonstrate its properties, three semi-technical reactors were linked in series. The test ran without significant disturbance for 1500 hours, and demonstrated that a catalyst had been developed with the activity and stability needed to methanate a variety of process gases at temperatures up to 750°C. By proving the catalyst under simulated process conditions it was shown that deactivation rates were low enough to be industrially acceptable, but large-scale plants have yet to be built.

7.5. Heat Transfer Applications

There are many examples of large qauntities of heat being available in areas remote from places of high energy consumption. Various methods of transporting energy have been considered, including the use of reversible chemical reactions, and the highly endothermic steam reforming/exothermic methanation reactions have been evaluated for this purpose. In this section one such application is briefly discussed.

7.5.1. The EVA–ADAM Project

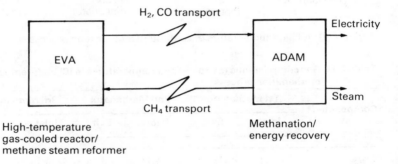

Figure 7.21. The EVA-ADAM system for long-distance energy transfer.

Beginning in the mid-1970s, Kernforschungsanlage Jülich GmbH (KFA) and Rheinische Braunkohlenwerke AG of Cologne, West Germany collaborated on the NFE project (long-distance nuclear energy transport). The basic concept, outlined in Figure 7.21, was to use heat from a high-temperature gas-cooled reactor (HTGR) to steam reform methane as in the equations (17) and (18) and to transport chemical energy in the form of the hydrogen and carbon oxides produced with subsequent release of the energy for consumption[223] via methanation. A HTGR can provide heat at about 950°C, so it is a suitable energy source for steam reforming.[224] The methane produced where the heat is liberated would be returned to the energy source for steam reforming or as an alternative to this closed system, the product methane could be used as fuel. In this configuration, this system could be regarded as a SNG plant with a long distance separating the synthesis gas generator from the methanator.

$$CH_4 + H_2O \rightarrow CO + 3H_2 \qquad (17)$$

$$CO + H_2O \rightarrow CO_2 + H_2 \qquad (18)$$

The EVA-1 pilot plant (*E*inzelrohr-*V*ersuchs-*A*nlage) was operated from 1972 to study the behaviour of a single full-sized reformer tube heated with helium at 950°C at 40 bar pressure. The ADAM-1 methanation pilot plant, commissioned in 1979, was subsequently linked with EVA-1 in a closed loop. In spring 1980 a larger demonstration plant, EVA-2/ADAM-2, with a 30-tube reformer and a power input of 10 MW was also put into service at Jülich. The methanation process aimed to produce superheated steam at about 540°C at 110 bar, necessitating temperatures of up to 700°C for methanation. This and other requirements led to the selection of a methanation process with three consecutive adiabatic fixed beds with decreasing exit temperatures (cf. Section 7.4.2.3). The temperature in the first methanator was controlled by hot gas recycled from the first or second reactor. Recycle from the first methanator is illustrated in Figure 7.22, and typical operating conditions are shown in Table 7.14.

Table 7.14. Typical conditions for ADAM-1

Composition*	1st Methanator		2nd Methanator		3rd Methanator	
	Inlet	Exit	Inlet	Exit	Inlet	Product[†]
CH_4/%	22.14	32.88	32.88	40.26	40.26	80.42
CO/%	5.25	1.65	1.65	0.05	0.05	0
CO_2/%	6.10	4.02	4.02	1.27	1.27	0
H_2/%	44.67	26.04	26.04	12.67	12.67	9.88
N_2/%	4.08	4.57	4.57	4.95	4.95	9.72
H_2O/%	17.76	30.86	30.86	40.81	40.81	0
T/°C	309	658	270	486	261	20
P/bar	27.1	27.1	26.9	26.9	26.7	1.0
Flow rate/Nm^3 h^{-1}	1086	944	324	299	299	148

*Wet gas basis, [†]dry product.

Chapter 7. Methanation

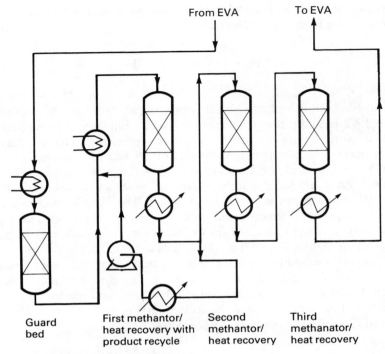

Figure 7.22. Methanation Plant ADAM-1.

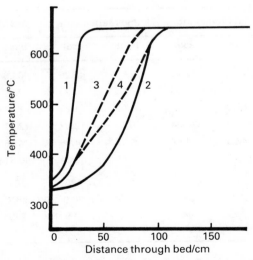

Figure 7.23. Methanator temperature profiles in the first reactor in ADAM-1. 1, 0 h; 2, 8500 h; 3, 8510 h; 4, 10 000 h.

7.5. Heat Transfer Applications

The catalyst in the first reactor must have good low temperature activity at the reactor inlet (i.e. at ~300°C) and also good thermal stability because the outlet temperature is ~650°C. No commercial catalysts were then available for such a duty, and so numerous nickel-based catalysts were developed and tested. Figure 7.23 shows the results of one test on an experimental catalyst. In this example it was estimated that of the catalyst deactivation observed over one year about 65% was caused by irreversible sulphur-poisoning and sintering, while 35% was due to carbon formaton on the catalyst. It was shown that the carbon could be removed by hydrogen treatment at high temperature, and Figure 7.23 shows temperature profiles before (profile 2) and after (profile 3) regeneration by hydrogen treatment at 500°C. Figure 7.24 shows typical[225] catalyst temperature profiles for the three stages of ADAM-1.

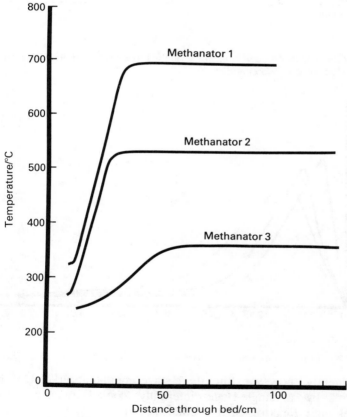

Figure 7.24. Typical temperature profiles in ADAM-1 methanators.

381

Chapter 7. Methanation

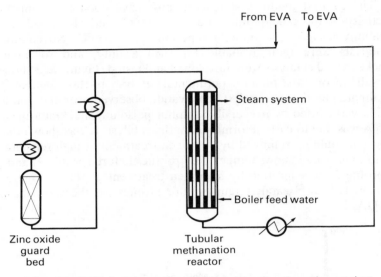

Figure 7.25. The IRMA flowsheet, showing the tubes of methanation catalyst surrounded by boiler water and steam.

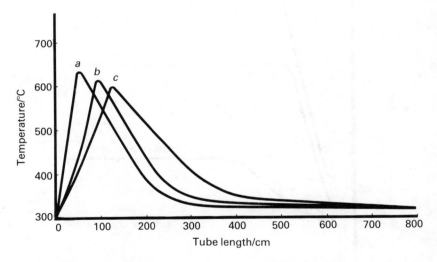

Figure 7.26. Temperature profiles in IRMA for increasing throughputs $c > b > a$.

Subsequently, however, the three-stage methanation system was replaced[226] by a single-stage internally cooled reactor—IRMA (*I*nnenbekühter *R*eaktor einer *M*ethanisiezung *A*nlage). This reactor contained five reaction tubes in a pressure vessel, and the reaction heat was used to raise high-pressure steam. The flowsheet is shown in Figure 7.25, and the temperature profile, measured for different flow rates of synthesis gas, in Figure 7.26. Although the peak temperature in the methanator is 600–650°C, the temperature at the exit was 310–320°C, with a final methane content of over 83%. Details of all the NFE work have been reported,[227–230] and although the commercial realization of long-distance energy transfer may be many years in the future, the project illustrated that methanation catalysts can be developed for a very wide range of duties and process conditions.

Chapter 8

Ammonia Synthesis

8.1. Introduction

At the beginning of this century the use of nitrogenous fertilizers was already well-established, with the principal sources being sodium nitrate from Chile and by-product ammonium sulphate from the destructive distillation of coal, the basis of the town-gas industry. However, it was recognized[231] that supplies would not be able to keep pace with demand as populations increased and as alternative uses, such as in the explosives industry, developed. Contemporary estimates indicated that even at the relatively low levels of usage in the early part of the century the Chilean deposits would be exhausted within 50 years, and efforts were therefore devoted to finding other sources of fixed nitrogen. As a result, during the period 1900–1920 the technology of the chemical fixation of atmospheric nitrogen was born, and underwent rapid development. By 1910 two processes had been commercially established, the cyanamide process and the electric arc process, and indeed up to 1912 general opinion seemed to agree that future expansion of the fixed-nitrogen industry lay with further development of these processes rather than with the development of a direct ammonia synthesis process. However, an extensive programme of work had been carried out by BASF in Germany on the Haber process, the results of which had been kept secret.

The cyanamide route to ammonia is shown in equations (1), (2) and (3). Calcium carbide (CaC_2), obtained by fusion of lime and carbon, reacts at 1000°C with nitrogen to form calcium cyanamide ($CaCN_2$). Hydrolysis of the cyanamide yields ammonia, with simultaneous precipitation of calcium carbonate. The first plant was commissioned in Italy in 1906, and other plants using the process were built in Germany and the USA. Somewhat surprisingly, in Germany production by the cyanamide route had reached 140 000 tonnes per year by 1915. However, the process was inefficient in its use of energy, and required 230 GJ tonne^{-1} of nitrogen, but such was the need for explosives that this extreme cost was apparently justified.

$$CaO + 3C \xrightarrow{2000°C} CaC_2 + CO \qquad (1)$$
$$CaC_2 + N_2 \xrightarrow{1000°C} CaCN_2 + C \qquad (2)$$
$$CaCN_2 + 3H_2O \rightarrow 2NH_3 + CaCO_3 \qquad (3)$$

8.1. Introduction

The electric arc process was established using cheap electricity in Norway in 1905 and in the USA in 1917. In this process, air was passed through an electric arc and attained a temperature of 3000°C, conditions under which oxygen and nitrogen combine directly to form nitric oxide. The energy consumption for this process is 720 GJ tonne^{-1} of nitrogen fixed. Indeed, it was the inefficiency and high cost of this process and the cyanamide route that provided the incentive for the development of the direct synthesis of ammonia from hydrogen and nitrogen. The significance of the development is illustrated in Figure 8.1 which traces the energy required for the fixation of nitrogen during this century. The introduction of the Haber process, using coal, gave rise to a substantial improvement in the energy requirement. Subsequently the improvement obtained was small as the coal based process was refined, and it was not until the introduction of natural gas and naphtha based plants followed by the large single stream plants that further significant reductions in energy consumption were achieved. More recently the development of the high efficiency processes has brought the energy consumption close to the theoretical minimum for processes based on hydrocarbon feedstocks.

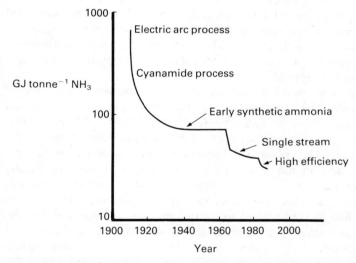

Figure 8.1 Improved efficiency of nitrogen fixation during this century.

The development of the science required for ammonia synthesis started in the 19th century, and its subsequent rapid progress owes much to Haber,[232, 233] who first determined the equilibrium constant for the synthesis reaction in 1904. He conducted his experiments at 1020°C and atmospheric pressure over finely divided iron catalyst, and he was able to extrapolate the data obtained to lower temperatures using newly developed theories in physical chemistry. He concluded that it would not be possible to develop a viable process because conversions were so low. In 1906 Nernst[234] applied his new heat theory, which we now know as the Third Law of Thermodynamics, to the synthesis reaction. From heats of reaction he was able to calculate the equilibrium constant at any temperature and pressure. Moreover, he also determined a value at 75 bar, which appeared to substantiate the value he had obtained from his theory and to refute the earlier data obtained by Haber. In the ensuing debate Haber and Le Rossignol[235] undertook further experimental work, this time at pressures up to 30 bars, and showed that the equilibrium constant was higher that that predicted by Nernst. He attributed this to the uncertainty of the value used by Nernst for the specific heat of ammonia at high temperatures. More importantly, Haber decided that even though conversions were low, a process to synthesize ammonia was feasible provided that: (1) synthesis and removal of the ammonia formed were carried out at high pressure; (2) synthesis gas was recirculated over the catalyst and (3) a satisfactory catalyst could be developed. Haber and his co-workers developed a catalyst based on osmium, and in 1909 built a semi-technical plant capable of producing 90 g h^{-1} of ammonia. Following the successful demonstration of the unit, BASF decided to undertake large-scale development work and, with Carl Bosch as project leader, a plant was built at Oppau to produce 30 tonnes day^{-1} of ammonia, which was commissioned in September 1913.

The successful development of the ammonia synthesis process resulted from the application of the then recently developed thermodynamic, kinetic and chemical engineering principles. The importance and magnitude of the achievement cannot be overestimated. The plant used entirely new concepts in process technology, and in many instances new equipment was specifically designed for the purpose. Never before had plant been built on this scale to handle such high pressures and high temperatures. Instruments and valves had to be designed to control process streams, while in the chemical area a massive research effort was undertaken that culminated in the development of an iron-based catalyst as an alternative to the original osmium catalyst, which was not only rare but difficult to handle. These impressive achievements were accomplished in less than 5 years from the start of process development.

Figure 8.2 shows the growth in synthetic ammonia production from early in the 20th century to the present time. The current annual world capacity of ammonia is in excess of 110 million tonnes (expressed as nitrogen). There are now more than 600 plants in operation, with most

8.1. Introduction

modern plants having an output of about 1000–1500 tonnes day^{-1}, but as shown in Table 8.1, a substantial amount of ammonia is still produced by plants which have a smaller capacity than this. The design and operation of these plants depends on a large number of interrelated variables, and these factors are discussed in this chapter.

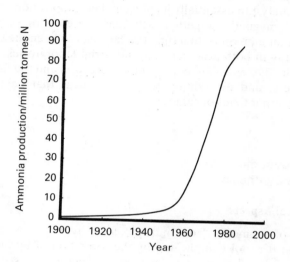

Figure 8.2 The growth of world nitrogen production. (Constructed from data supplied by the British Sulphur Corporation.)

Table 8.1. World ammonia capacity (1984–1985)

Individual plant capacity/ thousand tonnes N year^{-1}	Share of world capacity[a]/ %
<100	17.1
100–200	20.6
201–300	33.5
301–400	23.7
401–501	5.1

[a]World ammonia capacity = 113 million tonnes N.

Chapter 8. Ammonia Synthesis

The commercial ammonia synthesis catalyst is an excellent example of a heterogeneous catalyst, with one major component generating the activity, but requiring the addition of other compounds as promoters and stabilizers to increase its effectiveness and stability over long periods. In the catalyst research programme up to 1911, Mittasch of BASF investigated more than 2500 different formulations and optimized the catalyst to essentially its present day composition, namely alkali-promoted magnetite together with the presence of oxides of calcium and aluminium. Since that time the catalyst has been the subject of continual study in both academic and industrial laboratories. These studies have helped to elucidate both its mode of action and the function of the promoters, and in addition so provide ideas that have been applied in many other fields of catalysis.

8.2. Thermodynamics of Ammonia Synthesis

8.2.1. Theoretical Aspects

The synthesis of ammonia from nitrogen and hydrogen is a clean reaction, in that it is not complicated by the formation of by-products such as hydrazine, and moreover the thermodynamics are seemingly straightforward. However, the non-ideality of some of the gases under normal operating conditions, and the presence of the inert gases methane and argon, complicates matters. The synthesis reaction is shown in equation (4), together with some associated thermodynamic values.

$$\frac{1}{2}N_2 + \frac{3}{2}H_2 \rightarrow NH_3 \qquad \begin{aligned} \Delta H°_{700K} &= -52.5 \text{ kJ mol}^{-1} \\ \Delta G°_{700K} &= 27.4 \text{ kJ mol}^{-1} \\ \Delta S°_{700K} &= 288 \text{ J mol}^{-1} \text{ K}^{-1} \end{aligned} \qquad (4)$$

The reaction is exothermic and is accompanied by a decrease in volume at constant pressure. The value of the equilibrium constant (K_p) therefore increases as the temperature is lowered, and the equilibirum ammonia concentration increases with increasing pressure. The determination of equilibrium constants for this reaction has been a subject of intense research, partciularly at the beginning of this century. Indeed, the resolution of discrepancies between the respective results obtained by Haber and Nernst provided the impetus for the pioneering work on catalysts for the reaction. Thermodynamic data have been

8.2. Thermodynamics of Ammonia Synthesis

published by Haber et al.,[236] Larson and Dodge,[237] and Larson,[238] covering the pressure range of industrial interest. These data have been analysed by Gillespie and Beattie,[239] who developed a method for calculating the equilibrium composition of hydrogen, nitrogen and ammonia in the presence of inert gas. In Figure 8.3 the equilibrium concentration of ammonia is shown as a function of temperature and pressure with a 3 : 1 hydrogen/nitrogen gas mixture for the two conditions where there are either no inerts present at all, and where the synthesis gas contains 10% inert components. The early work on equilibria in the N_2–H_2–NH_3 system has been comprehensively reviewed by Nielsen.[240]

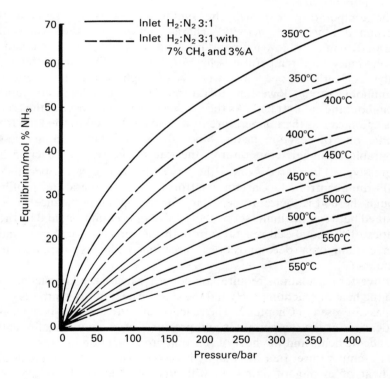

Figure 8.3 Effect of pressure, temperature and inert gas on equilibrium ammonia concentration.

8.2.2. Process Consequences

The formation of ammonia is favoured by operation at high pressure and low temperature. Over the past 50 years the optimum pressure for economic operation with the available catalysts has been in the range 150–350 bar. Processes operating at higher pressures have also been used, and the Ruhrchemie plant at Oberhausen in Germany, which operated at 750 bar with a capacity of 150 tonnes day^{-1}, was shut down only as recently as 1984. Normally, however, the advantages of the higher equilibrium concentration of ammonia at very high pressure are more than offset by the higher costs of both gas compression and additional plant capital. In new plants which are now being built using the latest technology such as ICI's AMV process, the optimum pressure can be as low as 70 bar.[241]

The temperature at which the synthesis process is operated is determined by the activity characteristics of the catalyst. Thermodynamically, low temperature is advantageous, but for kinetic reasons fairly high temperatures have to be used. The most effective catalyst is clearly the one which will give the highest rate of conversion of ammonia at the lowest temperature. These considerations apply throughout the catalyst bed. As the synthesis reaction proceeds the heat of reaction causes the temperature to rise down the bed, so making the specific rate of reaction faster. Since the equilibrium becomes less favourable at higher temperatures, the rate of the reverse reaction is progressively increased and the overall conversion becomes equilibrium-controlled. Careful control of the temperature profile through the bed is therefore necessary for the optimum balance to be obtained between the limits set by thermodynamic equilibria, and by the kinetics of the catalysed reactions in both the forward (synthesis) and reverse (ammonia decomposition) directions. This is discussed in Section 8.6.3.

Converter calculations require, among other factors, a knowledge of both the heat of reaction ($\triangle H$) and the specific heat of the synthesis gas (c_p), as discussed in Chapter 2. The reaction mixture does not behave as an ideal gas at the temperatures and pressures used commercially, and the specific heats and the heat of reaction are a function of pressure as well as temperature. In a strict analysis account should also be taken of the heat of mixing of ammonia with unconverted synthesis gas. A summary of values of heats of reaction at 500°C from several authors is given by Nielsen[240] and is shown in Table 8.2. The actual values depend on the correction applied for the heat of mixing. In practice it is convenient to work with a standard heat of reaction of 54 kJ mol^{-1} at 450°C, which is a reasonable average temperature used in commercial synthesis. A value of c_p (expressed as kJ mol^{-1} K^{-1}), for the inlet gas to the bed may be estimated from equation (5), in which it is assumed that the hydrogen/nitrogen ratio of the synthesis gas is 3 : 1

$$c_p = 1.632(1 + a_i) + 1.551b_i - 0.517c_i \tag{5}$$

where a_i is the mole fraction of inlet ammonia, b_i is the mole fraction of inlet methane and c_i is the mole fraction of argon and helium.

When calculating temperature rises through adiabatic beds, it may also be assumed that at temperatures above 250°C and pressures in excess of 100 bars the mass specific heat is constant. Taking the heat of reaction at 450°C as 54.13 kJ mol^{-1}, a temperature rise factor $\triangle H/c_p$ may be calculated which, when multiplied by the fraction of inlet gas converted to ammonia, gives the adiabatic temperature rise. This is represented by equation (6)

$$\triangle T = (\triangle H_{450}/c_p)\{[a_o(1 + a_i)/(1 + a_o)] - a_i\} \tag{6}$$

in which a_o is the mole fraction of ammonia in the exit stream.

8.2.3. The Synthesis Loop

The principle of circulating gases over the catalyst, which was first appreciated by Haber in 1908, is still an important feature of modern ammonia plants, as illustrated in Figure 8.4. This shows the synthesis loop for a typical 1000 tonnes day^{-1} plant operating at 220 bar using a three-bed quench converter (see Section 8.8.2). Synthesis gas of the appropriate composition passes through the catalyst beds, and the ammonia produced is condensed and recovered. Unreacted gas, to which fresh make-up gas is added, is then recirculated through the catalyst. Using heat exchangers the temperature of the recirculating gas is raised in two stages to a reaction temperature of about 400°C and, at the same time, the temperature of the converter effluent gas is reduced. As shown in Figure 8.4, the heat exchanger immediately downstream from the catalyst basket is also contained within the high-pressure converter, as discussed further in Section 8.8. To prevent accumulation of inert gases generally present in the synthesis gas, part of the circulating gas is purged. The residual ammonia in the purge gas is usually recovered, and the hydrogen content is either used as fuel in the primary reformer or recovered and recirculated. (For further discussion see Section 8.7.4.)

Chapter 8. Ammonia Synthesis

Table 8.2. Values of $\triangle H$ (kcal mol^{-1}) for ammonia synthesis over a range of pressures at 500°C

Pressure/ atm	Gillespie and Beattie[a]	Kazarnovskii and Karapet'yants[b] without correction for heat of mixing	Kazarnovskii and Karapet'yants[b] corrected for differential heat of mixing (dilute solution)	Kazarnovskii[c] corrected for heat of mixing 17.6% NH$_3$ 20.6% N$_2$ 61.8% H$_2$
1	−12.660	−12.893	−12.893	−12.895
100	−12.920	−13.149	−13.121	−13.040
200		−13.413	−13.293	−13.182
300	−13.450	−13.708	−13.411	−13.210
400		−14.023	−13.555	−13.045
500		−14.275	−13.650	−12.895
600	−14.240	−14.493	−13.741	−12.940
700		−14.702	−13.842	
800		−14.903	−13.943	−13.257
900		−15.098	−14.048	
1000	−15.290	−15.280	−14.145	−13.595

[a] L. J. Gillespie and J. A. Beattie, *Phys. Rev.*, **36**, 1008 (1930).
[b] Y. S. Kazarnovskii and M. K. Karapet'yants, *Zh. Fiz. Khim.*, **15**, 966 (1941).
[c] Y. A. Kazarnovskii, *Zh. Fiz. Khim.*, **19**, 392 (1945).

8.3. Ammonia Synthesis Catalysts

All commercial ammonia synthesis catalysts are currently based on metallic iron promoted with alkali (potash), and various metal oxides, such as those of aluminium, calcium or magnesium. The principal material used to make these catalysts is usually magnetite (Fe_3O_4), with some of the components in the catalyst originating as impurities in the magnetite. A typical catalyst, such as ICI Catalyst 35-4, contains approximately 0.8% K_2O, 2.0% CaO, 0.3% MgO, 2.5% Al_2O_3 and 0.4% SiO_2, as well as traces of TiO_2, ZrO_2 and V_2O_5. In developing the process to manufacture catalysts of this sort, it was recognized that these minor components could have a large effect on the performance of the final catalyst, since they may also interact with each other, giving rise to both harmful and beneficial effects. In modern catalysts these factors have been taken into account, resulting in optimized performance in terms of high activity and long life.

Almost all ammonia synthesis catalysts are manufactured by fusing

8.3 Ammonia Synthesis Catalysts

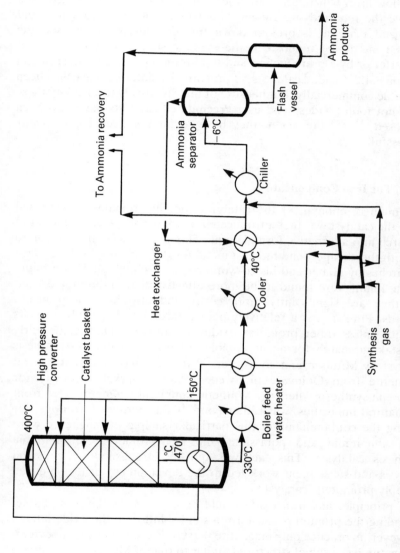

Figure 8.4 Ammonia synthesis loop. This layout is typical of a loop for a large capacity (1000 tonne day^{-1}) plant. Preheat of the converter feed gas to 150°C allows high grade heat recovery from the converter exit gas.

magnetite together with the required amounts of promoters to generate a homogeneous melt. The molten mixture is cooled by pouring it out as a shallow layer which, after solidification, is broken down and screened to give the required size range. Undersized material is recycled, and oversized material is broken down further. Before being used the catalyst has to be reduced to the metallic state, either in the plant converter or by a prereduction and stabilization process as described in Section 8.4.2. Shaped catalysts,[242] prepared by sintering, have also been available commercially, but because of their relatively low activity these have not found widespread use. Precipitated catalysts have also been proposed,[243, 244] but again, they have so far not been commercially successful.

8.3.1. The Iron Component

The main component of the catalyst, iron, has remained unchanged since the catalyst was first introduced in 1913, despite a large amount of research into alternative formulations. This work has confirmed iron to be both the cheapest and the best metal for the purpose. In the earliest researches by Haber and his co-workers, other metals such as osmium and uranium were found to be more effective than unpromoted iron, but they are significantly more costly and also introduced health hazards. Pure iron is a relatively poor catalyst which quickly loses its activity unless other promoter oxides are present. During the early painstaking search for an acceptable commercial ammonia synthesis catalyst by Mittasch and co-workers[245] it was discovered that natural magnetite from Gollivare in Sweden, had remarkable activity for ammonia synthesis whereas a synthetic sample of magnetite, free from the natural impurities, was rather poor. It was perhaps fortuitous, for among the contaminants in this particular sample of magnetite were both aluminium and potassium—the key promoters in ammonia synthesis catalysts. This fact was quickly established by chemical analysis and subsequent work led to the development of the so-called "doubly-promoted" catalyst.

In principle, any iron oxide would be effective as the iron source, providing the promoters can form a solid solution in the oxide lattice. However, in practice only magnetite, Fe_3O_4, is used to any great extent. Magnetite has a spinel structure (similar to that of $MgAl_2O_4$) consisting of a cubic packing of oxygen ions, in the interstices of which Fe^{2+} and Fe^{3+} ions are distributed. As shown in Figure 8.5, the crystals of magnetite in the unreduced catalyst are relatively large by catalysis standards, being up to a micron or even more. However, during reduction oxygen is removed from the crystal lattice without shrinkage, so the metallic iron is obtained as a pseudomorph of the original magnetite. Metallic iron produced in this way is therefore extremely porous, and the way in which porosity is developed is an important factor affecting the activity of the final catalyst. Another major factor is the size of the individual crystals of iron produced during reduction. This is largely determined by the nature and amounts of promoters present, and of course the actual conditions during reduction. Scanning electron microscope (SEM) photographs at increasing magnifications, showing both the porosity and the pseudomorphic nature of the reduced catalyst, are also shown in Figure 8.5.

The early discoveries, though essentially empirical, are still very relevant to modern ammonia synthesis catalyst. Research since Mittasch's time has concentrated largely on kinetics and mechanism, in explaining the promoting action of the minor components, and the effect of poisons. It is in this context that the theoretical considerations of Chapter 1 have proved valuable. A fairly comprehensive picture of the working catalyst has been put together as a result of X-ray and microscopy studies, from which the different crystalline phases have been identified. Selected gas absorption studies have also provided additional information about the distribution of the components within the reduced catalyst, and recent studies using modern surface science techniques have added further to our understanding of the reaction mechanism and modes of promotion. These aspects are discussed in Sections 8.3.2.2 and 8.6.4.

8.3.2. Promoters

The key to the generation of a successful commercial iron-based ammonia synthesis catalyst lies in the incorporation of several metal oxides within the magnetite structure that promote activity and improve stability of the operating catalyst. The most important of these are alumina and potash, which generate the so-called "doubly-promoted" catalyst, but several other oxides may also be added. They include calcium oxide, silica, magnesia and, to a lesser extent, the oxides of manganese, chromium, zirconium and vanadium. Such promoters are conveniently classified as "structural" or "electronic", depending on their accepted mode of action, and they are discussed below using this classification.

8.3.2.1. Structural promoters

The production and preservation of a porous structure during the

Chapter 8. Ammonia Synthesis

Figure 8.5. Scanning electron micrographs of ammonia synthesis catalyst: (a) unreduced catalyst showing well-formed crystals of magnetite 230 X; (b), (c) and (d) reduced catalyst at different magnifications showing porous iron pseudomorphs at magnifications of 2300, 7700 and 23,000.

8.3 Ammonia Synthesis Catalysts

reduction of ammonia synthesis catalyst is of fundamental importance in obtaining high activity, and the prime role of structural promoters such as alumina, magnesia and chromium sesquioxide is to facilitate the formation of porous, high-area, metallic iron. During catalyst manufacture the components are fused together and a reasonably homogeneous melt is obtained. On cooling, the melt solidifies and magnetite crystals are formed with the aluminium and magnesium ions incorporated into the lattice as a solid solution. Calcium, on the other hand, occurs mainly as mixed ferrites and aluminates because Ca^{2+} is a significantly larger divalent cation than either magnesium or ferrous cations, and it is unable to enter the magnetite lattice without causing severe distortion. In early work[246] on the solubility of alumina in magnetite, it was suggested that below 500°C the limit of solubility corresponded to about 3% alumina, a level comparable with the maximum alumina content in most commercial ammonia synthesis catalysts.

During the catalyst activation process, reduction starts on the outside of the fused magnetite granules, and the dissolved alumina and magnesia separate out of solution in the forming pores between the iron crystallites. They hinder further growth of the iron crystallites during both reduction and subsequent use. The iron crystallites are typically between 20 and 40 nm, with the pores being of similar size. This is illustrated in the series of SEM photographs shown in Figure 8.5, which shows unreduced and reduced catalyst over a range of magifications. At low magnification the surface appears very smooth, resembling the original magnetite, but at higher magnification separate iron crystallites can be distinguished, revealing the open-pore structure. Despite the presence of skeletal alumina, however, the iron crystallites can grow to around 80–100 nm during subsequent use in ammonia synthesis. The BET nitrogen surface area of freshly reduced catalyst can be as high as 15–20 $m^2\ g^{-1}$, compared with less than 1.0 $m^2\ g^{-1}$ for unreduced catalyst. In contrast, unpromoted iron, made by reduction of pure magnetite, has a surface area of less than 1 $m^2\ g^{-1}$. Promoters such as alumina therefore assist in both the formation and the preservation of small iron crystallites.

Silica and other acidic components are common impurities in natural magnetite used for catalyst production, and consequently they become incorporated into the structure of the finished catalyst. They have the effect of "neutralizing" the electronic promotional effect of basic components such as K_2O and CaO, and their presence in excess can result in lower catalyst activities. However, silica also has a stablizing effect like alumina, and high-silica catalysts tend to be more resistant to both water-poisoning and sintering. There is obviously considerable interaction between the various components of the catalyst, and the optimum quantities have to be determined for each combination, taking into account the kind and amount of impurities present. Further more, the formulation should be optimised for the particular conditions under which the catalyst will be operated. The method of manufacture also has to be considered, because some components can be lost by volatilization, and other factors such as the rate of cooling the melt can affect the crystallinity of the catalyst.

The distribution of elements within the different crystalline phases in the unreduced and reduced catalysts can be determined by electron-probe micro-analysis. These studies show that the promoters, particularly potassium, are intimately mixed with the iron, at least on the 100-nm scale, and that at least 90% of the iron surface is covered by them. Presumably only the remaining 10% of the iron surface area is available for catalysis. However, the available surface area of iron and the activity of the promoted surface are both significantly greater than that of unpromoted iron.

8.3.2.2. Electronic promoters

The presence of alkali-metal species in the ammonia synthesis catalyst is essential to attain high activity. Although all of the alkali metals are effective to some extent, the best are potassium, rubidium and caesium, with potassium being the most cost-effective. The light alkali metals, lithium and sodium, are relatively poor promoters, and are not used commercially.

Potassium is added in the form of potassium carbonate during catalyst manufacture, and the principal compounds which result are mixed ferrites,[247] though the presence of potassium aluminate and potassium silicate phases have also been reported. On reduction of the catalyst much of the potash remains associated with the somewhat acidic support phase, though some interacts with the iron particles and greatly increases their intrinsic activity. An undesirable side-effect of potassium promotion is an increased growth in the size of the iron crystallites during reduction, with a consequent loss of metal surface area. Nevertheless, this effect is vastly outweighed by the large increase in intrinsic activity of the promoted surface.

Calcium oxide reacts with iron oxides to form complex calcium ferrites,[247] and some reacts with alumina and silica to form glassy alumino-silicates. This neutralizes some of the acidity of the amphoteric support components, and leaves more of the K_2O available to activate the iron. Possibly as a result of forming these compounds, calcium oxide is said to enhance the action of alumina in stabilizing the iron surface area, and reduces the level of sintering during reduction. However, there is a penalty in that the catalyst becomes more difficult to reduce and the consequent higher temperatures required during reduction can lead to some sintering. The presence of calcium ions also makes the catalyst rather more resistant to poisoning by sulphur and chlorine, a feature more important in earlier days when high sulphur feedstock was used. A minor proportion of the calcium oxide also dissolves in the magnetite lattice, and this tendency is most marked when the ferrous content of the magnetite is high; that is, when the ferric/ferrous ratio is less than 2 : 1.

Despite the enormous amount of research on the mechanism of potassium promotion, there is still considerable debate around this subject. In one of the early proposed mechanisms,[248] it was suggested that strongly basic alkali oxides or hydroxides would react with weakly acidic sites on the catalyst which would otherwise react with the basic ammonia product, leading ultimately to blockage of active sites on the catalyst surface by ammonia or its precursors. Another view[249] was that alkali metals on the iron surface promote electron release from the iron surface to the adsorbing nitrogen molecule, consequently promoting adsorption of N_2 and simultaneously weakening the nitrogen–nitrogen bond. Modern ideas on alkali-metal promotion stem from the more recent surface science studies on single iron crystals. Ertl and co-workers[250] showed that the dissociative chemisorption of nitrogen on

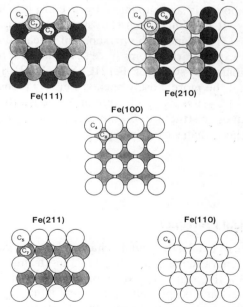

Figure 8.6. Representations of the structures of various iron crystal faces indicating the different sites available on each face.

low-index iron surfaces is extremely structure-sensitive, confirming conclusions from earlier work.[251] The close-packed Fe(110) plane was found to be the least active, while the open Fe(111) surface was considerably more active. The structures of the relevant crystal faces are shown in Figure 8.6, where the differences in the packing of the Fe(111) and Fe(110) faces can clearly be seen.

This work was closely paralleled by that of Somorjai and co-workers,[252] who subsequently measured the rate of ammonia synthesis at 20 bar over the same crystal planes. The relative rate of reaction for the Fe(111), Fe(100), and Fe(110) planes was found to be 418 : 25 : 1. In the presence of potassium co-adsorbed with oxygen, Ertl and co-workers[253, 254] then showed that the dissociative sticking probability of nitrogen on iron was increased by an order of magnitude, a figure corresponding well with the rate-enhancement observed under industrial conditions. In keeping with the thermodynamics of the bulk catalyst, Somorjai and co-workers[255] have since shown that elemental potassium is not stable on the iron surface during synthesis. However, some potash can be retained on the surface for a longer period when co-adsorbed with oxygen, and a substantially higher level when alumina is present. Presumably this is due to compound formation. An interesting observation made during this work was that reorganization of the Fe(110) to the Fe(111) surface is induced by alumina, and that this, as noted above, is accompanied by a considerable *increase* in synthesis activity. It has since been shown[256] that the most important feature on the Fe(111) plane is the presence of so-called C_7 sites—that is, sites in which exposed iron atoms have seven nearest neighbours. In an elegant comparative study of Fe(211) and Fe(210) faces it was demonstrated that the more closely packed Fe(211) plane, which has a high population of C_7 sites, is significantly more active than the Fe(210) face which has an open structure. C_7 sites are therefore more important in ammonia synthesis than surface roughness.

8.4. Catalyst Reduction

8.4.1. Typical Plant Procedure

Ammonia synthesis catalyst is usually charged into the converter in the oxidized form, generally magnetite (Fe_3O_4), and it must be reduced to bring it into the active state. The reaction is slightly endothermic with evolution of water, and is generally done using ammonia synthesis gas. It is important that the reduction procedure leads to optimum catalytic performance, and hence it must be performed in a way which will give

8.4. Catalyst Reduction

minimum sintering of the iron crystallites and the most effective dispersion of the potash. The effects of temperature and oxygenated molecules, including water, on the catalyst structure determine the way in which the catalyst should be reduced and operated. As far as possible, metallic iron produced in one part of the bed should not be exposed to water formed during reduction of catalyst in other parts of the bed. Within a single particle this cannot be avoided, because metallic iron produced on the surface of the particles is exposed to water produced by the subsequent reduction of oxide in the interior of the same particle. Indeed, as a result of this effect larger catalyst particles tend to have a lower intrinsic activity than smaller catalyst particles that are exposed to less water vapour during reduction. Incidentally, smaller particles are also more effective because they are less affected by gaseous diffusion (see Section 8.6.2). The gas circulation rate during reduction should therefore be controlled to ensure that water produced in the lower (exit) parts of the bed does not come into contact with the upper (inlet) reduced catalyst as a result of back-diffusion, and the water formed during reduction must be removed from the exit gas before it is recycled. In practice, once ammonia synthesis has begun the circulation rate is kept as high as possible to ensure that the water concentration is minimized. Reduction is normally carried out at temperatures and pressures which offer a compromise between bringing the plant on-line as quickly as possible, and operating at such a rate that the average partial pressure of water throughout the converter is not detrimental to catalyst activity.

When synthesis gas is used to carry out the reduction (which in commercial practice is almost always the case) ammonia synthesis starts as soon as some metallic iron is produced, albeit at a low rate, and the heat of reaction raises the catalyst temperature, which in turn increases the rates of both reduction and synthesis. This not only increases the concentration of water in the gas, but also gives rise to the production of aqueous ammonia solution.

The formation of aqueous ammonia during reduction is generally an inconvenience to the operator, since it requires disposal. The amount produced depends on the conditions, and the rate at which it is formed gradually increases as the reduction progresses through the bed. The extent of the synthesis reaction must therefore be controlled by keeping the pressure low, and by temperature-control methods such as admission of cold gas. Typically reduction is carried out at 70–100 bar. The progress of reduction is indicated by the quantity of water in the exit gas. In general, the lower the water concentration the better, and although experience has shown that with ICI Catalyst 35-4, water levels of 5000–10 000 ppm do not appear to result in low activity, however an

Chapter 8. Ammonia Synthesis

exit water concentration of 5000 ppm is a good operating limit and 10 000 ppm should be regarded as an absolute maximum. These levels are equivalent to water being evolved from the catalyst at rates of 14 and 28 kg h^{-1} tonne^{-1} of catalyst, respectively, at a space velocity of 10 000 h^{-1}.

Various catalysts with different formulations are sensitive to water to differing extents, and the reduction conditions should be optimized for each situation. It is also important that the catalyst is not overheated during reduction, otherwise sintering may occur and the activity will be adversely affected. The catalyst should not be heated above the temperatures at which it is to be operated (about 450–500°C) and hot-spots within the bed should be avoided. This can be achieved by operating the plant during reduction at high circulation rates, thus ensuring good gas distribution through the bed. Temperature control is also facilitated by operating at low pressure, which minimizes the amount of ammonia produced, which in turn limits the temperature rise. If gas distribution is poor, then hot-spots can result, and catalyst in parts of the bed subjected to low gas flow will be reduced in the presence of high water levels, which will lead to loss of some activity. It is usually necessary to use a low circulation rate in the early stages of reduction, because of limits imposed by the capacity of the start-up heater. However, once synthesis starts, and heat is recovered from the exit of the catalyst bed, the circulation rate can be increased.

A recent study[257] on the reduction of commercial synthesis catalysts using X-ray diffraction techniques generated some interesting results. Reduction of small particles of catalyst at 450°C and atmospheric pressure using high space velocities to minimize the water vapour partial pressure gives rise to an amorphous phase rather than the usual crystallites of α-iron. It is this phase that is present in all samples of discharged synthesis catalyst. The amorphous phase was reported to be very active in ammonia synthesis, but its chemical nature was not characterized. It is thought[258] that the structure could be interpreted in terms of a highly defective lower oxide of iron. Such an interpretation is consistent with some observations using Auger electron spectroscopy, which has shown that even in the working state the catalyst surface still contains some residual oxygen atoms. However, at present this new phase has little practical significance, since the conditions under which it was produced in the laboratory could not be attained economically on a commercial scale.

8.4.2. Prereduced Catalysts

While ammonia synthesis catalyst is normally supplied in the oxidized

8.4. Catalyst Reduction

form, it may also be reduced by the catalyst manufacturer to give a prereduced catalyst. In the reduced state the synthesis catalyst is pyrophoric, and direct exposure to air would result in vigorous oxidation with the catalyst becoming incandescent with the heat of reaction. Not only would this be extremely hazardous, but the high temperature associated with severe oxidation would lead to extensive sintering, and to a marked reduction in catalytic activity. To facilitate storage, transport and catalyst charging, prereduced catalyst is always "stabilized" by the supplier. This is usually done by partial re-oxidation of the reduced catalyst by gradual exposure to low but increasing concentrations of oxygen in an inert carrier gas at low temperature.[258a] The so formed passivated iron surface can then be exposed to air at ambient temperatures without further oxidation. Prereduced catalyst should, however, be stored carefully in sealed drums in a dry cool place and not handled roughly, because in an extreme case this could initiate further re-oxidation. Stabilized prereduced catalyst can be charged to plant converters and re-reduced in the usual manner.

On stabilization the catalyst is usually less than 10% oxidized, and the oxidation "halo" around the iron crystallite, as seen by transmission electron microscopy, is generally less than 3 nm thick. Reductive activation of prereduced catalyst starts at about 330°C compared with a temperature of about 400°C for normal catalyst. This lower temperature can be of particular benefit where the size of the start-up heater may be a limiting factor, and the use of prereduced catalyst can shorten the time taken for reduction by a factor of two or more, which usually justifies the higher cost of the prereduced catalyst. Most advantage is gained when the whole charge is prereduced, but prereduced catalyst can also be used with advantage as a part-charge at the inlet of a converter. This is because reduction of the surface oxidation film is rapid and, once the ammonia synthesis reaction is initiated, it provides a source of heat to supplement the start-up heater. This is beneficial as it reduces the heat load on the start-up heater and permits a high gas space velocity to be used.

Prereduction and stabilization is carried out by the catalyst supplier and careful control of the process is essential, otherwise excessive loss of activity results when the catalyst is re-reduced. As a consequence it is expensive. Laboratory microreactor tests under controlled reduction conditions indicate that prereduced catalysts are generally somewhat less active than conventional catalysts, but in practice this does not cause any appreciable loss in plant capacity. This may be because the overall production of ammonia is rarely limited by the activity of the synthesis catalyst, except towards the end of life when the catalyst has become deactivated.

8.4.3. Economics of Prereduced Catalyst

The reduction of a typical charge of ammonia synthesis catalyst can take up to seven days. During this period not only is production lost while fuel is being consumed, but the aqueous ammonia solution generated can present a disposal problem. The costs associated with on-line reduction can be substantially lower when prereduced catalyst is used, because it is only about 10% oxidized; on re-reduction it gives only 10% of the water obtained with conventional catalyst, and consequently much less ammoniacal liquor is produced. The principal advantage is that the whole reduction procedure is shortened and the converter can be commissioned much more quickly. Typically this can take about one day when a full charge of prereduced catalyst is used. The extra ammonia production which then results, together with the savings in feedstock cost which accrue, makes a powerful argument for the use of the prereduced catalyst, although it is significantly more expensive.

A part-charge of prereduced catalyst can also be used with advantage, since the lower initiation temperature (typically 330°C) for the prereduced catalyst can be of particular benefit when the output from the start-up heater is a limiting factor in the reduction procedure. The operator can then use the prereduced catalyst in the top bed (inlet) of a multibed converter, or perhaps in the top 20% of a single-bed converter. The low strike temperature of the prereduced catalyst enables the start-up heater capacity to be quickly supplemented by the heat of the synthesis reaction. This allows the gas circulation rate to be increased more rapidly, so the activation of the remaining charge becomes faster than otherwise would be the case. Using this procedure the total time taken is typically only four days, and as a result significant cost savings are achieved.

8.5. Poisoning and Deactivation

8.5.1. Introduction

The ammonia synthesis catalyst generally has a much longer life than other catalysts used in an ammonia plant, and many plants are designed so the catalyst is only changed every 5–10 years. Nevertheless, loss of activity does occur by various mechanisms which are common to all catalysts. Deactivation may occur by thermal sintering, which is a slow, progressive process gradually leading to loss of iron surface area and hence activity, as shown in Figure 8.7. It may also occur as a result of poisoning by oxygenated compounds such as water, carbon monoxide

8.5. Poisoning and Deactivation

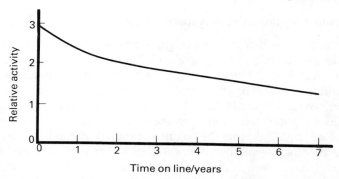

Figure 8.7 Typical activity die-off curve for an ammonia synthesis catalyst.

and carbon dioxide. However, loss of actvity resulting from exposure to these compounds is largely reversible, provided that the partial pressure of the oxygenate is low and that the period of exposure is no more than a few days.

The more traditional poisons, such as sulphur and chlorine affect the catalyst irreversibly. During the reduction process the overall surface area of the catalyst increases steadily, but the area of the metal component is said to increase at a faster rate towards the end of the reduction, with a corresponding increase in catalytic activity. This suggests that the catalyst surface is rather heterogeneous, and that the most active iron sites are the last to be reduced, being made available by the removal of the residual oxygen atoms. The most active part of the catalyst may therefore constitute only a small proportion of the total iron surface. This is consistent with the large structural sensitivity for ammonia synthesis reported by many groups of research workers (see Section 8.6.4), though the presence of potash in the commercial catalyst somewhat complicates the situation.

Further evidence that the proportion of active surface is small is provided by the sensitivity of the catalyst to relatively low levels of gaseous poisons such as oxygen, sulphur, arsenic, phosphorus and chlorine. Very small amounts of these compounds have a drastic effect on catalyst activity. Catalysts in operating plants that have lost activity as a result of poisoning have been found on discharge to contain less than 0.1% sulphur. However, significant effects are produced by much smaller amounts of sulphur, for example around 0.01% sulphur, which on a freshly reduced catalyst with a crystallite size of 40 nm is equivalent to less than 1% of the iron surface having been sulphided.

Poisoning of ammonia synthesis catalysts is normally considered in terms of permanent and temporary deactivation, depending on the nature of the poison and on the conditions under which it was administered. This classification is used in the following sections.

8.5.2. Temporary Poisoning in Ammonia Converters

Oxygen and oxygen-containing compounds such as water, carbon monoxide and carbon dioxide are well-known poisons for ammonia synthesis catalyst. Under ammonia synthesis conditions, at temperatures in excess of 450°C and in the presence of a large excess of hydrogen, oxygen and oxides of carbon are thought to be rapidly converted to water. Ammonia synthesis catalysts, while inferior to nickel-based catalysts for the methanation of carbon oxides, are nevertheless effective methanation catalysts under synthesis conditions, and indeed have even found application in this role in some early coal-based plants (see Section 7.1). It is therefore not surprising that experience has shown that these poisons are about equivalent on an oxygen content basis, which is also consistent with some radioactive tracer studies.[259] Thus, 100 ppm oxygen has about the same effect as 100 ppm carbon dioxide, or 200 ppm carbon monoxide or water. If ammonia synthesis is carried out in the presence of oxygenates at low levels for comparatively short periods, then the deactivation effect is temporary and reversible, so that the overall conversion of synthesis gas to ammonia is restored completely by reverting to pure, oxygen-free synthesis gas. With catalyst operating at 450°C and 300 bar pressure and with a space velocity of about 15 000 h^{-1}, 100 ppm carbon monoxide in the inlet gas reduces the ammonia concentration in the exit gas by some 25% over six days, the activity remaining constant at this low level while the poison concentration is maintained. However, activity is completely restored after operating for one day with pure gas. Gas containing 500 ppm carbon monoxide reduces the exit ammonia concentration by about 67% over three days and, again, the activity completely recovers after operating for four days with pure gas. At 500°C, on the other hand, 50 ppm oxygen reduces the exit ammonia concentration by only 4%, but this effect is permanent, and no recovery is subsequently obtained when pure gas is used.

Some permanent damage almost certainly takes place even during so-called "temporary" poisoning, due to sintering or reorganization of the iron surface. This is normally not seen as a loss in production, because of the universal conservatism in converter design, but laboratory microreactor test results suggest that even low levels of carbon monoxide (e.g. 10 ppm) produce a measurable permanent deactivation, particularly at synthesis temperatures below 400°C. This effect is less pronounced at low temperature with carbon dioxide, and this probably reflects the lower rate of methanation compared with that for carbon monoxide. Water is produced at a significantly slower rate from carbon dioxide than from carbon monoxide, suggesting that water is the active poison.

There is evidence[260] that the amount of oxygen the catalyst adsorbs is proportional to $(P_{H_2O}/P_{H_2})^{1/2}$ at the low partial pressures of water present in the synthesis loop, which suggests that only a comparatively small proportion of the iron surface is active for ammonia synthesis. In another study[261] on the effect of oxygen and water on ammonia synthesis, it was found that the synthesis rate was proportional to P_{H_2}/P_{H_2O}. The water levels required to reduce catalyst activities are very much less than those required thermodynamically to oxidize bulk iron to Fe_3O_4. For example, at 450°C, the ratio P_{H_2O}/P_{H_2} must exceed 0.16 for bulk oxidation to take place, whereas in practice significant deactivation occurs at a value of only 50×10^{-6} or less. It is therefore probable that the deactivation involves reversible oxidation of the most highly active parts of the iron surface that are expected to have a higher free energy than bulk iron. It therefore appears that the surface of the catalyst contains crystallites of iron covering a range of free energies and activities.

8.5.3. Permanent Poisoning in Ammonia Converters

When poisoning by oxygen-containing molecules is continued for several weeks (rather than a few days) permanent deactivation results. Activity does not then completely recover when pure synthesis gas is subsequently used, and the extent of permanent deactivation is exacerbated by operation at high temperature. It appears that when the reduced catalyst is exposed to moderate partial pressures of water vapour, partial and reversible surface oxidation takes place rather than bulk oxidation, and in contrast with fresh catalyst, the transient oxidic phases probably do not contain alumina in solid solution. Thus, no support-precursor phase is present in the oxide structure to prevent sintering during re-reduction of the reforming iron crystallites. This continuous oxidation/reduction mechanism for "chemical" sintering is progressively enchanced at high partial pressures of the poison, particularly as the temperature is increased. A similar oxidation/reduction mechanism could also result in reorganization of the iron surface, decreasing the population of the highly active open-packed Fe(111) faces in favour of the close-packed and less-active Fe(110) ones. The increase in the size of the iron crystallites is not reversible, so permanent loss of iron surface area and activity results.

Irreversible poisoning by sulphur compounds has long been a problem in ammonia synthesis, particularly in the days when the feedstock was coal, and the synthesis gas had a relatively high sulphur content. The move to low-sulphur natural gas and petroleum-based feedstocks, coupled with the recognition that sulphur-containing compressor oils used in the make-up gas circulation system were detrimental, has largely removed this problem. Arsenic, antimony and phosphorus compounds are also severe poisons, and in some plants there is potential for the inadvertent introduction of arsenic from the carbon dioxide removal stage. Even very low levels of sulphur are detrimental and, unlike oxygen, sulphur tends to remain on the catalyst surface although thermodynamically reduction of bulk ferrous sulphide should be favourable under ammonia synthesis conditions. The use of calcium promoters in synthesis catalysts leads to enhanced stability towards sulphur, perhaps through formation of "calcium sulphide" on the surface.

Chlorine compounds are also severe poisons because metal chlorides are relatively volatile, and their transient formation can lead to irreversible sintering and loss of activity. An additional mechanism commonly quoted for chlorine poisoning is the removal of the potassium promoters from the active sites through formation of the relatively volatile potassium chloride.

Copper contamination, arising from ammoniacal copper liquor used in older plants for carbon monoxide removal, has also been found to lower the catalyst activity irreversibly, though this no longer presents any problem in modern plants that use low temperature carbon monoxide shift conversion and methanation.

In addition to chemical poisoning of the iron surface, permanent loss of activity can also result from physically covering the catalyst surface with otherwise inert material that obstructs access of reactants to the catalytically active sites. Such fouling occurs when amorphous carbon is deposited by the thermal or catalytic cracking of higher hydrocarbons from compressor lubrication oil. This effect can be partly reversible, provided the problem is recognized quickly and rectified because over a period of time small levels of carbon are removed from the catalyst as methane. Unfortunately, minor leakage of oil is not easily recognized, and permanent damage to the catalyst sometimes occurs. Sulphur-free lubricants must obviously be used to avoid sulphur poisoning. In practice, the well-maintained centrifugal compressors and circulators used in modern plants should not introduce significant amounts of oil into the circulation loop, although slugs of oil may be introduced as a result of operational upsets.

Even in the absence of poisons, recrystallization of the iron surface occurs slowly. This process proceeds even at the normal operating temperature, but is significantly faster at higher temperatures. In general 540°C should be regarded as an upper limit for ammonia synthesis catalysts such as ICI 35-4 used in conventional plants, though this does depend on the actual duty. For example, while ammonia converters are usually conservative in design, prolonged operation at excessive temperatures reduces activity of the catalyst to the point where high temperature becomes *necessary* to maintain the reaction rate. This effect, of course, is exacerbated by even quite small levels of oxygenated compounds. However, associated with increasing temperature is the penalty that the thermodynamic equilibrium concentration of ammonia becomes less favourable.

8.6. Kinetics and Mechanism

Given a satisfactory understanding of the thermodynamics of ammonia synthesis, the key to efficient converter design lies in the generation of an expression which correctly predicts the rate of ammonia synthesis as the partial pressures of reactants, products and inerts vary throughout the converter at different temperatures and total pressures. While high temperature and pressure increase the rate of reaction, high temperature also reduces the value of the equilibrium constant, and hence the maximum ammonia concentration attainable decreases as the operating temperature is increased. Thus, at high temperatures whereas the initial rate of reaction is high the rate soon falls as the relatively low equilibrium ammonia concentration is approached. The optimum yield of ammonia from a converter at a given pressure is therefore obtained when the temperature profile falls continuously through the bed, as the ammonia concentration increases. This is discussed later in this Chapter in Section 8.6.3. Within normal operating parameters the temperature for maximum rate is approximately 70°C below the temperature at which a given synthesis-gas composition would be in thermodynamic equilibrium. The most useful published work on ammonia synthesis kinetics is that of Temkin and his co-workers, and this is considered in the following section.

8.6.1. Temkin Kinetics

Many reaction mechanisms have been proposed for ammonia synthesis but, because of their almost universal application to plant design, only those of Temkin and his co-workers are considered here. The first kinetic equation to give reasonable agreement with observed rates was that due to Temkin and Pyzhev,[262] published in 1940. This equation was based on the assumption that the absorption of nitrogen on a non-uniform surface is the rate-controlling step, and it led to the now well-known equation (7) for the intrinsic reaction rate in the absence of diffusion. In this equation r is the rate of reaction and K_p the equilibrium constant for the synthesis reaction. The constant α has a value between 0 and 1, and k_2 is given by equation (8). The value of $\triangle E_{k_2}$ is about 150 kJ mol^{-1}.

$$r = k_2 \{K_p P_{N_2}(P_{H_2}^3/P_{NH_3}^2)^\alpha - (P_{NH_3}^2/P_{H_2}^3)^{1-\alpha}\} \tag{7}$$

$$k_2 = k_{2(0)} \exp-\{(\triangle E_{k_2}/R)[(1/T)-(1/T_0)]\} \tag{8}$$

Equation (7) has been the basis of industrial converter design for the past 30 years. Most workers, including those at ICI[263] use the value of α found by Temkin (i.e. α = 0.5). Others, notably Nielsen,[240] have found their results best supported a figure of 0.75. In general it has been found necessary to allow k_2 to decrease with increasing pressure, although again Nielsen (using α = 0.75), and Livshits and Siderov[264] (using α = 0.5) claim that k_2 is substantially pressure-independent if fugacities rather than partial pressures are used to allow for non-ideality. Inspection of equation (7) shows that it cannot apply when the ammonia concentration is zero, since it then predicts infinite reaction rate. More-recent work[265] has established that under these conditions the rate is best given by equation (9).

$$r = k P_{H_2}^\alpha P_{N_2}^{1-\alpha} \tag{9}$$

In 1963 Temkin et al.[266] proposed a mechanism which incorporated, as an important step, the addition of the first hydrogen molecule to the absorbed nitrogen. They obtained equation (10), in which k_* and l are given by equations (11) and (12).

$$r = \frac{k_* P_{N_2}^{1-\alpha}(1 - P_{NH_3}^2/K_p P_{N_2} P_{H_2}^3)}{[(l/P_{H_2}) + (1/K_p)(P_{NH_3}^2/P_{N_2} P_{H_2}^3)]^\alpha [1 + (l/P_{H_2})]^{1-\alpha}} \tag{10}$$

$$k_* = k_{*(0)} \exp-\{(\triangle E_k/R)[(1/T)-(1/T_0)]\} \tag{11}$$

$$l = l_{(0)} \exp-\{(\triangle E_l/R)[(1/T)-(1/T_0)]\} \tag{12}$$

It can be shown that under the two extreme conditions of close to equilibrium and far from equilibrium, equation (10) becomes equivalent to equations (13) and (14), respectively. If k_* is pressure-independent, then equation (9) suggests the pressure dependence of k_2 through the factor $K_p^{(\alpha-1)}$.

$$r = (k_*/K_p^{(1-\alpha)})\{K_p P_{N_2}(P_{N_2}^3/P_{NH_3}^2)^\alpha - (P_{NH_3}^2/P_{H_2}^3)^{1-\alpha}\} \tag{13}$$

$$r = (k_*/l^\alpha) P_{H_2}^\alpha P_{N_2}^{1-\alpha} \tag{14}$$

A considerable amount of work on the kinetics of ammonia synthesis was carried out in ICI laboratories during the 1950s and 1960s and catalysts were tested in both integral and differential reactors. It was found, from a large number of differential rate determinations over a wide range of conditions, that the model of Temkin et al. given in equation (10) gave a much better fit than equation (7) for non-diffusion limited situations. Under most conditions of commercial interest equation (10) corresponded quite closely with equation (7), with $k_2 = k_*/K_p^{(1-\alpha)}$. The "best fit" values, $\triangle E_{k_*} = 110.8$ kJ mol^{-1} and $\alpha = 0.46$, agree quite well with those of Temkin et al.[265] ($\triangle E_{k_*} = 104.5$ kJ mol^{-1} and $\alpha = 0.4$), and k_* was found to be independent of pressure. Thus, it is possible to calculate the pressure and temperature dependence of k_2. At 200 bar and 450°C, K_p varies as $P^{0.44}$, and if $\alpha = 0.46$, k_2 varies as $P^{-0.24}$. At low pressures K_p is a function of only temperature, and consequently k_2 is independent of pressure. Similarly, the activation energy associated with k_2 can be predicted from equations (15) and (16) in which $\triangle H_R$ is the enthalpy of reaction. Hence $\triangle E_{k_2} = 110.8 + 58.1 = 168.9$ kJ mol^{-1} ($\triangle H_R \simeq 109$ kJ mol^{-1}), which is close to the normally quoted value of 158.8 kJ mol^{-1} (38 kcal mol^{-1}).

$$\triangle E_{k_2} = \triangle E_{k_*} + (1 - \alpha)\triangle H_R \qquad (15)$$

$$\triangle H_R = -RT^2(\delta \ln K_p/\delta T)_P \qquad (16)$$

8.6.2. Effect of Catalyst Size

The kinetic expressions discussed in Section 8.6.1 relate to conditions where the rate of reaction is not limited by diffusional effects. Rate measurements made using catalysts of different sizes show that diffusion does, indeed, have a marked effect, particularly at high temperature. This is illustrated in Table 8.3, where reaction rates obtained using catalyst in the size ranges 0.6–1.2 mm, 3.0–4.5 mm and 6.0–9.0 mm are compared. The measurements were made in a differential reactor at 500°C and 100 bar, with a 3 : 1 hydrogen/nitrogen gas mixture containing 4% ammonia. It is clear that the larger size catalyst particles are considerably less active than the smaller particles. This is mainly due to mass transfer limitations within the catalyst pores. Another contributory factory is the lower intrinsic activity of larger catalyst particles, which is due to their outer regions being more exposed to the sintering action of water vapour than smaller particles during the reduction process (see Section 8.4.1).

Under conditions of low linear gas velocity the rate of reaction can also be limited by the rate of transfer of reactants and products through the thin gas film around the catalyst particles. This film diffusion phenomenon is most commonly encountered in small-scale laboratory reactors which are characterized by low turbulence around the catalyst

Chapter 8. Ammonia Synthesis

particles as described by Reynolds' numbers in the range 0–10. On the other hand, industrial reactors generally operate with much higher linear velocities, with Reynolds' numbers greater than 100. Under these conditions film diffusion limitations are insignificant, but at high reaction rates, particularly at high temperatures and pressures, diffusion of reactants or products throughout the particle can then become rate-limiting. This pore diffusion phenomenon is observed principally at the inlet of an ammonia converter, where the concentration of ammonia is low and the rate of synthesis relatively high. A detailed discussion of effects in ammonia synthesis is given by Nielsen.[240]

Table 8.3. Variation of synthesis rates with catalyst granule size

Normal size range/mm	Rate/kmol N_2 h^{-1} m^{-3} of catalyst
0.6–1.2	300
3–4.5	112
6–9	61

8.6.3. Implications on Process Design

The overall rate of reaction to produce ammonia from nitrogen is dependent on the relative rates of the forward and reverse reactions. Far away from equilibrium the forward reaction predominates and the rate increases as the temperature is raised. Close to the equilibrium—that is, in the presence of substantial amounts of ammonia—the reverse reaction becomes increasingly significant.

The results of these effects are illustrated in Figure 8.8, which shows the equilibrium concentration of ammonia as a function of temperature, together with a series of constant reaction rate contours. The contours are expressed in terms relative to the overall reaction rate at 350°C in the presence of 20% ammonia. For any ammonia concentration in the reaction gas there is a temperature at which the rate of reaction has a maximum value, which falls sharply at higher temperatures as the equilibrium curve AB is approached. The loci of the maxima of the constant rate contours generates the curve CD, which may be regarded as representing the ideal temperature profile for an ammonia converter, and its position depends on catalyst activity and particle size. Such a profile cannot be achieved in practice, but it is nevertheless the aim of the converter designer to approach it as closely as possible.

The intrinsic activity of an ammonia synthesis catalyst gradually falls

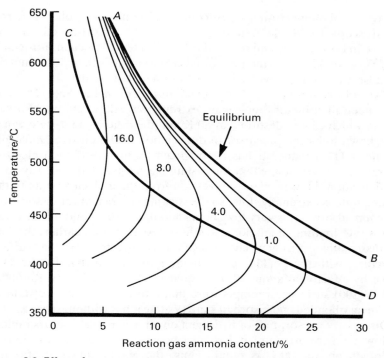

Figure 8.8 Effect of temperature and ammonia concentration on reaction rate. The line CD is the maximum rate curve and is in effect the ideal temperature profile for an ammonia converter.

during use, initially relatively rapidly, followed by a slow decline to a steady level approximately half of the initial activity. The extent and rate of this deactivation is very dependent on the operating conditions, and reactors must be designed to operate at this steady-state level. The reaction rates given in Table 8.3 are for freshly reduced catalyst and are thus not suitable for designing a converter.

8.6.4. Reaction Mechanism

The sequence of events that takes place at the molecular level on the catalyst surface during ammonia synthesis has been studied by several research groups over many years. It was established by the combined efforts of several early workers[267, 268] that the rate-determining step involves the initial interaction of the nitrogen molecule with the iron surface, though the exact nature of this interaction was not known. Evidence obtained from modern surface science techniques has helped to substantiate and to clarify this picture. It has been shown by ultraviolet photoelectron spectroscopy (UPS) and X-ray photoelectron spectroscopy (XPS) data that nitrogen can be adsorbed on an iron surface in both atomic and molecular[269-272] states. Molecular nitrogen is only very weakly held (the heat of adsorption is less than 40 kJ mol^{-1}) but adsorption is rapid with a sticking coefficient of about 0.01.

Two adsorption states of molecular nitrogen on the Fe(111) plane have been identified:[273] one is an extremely weakly held physisorbed γ-state which can be desorbed at 80 K and the other is an α-state which was shown to be the precursor of dissociation of the nitrogen molecule on the Fe(111) plane. It has been shown by high-resolution electron energy-loss spectroscopy that the α-state has a π-bonded "sideways-on" configuration. However, the conversion from the molecular state to the atomic state is extremely slow, so the sticking coefficient for dissociative nitrogen adsorption is very low, being about 1×10^{-7}, and it is this step that is rate-limiting in ammonia synthesis. As indicated earlier, the rate of dissociative nitrogen chemisorption is affected by the surface structure, with the Fe(111) plane being more active than the Fe(110) plane by a factor of about 20. In similar work[274] on Ni(100), Re(0001) and W(100) surfaces, attempts were made to find a precursor state to the lineary bonded chemisorbed state (γ-state), but without success.

Dissociative adsorption of hydrogen on an iron surface occurs rapidly at low temperature, and it has also been shown that desorption of hydrogen above 200°C is rapid. Thus, the equilibrium state for the adsorption and desorption of hydrogen on iron will always be maintained under ammonia synthesis conditions. Since the onward hydrogenation of dissociatively adsorbed nitrogen is extremely rapid, direct observation of intermediates is difficult. In order to obtain information on the nature of the hydrogenated[275, 276] species the reverse reaction—that is, the decomposition of ammonia on iron surfaces—was studied by Ertl and co-workers. Ammonia adsorbs at low temperature and is rapidly desorbed above 100°C. When it decomposes on an iron surface to chemisorbed hydrogen and nitrogen, atoms are generated. Exchange reactions with deuterium led to the production of NH$_2$D, suggesting that reversible dissociation takes place as shown in equation (*17*)-(*19*).

$$D_2(ad) \rightleftharpoons 2D(ad) \qquad (17)$$

$$NH_3(ad) \rightleftharpoons NH_2(ad) + H(ad) \qquad (18)$$

$$NH_2(ad) + D(ad) \rightleftharpoons NH_2D(ad) \qquad (19)$$

Studies on the interaction of ammonia with Fe(110) by UPS have shown that at 500 K the stable species is N(ad) but evidence was also presented for a partially hydrogenated phase, stable at 340 K. This intermediate was, however, shown to be NH(ad) rather than NH_2(ad) by secondary ion mass spectrometry.[277] Direct evidence for the presence of a wide range of intermediates in the interaction of nitrogen, hydrogen and ammonia with iron surfaces has therefore been obtained, and this has been compiled into an overall reaction mechanism shown in Scheme 1. Other similar reaction sequences had also been proposed[278, 279] by earlier workers, but without characterization of the surface intermediates.

Scheme 1.

$$H_2 \rightleftharpoons 2H(ad)$$
$$N_2 \rightleftharpoons N_2(ad)(\gamma) \rightleftharpoons N_2(ad)(\alpha) \rightleftharpoons 2N(ad)$$
$$N(ad) + H(ad) \rightleftharpoons NH(ad)$$
$$NH(ad) + H(ad) \rightleftharpoons NH_2(ad)$$
$$NH_2(ad) + H(ad) \rightleftharpoons NH_3(ad) \rightleftharpoons NH_3$$

The promotional effect of potassium in the catalyst must arise by influencing the rate-limiting step in ammonia synthesis, which is the dissociation of the α-state of adsorbed molecular nitrogen into the atomic state. It has been shown[253] that the heat of adsorption of molecular nitrogen on the iron surface is increased by about 45 kJ mol^{-1}, with a consequent lowering of the activation energy for dissociation when the nitrogen molecule is in close proximity to a potassium atom. This results in a large increase in the sticking coefficient, a feature more pronounced on the less active Fe(100) face than on the open-packed and more active Fe(111). This leads to a situation in which all of the iron surface planes have a similar activity in the presence of potassium. It is suggested that the increase in the nitrogen molecule adsorption energy is due to charge transfer from the potassium to the iron surface, allowing much stronger π-backbonding from the iron surface to the nitrogen molecule. However, further work is still needed to link together these elegant surface studies with "real" catalysts, operating under industrial conditions.

8.7. Plant Operation

8.7.1. General Considerations

It is apparent from the thermodynamics of the exothermic synthesis reaction that ammonia production should be carried out at low temperature and elevated pressure if the best conversion is to be obtained. In practice the actual operating temperature and pressure

Chapter 8. Ammonia Synthesis

used are determined by catalyst activity. Many variations of the synthesis process have been developed, with most being designed to operate at pressures governed by the material limits of the equipment, particularly the synthesis reactor vessel and the compressor. Plants designed before about 1960 usually consisted of multiple parallel streams with numerous interconnections between the main process steps, and with the ammonia synthesis being carried out at pressures of 250–350 bar. At such pressures the conversion to ammonia is sufficiently high that it is possible in temperate climates to operate a basic synthesis loop consisting of converter, water- or air-cooler, liquid-ammonia separator and gas-circulator. Such a loop configuration, with simple cooling rather than refrigeration of the circulating gases, returns gas to the ammonia converter with an ammonia concentration of about 5–6%. The ammonia concentration in the gas phase in equilibrium with liquid ammonia in a 3 : 1 hydrogen/nitrogen mixture may be estimated from the curves given in Figure 8.9, or derived from the expression given in

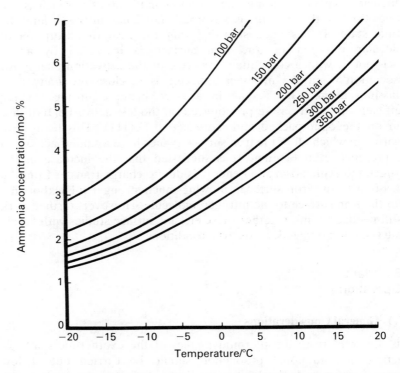

Figure 8.9 Concentration of ammonia vapour in equilibrium with liquid ammonia in a 3:1 hydrogen/nitrogen mixture.

8.7. Plant Operation

equation (20) where a is the mole fraction of ammonia in the vapour, P is the pressure (bar) and T is the absolute temperature. These relationships have been derived for a 3 : 1 hydrogen/nitrogen gas mixture containing ammonia at pressures above 100 bar, but they do not take into account the effect of methane and argon on the vapour/liquid equilibrium.

$$a = (0.0035 + 1.0138/P)10^{(4.84 - 1144/T)} \qquad (20)$$

The ammonia concentration obtained from a converter operating at a pressure of 350 bar is typically about 20%. Thus, an adequate incremental conversion per pass can be obtained at 350 bar with water- or air-cooling alone. However, the incorporation of a refrigeration unit to condense the ammonia to below the 5% level significantly reduces the circulation rate required for a given production rate. The complex interactions between loop operating pressure, refrigeration level, circulation rate, the power required to drive the major loop machinery

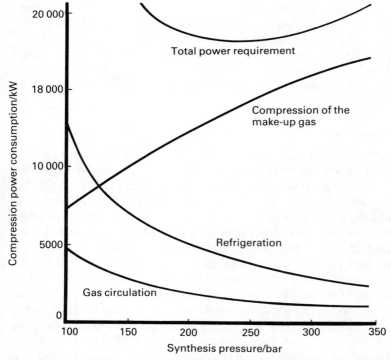

Figure 8.10 The effect of loop pressure on major power requirements for a typical 900 tonne/day ammonia plant. The total power requirement shows a shallow minimum under the condition considered in the diagram. (Courtesy of British Sulphur Corporation.)

Chapter 8. Ammonia Synthesis

and capital costs are of critical importance in the design of efficient, high-capacity, single-stream ammonia plants. A number of studies have been carried out to investigate optimum loop pressure[280] using typical commercial catalysts, and in Figure 8.10 an estimate of the power required to drive the main loop machinery in a 900 tonnes day^{-1} plant for a range of pressures from 100 to 350 bar is shown. This figure has been constructed on the assumption that a high degree of refrigeration is required when the synthesis loop is operated at low pressures to compensate for the lower equilibrium ammonia concentration.

The additional cost of refrigeration, together with the further cost of higher circulation rate, can exceed the benefits from operating at reduced pressure, and in this particular instance energy consumption is lowest around 250 bar. The equilibrium concentration of ammonia can be increased by lowering the operating temperature, but this would require either additional catalyst volume or high-activity catalysts. The generation of new high-activity synthesis catalysts have allowed the development of the ICI AMV Process,[241] in which very low refrigeration temperatures are not required. In this process the total loop power requirements are reduced, typically to about 12 MW for a 900 tonnes day^{-1} plant compared with about 20 MW for the conventional loops included in Figure 8.10. The significant improvement in the overall energy efficiency of the AMV Process is also partly due to the lower fuel demands of the synthesis-gas preparation section of the process.

8.7.2. Circulation

The ammonia production rate as a function of circulation rate through the converter and conversion per pass can be readily derived from the mass balance expression shown in equation (*21*) where V is the converter inlet rate in N m^3 h^{-1}, M is the ammonia production rate in tonnes day^{-1}, and a_i and a_o are the inlet and exit ammonia mole fractions, repectively. If other operating parameters are kept constant, as the circulation rate is increased the composition of the reaction mixture moves further away from equilibrium, and the exit ammonia concentration is decreased. However, the higher throughput more than compensates for the lower ammonia concentration, and the overall effect is an increase in ammonia production. The effect is illustrated in Figure 8.11 which shows a wide range of circulation rate and assumes that pressure is constant. In a practical situation the synthesis make-up gas rate is often fixed and a change in circulation rate is reflected in a change in loop pressure. In addition the range of circulation rate over which control can be effected is quite small. Some centrifugal machines provide the facility of inlet guide vanes, by means of which the circulation rate can be adjusted over a limited range without loss of efficiency. It is then possible to control circulation rate and in this way optimize output and heat recovery.

$$1/M = (58.83/V)[(1 + a_o)/(a_o - a_i)] \qquad (21)$$

The measurement of circulation rate is important if any investigative work on ammonia-loop performance, such as the estimation of catalyst efficiency, is to be undertaken. Direct measurement is not always

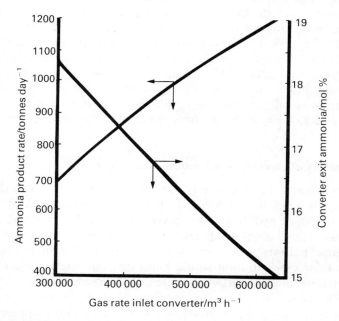

Figure 8.11 Effect of circulation rate on ammonia production rate and conversion. Although conversion per pass is higher at lower rates, the maximum output is obtained at high circulation rate.

possible, but a satisfactory estimate can often be made from a heat and mass balance across the converter, given the overall ammonia production rate together with inlet and exit temperatures. Precise enthalpy data are required. Equation (22) can be used to obtain an approximate value of the inlet gas rate V (in $Nm^3\ h^{-1}$) that is sufficiently accurate for checking the calibration of flow meters. T is the rise in temperature over the converter. This equation assumes heat is not removed from the converter by external cooling.

$$V = 95\ 000M/T(1 + a_i) \qquad (22)$$

8.7.3. Hydrogen/Nitrogen Ratio

The ratio of hydrogen to nitrogen is one of the more important parameters to be considered in the design and operation of an ammonia-synthesis loop. As shown in Section 8.6.4, the rate-determining step in the synthesis reaction is the dissociative chemisorption of nitrogen on the activated catalyst surface, suggesting that the use of nitrogen-rich synthesis gas would enhance the reaction rate. In practice, within the limits imposed by overall optimization, a ratio in the range 2.2 : 1 to 3.0 : 1 gives an efficient operating point. The manner in which the hydrogen/nitrogen ratio in the loop is affected by changes in the amount of inert gas in the synthesis gas is shown in Figure 8.12. The effect is most pronounced when operation with low concentrations of inerts—as, for example, where synthesis gas is obtained by partial oxidation and purified by a nitrogen wash. Such loops require little or no purge to control the concentration of inerts, and therefore small changes in synthesis-gas composition have a profound effect on the hydrogen/nitrogen ratio in the loop and on the operability of the plant.

8.7.4. Influence of Inert Gas Concentration and Purge Rate

Inert gases ("inerts") are invariably present in synthesis gas, regardless of the way in which it is produced. Typically, gas obtained by the steam-reforming of hydrocarbons will contain about 1% methane and 0.3% argon. These components take no part in the reaction, nor are they catalyst poisons, but they must be removed from the loop otherwise their concentration would build up and quickly affect the process by lowering the partial pressures of the reacting gases. Their concentration in the loop is controlled by purging—the higher the purge rate is, the lower the concentration in the loop. The removal of inerts is achieved in two ways: (a) a fixed involuntary purge via methane and argon dissolving in the liquid ammonia product and (b) deliberate, controlled venting, direct from the synthesis loop. The purge gas contains not only the "inerts", but also some ammonia, hydrogen and nitrogen. The ammonia is generally recovered, and the residual hydrogen-containing gas is used as fuel. In such circumstances the loss of reactants is significant and can govern the amount of ammonia produced.

The effect is illustrated in Figure 8.13, in which curve 1 shows the maximum amount of ammonia that can be produced from a fixed rate of production of synthesis gas, without recovery of the hydrogen from the purge. The data cover a range of concentrations of "inerts", and it is assumed that the circulation rate can be increased to keep the pressure

8.7. Plant Operation

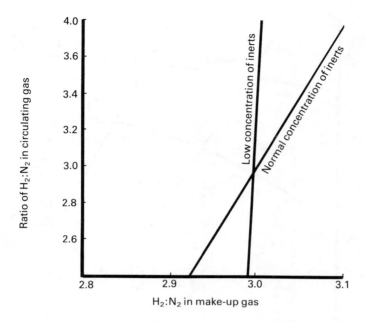

Figure 8.12 Relationship of H_2 to N_2 ratio in make-up gas and circulating gas. The normal inerts line represents the situation where the loop make-up gas contains 1–1.5% of methane and argon. In the low inerts case, the loop make-up gas contains only a few parts per million inerts and little or no purge is required to control the loop inerts. Here, small deviations from 3:1 in the make-up gas H_2 to N_2 ratio produce a large deviation in the loop H_2 to N_2 ratio.

constant as the "inerts" level is increased. It can be seen that the maximum ammonia production occurs at the highest "inerts" levels. This rather anomalous conclusion arises as a consequence of the smaller losses of reactants, since the high level of inerts results from a lower purge rate.

It has, however, become increasingly common to recover the high-value hydrogen from the purge gas and to recycle it to the synthesis loop. This can be accomplished in a cryogenic plant, where typically about 90% of the hydrogen in the purge gas may be recovered at a purity of 90–95%. Other systems, notably those using permeable membranes and molecular sieves, are also becoming more common.

Whichever system is used, the methane, argon and nitrogen together with some hydrogen are separated from the purge as a tail gas stream and may be incorporated into the plant fuel gas system. A

Chapter 8. Ammonia Synthesis

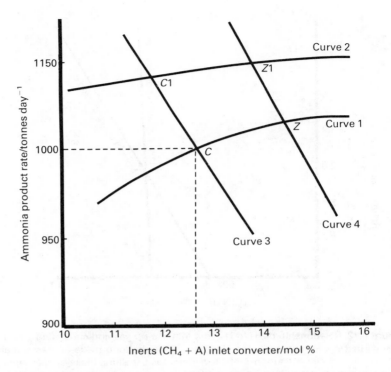

Figure 8.13 Effect on converter operation on loop output. Curve 1 represents the mass balance of an ammonia loop where hydrogen lost in the higher purge rate required to maintain a low loop inerts level results in a low ammonia production rate. This effect is less marked in curve 2 which represents the recovery of a substantial proportion of the hydrogen purged from the loop in curve 1. Curves 3 and 4 are two examples of different converter operating conditions. The intersection of a converter operating line with a loop mass balance line, for example at C gives the operating point of the loop and converter, in this case 1000 tonne day^{-1} at 12.7% inerts. Improvement in converter operating condition to curve 4 allows operation at point Z which gives an increase in output at the same make-up gas rate. A change in converter operating conditions from curve 3 to curve 4 may be brought about in a number of ways, for example, by increased circulation rate, improved catalyst activity, or improved converter temperature profile.

representation of the situation where hydrogen is recovered from the purge and recycled to the loop is shown in Figure 8.13 (curve 2), and it can be seen that as a result of this the production rate of ammonia is increased significantly. The shape of the curve depends on the capacity and efficiency of the hydrogen recovery unit as well as the purity of the recovered hydrogen stream. Typically, where the synthesis-gas production rate is fixed, and where the concentration of inerts in the loop is controlled at 12–13%, the addition of a hydrogen recovery unit makes it possible to increase the ammonia output by 4–5%. Since the nitrogen in the purge gas is not usually recovered for recycle, it is necessary to adjust the composition of the synthesis gas by addition of a little more air in the secondary reformer than is required for 3 : 1 stoichiometry. The fresh make-up gas has a hydrogen/nitrogen ratio of about 2.8 : 1 before addition of the recovered hydrogen stream. The ICI AMV process incorporates a large efficient hydrogen recovery unit operating at synthesis loop pressure. The hydrogen to nitrogen ratio in the make-up gas here is typically 2.5:1.

A general rule for converter operation is that as the concentration of "inerts" falls, the partial pressure of reactants increases with a consequent increase in both reaction rate and conversion. Thus, for a fixed circulation rate, provided sufficient synthesis gas is available to maintain the operating pressure, output will increase as the inerts level is reduced. This is represented as the converter performance line (curve 3) in Figure 8.13. The intersection of this curve with the mass balance line (curve 1) defines the optimum operating point (C or $C1$) depending on whether hydrogen recovery is used or not. This fixes the output of the converter and the concentration of inerts in the circulating gas. If the circulation rate were to be increased, then the new converter performance line would be represented by curve 4 and the new operating point would then be either Z or $Z1$. Similar performance lines can be constructed to examine the effect of variation of any of the converter operating parameters.

The use of performance lines forms the basis of a method for optimizing the operation of a given loop and converter configuration. Curves 3 and 4 in Figure 8.13 could represent non-optimal and optimal temperature profiles respectively, point Z therefore representing the maximum output possible under this set of operating conditions. It can also be seen that operation with hydrogen recovery, particularly in this case, is less sensitive to converter performance. In drawing the curves, it was assumed that the hydrogen recovery unit was of sufficient capacity to deal effectively with the higher purge gas rates.

8.8. Commercial Ammonia Converters

8.8.1. General Considerations

In its simplest form the ammonia converter consists of a pressure vessel into which is fitted a cartridge consisting of a catalyst section and a heat exchanger. The cartridge is usually fabricated of stainless steel, and in many designs it can be withdrawn completely from the pressure vessel.

Chapter 8. Ammonia Synthesis

Reaction conditions lie in the regime where most alloy steels would suffer embrittlement due to nitriding or hydrogen attack, and it is usual for ammonia converter pressure vessels to be protected from exposure to the reaction temperature by passing cool inlet gas (or in some designs, exit gas) through an annular space between the vessel and the cartridge. As the cartridge is contained within the pressure vessel, it is not itself required to withstand the full differential synthesis loop pressure. It can therefore be constructed from relatively thin stainless steel rather than low-alloy steel, which minimizes the problems associated with embrittlement.

The synthesis converter is undoubtedly the most complex catalyst containing vessel in the ammonia plant. Even given the overall broad operating conditions of the plant and synthesis loop, the designer is still faced with a range of converter types, the chacteristics and philosophies of which not only affect the operation and detailed design of the synthesis loop, but can also be reflected in the design and economics of the plant as a whole. In general, converters can be classified under two main headings: the type of flow used (axial, radial or cross flow), and the methods used to control temperature (quench or indirect cooling) and recover the heat of reaction.

8.8.1.1. Flow type

Only a limited conversion to ammonia is achievable on a single pass of reaction gas through any ammonia converter, and it is therefore necessary to recycle unreacted gas back to the converter. Typically the circulation rate is four to five times the rate of addition of synthesis make-up gas to the loop. As the diameter of a pressure vessel may be limited by shell thickness, closure size and cost, high pressure ammonia converters tend to be tall vessels of relatively small diameter. A high gas rate flowing axially through a catalyst bed in such a vessel will therefore create a high pressure-drop, and consequently the circulator power consumption will be high. The cartridge and associated loop feed/effluent heat exchangers have to be designed to withstand this relatively high differential pressure, but nevertheless an axial flow converter does provide a simple design.

The high pressure-drop penalty incurred in a high length/diameter ratio axial-flow converter may be overcome in a radial-flow design. In such a design the gas is introduced in an annular space surrounding the cylindrical bed. The gas then flows radially through the bed to a central exit-gas collector. The reverse of this flow is also possible; in either case the result is, in effect, a bed of shallow depth and large cross-sectional area and hence low pressure-drop. An alternative is a cross-flow pattern which has the same effect. These low pressure-drop beds need to be carefully designed to ensure even flow and to avoid bypassing of the bed. The cartridge is necessarily more complex than that of an axial-flow converter, so is often limited to two beds. However, such converters are attractive where energy costs are high and circulation power consumption has to be minimized.

8.8.1.2. Temperature control and heat recovery

Since the ammonia synthesis reaction is exothermic, the equilibrium concentration of ammonia is lower at higher temperatures, and one of the fundamental aspects of converter design is the provision of an effective means of catalyst temperature control. The process designer can choose from several methods of temperature control; the method actually employed has important consequences which affect the design of the converter, the loop and the plant as a whole. Basically two methods of temperature control are available.

The first method involves adding part of the cool reaction gas to the catalyst bed at intervals along its length; the so-called quench converter. In the second method the catalyst bed is cooled by heat exchange. The cooling medium is usually the inlet gas to the catalyst bed, but in some designs steam generators have been used. Heat exchange with the catalyst bed can be applied at intervals down the length of the bed, or continuously by means of cooling-tubes passing through the bed. The indirectly cooled converter, as opposed to the directly cooled quench reactor, is generally preferred when it is required to recover reaction heat efficiently at high temperature. This avoids the degradation of high-grade heat, inevitable in quench designs, consequent upon the mixing of hot and cold gas streams.

When temperature control is applied at intervals along the catalyst bed, the catalyst is effectively subdivided into a number of adiabatic beds separated by the cooling means, either quench or heat exchange. Such multibed converters can utilize axial-, radial- or cross-flow configurations, depending on the pressure-drop requirements of the synthesis loop. However, continuously cooled (tube-cooled) converters are generally restricted to axial flow. In recent years there has been a tendency to increase the temperature of the converter exit gas in order to recover reaction heat at a high enough temperature to make a useful contribution to the overall plant energy requirements. A typical synthesis' loop with heat being recovered at the converter exit is illustrated in Figure 8.4. In this example the converter feed gas (including the quench gas) is preheated to 150°C in the external feed effluent exchanger, while the gas from the outlet of converter internal exchanger is at a temperature of 330°C. This temperature is sufficiently high to preheat boiler feedwater for 100-bar steam production, but in

Chapter 8. Ammonia Synthesis

order actually to generate steam at this pressure it is necessary to have a converter exit temperature of about 450°C. To achieve this an indirectly cooled multibed converter is normally used. As the required exit temperature is similar to that of the final bed exit temperature, such converters do not usually have a heat exchanger immediately after the final bed, because all the internal heat exchange is accomplished by the inter-bed exchanger(s).

In the following sections some of the principles of the main types of ammonia converters are reviewed, together with examples of some of the commercially available converters. The review is not intended to be exhaustive, but rather to highlight the principles and main features of currently available equipment.

8.8.2. Quench Converter

In this type of converter, catalyst bed temperatures are controlled by injection of cold synthesis gas between beds of catalyst. Earlier in this Chapter (see Section 8.6.3) the concept of the maximum rate curve was discussed, and it was stated that it was impossible to design a converter which followed that temperature profile exactly. A satisfactory approximation to the ideal profile is achieved by the use of a multibed converter, and the temperature profile of a typical three-bed quench converter designed for operation at 220 bar is shown in Figure 8.14. In this particular example the temperature rise over the successive beds is approximately 95, 45 and 40°C, respectively, while the concentration of the ammonia increases from 3% to 15% across the converter. In the design of a multibed converter it is possible to minimize the total catalyst volume, and hence converter size, by determining the optimum temperature profile for the converter. Typical catalyst volumes for the three beds, in a 1000 tonnes day^{-1} ammonia plant (at 220 bar), using catalyst in the size range 6–9 mm are: Bed one, 7 m^3; Bed two, 12 m^3; and Bed three, 22 m^3.

In determining the necessary catalyst volume the approach to equilibrium is seldom selected to be less than 15°C, since a lower approach than this without a significant increase in catalyst activity would substantially increase the catalyst volume for little increase in conversion. In a multibed converter the use of four beds rather than three to achieve the same conversion per pass would result in a lower catalyst volume. In such a case the temperature rise over the individual beds will be reduced and the overall profile would be slightly closer to the maximum rate curve. However, the use of more than four beds to increase conversion per pass is not usually considered to be practicable, because of the increased mechanical complexity of the converter.

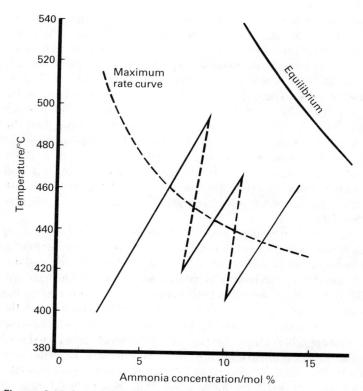

Figure 8.14 A typical quench converter profile. The converter catalyst bed temperatures lie on either side of the maximum rate curve. Note the reduction in ammonia concentration, due to dilution by the quench gas, between the exit of one bed and the inlet of the next.

Temperatures in the final catalyst bed, where the ammonia concentrations are highest, must be kept low to avoid equilibrium limitations. However, lower converter temperatures may affect operability, as converters of this type are designed to operate autothermally, with the first bed inlet temperature being maintained by heat exchange with the exit gas from the final bed. A temperature profile for a multibed quench converter could be produced in which the exit temperature of the final bed is less than that required to sustain reaction in the first bed. Such a converter would, of course, be inoperable and it is usual to design quench converters with an overall $\triangle T$ of about 60°C, so that the internal heat exchanger will be of reasonable size.

Many different forms of axial-flow quench converters are in use, and the main differences between the designs lie in the catalyst bed support systems, and the means of quench-gas injection. The design illustrated in Figure 8.15 is an ICI lozenge quench converter in which the continuous catalyst bed is divided by lozenge distributors through which the quench gas is introduced and mixed. Two other common quench types are the Kellogg four-bed axial flow converter and the Topsøe two-bed radial flow. The Kellogg converter, shown in Figure 8.16, consists of four beds held on separate grids. Quench-gas distributors/mixers are placed in the spaces between the beds, and the heat exchanger is located at the top of the vessel. A version exists in which the shell-cooling sheath is provided by gas from a point downstream of the converter. The cartridge is thus under internal pressure rather than external pressure, as in the more common arrangement using converter inlet gas, thus allowing the use of a lighter cartridge construction.

The Topsøe converter uses two radial beds with an injection of quench gas between them, and thus offers a lower pressure drop than is generally possible with axial-flow types. The layout of the quench converter is similar to the indirectly cooled S200 version illustrated in Figure 8.22, below, the main difference between the two types being the provision of an inter-bed heat exchanger in the S200. In general the design of radial beds requires care to ensure even gas distribution and to avoid gas bypassing direct from inlet to exit. A relatively new radial flow design is the Casale converter, illustrated in Figure 8.17. The flow pattern is actually a mixture of axial and radial; part of the gas enters the top of each bed in predominantly axial flow before axial flow is established lower down the bed. The flow pattern is designed to avoid problems associated with possible bypassing of the catalyst bed. The Casale converter also features the use of a cartridge in which each bed is contained in a module. Modules are stacked one on top of the other as the cartridge is assembled in the pressure vessel.

Radial-flow designs often result in tall vessels of relatively small diameter. A consequence of this can be problems associated with providing the high lift necessary to dismantle the converter during overhauls. An interesting variation using a cross-flow principle which avoids excessive pressure vessel height is the Kellogg horizontal converter illustrated in Figure 8.18. The gas flow through the catalyst is perpendicular to the axis of the vessel, the catalyst being contained in shallow beds of rectangular cross-section. The version illustrated is a quench type, although indirectly cooled versions have also been developed. Access to the cartridge for catalyst charging is obtained by withdrawing the cartridge on tracks through the full-bore opening in the converter pressure vessel. The original quench version was started up in 1971 at the Higashi Nihon ammonia plant in Japan, which was then the

8.8. Commercial Ammonia Converters

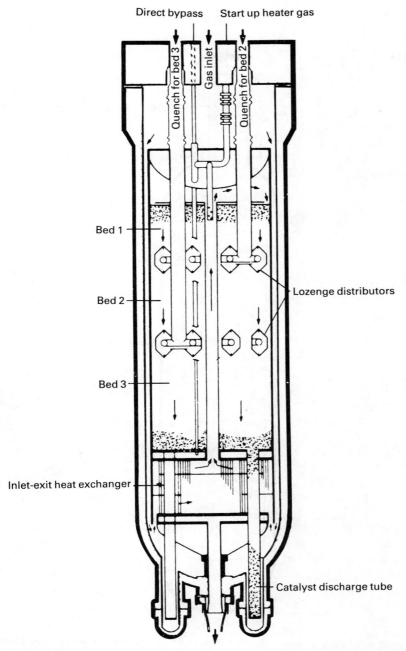

Figure 8.15 ICI quench converter. Quench gas is injected and mixed by means of lozenges which also allow gravity discharge of the catalyst bed.

Chapter 8. Ammonia Synthesis

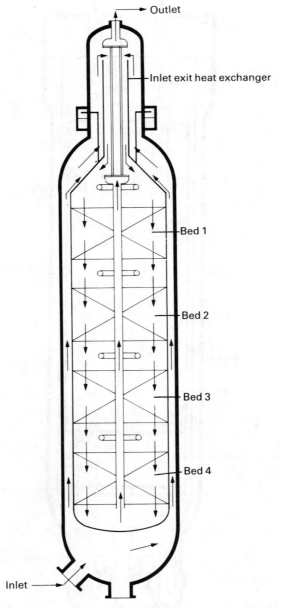

Figure 8.16 Kellogg quench converter. The four beds are supported on separate grid plates. The internal heat exchanger is located at the top of the vessel. (Courtesy of British Sulphur Corporation.)

8.8. Commercial Ammonia Converters

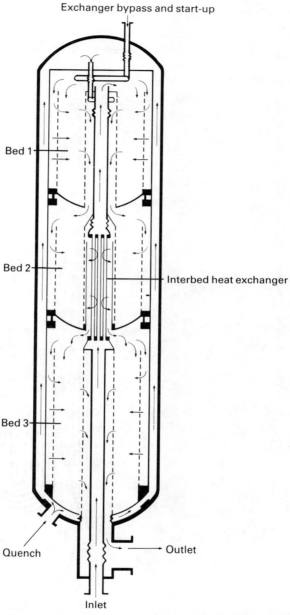

Figure 8.17 Casale axial-radial converter. The majority of the gas passes through the beds in radial flow. Complicated sealing arrangements at the tope of each bed are avoided by allowing a portion of the gas to pass in axial flow. The cartridge is assembled in the pressure vessel in stackable modules. (Courtesy of British Sulphur Corporation.)

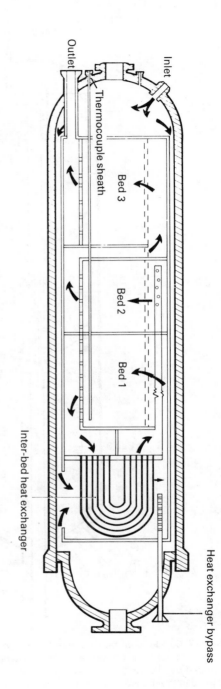

Figure 8.18 Kellogg horizontal converter. This version is a quench converter although indirectly cooled types are possible. (Courtesy of British Sulphur Corporation.)

largest ammonia plant in the world. A small particle size version of ICI Catalyst 35-4 was used in this converter. The horizontal converter was not immediately widely adopted, but developed versions including those with indirect inter-bed cooling have come into prominence in the mid-1980s for use in large, high-efficiency plants.

8.8.3. Indirectly Cooled Multibed Converter

The quench converter is not ideal for recovering reaction heat at the highest temperatures possible. Such a design would require the feed gas to the converter (including the quench gas) to be at a high temperature. The relatively high temperature of the quench gas would then require a large proportion of the converter feed to be introduced as quench, to sustain the converter temperature. This would reduce the overall ammonia concentration and hence overall conversion per pass. This is typical of the poor efficiency that can result from the cooling of a hot gas stream by mixing with colder gas. Such a disadvantage can generally be overcome by replacing the cold shot quench with conventional heat exchangers. Figure 8.19 shows a temperature profile for a two-bed converter with indirect cooling between the beds. In this example the same conversion per pass as in the three-bed quench converter shown in Figure 8.14 is achieved. The addition of a further bed would increase the conversion, at the expense of a higher catalyst volume and a more complex converter. An example, of a two-bed indirectly cooled converter configuration is shown in Figure 8.21, and a converter in Figure 8.20. This is the C. F. Braun design, in which separate vessels are used for each bed. While this design makes efficient use of available volume for catalyst containment, special precautions are necessary in designing high-temperature nozzles and external connecting pipework.

The Topsøe S200 converter, shown in Figure 8.22, also uses two beds with an inter-bed heat exchanger, this time contained within a single vessel. The catalyst beds use radial, rather than axial, flow. Like the quench version, this arrangement offers a substantial reduction in pressure drop through the catalyst bed compared with axial-flow designs, and allows smaller catalyst particles to be used. The pore-diffusion limitation effects are thus reduced and higher activities per unit catalyst volume are obtained.

8.8.4. Tube-cooled Converter

A typical tube-cooled converter is illustrated in Figure 8.23. Cooling tubes pass through a single bed of catalyst, and the internal

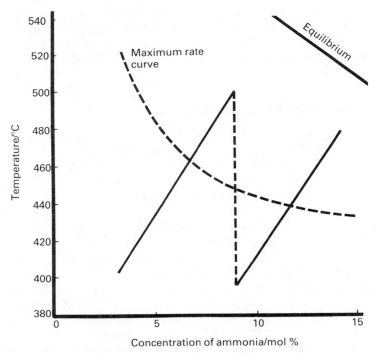

Figure 8.19 Temperature profile of a two-bed indirectly cooled converter. Because interbed cooling is achieved without reduction of ammonia concentration, such a converter achieves a similar conversion per pass as the three bed converter illustrated in Figure 8.18.

heat-exchangers are all contained wthin the cartridge. The cool inlet gas flows through the annulus between the pressure vessel and cartridge. It then passes through an internal start-up heater and the cooling tubes before flowing into the catalyst bed at reaction temperature. Although an electric heater is shown in Figure 8.23, an external fired heater can be used. The temperature profile in the catalyst bed reaches a maximum at a position corresponding approximately to one-third of the bed depth, as shown in Figure 8.24. The converter is equipped with two bypasses, one controlling the temperature at the exit of the internal heat exchanger (the inlet to the cooling tubes), and the other (the direct bypass) providing the means for a cold quench at the inlet to the catalyst bed. The mode of operation of the bypasses depends on the temperatures at the inlet and exit of the bed, as well as the peak temperature to be controlled.

As with multibed converters, the aim with tube-cooled converters is to operate the catalyst bed so that the temperature profile lies as close as

8.8. Commercial Ammonia Converters

possible to the maximum rate curve. The tube-cooled type of converter was originally developed by the Tennessee Valley Authority, and is often referred to as a TVA converter. Many mechanical variations have been designed and operated, with examples in ICI producing over 600 tonnes day^{-1} of ammonia. However, this type of converter has not been favoured for larger capacities because of the increased mechanical complexity. Nevertheless, it does provide another means of controlling the catalyst bed temperature by direct heat exchange rather than by quench, and is therefore free from the inefficiencies inherent in that type of design.

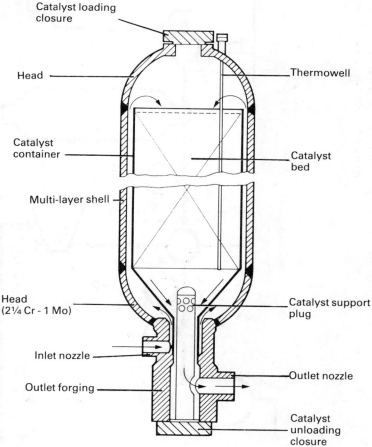

Figure 8.20 The C. F. Braun single bed ammonia converter as used in an indirectly cooled configuration illustrated in Figure 8.21. (Courtesy of British Sulphur Corporation.)

Chapter 8. Ammonia Synthesis

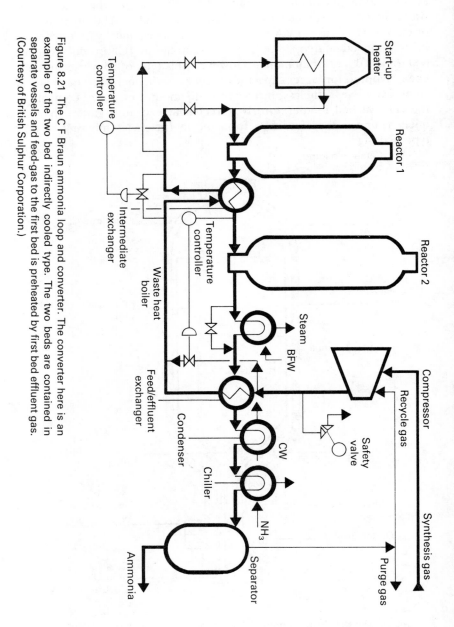

Figure 8.21 The C F Braun ammonia loop and converter. The converter here is an example of the two bed indirectly cooled type. The two beds are contained in separate vessels and feed-gas to the first bed is preheated by first bed effluent gas. (Courtesy of British Sulphur Corporation.)

436

8.8. Commercial Ammonia Converters

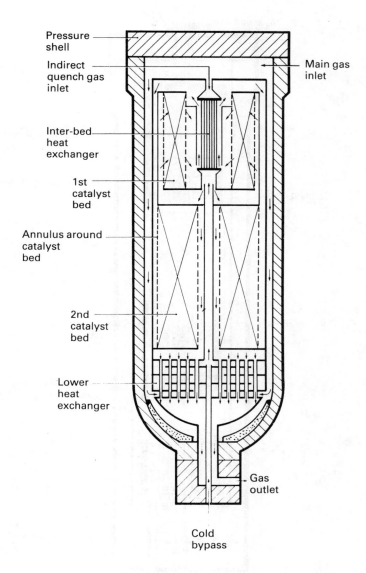

Figure 8.22 The Topsøe S 200 converter. This converter combines two radial flow beds with indirect interbed cooling. (Courtesy of British Sulphur Corporation.)

Chapter 8. Ammonia Synthesis

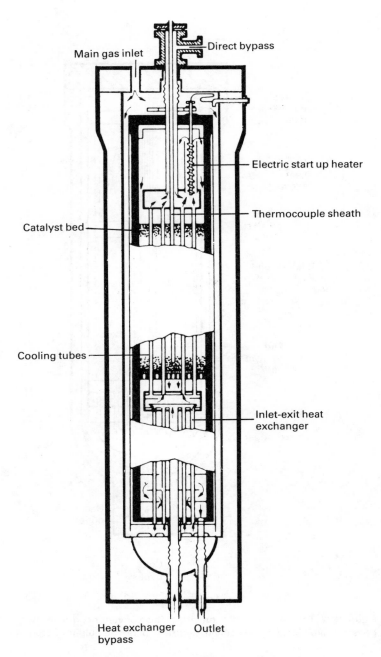

Figure 8.23 Tube cooled converter.

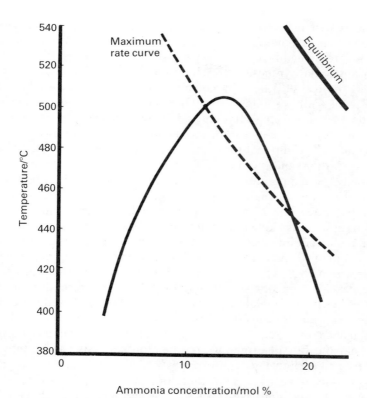

Figure 8.24 Typical temperature profile in a tube cooled converter.

8.9. The Future

Ammonia synthesis catalyst has quite a long history of full-scale use, development and academic research that is culminating in highly sophisticated surface science studies. Its introduction led the way to high-pressure catalytic processes, and thereby opened up much of the chemical industry taken for granted today. Research on the catalyst itself, which has been empirical, developmental and fundamental, has revealed many aspects of catalysis—for example, the function and mode of action of promoters and activators. However, despite its place as the oldest catalyst in the ammonia process, the ammonia synthesis catalyst has by no means reached the limit of its development. Demands, in recent years, for larger plants and for use of different feedstocks as well as the rapid increase in energy costs, have required even closer investigation of the role played by this important catalyst.

Chapter 8. Ammonia Synthesis

In the search for improved energy efficiency, the need to reduce compression and circulation costs has been well-recognized, and this has placed severe demands on the catalysis chemist to improve the activity of the ammonia synthesis catalyst. In ICI's AMV Process,[241] for example, the use of a new ammonia synthesis catalyst in a relatively low-pressure synthesis loop of 70–80 bar allows attractive economies to be made in energy consumption and in overall capital cost. Other developments also give indications to the future. Scientists from British Petroleum[281, 282] claim that ruthenium catalysts supported on particular forms of carbon and promoted by barium offer high activities at surprisingly low temperatures. This could enable advantage to be taken of the higher thermodynamic equilibrium concentration of ammonia at low temperature. This catalyst has not yet been commercially proven and its acceptance is likely to be affected by the price and large amounts of ruthenium required. However, this is further demonstration of the drive to both lower-pressure and and lower-temperature operation. The Kellogg Company,[283, 284] with whom British Petroleum has agreements for the development of the ruthenium catalysts, has recently claimed processes in which the inlet catalyst is conventional iron-based material, and the exit catalyst consists of ruthenium on carbon. The work of Somorjai and co-workers[285–287] and Grunze et al.[288] on rhenium catalysts shows that the rate-determining step on iron catalysts can be moved from the dissociative chemisorption of nitrogen atoms to the hydrogenation stage. There is therefore still potential to develop systems which may improve on both the iron and rhenium catalysts.

In the context of future developments, the work of the group at the Nitrogen Fixation Unit in the University of Sussex is also worthy of note.[289] While clearly not competitive at present with conventional fossil-fuel based, large-scale ammonia manufacture, the recent development of an electrochemical route[290, 291] to ammonia using diphosphine complexes of tungsten and molybdenum has longer-term potential for small-scale ammonia manufacture in remote areas. However, in the context of the established ammonia production technology, developments in amonia synthesis catalysts have the greatest potential to improve the efficiency of the process.

Chapter 9

Methanol Synthesis

9.1. Introduction

The introduction by BASF in 1923 of the process to synthesize methanol from carbon monoxide and hydrogen at pressure was the second large-scale application of catalysis and high-pressure technology to the chemical industry. It followed some 10 years after the introduction of the catalytic synthesis of ammonia, and resulted from the understanding of the thermodynamic and kinetic principles controlling chemical reactions which had come from the work of Ostwald, Nernst, and others. The process was the outcome of extensive research of the hydrogenation of carbon monoxide, which had shown that, depending on the catalyst used and the process conditions, carbon monoxide and hydrogen could be reacted at a pressure of 100–300 bar to give products ranging from methanol to higher alcohols, other oxygenated compounds and hydrocarbons. The work of Fischer and Tropsch subsequently led to the establishment in the 1930s of the process for the manufacture of synthetic fuels, which bears their names.

Like the ammonia process, methanol synthesis was dependent on the development of an effective catalyst, but unlike the ammonia synthesis catalyst the methanol catalyst had to be selective as well as active. It had not to give significant amounts of methane or other Fischer–Tropsch products. One catalyst, containing zinc oxide and chromia, was found which met these requirements so well that its composition remained virtually unchanged for 40 years. During that time several attempts were made to improve the catalyst, but without success, until 1966 when ICI introduced a copper/zinc oxide/alumina catalyst with much higher activity, which enabled methanol synthesis to be carried out at lower temperatures, less than 300°C, and which permitted operation at lower pressures, 50–100 bar. The low-pressure process is more efficient, has a lower capital cost and is cheaper to operate than the early high-pressure process, which is now obsolete. Although other new methanol processes and catalysts have been proposed in recent years, none has reached industrial practice and the low-pressure process remains the only economical route. Recent advances in methanol production technology have come from improvements in catalysts and process design. Several

Chapter 9. Methanol Synthesis

recent reviews[292–297] of the processes and catalysts for methanol synthesis supplement this chapter.

9.2. Thermodynamic Aspects

9.2.1. Methanol Formation

The synthesis gas from which methanol is made used to be manufactured by coke gasification, but is now almost invariably produced by steam reforming or partial oxidation of hydrocarbons, usually natural gas (see Chapter 5). It is a mixture of carbon monoxide, carbon dioxide and hydrogen, together with some nitrogen and other inert gases. The reactions involved in the synthesis process are shown in equations (1)–(3) with associated thermodynamic values for reactions of gas-phase components.

$$CO + 2H_2 \rightleftharpoons CH_3OH \quad \Delta H°_{298} = -90.64 \text{ kJ mol}^{-1} \quad (1)$$
$$\Delta G°_{298} = -25.34 \text{ kJ mol}^{-1}$$

$$CO_2 + 3H_2 \rightleftharpoons CH_3OH + H_2O \quad \Delta H°_{298} = -49.47 \text{ kJ mol}^{-1} \quad (2)$$
$$\Delta G°_{298} = +3.30 \text{ kJ mol}^{-1}$$

$$CO + H_2O \rightleftharpoons CO_2 + H_2 \quad \Delta H°_{298} = -41.17 \text{ kJ mol}^{-1} \quad (3)$$
$$\Delta G°_{298} = -28.64 \text{ kJ mol}^{-1}$$

Carbon dioxide produces water as well as methanol, reaction (2). Both carbon monoxide and carbon dioxide take part in reaction (3), the water-gas shift reaction which is catalysed by the synthesis catalysts. Reactions (2) and (3) combined are equivalent to reaction (1), so that either, or both, of the carbon oxides can be the starting point for methanol synthesis, an aspect which has been the subject of several investigations in recent years (see Section 9.6 and Chapter 1).

Reactions (1)–(3) are exothermic, and reactions (1) and (2) are accompanied by a decrease in volume. Hence, the value of the equilibrium constant, K_p, for synthesis from carbon monoxide

$$K_p = p_{CH_3OH}/p_{CO}p^2_{H_2} \quad (4)$$

decreases with temperature, as shown[298] in Figure 9.1. The increase in K_p with pressure follows[295] from non-ideality. Thus, high conversions to methanol, given a sufficiently active catalyst, will be obtained at high pressures and low temperatures. The equilibrium constant for reaction (1), i.e. methanol formation from carbon monoxide, is given[295] by

$$K_1 = 9.740 \times 10^{-5} \times \exp[21.225 + (9143.6/T)$$
$$- (7.492 \ln T) + (4.076 \times 10^{-3})T - (7.161 \times 10^{-8})T^2] \quad (5)$$

9.2. Thermodynamic Aspects

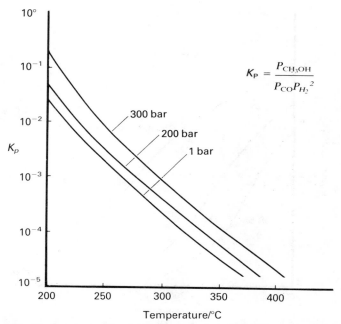

Figure 9.1. Thermodynamics of methanol synthesis. Variation of equilibrium constant K_p with temperature and pressure.

in which K_1 is in kPa^{-2}. Wade et al.[295] point out that there are discrepancies in the literature on the value of K_1. The equilibrium constant for the water-gas shift reaction, K_3, is better established[299], equation (6),

$$K_3 = \exp[-13.148 + (5639.5/T) + (1.077 \ln T) + (5.44 \times 10^{-4})T - (1.125 \times 10^{-7})T^2 - (49\,170/T^2)] \qquad (6)$$

(see also Appendix 7). Corrections for non-ideality are significant in methanol synthesis reactions, and the use of appropriate fugacity coefficients is discussed in the literature.[295, 297]

The use of high pressure, particularly in very large modern plants with capacities of more than 1000 tonnes day^{-1} of methanol is expensive both of capital and gas compression. Therefore, with a very active catalyst, it is most economical to use lower pressures and obtain the same conversions as at high pressures but at lower temperatures. Figure 9.2 shows the thermodynamic equilibrium conversion of carbon monoxide and hydrogen to methanol at different temperatures and pressures. It shows that theoretically a catalyst which is only active at 350°C or above (such as the early zinc oxide/chromia catalyst) requires an operating pressure of 300 bar to achieve a conversion to methanol of

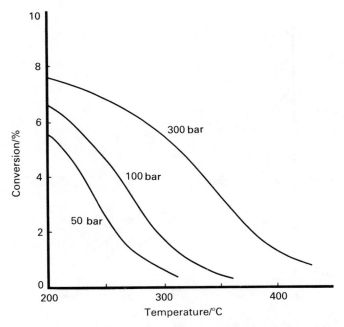

Figure 9.2. Effect of temperature on equilibrium concentration of methanol at different pressures. (Inlet gas composition 8% CO, 8% CO_2, 5% H_2, 27% inerts).

3%. Given a catalyst which is active at temperatures as low as 250°C (such as the modern copper/zinc oxide/alumina catalyst) the same conversion is obtainable at 50 bar. Further details of the limits to process operation imposed by equilibrium considerations are given by Wade et al.[295]

9.2.2. Selectivity

In addition to the synthesis of methanol, both carbon monoxide and carbon dioxide can take part in other hydrogenation reactions, producing by-products such as hydrocarbons, ethers and higher alcohols, as illustrated by examples in reactions (7)–(9) with associated thermodynamic values for reactions of gas-phase components.

$$CO + 3H_2 \rightleftharpoons CH_4 + H_2O \quad \Delta H°_{298} = -206.17 \text{ kJ mol}^{-1} \quad (7)$$
$$\Delta G°_{298} = -142.25 \text{ kJ mol}^{-1}$$

$$2CO + 4H_2 \rightleftharpoons CH_3OCH_3 + H_2O \quad \Delta H°_{298} = -204.82 \text{ kJ mol}^{-1} \quad (8)$$
$$\Delta G°_{298} = -67.20 \text{ kJ mol}^{-1}$$

$$2CO + 4H_2 \rightleftharpoons C_2H_5OH + H_2O \quad \Delta H°_{298} = -255.58 \text{ kJ mol}^{-1} \quad (9)$$
$$\Delta G°_{298} = -122.55 \text{ kJ mol}^{-1}$$

9.2. Thermodynamic Aspects

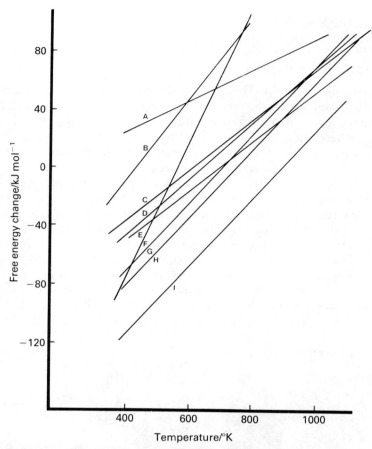

Figure 9.3. Standard free energies of formation of hydrocarbons and alcohols from carbon monoxide and hydrogen with water as by-product. A, ethanol; B, methanol; C, acetylene; D, benzene; E, propylene; F, ethylene; G, propane; H, ethane; I, methane.

These reactions are much more exothermic than the methanol synthesis reactions, and formation of by-products involves even larger negative changes of free energy. This is illustrated in Figure 9.3. Methanol is thermodynamically less stable and thermodynamically less likely to be formed from carbon monoxide and hydrogen than the other possible products, such as methane. Which of the products is formed in practice is controlled by kinetic factors; that is, by the catalyst being selective in favouring a reaction path leading to the desired product with a minimum of by-products. Some fundamental aspects of selectivity are discussed in Chapter 1.

445

Chapter 9. Methanol Synthesis

9.3. The Methanol Synthesis Process

9.3.1. The Synthesis Loop

Both the modern 50–100 bar process and the earlier 350 bar process use the same principle of the recirculation loop used in ammonia synthesis. Synthesis gas of an appropriate composition (i.e. ideally $p_{H_2} = 2p_{CO} + 3p_{CO_2}$, where there is sufficient hydrogen to react with the oxides of carbon by reactions (*1*) and (*2*)) is supplied to the loop and circulated continuously so that unreacted gas is recycled over the catalyst. From the converter the gas passes to a condenser, which removes the crude methanol liquor, and then to the circulator, which takes the gas back to the converter. Fresh "make-up gas" is supplied continuously to maintain the pressure in the loop as the synthesis proceeds. This gas always departs from the ideal composition to some degree: methane and other inert gases are present and frequently there is excess hydrogen, depending on the feedstock and processes used to manufacture the make-up gas (Section 9.3.2). To prevent these gases from building up in the loop and diluting the reactants, a continuous purge is taken off. Depending on process economics, the purge may be used as fuel for the

Figure 9.4. General view of the Methanol 2 plant at Billingham, U.K.

9.3. The Methanol Synthesis Process

reformer or, as is becoming more common, it gives a supply of hydrogen after passage through a pressure-swing adsorption unit.

The crude methanol is distilled to separate the methanol from water and impurities such as higher alcohols, ethers, etc., that are present in low concentrations. Figures 9.4 and 9.5 show a general view and a flowsheet for typical low-pressure plants and the process is described in detail in References 292 and 293. The use of low conversion per pass together with the recycling of the unreacted gas around the loop facilitates the control of temperature in the catalyst bed. Nevertheless the highly exothermic nature of the reaction requires the use of special reactor designs.

Quench reactors are ICI's standard reactor[300] for large plants, combining a near approach to equilibrium with reliable performance. A typical example is shown in Figure 9.6(a).

Part of the circulating gas is preheated and fed to the inlet of the reactor. The remainder is used as quench gas and is admitted to the catalyst bed through lozenge distributors in order to control bed temperatures. Temperature control within the catalyst bed can also be achieved using tube-cooled or steam-raising reactor designs.

The tube-cooled reactor[300] combines a lower capital cost than the quench reactor with simplicity of operation and a very flexible design. Circulating gas is preheated by passing it through tubes in the catalyst bed, as illustrated in Figure 9.6(b). This removes the heat of reaction from the catalyst bed. Temperature control is achieved by using a gas bypass around the catalyst bed and by controlling heat recovery in a feed–effluent heat exchanger immediately downstream. Two configurations are possible with the steam-raising type of reactor.[300] The catalyst may be contained within the tubes of a shell-and-tube heat exchanger, with boiling water acting as coolant on the shell side. Alternatively the catalyst may be contained within the shell with boiling water in the tubes, and this arrangement is similar to that used in the tube-cooled reactor design, as in both reactors the catalyst bed is cooled by fluid flow through tubes in the bed. It is illustrated in Figure 9.6(c). Since the design with catalyst on the shell side allows the heat transfer surface area to be reduced to as little as one-seventh that of the design with catalyst in the tubes, the former is the preferred configuration.[300] In the ICI steam-raising reactor design illustrated in Figure 9.6(c) circulating gas enters through a vertical distribution plate and flows transversely across the catalyst bed. This allows the steam coils to be placed asymmetrically within the catalyst bed, so optimizing heat removal, and giving a small pressure drop. In plants based on natural gas, steam-raising reactors are less efficient at recovering the heat of reaction than quench or tube-cooled designs. This is because heat can be

Chapter 9. Methanol Synthesis

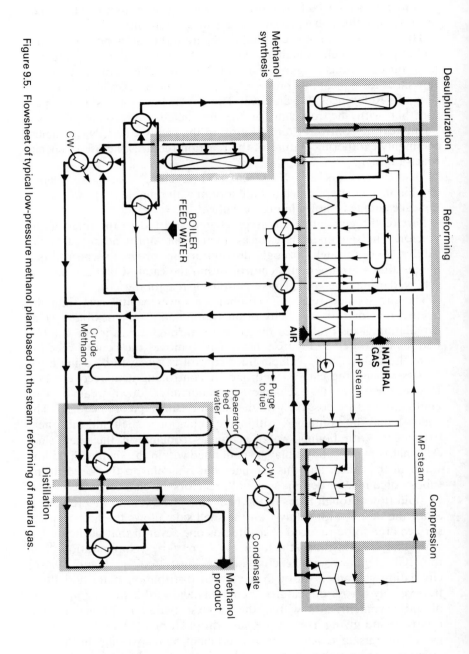

Figure 9.5. Flowsheet of typical low-pressure methanol plant based on the steam reforming of natural gas.

9.3. The Methanol Synthesis Process

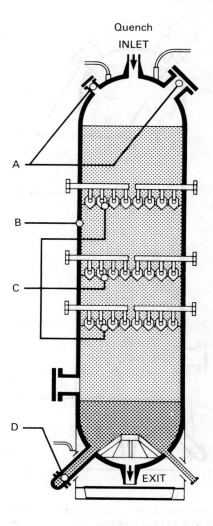

Figure 9.6. (a) A modern low-pressure methanol synthesis converter, showing quench distributors for catalyst temperature control. A, catalyst is charged and inspected through these man-holes; B, the pressure vessel is of a simple design—no internal catalyst basket is required; C, the ICI lozenge quench distributors ensure good gas distribution and allow the free passage of catalyst for charging and discharging; D, gravity discharge of catalyst permits rapid preparation for maintenance or recharging.

Chapter 9. Methanol Synthesis

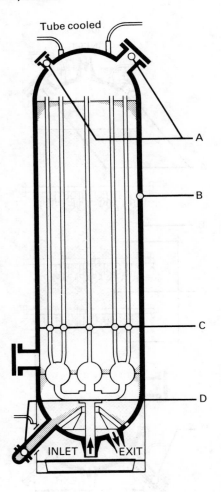

Figure 9.6. (b) A modern tube-cooled low-pressure methanol synthesis reactor. A, catalyst is charged and inspected through these man-holes; B, the pressure vessel is of a simple design—no internal catalyst basket is required; C, thin walled cooling tubes are welded to a simple header system embedded in the catalyst; D, gravity discharge of catalyst permits rapid preparation for maintenance or recharging.

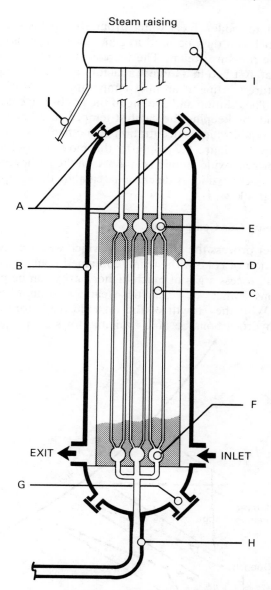

Figure 9.6. (c) The ICI-steam-raising reactor for low-pressure methanol synthesis. A, catalyst is charged and inspected through these manholes; B, the pressure vessel is of simple design; C, cooling tubes are welded to a simple header system, F, embedded in the catalyst bed; D, perforated catalyst support grids allow gas to be distributed along the length of the catalyst bed; E, an upper header system collects steam from the tube bundle, which then passes to the steam drum, I; H, boiler feed water enters the base of the converter; G, gravity discharge of catalyst permits rapid preparation for maintenance or recharging.

Chapter 9. Methanol Synthesis

recovered into boiler feed water from the quench or tube-cooled designs, which can then be used to generate high-pressure rather than intermediate-pressure steam. The increase in methanol concentration down the catalyst bed in a quench reactor and the consequent increase in temperature starting at an inlet temperature of 200°C are shown in Figure 9.7. The dilution and cooling of the gas by the cold quench gas is apparent and, by keeping the catalyst temperature below 250°C at the exit, a high equilibrium concentration of methanol is approached. The temperatures in Figure 9.7 apply to the modern 50–100 bar process using copper/zinc oxide/alumina catalysts. In the earlier 350-bar process with the less-active zinc oxide/chromia catalyst the temperature range would be 350–400°C.

9.3.2. Make-up Gas Composition

In the earlier process the make-up gas, which was usually produced by the gasification of coke with steam, contained many impurities, but in the modern process a gas of much higher purity can be produced from hydrocarbons either by catalytic steam reforming or by partial oxidation. With the growth of the oil industry before 1960, suitable feedstock hydrocarbons became available for steam reforming, and it

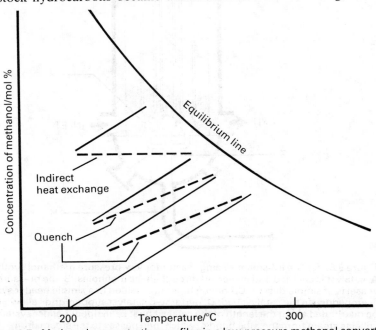

Figure 9.7. Methanol concentration profiles in a low-pressure methanol converter. Inlet temperature 200°C, exit temperature 240°C.

was the consequent increasing availablity of reformer gas of high purity which made it feasible for ICI to embark upon the research in the early-1960s which led to the introduction of the world's first successful low-pressure methanol process.

With methane as the reformer feedstock, a methanol synthesis gas with an excess of hydrogen is produced as indicated in the following equations:

$$CH_4 + H_2O \rightarrow CO + 3H_2 \rightarrow CH_3OH + H_2 \qquad (10)$$

$$CH_4 + 2H_2O \rightarrow CO_2 + 4H_2 \rightarrow CH_3OH + H_2O + H_2 \qquad (11)$$

The excess hydrogen can be used for synthesis by adding more carbon dioxide to the synthesis gas (if available from another source, such as an ammonia plant). Alternatively it can be purged from the synthesis loop and used either as fuel on the reformer or as a source of hydrogen after removal of the components of purge gas, usually by pressure-swing adsorption. In economic terms it is usually better to add carbon dioxide, and so use the excess hydrogen to make more methanol. Naphtha, on the other hand, with a higher C/H ratio approximating to CH_2, produces a synthesis gas of the correct stoichiometry for methanol synthesis when steam reformed as in equation (12). However, few methanol plants are based on a naphtha feedstock, despite the favourable stoichiometry, because at present naphtha is a relatively expensive feedstock. In future, when natural gas and naphtha are likely to be less readily available, synthesis gas will be derived from coal gasification and partial oxidation of heavy petroleum fractions, the raw gas will have an even higher C/H ratio, which will have to be adjusted to the stoichiometric requirement by water-gas shift followed by removal of the excess of carbon dioxide before methanol synthesis.

$$(1/n)(CH_2)_n + H_2O \rightarrow CO + 2H_2 \rightarrow CH_3OH \qquad (12)$$

9.4. Methanol Synthesis Catalysts

9.4.1. High-pressure Catalysts

The catalyst used in the original methanol synthesis process was derived by empirical methods. It contains zinc oxide and chromia (typified by ICI Catalyst 26-1) and was used in the high-pressure process for 40 years. A list of catalysts used or proposed for methanol synthesis is given in Table 9.1. The development work on methanol synthesis catalysts from 1920 to 1955 has been summarized by Natta.[301] Zinc oxide alone was a good catalyst for methanol synthesis at high pressure and

Chapter 9. Methanol Synthesis

temperatures above 350°C, but it was not stable, and in use quickly lost its activity. It was found that die-off could be retarded by the incorporation of chromia, which acted as a stabilizer preventing the growth of the zinc oxide crystals. This demonstrated an appreciation at that time of the significance of the surface area of the active zinc oxide component. Today zinc oxide, still a major component in synthesis catalysts, is known to have a defect structure with a non-stoichiometric oxide lattice which is probably responsible for its catalytic activity. Stabilizers such as chromia probably prevent recrystallization of the zinc oxide and preserve the defect structure, as well as preventing crystal growth and loss of surface area. The most active zinc oxide was obtained from the zinc carbonate mineral smithsonite, containing traces of various oxide impurities which acted as "promoters", probably by forming solid solutions. The promoters may have both increased the specific activity of the zinc oxide, e.g. by inducing lattice defects, and

Table 9.1. Catalysts proposed or used for industrial methanol synthesis

Catalyst composition	Active phase in methanol synthesis[a]	Properties and use
ZnO	ZnO	Original synthesis catalyst, short life
ZnO/Cr$_2$O$_3$ (ICI Catalyst 26-1)	ZnO	Standard high-pressure catalyst
ZnO/MnO/Cr$_2$O$_3$ + alkali	alkalized ZnO (+ MnO?)	Standard high-pressure catalyst for methanol/higher alcohol mixtures
Cu/ZnO Cu/ZnO/Cr$_2$O$_3$	Cu	Early low-pressure catalysts, short life
Cu/ZnO/Al$_2$O$_3$ (ICI Catalyst 51 series)	Cu	Industrial low-pressure catalyst
Pd/SiO$_2$ Pd/basic oxides	Pd	Active; poorer selectivity than copper catalysts[b]
Rh/SiO$_2$ Rh/basic oxides	Rh	Active; poorer selectivity than copper catalysts[b]
Rh complexes	Rh complex	Low activity; poorer selectivity[c]

[a]There is evidence that the active phases in the Cu, Pd and Rh catalysts, essentially the metals, are partially oxidized, either bulk or surface.

[b]Catalysts based on Group VIII metals are discussed in Section 9.7. The by-products are mainly hydrocarbons.

[c]Homogeneous catalysis with rhodium complexes gives ethylene glycol + methanol. No hydrocarbons are formed.

stabilized the zinc oxide by the inhibition of sintering. Later work[302] has shown the importance of the defect state and crystal habit of zinc oxide in methanol synthesis.

The zinc oxide/chromia catalyst was tolerant of the impure synthesis gas, and could have a plant life of several years. It was not very selective and depending on synthesis conditions as much as 2% of the inlet carbon oxides could be converted to methane, with a similar proportion to dimethyl ether. Because these side-reactions are very exothermic, careful control of catalyst temperatures was necessary. The catalyst was made by precipitation from zinc and chromium solutions, or by impregnation of, for example, zinc carbonate with chromic acid or dichromate solution. In the catalyst produced by impregnation the chromium is present in the hexavalent form, as CrO_3 rather than Cr_2O_3 as in the precipitated catalyst. Cr(VI) is particularly toxic and presents a health hazard in the handling of the catalyst. The chromate also has to be reduced with great care to avoid a temperature runaway, because the reduction of CrO_3 to Cr_2O_3 with hydrogen

$$2CrO_3(s) + 3H_2(g) \rightarrow Cr_2O_3(s) + 3H_2O(g), \qquad (13)$$
$$\triangle H°_{298} = -695 \text{ kJ mol}^{-1}$$

is extremely exothermic. Although the low-pressure process based on high-activity cataysts has almost entirely taken the place of the high-pressure process, there have been occasions when a high-pressure plant being supplied with pure synthesis gas (e.g. from a steam reformer) could, to advantage, use a catalyst of higher activity. The copper/zinc oxide/alumina type (described below for low-pressure operation) was found to be very satisfactory, although at the synthesis pressure of around 350 bar, with high partial pressures of water, sintering processes are accelerated.

9.4.2. Low-pressure Catalysts

After the introduction of the high-pressure process in 1923, work continued on the catalyst and the efficiency of many different elements and compounds as synthesis catalysts were tested.[301] Among them was copper oxide which, while it appeared to have little synthesis activity itself, was very effective when added to zinc oxide. This was also the case with the zinc oxide/chromia catalyst, and when copper oxide was added to it, its activity increased, so it could be used at temperatures as low as 300°C. However, the catalysts containing copper were not stable and lost activity. For example, a Zn/Cu/Cr catalyst in atomic proportions of 6 : 3 : 1 lost 40% of its activity in 72 hours. The instability of the copper-containing catalysts was borne out in

Chapter 9. Methanol Synthesis

unsuccessful research work during the 1930s aimed at producing a high-activity catalyst based on copper.[303, 304] However, as a result of the ICI work on methanol catalysts, stable copper catalysts were produced. During this work an extensive study was made of possible catalysts (for example, by X-ray diffraction and oxygen-chemisorption) to identify the phases which were present, to determine crystal sizes and to measure catalytic surface areas. The identity and distribution of poisons and waxes which might accumulate in the catalyst pores and on the active surfaces were also determined. Much of the work was carried out in parallel with development of similar copper/zinc oxide catalysts (the ICI Catalyst 52 series) for the low-temperature shift process (see Chapter 6).The loss of activity of the early copper catalysts in use was almost certainly due to loss of copper surface area, despite the presence of zinc oxide and other "stabilizers". The copper metal produced by reduction of the oxide, is in the form of crystals 5–7 nm in size which readily sinter at synthesis temperatures with consequent loss of activity. One of the functions of the stabilizer is to act as a dispersant, separating the copper crystallites physically so that the sintering process is hindered. However, it is desirable to minimize the amount of stabilizer, because if it has negligible synthesis activity itself, it dilutes the effective catalyst component. Although zinc oxide has activity for methanol synthesis, this is so much less than that of copper that it is in effect an inert diluent in copper/zinc oxide catalysts. The other possible roles of zinc oxide are discussed in Section 9.6.1, and Andrew[305] has described the complex criteria by which suitable supports can be selected for both methanol synthesis and low-temperature water-gas shift catalysts.

In an oversimplified model the catalyst can be considered[305] as particles of copper metal surrounded by and kept apart by stabilizer particles. There is a rough geometric relationship shown[306, 307] between the diameter of the metal and stabilizer particles and the volumes of metal and stabilizer, which has been used to calculate the curves in Figure 9.8. It is clear that the smaller the diameter of the refractory spacers, the higher is the practical copper loading and the higher the activity. Only when the spacer particles are less than about 2 nm in diameter can any significant advantage in activity be obtained by a decrease of copper crystallite size from 15 to 5 nm. High metal concentrations are used in industrial catalysts: in use both metal and spacer dimensions increase, and the rough relationship still holds. While such calculations are not precise, they provide useful guidelines during catalyst development. In Figure 9.9 the ideally-stabilized catalyst A would show no loss of activity with time on-line. Catalysts B and C, with higher metal concentrations, have higher initial activities, but because they contain less stabilizer some sintering would take place and they

9.4. Methanol Synthesis Catalysts

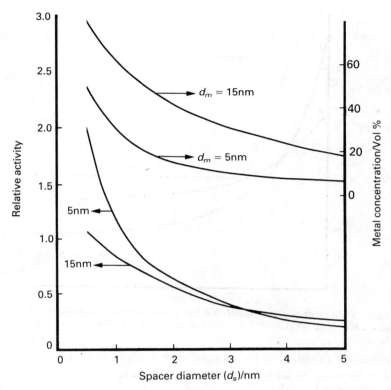

Figure 9.8. Effect of stabilizer crystal size on optimum metal content and relative catalytic activity.

would die-off to activity levels below that of the stable catalyst. As the overall production from a plant (under otherwise constant conditions) is determined by the area under the activity–time plot in Figure 9.9, the optimum catalyst is not necessarily A, but usually a formulation lying between A and B.

In the modern copper/zinc oxide/alumina synthesis catalysts high activity and stability are obtained by optimizing the composition and producing very small particles of the components in a very intimate mixture. By precipitation at a controlled pH, in which the acidic and alkaline solutions were mixed continuously, a catalyst of optimum composition and particle size is obtained (see Chapter 1). Although chromia had been effective as a stabilizer in the high-pressure catalyst, alumina is superior to it as the third component in the low-pressure catalyst. The alumina, present as a high-area, poorly crystalline phase, is more effective than zinc oxide in preventing the sintering of copper

Chapter 9. Methanol Synthesis

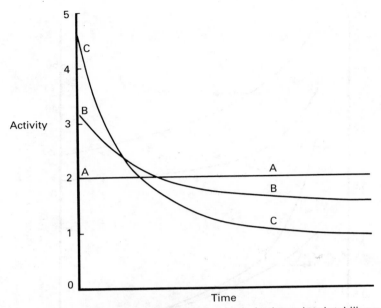

Figure 9.9. Stability of catalysts. Catalyst A has optimized metal and stabilizers contents. Catalysts B and C with high metal, lower stabilizer contents, are less stable.

crystallites. It might then be thought that a copper/alumina catalyst would be superior to the ternary catalysts, but there are several reasons why this is not so.

1. The chemistry of the metal and support precursors, formed in the precipitation process, gives smaller copper crystallites with zinc oxide support than with alumina. Hence, the optimum combination of high initial activity and catalyst stability is obtained with the ternary $Cu/ZnO/Al_2O_3$ catalyst.

2. High-area aluminas have acidic sites on the surface, which catalyse the parasitic reaction of methanol to dimethyl ether, so Cu/Al_2O_3 catalysts have poor selectivity. In the ternary catalysts, formation of dimethyl ether is very low, showing that the acidic sites are neutralized by the basic zinc oxide (either as a surface reaction only or in bulk reaction to zinc spinel).

3. Although poisoning is not usually a significant problem in well-run methanol plants, the presence of a "poison-soak" in the support phases can be useful. Zinc oxide is much more effective than alumina in picking up and holding typical poisons, such as sulphur and chlorine compounds (see Section 9.5).

9.4. Methanol Synthesis Catalysts

The design of methanol synthesis catalysts is also discussed in Chapter 1 and, in more detail, by Andrew[305] and Chinchen et al.[297] The results of two typical life tests are given in Figure 9.10, where loss in activity is indicated by the increase in catalyst temperature necessary to maintain a constant synthesis rate. The higher activity of the alumina-containing catalyst (lower operating temperature) and its higher stability (less change of temperature with time) are very apparent. It can be seen that the formation and maintenance of high activity, i.e. a high surface area of copper metal, is a function of the oxidic support, as discussed by Andrew.[307]

Catalysts made by the continuous procedure were used in the first low-pressure synthesis plants, operating at 50 bar, which were started up in 1966. They had lives of more than three years and produced methanol of a higher purity than the high-pressure process. Continued development resulted in catalysts suitable for operation at 100 bar, around the optimum operating pressure for high-capacity plants producing more than 1000 tonnes day^{-1} of methanol. Initial tests showed that the rate of catalyst die-off was higher at the higher pressure. Zinc oxide/alumina is not sufficiently refractory as a support at 100 bar, but a more refractory support is provided by introducing some of the zinc component as a compound with the alumina component; as zinc spinel ($ZnO.Al_2O_3$), which is produced in a finely divided state.

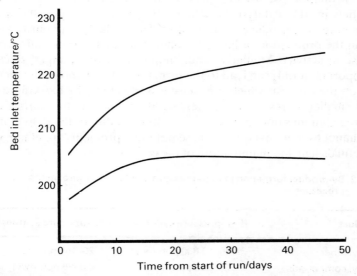

Figure 9.10. Activities of low-pressure synthesis catalysts containing alumina and chromia. Activity is inversely related to temperature. (Constant equal methanol synthesis rates; 50 bar; Cu/Zn/Cr, 6 : 3 : 1.)

9.5. Selectivity and Poisons

As indicated earlier (Figure 9.3), hydrocarbons and higher alcohols are thermodynamically more stable than methanol, and could be formed from carbon oxides and hydrogen. The extent to which this occurs is controlled by the selectivity of the catalyst. In this respect the copper catalysts are greatly superior to the zinc oxide/chromia catalyst, as shown in Table 9.2. With the latter, up to 5% of methane and dimethyl ether are formed, a high level of inefficiency which is almost eliminated in the low-pressure process. Methanation needs to be minimized, both because it is a loss of feedstock and, as a highly exothermic reaction, it can cause temperature runaways. The high selectivity of copper catalysts arises in two ways. As the by-products are formed in parasitic, i.e. parallel, reactions and not by any intrinsic inefficiency of the synthesis reaction itself, improved activity for methanol synthesis alone also gives improved selectivity. Thus, the copper catalysts are more selective than zinc oxide catalysts under the same process conditions, but there is a further benefit. The higher activity allows the use of lower process temperatures (Section 9.3), so by-product formation is further decreased relative to methanol synthesis, because most of the parasitic reactions have higher activation energies.

The formation of by-products is influenced markedly by any impurities in the catalyst,[305] either left in the catalyst during its manufacture or introduced during use. Thus, alkaline impurities can result in the production of higher alcohols and, in addition, cause some decrease in activity. Similarly, acidic impurities (for example, silica in the support material) can lead to the formation of high molecular weight waxes on the catalyst, which can cause loss of activity by blocking some of the smaller pores. The weaker acidity found on the surfaces of high-area aluminas does not give waxes, but it catalyses the dehydration of methanol to dimethyl ether. These potential problems are eliminated by careful design and manufacture of the catalyst.

Table 9.2. By-product formation in high-pressure and low-pressure methanol processes

By-product	High-pressure synthesis	Low-pressure synthesis
Dimethyl ether	5000–10 000 ppm(w/v)	20–150 ppm(w/v)
Carbonyl compounds	80–220 ppm(w/v)	10–35 ppm(w/v)
Higher alcohols	3000–5000 ppm(w/v)	100–800 ppm(w/v)
Methane	2% Input carbon	None

9.5. Selectivity and Poisons

Table 9.3. Effects of possible contaminants and poisons on Cu/ZnO/Al$_2$O$_3$ catalysts for methanol synthesis

Contaminant or poison	Possible sources[a]	Effects
Silica, other acidic oxides	Transport in steam in plant gases	Waxes, other by-products formed
γ-Alumina	Catalyst manufacture	Dimethyl ether formed
Alkali	Catalyst manufacture	Decreased activity; higher alcohols formed
Iron	Transport in plant as Fe(CO)$_5$	Methane, paraffins, waxes formed
Nickel	Transport in plant as Ni(CO)$_4$	Methane formed; decreased activity
Cobalt	Catalyst manufacture	Methane formed; decreased activity
Lead, heavy metals	Catalyst manufacture	Decreased activity
Chlorine compounds	Transport in plant gases	Permanent decrease in activity
Sulphur compounds	Transport in plant gases	Permanent decrease in activity

[a]Raw materials for catalyst manufacture, if not of sufficient purity, can also be a source of contamination

The presence of Group VIII metals, such as iron, is particularly undesirable, as they increase hydrogenation activity and promote the dissociation of carbon monoxide and dioxide, leading to formation of methane and of long-chain paraffins and/or waxes by Fischer–Tropsch type reactions. Methanation was always a problem with the high-pressure catalyst (Table 9.2), and it is thought that this was mainly caused by iron impurities. Iron is sometimes deposited on a synthesis catalyst during use by the decomposition of the gaseous iron pentacarbonyl, Fe(CO)$_5$, formed from rust which may be present in the make-up gas system. The effects of a given level of iron impurity in a catalyst are strongly dependent on its form, oxidation state and distribution. Other Group VIII metals, e.g. cobalt and nickel, can also catalyse methanation but they have been found[308] to have a further deleterious effect. Methanol synthesis itself is inhibited, probably by surface coverage of copper crystallites with support oxide. Thus, alkalis, acidic species and Group VIII metals can be regarded as poisons for synthesis catalysts. In addition, traces of sulphur and chlorine can reduce the activity of the copper catalyst by reacting with the active surface. Chlorine is particularly undesirable because, in addition to poisoning the copper surface, the copper chloride produced is mobile and causes rapid sintering of the metal. Chlorine and sulphur also react with the free zinc oxide in the catalyst, so that much of the uptake occurs on the catalyst near the inlet to the bed, which therefore acts as a guard for the rest of the bed. Poisoning of synthesis catalyst is not normally a problem except as a result of maloperation. The diverse effects of various contaminants and poisons on $Cu/ZnO/Al_2O_3$ catalysts are summarized in Table 9.3.

9.6. Mechanism and Kinetics

9.6.1. Reaction Mechanism

The methanol synthesis reaction has been the subject of many mechanistic studies since the process was first introduced.[292–294, 297, 301] The investigations have been concerned with catalysts ranging from zinc oxide alone, through the zinc oxide/chromia high-pressure catalyst, to copper/zinc oxide/alumina types of catalyst. Most recent work has been with copper catalysts. The results of these studies are often conflicting for several reasons.

1. The properties of the catalysts, especially $Cu/ZnO/Al_2O_3$ catalysts, depend very much on details of preparation, and considerable differences are found in catalysts of the same nominal composition but made in different ways. For example, Klier[292] found a maximum in synthesis activity with a 30/70 mol ratio of CuO/ZnO catalyst, but this is a much lower copper content than the optimum used in many industrial catalysts.

2. The $Cu/ZnO/Al_2O_3$ is both more complex and more variable than most of the supported metal catalysts which have been studied in detail. In the reduced catalyst there are at least three bulk phases present: metallic copper, zinc oxide and an oxidic phase containing aluminium. Moreover, each of these phases can, and frequently does, contain elements from the other phases. The presence of water and carbon dioxide in the reaction gases prevents the reduction of zinc oxide to metallic zinc, but the reduction potential on the system is sufficient to allow the formation of dilute solid solutions of zinc in copper metal, i.e. dilute α-brasses.[309] Zinc oxide has been shown to take copper ions in the lattice.[292] The alumina-containing phase must also contain zinc oxide in a practical catalyst (Sections 9.4.2 and 9.5). The oxidation state of copper (clean metal → surface-oxidized metal → Cu(I) oxide → Cu(II) oxide), the defect state of zinc oxide and the

formation of the various possible solid solutions are all functions of catalyst activation procedures and reaction conditions.

3. As the nature of the catalyst varies with reaction conditions, the many investigations done under conditions far removed from industrial practice are of uncertain relevance to the full-scale process.

4. Although poisoning is rarely a problem in plant operation, it can cause severe difficulties in small-scale testing (see Section 1.3), and some investigators do not appear to take sufficient precautions to avoid this.

There is still much controversy over the nature of the active components of the catalysts and the reaction steps that take place on them. Chinchen et al.[310] set the discussion as the answers to five questions.

(a) Is methanol synthesized from carbon monoxide or carbon dioxide?

(b) What is the state of the copper in a working catalyst?

(c) What roles are played by the zinc oxide and alumina components in the commercial catalyst?

(d) What is the mechanism and which reaction step is rate-determining?

(e) What are the active sites for methanol synthesis on a $Cu/ZnO/Al_2O_3$ catalyst?

Much work has still to be done to remove remaining ambiguities, but nevertheless a consistent picture can be drawn from the answers as detailed below.

(a) The Role of Carbon Dioxide
Many workers have shown that methanol can be made from CO_2/H_2 mixtures over supported copper catalysts. The reaction with a CO_2/H_2 mixture is usually faster and starts at a lower temperature than with a CO/H_2 mixture.[311, 312] All industrial plants operate with $CO/CO_2/H_2$ mixtures, and it is not obvious which carbon oxide is the source of methanol. In most studies it was concluded or assumed that adsorption of carbon monoxide is the starting point,[292] but in recent years evidence has been obtained indicating that methanol is synthesized directly from carbon dioxide.[313] Certainly the presence of carbon dioxide in the reacting gas has a marked effect in increasing synthesis rate,[312, 314–316] and the effect is reversible. In contrast with this observation, other workers[317] using a zinc/chromia catalyst found that carbon dioxide decreased the synthesis rate.

Studies[310, 313] using $^{14}CO_2$ and ^{14}CO with a copper-containing catalyst

Chapter 9. Methanol Synthesis

show that methanol comes from carbon dioxide rather than carbon monoxide. In one of these experiments a synthesis gas (10% CO, 10% CO_2, 80% H_2) containing $^{14}CO_2$ was passed at 50 bar pressure over a $Cu/ZnO/Al_2O_3$ catalyst. The results obtained at different gas flow rates are summarized in Figure 9.11. At high flow rate (short residence time) the specific radioactivity of the product methanol is the same as that of the inlet radioactive $^{14}CO_2$. This indicates that the methanol has been produced from $^{14}CO_2$ and not from the non-radioactive carbon monoxide either directly, or indirectly by the water-gas shift reaction, which is slower than the synthesis reaction under these conditions.

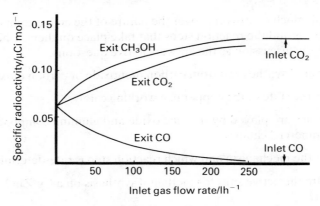

Figure 9.11. Effect of gas flow rate on exit gas composition and distribution of radioactivity in products.[310] Inlet gas contains $^{14}CO_2$.

(b) The State of Copper in a Working Catalyst

Accurate measurements[318] of copper metal surface areas of reduced $Cu/ZnO/Al_2O_3$ catalysts, especially *in situ* before and after synthesis, have given much useful information. A number of $Cu/ZnO/Al_2O_3$ catalysts of different compositions and covering a range of copper particle sizes, together with other supported copper catalysts, were found[310, 319, 320] to have the same linear dependence of catalyst activity on total copper surface area (see Figure 9.12). This has also been reported by other workers.[321, 322] Thus, the critical steps in the reaction mechanism must occur on the surface of the copper crystallites. Earlier interpretations[292, 293] in which it was concluded that carbon monoxide was hydrogenated over Cu(I) ions in zinc oxide now appear to be mistaken. The oxide support plays a minimal role in the synthesis mechanism. The critical test of synthesis over pure copper has given conflicting results: traces of impurities can have powerful effects, but it seems more likely that the reported[292] inactivity is due to poisons than the observed[323] activity due to promoters. About 30% of the initial

9.6. Mechanism and Kinetics

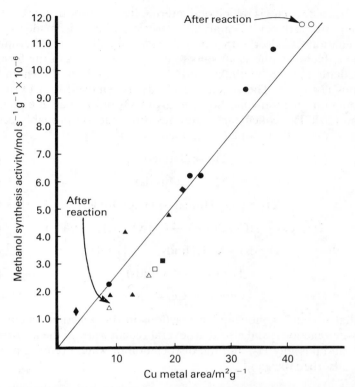

Figure 9.12. Relationship between methanol synthesis activity and copper metal surface area for Cu/ZnO/Al$_2$O$_3$ and other supported copper catalysts.[319] ●, ◆, CuO/ZnO/Al$_2$O$_3$ (60 : 30 : 10 and 45 : 37 : 18, respectively); ○, CuO/SiO$_2$; ▲, CuO/Al$_2$O$_3$; △, CuO/MgO; □, CuO/MnO; ■, CuO/ZnO.

copper surface area of a typical industrial catalyst is covered with adsorbed oxygen under working conditions.[310, 317, 320, 324] This coverage has been shown[319, 324] to be a function of the CO$_2$/CO ratio, so the catalyst surface is in a dynamic state.

(c) The Role of Support Oxides
This has already been covered in Sections 9.4.2 and 9.5. The support oxides play a minor role in the reaction mechanism, but it has been observed[310] that the coverage of adsorbed oxygen varies with different support oxides.

(d) Reaction Steps in Methanol Synthesis
The techniques of temperature-programmed desorption (TPD) and temperature-programmed reaction (TPR) spectroscopy have shown

Chapter 9. Methanol Synthesis

that a surface formate is the pivotal intermediate over both zinc oxide[302] and supported copper[310] catalysts. Surface formate is made by the hydrogenation of adsorbed carbon dioxide, and the rate-determining step in methanol synthesis appears to be hydrogenolysis of the formate intermediate first to methoxy and then methanol as shown in the equations (*14*)-(*18*). The remaining O(ads) is removed by CO or H_2, depending on reaction conditions, to give CO_2 or H_2O as in equations (*19*) and (*20*). The concurrent water-gas shift reaction probably occurs via O(ads) rather than formate intermediates:[324]

$$H_2 \rightarrow 2H(ads) \quad (14)$$

$$CO_2 \rightarrow CO_2(ads) \quad (15)$$

$$CO_2(ads) + H(ads) \rightarrow HCOO(ads) \quad (16)$$

$$2H(ads) + HCOO(ads) \rightarrow CH_3O(ads) + O(ads) \quad (17)$$

$$H(ads) + CH_3O(ads) \rightarrow CH_3OH \quad (18)$$

$$H_2O \rightleftharpoons H_2 + O(ads) \quad (19)$$

$$O(ads) + CO \rightleftharpoons CO_2 \quad (20)$$

Note that several of the steps in reactions in this reaction scheme are composite reactions, and also that several reactions (e.g. the adsorption of CO_2 and possibly H_2, and the dissociation of water) require promotion by adsorbed oxygen.[324]

(e) The Active Site
It is clear from the above results that the active sites are on the surface of the copper metal crystallites, but it is incorrect to envisage fixed sites. The surface of copper is mobile under reaction conditions[309] and is partially covered by a mobile layer of adsorbed oxygen. Adsorbed hydrogen mostly resides on the uncovered copper metal (even if requiring O(ads) for adsorption), whereas carbon dioxide adsorbs on the oxidized surface. Thus, the site of reaction consists of bare copper metal atoms next to an oxide surface site, i.e. a copper(0)/copper(I) site, but this occurs in different parts of the surface as the reaction proceeds. The mechanism of the synthesis of methanol on copper catalysts from CO_2-free gas mixtures is less well understood. Formate intermediates, made from residual O(ads), may be involved[325] again or successive additions of H(ads) to adsorbed carbon monoxide may give first adsorbed methoxy and then methanol.

9.6.2. Kinetics

Rate equations have been derived giving methanol synthesis rate as a function of gas composition, pressure and temperature.[294–297] In some cases the reaction mechanism has been deduced from kinetic measurements by assuming possible mechanisms involving adsorption, reaction of surface species and desorption, and developing a rate expression for each mechanism, and then assuming the one which gave the best fit to the measured kinetics indicated the actual reaction path. This approach is notoriously unreliable, partly because experimental results often fit more than one model and the expression may be only applicable over a limited range of conditions, and in most cases not relevant to the full-scale process. The complications of the concurrent water-gas shift reaction and of the variable state of the catalyst further vitiate any mechanistic conclusions.

The rate expressions which have been evolved[294–297] are of the type

$$\text{rate} = k p^x_{CO} p^y_{H_2} (1 - p_{ML}/K p_{CO} p^2_{H_2}) \tag{21}$$

in which p is the partial pressure, or fugacity, of the indicated reactant or product. The subscript "ML" refers to methanol. The negative term containing K, the equilibrium constant for the synthesis reaction (1), recognizes that the overall rate is affected by the rate of the back-reaction (methanol decomposition), which increases as equilibrium is approached. Industrial processes are designed to come close to equilibrium, so this factor has to be taken into account for full-scale plant design. The indices x and y vary, although the different expressions agree that hydrogen has a larger effect on the rate than carbon monoxide does; thus x ranges from 0.2 to 1.0, and y from 1.0 to 2.0. It should be noted that no term in p_{CO_2} appears in equation (21), even though carbon dioxide is the source of carbon for methanol under industrial conditions. Other forms of equations (21), with added terms for the effects of CO_2 and inhibition by water and methanol, have been used and are discussed by Chinchen et al.[297] The empirical kinetic equations used for the design of methanol synthesis reactors usually remain the commercial property of the industrial organization and have not been published.

9.7. Recent Developments

In recent years some new catalysts and process configurations have been proposed. Although none of these has yet reached commercial practice, some aspects may form the basis of future methanol processes. Two

Chapter 9. Methanol Synthesis

novel supported copper catalysts with high activity for methanol synthesis have been reported. Raney copper, prepared by leaching ternary copper/zinc/aluminium alloys with strong aqueous sodium hydroxide, has activity comparable with industrial precipitated catalysts.[326–329] Both zinc and aluminium were removed, but not completely, from the alloys to give catalysts with maximum activity at about 98% copper. These results support the conclusions in Section 9.6.1 that copper metal is the catalytically-active phase.

A more unusual route to supported-copper catalysts involves the careful oxidation of suitable intermetallic compounds. The thorium/copper compounds, $ThCu_6$, $ThCu_{3.6}$, $ThCu_2$ and Th_2Cu, all react directly with synthesis gas to give a form of copper supported on thoria which is extremely active for methanol synthesis.[330, 331] Carbon dioxide is not essential for these catalysts, so there is potential for a low-pressure process in which anhydrous methanol is made directly from a CO/H_2 feed. Since the discovery of the copper/thoria catalysts, similar copper catalysts made from non-radioactive rare-earth metals have been shown[332, 333] to be surprisingly active in methanol synthesis, operating at temperatures as low as 70°C in gas mixtures that are essentially free of carbon dioxide. In contrast, catalysts based on the platinum group metals have generally shown poor selectivity for methanol (Section 9.5) because of the ease with which C–O bonds are broken. The discovery[334] that palladium, on some supports and under certain conditions, could be selective in methanol synthesis has been followed by much work[335, 336] to identify the basis of the selectivity. So far there is no agreed interpretation. However, practical problems associated with the low rate of reaction, the high cost of palladium, and the relatively high reaction temperature leading to some methanation (Section 9.5), have discouraged any industrial applications. All-round, the conventional precipitated copper/zinc oxide/alumina catalyst remains the best available.

A three-phase process, in which fine particles of a copper catalyst are suspended in a high-boiling hydrocarbon, has been designed by Chem Systems.[337, 338] The liquid medium facilities heat transfer and methanol removal, but the added complexity of a liquid phase has so far prevented its commercial exploitation. Future developments of the conventional process are likely to be mainly capital savings and improvements in energy efficiency. Catalysts of improved performance will doubtless enable process designers to circumvent chemical engineering constraints.

Chapter 10

Catalytic Oxidations

10.1. Introduction

The three oxidation processes discussed in this chapter are amongst the earliest examples of heterogeneous catalysed reactions to have been exploited on a large scale, and each uses air as an oxidant. The chief uses of nitric acid are in making inorganic nitrates and organic nitro-compounds. By far the largest of these is the manufacture of ammonium nitrate for fertilizer use, the production of which has grown considerably over the past two decades. The higher nitrogen content of ammonium nitrate has also helped it to displace the traditional "sulphate of ammonia". Another fertilizer-related use of nitric acid that is increasing is as a replacement for sulphuric acid in the acidulation of phosphate rock. The current annual worldwide production of nitric acid is some 71 million tonnes, a demand that is largely met by plants each producing 1000–1500 tonnes day^{-1}.

Like the oxidation of ammonia, the oxidation of methanol to formaldehyde is not thermodynamically controlled, and so requires a catalyst having good selectivity. Formaldehyde is the most important derivative of methanol, and accounts for some 40% of its usage. At present the typical output of a modern formaldehyde plant is about 40 000 tonnes year^{-1} expressed as a 37% aqueous solution (formalin). Principal uses of formaldehyde are in resins and adhesives, with a large proportion going into urea–formaldehyde type resins for "compressed wood" production.

The oxidation of sulphur dioxide, unlike the previous two processes, is thermodynamically controlled, and there are some gross similarities with ammonia synthesis, in that both are exothermic and do not rely on catalyst selectivity. Oxidation of sulphur dioxide for sulphuric acid manufacture is the earliest example of a catalytic process being worked on an industrial scale, and today sulphuric acid is still produced on a larger scale than any other chemical. Although sulphur dioxide oxidation has no direct link with the use of a hydrocarbon feedstock, as have ammonia and methanol, a discussion of the process is included in this chapter for the following reasons: historically sulphuric acid plants were associated with ammonia plants because the final fertilizer product

Chapter 10. Catalytic Oxidations

was frequently ammonium sulphate,* and today many integrated fertilizer complexes produce sulphuric acid as a stage in the manufacture of phosphates for incorporation in compound fertilizers.

10.2. Ammonia Oxidation

10.2.1. History of Nitric Acid Production

Nitric acid was certainly known in the 13th century, and probably much earlier.[339] In 1776 Lavoisier showed that it contained oxygen, and nine years later Cavendish proved it contained nitrogen and oxygen, but it was not until 1816 that it was completely characterized by Gay-Lussac and Berthollet. Early nitric acid manufacture was from saltpetre, derived from animal waste sources, which was reacted with sulphuric acid in heated retorts, with nitric acid being distilled over and condensed in the manner of Glauber's invention of 1648. In 1825 the discovery of Chile saltpetre assisted in the growth of the nitric acid manufacturing industry, but in turn the nitrate route was ousted in the early part of this century by the ammonia oxidation process.

10.2.1.1. Routes from atmospheric nitrogen

Processes other than ammonia oxidation have been used but with limited success. In 1902 an unsuccessful trial using electrical discharge was started by Bradley and Lovejoy at Niagara Falls, Canada. After many problems it was abandoned in 1904. Greater success was had by Birkland and Eyde, and their process operated from 1905 for about 25 years in Norway, also using cheap hydroelectric power. Air was passed through an electrical discharge and with the high temperature generated a small amount of nitrogen was oxidized directly to nitric oxide. It is the small quantity of nitric oxide formed which is the major drawback to the process. Chilton[340] illustrated this difficulty by examination of the equilibrium constants for the formation of nitric oxide from nitrogen and oxygen. At temperatures as high as 2400°C the amount of nitric oxide in the gas mixture is only about 2.5%, whilst at 1000°C it is less than 0.1%. The change of this equilibrium with temperature is shown in Figure 10.1. Nitric oxide formation reaction rates are also extremely unfavourable at lower temperatures. The rate of formation is quite slow at 1600°C, taking more than two minutes to reach half the equilibrium

---

*In the UK, Germany and other countries in Europe, most ammonium sulphate was manufactured by double decomposition of calcium sulphate and ammonium carbonate rather than by direct neutralization of sulphuric acid. The process used by ICI is described by G. I. Higson, *Chem. Ind.*, 750 (September 1951).

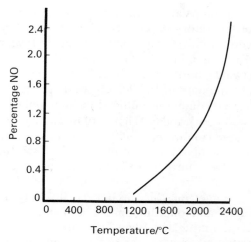

Figure 10.1. Percentage of nitric oxide at equilibrium in nitric oxide/air mixtures at different temperatures.

concentration, although only a small fraction of a second at 2400°C. Decomposition rates are rapid at equivalent temperatures, and to approach half-way to equilibrium from higher concentrations at 1600°C takes about one second. Consequently, the relatively small amounts of NO formed at very high temperatures must be rapidly chilled to lower temperatures, where decomposition rates are significantly slower and where the nitric oxide may be oxidized to nitrogen peroxide (a general collective term for the two allotropic forms nitrogen dioxide and dinitrogen tetroxide). Below 130°C nitric oxide is oxidized by air to nitrogen peroxide, and in the Birkland and Eyde process this was then absorbed in water to yield nitric acid. The huge power consumption of about 15 000 kW t^{-1} of HNO_3 and low yield precluded the proliferation of this process.

Thermal processes have also been used on a pilot scale where the high temperatures required for nitrogen oxidation are attained by passing air through refractory preheated to temperatures in excess of 2000°C by combustion of hydrocarbon gases. Refractory problems at these temperatures are serious and, as with the electrical discharge process, the economics placed this route to nitric acid with low nitric oxide yields and the required high absorption capacity at a disadvantage compared with ammonia oxidation.

10.2.1.2. Ammonia oxidation
The selective oxidation of ammonia to nitric oxide and water by heating an ammonia air mixture at elevated temperatures has been known for

Chapter 10. Catalytic Oxidations

almost two centuries. Various metals and oxides have been used as catalysts under laboratory conditions. The discovery that platinum is a particularly good catalyst was made about 150 years ago, and is attributed to Kuhlmann in 1839. Kuhlmann's discovery eventually bore fruit when Ostwald, shortly after 1900, developed ideas that resulted in its commercial exploitation. In 1902 Ostwald applied for a patent[341] in which his preferred conditions outlined the ammonia oxidation process essentially as we know it today. That is 10.6% ammonia in air, with the mixture preheated to around 300°C and short contact time over platinum giving a temperature rise of around 775°C, the exhaust gases then being cooled, further oxidation taking place followed by absorption of the nitrogen peroxide in water. Initially ammonia oxidation on a commercial scale was linked with coal coking, which gave rise to by-product ammonia liquor in reasonable quantities. The first ammonia oxidation nitric acid plant operating on by-product ammonia and using platinum sponge catalyst opened in 1908 at Gerthe, Westphalia. In the early years of industrial ammonia oxidation various plants were built operating with oxide catalysts using ammonia from ammonia liquor or generated from calcium cyanamide. However, the successful expansion of the use of the process followed from the equally successful and the then more technically difficult achievement of catalytic synthesis of ammonia (see Chapter 8). This process produced the large quantities of ammonia required for nitric acid manufacture, and moreover the ammonia was also free from catalyst poisons which sometimes caused problems in plants operating with ammonia from coal or cyanamide.

In the first quarter of this century, nitric acid plants were few, small and suffered problems, particulalrly corrosion in the absorption section. The development of acid-resistant steels, improved design and the impetus from increasing demand for nitric acid as a chemical and, more importantly, for manufacture of high-nitrogen fertilizers has led to a large expansion of the nitric acid industry. There are now large numbers of plants, of a variety of designs, in which the efficient use of energy is of paramount importance. Despite the evolutionary changes over the past 80 years giving rise to diversity of operation, the basic process and design remain recognizable with an ammonia burner section, nitric oxide oxidation section and absorber section (Figure 10.2). More significant is the fact that platinum still remains the only catalyst of importance although improvements to platinum catalyst have been made, notably the use of fine-mesh gauzes replacing foil, ribbon and sponge, and the incorporation of rhodium to strengthen the gauze. Other catalysts have, from time to time, been tried, both in the early days in Germany and in more recent years at various plants throughout

10.2. Ammonia Oxidation

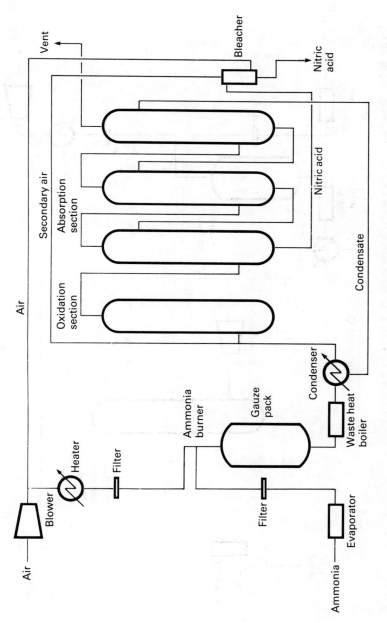

Figure 10.2a. Simplified flow diagram for atmospheric pressure nitric acid plant.

Chapter 10. Catalytic Oxidations

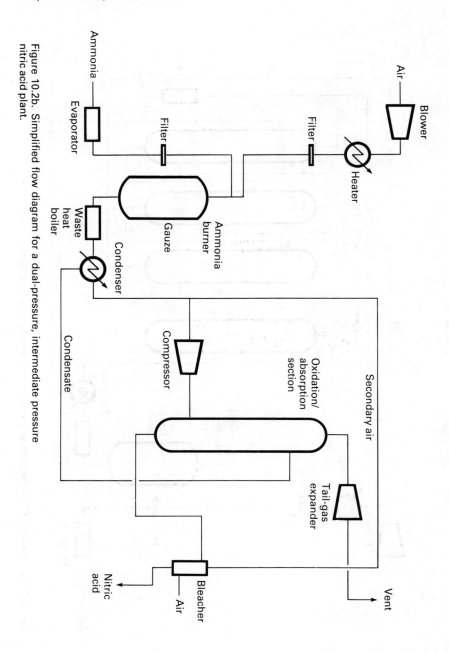

Figure 10.2b. Simplified flow diagram for a dual-pressure, intermediate pressure nitric acid plant.

10.2. Ammonia Oxidation

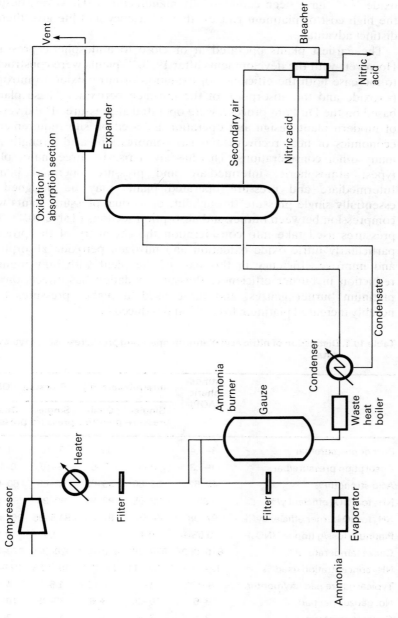

Figure 10.2c. Simplified flow diagram for a single pressure, pressure nitric acid plant.

475

Chapter 10. Catalytic Oxidations

the world. Recent catalysts based on promoted iron or cobalt oxide[342-344] have been commercially manufactured. However, despite the high cost of platinum gauzes, their efficiency and life give them a distinct advantage.

The earliest plants operated at or close to atmospheric pressure. However, with the developments after 1920,[345] plants were constructed to increase both the efficiency of oxidation of nitric oxide to nitrogen peroxide and the absorption of the nitrogen peroxide. These plants, based on the Du Pont process, were operated at pressure. The diversity of modern plant design and operation is based on site requirements, economics of heat recovery and gas compression, acid strength and many other considerations. This has given rise to three major plant types: atmospheric, intermediate and pressure operated plants. Intermediate and pressure operated plants may be designed as essentially single pressure throughout, or as dual pressure plants with compression between burner and absorption sections (Table 10.1). The pressures used take into consideration the chemistry of the process, particularly nitric oxide oxidation and nitrogen peroxide absorption, and improve efficiency in this part of the plant with only minimal reduction in burner efficiency. Pressure oxidation has effects on the platinum burner gauzes, and those used at higher pressures have notably increased platinum losses[346] and reduced life.[347]

Table 10.1. Distinction of nitric acid plants on operating pressures—some typical data

	Atmospheric (AOP)	Intermediate (IP)		Pressure (POP)	
		Single-pressure	Dual-pressure	Single-pressure	Dual-pressure
Burner pressure/bar	1–1.4	3–6	0.8–1	7–10	4–5
Absorption pressure/bar	1–1.3	3–6	3–5	6–10	0–12
Acid strength/%	49–52	53–60	55–69	52–65	60–62
NH_3 to HNO_3 efficiency/%	93	88–95	92–96	90–94	94–96
NH_3 to NO burner efficiency/%	97–98	96–96.5	97–98	94.5–95	96–96.5
Platinum loss/g (tonne HNO_3)$^{-1}$	0.05	0.1	0.05	0.3	0.1
Gauze temperature/°C	810–850	870–890	810–850	920–940	870–890
NH_3 concentration used/%	12–12.5	10.5–11	11–12.5	10–10.5	10.5–11
Typical gauze pad life/months	8–12	4–6	8–12	1.5–3	4–6
No. gauzes per pad	3–6	10–20	3–6	35–45	10–20
Gauze diameter/m	3–5	3–4	3–5	1.1–5	3–4

10.2.2. Chemistry of the Modern Process

The overall process shown in equation (*1*) is the exothermic oxidation of ammonia to nitric acid and water. Many sequences of reactions are involved in the overall process, which may be simplified into three equations; the burning of ammonia to nitric oxide (*2*), the oxidation of the nitric oxide (*3*) and the reaction (absorption) of nitrogen peroxide as dinitrogen tetroxide and water (*4*).

$$NH_3 + 2O_2 \rightarrow HNO_3 + H_2O, \quad \triangle H = -320.8 \text{ kJ mol}^{-1} \quad (1)$$

$$4NH_3 + 5O_2 \rightarrow 4NO + 6H_2O \quad \triangle H = -899.9 \text{ kJ mol}^{-1} \quad (2)$$

$$2NO + O_2 \rightarrow N_2O_4 \quad \triangle H = -171.6 \text{ kJ mol}^{-1} \quad (3)$$

$$2N_2O_4 + 2H_2O + O_2 \rightarrow 4HNO_3 \quad \triangle H = -73.6 \text{ kJ mol}^{-1} \quad (4)$$

Whilst there are at least eight oxides of nitrogen, only three are of major concern in nitric acid manufacture; NO, NO_2 and N_2O_4, the last two (nitrogen dioxide and dinitrogen tetroxide) being two different molecular forms of nitrogen peroxide. In ammonia oxidation important factors are the efficiency of forming nitric oxide from ammonia, known as the burner efficiency, the rate of oxidation of nitric oxide and the absorption of nitrogen peroxide in water to give dilute acid.

10.2.3. Chemistry of Absorption

In the absorption section dinitrogen tetroxide is reacted with cooled dilute nitric acid to yield nitric acid as in the overall reaction (*8*) which is net exothermic. The reaction takes place via a series of quite complex reactions involving other species, but may be conveniently described by equations (*5*)–(*7*).

$$4N_2O_4 + 4H_2O \xrightarrow{slow} 4HNO_3 + 4HNO_2 \quad \triangle H = +83 \text{ kJ mol}^{-1} \quad (5)$$

$$4HNO_2 \xrightarrow{rapid} 2H_2O + N_2O_4 + 2NO \quad \triangle H = +15 \text{ kJ mol}^{-1} \quad (6)$$

$$2NO + O_2 \xrightarrow{slow} N_2O_4 \quad \triangle H = -172 \text{ kJ mol}^{-1} \quad (7)$$

$$2N_2O_4 + 2H_2O + O_2 \rightarrow 4HNO_3 \quad \triangle H = -74 \text{ kJ mol}^{-1} \quad (8)$$

Nitric oxide formed in reaction (*6*) is further aerially oxidized to nitrogen peroxide (*7*), and further absorption occurs so that the exhaust tail gas has a very low content of nitric oxide—giving rise to the red plumes from the stacks where final oxidation to nitrogen peroxide occurs. The actual concentration of nitrogen oxides in the tail gas depends on the plant design, use of catalytic or combustion methods of

Chapter 10. Catalytic Oxidations

NO$_x$ removal and legislation. Typically the exhaust gas contains between 25 and 4000 ppm of oxides of nitrogen.

An important aspect of these reactions is that dinitrogen tetroxide (N$_2$O$_4$) is quite reactive towards water while nitrogen dioxide (NO$_2$) is not. The dissociation reaction of N$_2$O$_4$ shown in equation (9) is endothermic and is favoured by high temperature and low pressure (Figure 10.3), but because absorption is dependent on the concentration of N$_2$O$_4$ and since overall reaction (8) is exothermic, most plant designs incorporate good heat removal and are operated at pressure to enhance absorption. An added bonus is that for a given daily tonnage of nitric acid produced, higher pressure plants have a smaller absorption section, a significantly higher absorption efficiency and are capable of yielding higher acid strengths.

$$N_2O_4 \rightarrow 2NO_2 \qquad \triangle H = +57 \text{ kJ mol}^{-1} \qquad (9)$$

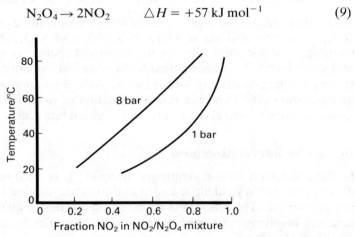

Figure 10.3. Effect of pressure and temperature on equilibrium dissociation of N$_2$O$_4$.

10.2.4. Nitric Oxide Oxidation Chemistry

Burner gases are rapidly cooled and mixed with secondary air to cause the oxidation of nitric oxide as in reaction (12). The reaction proceeds in two steps as shown in equations (10) and (11).

$$2NO + O_2 \xrightarrow{slow} 2NO_2 \qquad \triangle H = -115 \text{ kJ mol}^{-1} \qquad (10)$$

$$2NO_2 \rightarrow N_2O_4 \qquad \triangle H = -57 \text{ kJ mol}^{-1} \qquad (11)$$

$$2NO + O_2 \rightarrow N_2O_4 \qquad \triangle H = -172 \text{ kJ mol}^{-1} \qquad (12)$$

The dissociation of nitric oxide to nitrogen and oxygen does not occur at

10.2. Ammonia Oxidation

a significant rate at the temperatures used in the ammonia oxidation process, so the important factors are the rates of oxidation of nitric oxide, equation (*10*), and formation of dinitrogen tetroxide, equation (*11*). The dissociation of dinitrogen tetroxide (that is, the reverse of reaction (*10*)) occurs above about 130°C (Figure 10.4), so one requirement is to keep temperatures below 130°C and, since oxidation of nitric oxide is exothermic, cooling is required. As shown in Figure 10.5, even at moderately low temperatures conversion is low. The transformation of nitrogen dioxide to dinitrogen tetroxide desired for the absorption stage is also exothermic and is also favoured by low temperature and increased pressure, reinforcing the need for efficient heat removal and use of pressure.

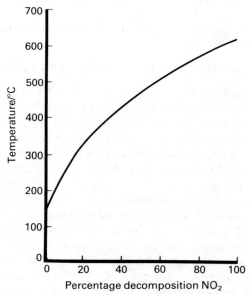

Figure 10.4. Change in equilibrium composition with temperature for decomposition of NO_2 to NO and O_2.

10.2.5. Ammonia Oxidation Chemistry

The desired reaction (*13*) is markedly exothermic and exceedingly rapid. However, at high temperatures the even more exothermic reaction (*14*) becomes significant, and ammonia is converted to undesirable nitrogen.

$$4NH_3 + 5O_2 \rightarrow 4NO + 6H_2O \qquad \triangle H = -903 \text{ kJ mol}^{-1} \qquad (13)$$

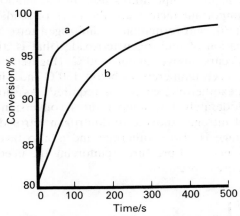

Figure 10.5. Rate of oxidation of nitric oxide in nitric oxide/air mixtures at 30°C with (a) 10% and (b) 2% initial NO.

$$4NH_3 + 3O_2 \rightarrow 2N_2 + 6H_2O \qquad \triangle H = -1261 \text{ kJ mol}^{-1} \qquad (14)$$

Both reactions involve a large number of molecules, and both must proceed by a sequence of steps, as discussed, for example, by Chilton.[340] However, since both reactions are exceedingly fast and occur at high temperature, extensive studies of the molecular species involved have not been made. Under the conditions used in normal practice there is high selectivity, the yield of nitric oxide being of the order of 95–98%. Whilst both reactions are favoured by pressure, reaction (14) is the more favoured. Consequently, increased burner pressures decrease the efficiency of conversion of ammonia to nitric oxide, and the highest maximum burner efficiencies are attainable by operation at atmospheric pressure. It becomes uneconomical to operate at significantly below atmospheric pressure, although some plants do operate under slightly reduced pressure. Since the efficiency of conversion decreases with increased pressure it also becomes uneconomical to operate above about 10 bar.

Another cause of drop in burner efficiency is leakage of unreacted ammonia through tears in the catalyst gauze, loose packing of gauzes, or due to poor sealing of the gauze periphery. This ammonia then reacts with nitric oxide to give nitrogen and water, so lowering the yield of nitric oxide as shown in equation (15).

$$4NH_3 + 6NO \rightarrow 5N_2 + 6H_2O \qquad \triangle H = -1803 \text{ kJ mol}^{-1} \qquad (15)$$

Reactions (13) and (14) are both highly exothermic, but ammonia oxidation is usually carried out with some preheating of the ammonia/air

10.2. Ammonia Oxidation

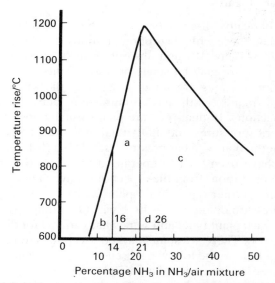

Figure 10.6. Adiabatic temperature rise for reacting ammonia/air mixtures showing: (a) stoichiometric region; (b) excess air—principal reaction $4NH_3 + 5O_2 \rightarrow 4NO + 6H_2O$; (c) excess ammonia—principal reaction $4NH_3 + 3O_2 \rightarrow 2N_2 + 6H_2O$; (d) explosive limits at 1 bar total pressure.

mixture (200–300°C) in order to maintain a burner temperature of 850–920°C with an efficiency around 96% or more (efficiency being defined as the percentage NH_3 converted to NO).[348] The adiabatic temperature rise for ammonia/air mixtures is shown in Figure 10.6. There are three regions:[349]

0–14% ammonia, where the main oxidation reaction is burning to nitric oxide leaving excess oxygen;

14–21% ammonia, the stoichiometric region in which all of the oxygen is used in reactions (*13*) and (*14*), reaction (*14*) becoming more prominant as the ammonia percentage is raised;

more than 21% ammonia, where the principal product of oxidation is nitrogen via reaction (*14*).

Most plants operate at below 12% ammonia to avoid explosive mixtures (lower limit 16% ammonia at 1 bar and about 12.4% at 10 bar)[350] although explosions due to the use of wrong proportions through maloperation are not unknown. Amounts used are also a compromise between increased NO from increased ammonia in the gas and decreased efficiency due to reaction (*14*); that is, to give maximum selectivity.

481

10.2.6. Modern Plants

There are basically three types of ammonia oxidation processes operated today, those with absorption at atmospheric pressure (AOP), absorption at 3–6 bar (IP) and absorption at 6–12 bar (POP). Plants operated at pressure may be single-pressure plants, where the burner and absorption sections are under essentially the same pressure, or dual-pressure plants, with gas compression between burner and absorption sections. IP dual-pressure plants usually operate the burner at atmospheric pressure, and POP dual-pressure plants operate the burner at intermediate 4–5 bar pressure (see Table 10.1). In recent years most plants have been built to operate at pressure. Regardless of the overall plant operation, from the aspect of catalytic ammonia oxidation there are three burner types; atmospheric- (1 bar), intermediate- (3–6 bar) and pressure-operated (7–10 bar). Operating parameters are different for each plant type, and somewhat variable for different plants of the same type. There are some marked exceptions to the average for any one plant type so the following discussion is of a general nature and cannot be regarded as a series of hard-and-fast rules.

In all cases ammonia and air are filtered to remove particulate matter before burning, and the mixture containing 8–13% (typically 10–11%) ammonia is usually preheated to between 120 and 300°C. The final gauze temperature depends on the ammonia content, the amount of preheat, the gas rate and the heat losses (mostly radiant). Typical examples for the increases in adiabatic temperature rise with increase in ammonia concentration are shown in Figure 10.7. These are measured rises of intermediate (I), atmospheric (A) and pressure (P) burners operating at or close to 96% efficiency and optimum gas rates, and show a temperature rise of about 40°C per 1% increase in ammonia near 8%, and about 60°C per 1% increase near 12% ammonia. For a fixed ammonia content, preheat and number of gauzes in the burner, there will be variation in efficiency with change in gas rate (Figure 10.8). Very low gas rates give poor efficiency and low gauze temperature with some nitric oxide loss that takes place via the decomposition reaction (16).

$$2NO \rightarrow N_2 + O_2 \qquad \triangle H = -173 \text{ kJ mol}^{-1} \qquad (16)$$

Increasing the gas rate increases the gauze temperature, but the efficiency rises to a peak then falls as selectivity falls at high temperatures when reaction (14) becomes more prominent and some ammonia slip may occur. Unreacted ammonia then consumes NO via reaction (15). Similarly, for fixed ammonia content gas rate and preheat the efficiency varies with number of gauzes. At low gas rates increasing the number of gauzes steadily decreases the efficiency, at medium gas

10.2. Ammonia Oxidation

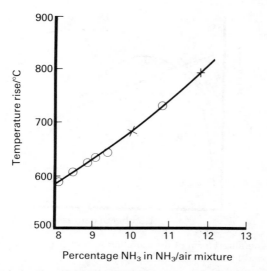

Figure 10.7. Variation in adiabatic temperature rise with ammonia content in the region 8–13% ammonia in air (measured values for +, 1 bar; ○, 4 bar and ×, 10 bar burners).

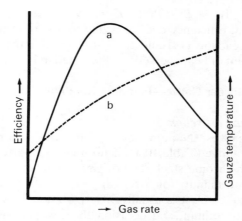

Figure 10.8. Generalized form of variation in: (a) the efficiency of ammonia–nitric oxide conversion and (b) change of gauze temperature with gas rate.

rates efficiency steeply rises to a peak then steadily falls, whereas at high gas rates the efficiency rises very rapidly then tends to plateau, as illustrated in Figure 10.9. In pressure plants, where the mass throughput of gas is greater, normally some 35–45 gauzes are used, while with atmospheric plants three to six gauzes are typical for optimum efficiency. Where burners are operated at pressure there is a decrease in

483

Chapter 10. Catalytic Oxidations

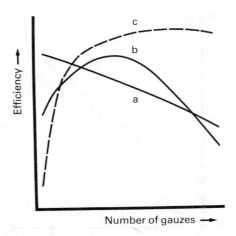

Figure 10.9. Generalized form of the variation in burner efficiency with number of gauzes at (a) low, (b) medium and (c) high gas rates.

burner efficiency, so that the best attainable efficiencies are about: AOP, 98.5%; IP, 96.5%; POP, 95.0%. In all cases, for fixed ammonia concentration, efficiency decreases due to loss of nitrogen in the more exothermic reaction (14) and the gauze temperature rises due to the greater amount of heat evolved, as illustrated in Figure 10.10.

10.2.7. The Burner Gauze—Platinum/Rhodium Catalyst

10.2.7.1. Gauze activation

The pad of gauzes, the number of which varies according to plant type as discussed above (Table 10.1), is normally supported on a large open mesh of high-chrome steel. The original catalyst of Ostwald was pure platinum, but mainly for strength reasons the composition of the modern gauze catalyst is usually 90% platinum/10% rhodium or 95% platinum/5% rhodium, and occasionally 85% platinum/15% rhodium. Rhodium is also reported to increase the conversion efficiency and reduce the rate of catalyst loss.[350]

A typical new gauze consists of a square woven mesh of side length 0.55 mm with a wire diameter of 75 μm. New unactivated gauze essentially consists of smooth wires with the principal surface roughness being draw marks, as illustrated in Figure 10.11(a). As the pad of gauzes (Figure 10.12) is made up, care is taken to ensure the gauzes are free from large holes or tears and that they are laid flat in close contact without creases or folds which would give gaps between gauzes. This

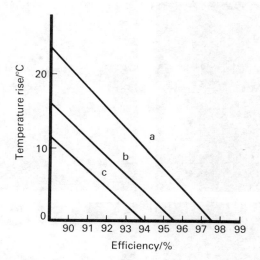

Figure 10.10. Temperature rise in gauze with decrease in efficiency due to nitrogen formation (equation (*14*)). (a) Atmospheric burner, 2.8°C per 1% efficiency loss; (b) intermediate-pressure burner, 2.8°C per 1% efficiency loss; (c) pressure-operated burner, 2.3°C per 1% efficiency loss.

careful construction of the burner pad is necessary to reduce slip of unreacted ammonia, which leads to reduced efficiency via reaction (*15*). The gauze pad is nowadays usually lit by a hydrogen torch playing in the upper surface of the gauzes through which the ammonia/air mixture is passed, although electrical heating is still used. In some plants it is the practice to hydrogen-flame the pad for several hours before use, which gives some activation of the surface of unused gauzes. Initially the activity of a new gauze is low, and the gauze requires activating. In order to reach high burner efficiency quickly, new gauzes are normally placed low in the pad (that is, downstream in the gas flow) of previously used cleaned active gauzes. However, activation of a new gauze starts immediately the burner is lit and is fairly rapid, taking only a few hours to reach a reasonable level of activity.

As activation proceeds the grain boundaries between the constituent crystals of the gauze take on an etched appearance and the crystals start to grow stepped faces, and etch pits develop (Figures 10.11(b)). Continued activation leads to highly roughened surfaces, and ultimately to dendritic excrescences of alloy which grow from the wire surface (Figure 10.11(c)). This increases metal area and reduces the hole size of the mesh. The gas therefore has to flow through smaller orifices with considerably rougher edges. The side facing into the gas flow activates

Chapter 10. Catalytic Oxidations

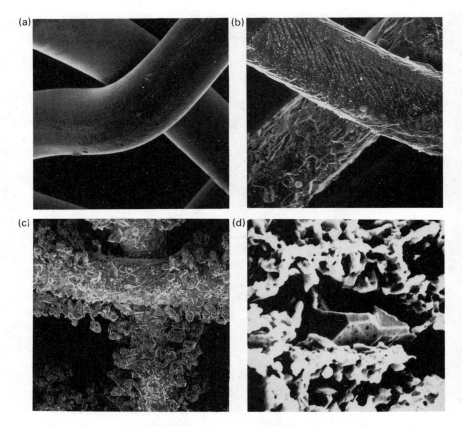

Figure 10.11. Scanning electron microscope (SEM) images of gauzes. (a) New unactivated gauze, showing draw marks (×200); (b) partially activated gauze, showing etching (×200); (c) well-activated gauze, showing excrescences (×200); (d) POP gauze with Rh_2O_3 crystals covering active alloy surface (×2500).

at a quicker rate and is roughened first, and top gauzes activate more quickly than lower ones. In a well-activated gauze pad the differences between surfaces and gauze position are not apparent.

Pielaszek[351] has shown that catalytic etching of the crystals in the ammonia/air mixture produces a variety of pits and islands, with the etch pits being related to the platinum alloy crystallographic surfaces exposed to the gas stream. Those pits related to the (100) planes have a four-fold symmetry, whilst those related to (111) planes have a three-fold symmetry. The (110) planes give rise to linear pits. As etching deepens, grain boundary etching also occurs and the wires thin and weaken. Some of the platinum volatilized from the gauze condenses on

10.2. Ammonia Oxidation

Figure 10.12. Installation of gauze pad—checking for wrinkles.

other portions and catches detached crystallites, leading to the growth of delicate excrescences. The effects are therefore increasing area and development of certain crystal faces, which increases activity, but during this growth the thinning imparts weakness, and as a result old gauzes tear easily on handling. A further effect of activation is that a lot of the alloy forms into delicate excrescences which are easily detached from the mesh. Some of the gauze loss can be attributed to this cause, since broken excrescences have been found down stream in sludges. However, the differences in platinum losses between atmospheric and pressure burners would suggest more platinum is lost by an additional mechanism in pressure burners, that is by vaporization. This loss of volatile platinum and its recovery are discussed in Section 10.2.7.3.

Chapter 10. Catalytic Oxidations

10.2.7.2. Gauze deactivation and cleaning

Nowadays new gauzes are supplied in a precleaned condition. Traces of lubricant and iron contamination from drawing and weaving machinery are carefully monitored and removed before despatch. Cleaning of used gauzes is carried out when they become obviously contaminated or broken, or when they lose efficiency. After cleaning, minor tears are patched by gently hammering a piece of gauze over the tear, generally after flaming to red heat. The highly malleable alloy usually adheres well. The cleaned and repaired gauzes are replaced in the burner or stored for later use. The cleaning and rejuvenation procedure varies according to the contamination or cause of deactivation involved. Loose debris such as refractory dust or iron oxide (frequently observed contaminants) causing physical blanketing is vacuumed or shaken off. The gauzes are then pickled in hydrochloric acid baths to dissolve attached or wedged material—again, usually iron oxide dust. Despite the precautions used to keep gases clean, iron oxide occurs on most gauzes after a long period of operation, and hydrochloric acid performs adequately in iron oxide dissolution, but other contaminants such as fused refractory are more difficult to remove. Gauzes may disintegrate by embrittlement and grain boundary attack by contaminants such as molybdenum disulphide. In addition to deactivation by physical blanketing, chemical deactivation by specific poisons such as iron, molybdenum, zinc and sulphur sometimes occurs. The origin of the iron, as its form frequently suggests, is plant scale. Sulphur may arise from compressor oils where compression is used, or may arise as atmospheric pollutant in the air intake. Filters, grills, refractories and other sources may provide small, but positively harmful, quantities of a wide variety of material.[352]

A common cause of loss of activity in burners operated at pressure is physical blanketing of the alloy gauze by rhodium oxide, Rh_2O_3[352, 353] (Figure 10.11(d)). This is insoluble in acid and can only be removed by thermal decomposition in nitrogen or air (in air, temperatures above about 1050°C are necessary), by reduction in hydrogen, or by slow decomposition at lower pressures of operation.[352] Obviously the product rhodium metal, being more-easily oxidized than the alloy, remains a problem unless it diffuses back into the wire by prolonged high-temperature annealing. The exact mechanisms whereby the rhodium in the alloy is oxidized are debatable.[347, 354] Certainly the increased partial oxygen pressure in pressure-operated burners shifts the thermodynamic equilibrium from rhodium dissolved in alloy to oxide. Whether, as in many ferrous alloys, the oxidation is surface oxidation of rhodium followed by diffusion outwards of rhodium from the alloy followed by further oxidation, or whether the loss of volatile platinum oxide, PtO_2, leaves a rhodium-enriched alloy which oxidizes more easily, is not immediately clear. Various theories have been advanced; however, high-pressure burners frequently suffer from rhodium oxide formation and also show enhanced platinum loss[347] believed to be due to formation of relatively volatile PtO_2. Comparisons between atmospheric and pressure-operated gauze compositions made by the present author[352] show distinct rhodium depletion of wire and metal excrescences in pressure-operated gauzes after several months on-line, suggesting that the rate of rhodium oxide formation is quicker than PtO_2 volatilization. Similarly, the extent to which nitric oxide might feature in either oxidation is not known.

10.2.7.3. Metal recovery

Metal losses by whatever cause—alloy particles, solid rhodium oxide or volatile platinum oxide—make a small but significant contribution to the cost of acid production. Several attempts have been made in the past to recover precious metal, either from absorber sludges or by filtration[355] of gases. Mechanical filters of glass wool or other ceramics placed some distance downstream of the burners are said to achieve up to 60% recovery of platinum as fine dust particles. However, they suffer from pressure-drop effects, and in addition efficiency is variable. Less often used are Raschig rings or marble chips just below the burner gauzes. Raschig rings are not as successful in high recovery and marble chips present problems, particularly in the rehydration of lime formed during use. Since about 1970 excellent recovery has been claimed by using a getter gauze system[346, 355] to capture particulate or volatile platinum and rhodium species. In recent years a gold/palladium gauze[346] just downstream of the burner has been tried. Significant recovery of platinum and rhodium occurs (60%), but the costs of fabrication and reprocessing the spent getter gauze are high. Universal acceptance of the use of getter gauzes has not occurred.

The dual aspect of increased catalyst loss and deactivation by rhodium oxidation on intermediate and pressure-operated plants have been given some attention in the past few years, particularly that of deactivation by rhodium oxide blanketing. It is claimed that by careful operation of the gauze temperature avoidance of rhodium oxide formation and increased life of gauzes is possible. As in all problems of this nature, a compromise must be made between loss of activity, loss of alloy and plant efficiency.[347, 356]

10.3. Methanol Oxidation

10.3.1. Introduction

Since its first commercial production at the beginning of this century formaldehyde has become a chemical of major industrial importance. Current world annual production capacity is estimated to be almost 5 million tonnes as 100% formaldehyde; present production quantities are difficult to assess. The main industrial use of formaldehyde is in the production of urea–phenolic and melamine resins which are used in the manufacture of chipboard (compressed wood) and plywood. Other well-established applications are in the production of paints, cosmetics, explosives, fertilizers, dyes, textiles and papers. At present, commercial formaldehyde manufacturing processes are based on methanol feedstock, other routes such as oxidation of hydrocarbons have had only limited success, and are no longer in operation.

The production of formaldehyde from methanol is based on reactions (*17*) and (*18*).

Oxidation

$$CH_3OH + \tfrac{1}{2}O_2 \rightarrow HCHO + H_2O \qquad \triangle H = -157 \text{ kJ mol}^{-1} \quad (17)$$

Dehydrogenation

$$CH_3OH \rightarrow HCHO + H_2 \qquad \triangle H = +85 \text{ kJ mol}^{-1} \qquad (18)$$

The world's formaldehyde production is split between silver-catalysed and metal oxide-catalysed routes, and to date neither technology has established a leading role. Oxidation of methanol is the basic mechanism of the metal oxide-catalysed route, whereas the silver-catalysed process is produced by a combination of oxidation and dehydrogenation reactions.

10.3.2. The Silver-catalysed Process

The silver-catalysed process for the commercial production of formaldehyde by the reaction of methanol vapour and air over a silver catalyst has been used since the beginning of this century. A typical silver catalyst based process is shown in Figure 10.13. It consists of a reactor containing a bed of extremely pure silver in a granular crystalline form or a silver gauze. Catalysts consisting of silver supported on a range of inert materials have been used at various times in the past, but these do not appear to have achieved acceptance, a possible reason for this is that spent metallic silver catalyst can easily be recycled, or regenerated at low cost.

A typical reactor contains the silver crystal catalyst supported above a steam-raising boiler so that the reaction products pass through the boiler tubes before passing to further cooling stages. The silver catalyst usually consists of electrolytically produced crystalline particles of varying sizes, which are typically from 5 mm to less than 0.5 mm. A layer of catalyst particles of selected sizes is supported above the boiler tube sheet, forming a permeable bed which may be typically up to 3 cm deep and up to 4 m in diameter.[357] The reactor is fed with an appropriate mixture of vaporized methanol, air and water vapour, and operates at a pressure slightly above atmospheric pressure. The catalyst is raised to the temperature at which reaction is initiated, usually above 300°C, and then operates at temperatures between 560°C and 680°C. The reacted gases containing formaldehyde pass from the catalyst bed, are cooled and then passed to a circulating absorber system, followed by a water-scrubbing system for the effluent gases. Aqueous formaldehyde is removed from the absorber. The scrubbed gases, consisting principally of nitrogen, hydrogen and carbon dioxide are vented.

Improvements to the silver process which are now well-established include the addition of diluent steam to the reactor feed gases, rapid cooling of the reactor effluent, and various methods of recovering process heat in order to maximize the efficiency of operation. In recent years there has been an increased demand for higher concentrations of formaldehyde, typically 50% formaldehyde with 1% methanol, so that the traditional formalin product containing 37% formaldehyde has been superseded. Silver-catalysed technology has been adapted to produce an intermediate product which, after distillation, contains up to 55% formaldehyde with 1% or less of methanol. Methanol separated from the intermediate product is returned to the process as reaction feed. Roughly half of the steam demand for distillation can be satisfied by burning the process vent gas, which contains approximately 20% hydrogen and small amounts of methanol, formaldehyde and methyl formate. In other processes higher concentrations of formaldehyde can be obtained without distillation by specialized absorber systems which may be allied with a process to obtain a urea formaldehyde solution.[358, 359]

The most significant development of silver process technology in recent years is the recycling of process tail gas to the reactor. This gas moderates catalyst temperatures and improves methanol conversions

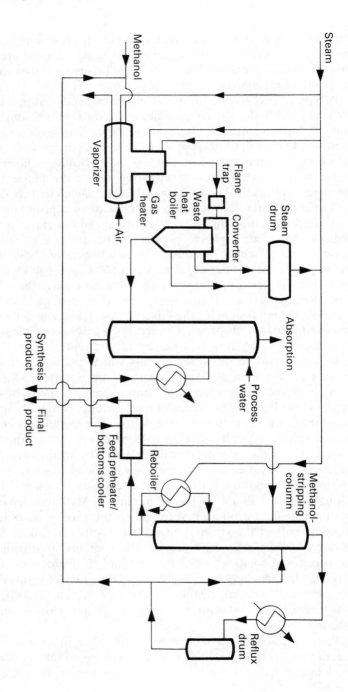

Figure 10.13. Flowsheet of a typical silver-catalysed formaldehyde process.

and process efficiencies; a product with up to 55% formaldehyde with less than 2% methanol is claimed. This is a significant improvement, as a concentrated formaldehyde product is achieved without distillation and its associated energy requirement.[360, 361]

10.3.2.1. Silver-catalysed reactions

The principal reaction in the silver-catalysed process is methanol dehydrogenation, and operation with a deficiency of oxygen in the reaction gas is essential in order to limit oxidation reactions. Formaldehyde production proceeds at reasonable rates above 500°C, but for an acceptable methanol conversion temperatures are normally in the range 560–680°C. At operating temperatures the equilibria for the decomposition reactions predict almost complete decomposition of methanol *and* formaldehyde. As a consequence of this, high space velocities are necessary, and these may be up to $3 \times 10^5 \text{ h}^{-1}$. Specification of the required space velocities is part of proprietary knowledge of each silver-catalysed process and is, in turn, dependent on precise operating conditions as well as the nature of the silver catalyst used. In the case of silver gauzes, too high a space velocity leads to bypassing and to low methanol conversions. The selection of silver crystal sizes and their distribution in the reactor are specific to each process, and are the subject of patents.[357] The endothermic dehydrogenation reaction is sustained by the exothermic reactions of hydrogen to form water, and the oxidation of methanol to formaldehyde and water. The contribution to the overall reaction made by the two exothermic reactions has been the subject of frequent investigations;[362, 363] however, the definitive mechanism is not yet established. Other significant reactions result in the formation of carbon dioxide, carbon monoxide and methyl formate:

$$CH_3OH + {}^3/_2 O_2 \rightarrow CO_2 + 2H_2O \quad \Delta H = -676.2 \text{ kJ mol}^{-1} \quad (19)$$

$$CH_2O \rightarrow CO + H_2 \quad \Delta H = +6 \text{ kJ mol}^{-1} \quad (20)$$

$$CH_3OH + {}^1/_2 O_2 \rightarrow H_2O + H_2 + CO \quad \Delta H = -150 \text{ kJ mol}^{-1} \quad (21)$$

$$CH_2O + CH_2O \rightarrow CH_3OCHO \quad \Delta H = -118 \text{ kJ mol}^{-1} \quad (22)$$

The process is largely explicable in terms of these reactions; for example, addition of water vapour to the reactant gases helps to lower reaction temperature and to suppress the oxidative reactions (*17*) and (*19*), producing formaldehyde and carbon monoxide. In practice water addition is limited by the requirements of the final product strength and of the additional water needed for tail-gas scrubbing.

Chapter 10. Catalytic Oxidations

Catalyst lives in commercial operation vary between three months and a year. Processes which use heavy catalyst loading and achieve high conversions normally experience shorter lives than those which impose less-severe conditions. As the catalyst "dies off" raising the air/methanol ratio in order to maintain conversion leads to higher temperatures which promote increased by-product formation. This, in turn, gives rise to further temperature increases which cause catalyst sintering and results in a build-up of pressure drop over the catalyst bed. A further factor affecting lives is loss of integrity of the bed with subsequent gas by-passing; this is likely to be aggravated by unsteady or irregular plant operation. When this happens the catalyst bed can normally be replaced or "resealed" with fresh catalyst without significant loss of production or catalyst cost. Figure 10.14 shows an electron micrograph of the silver catalyst before it is charged into the plant, and one of discharged catalyst.

10.3.2.2. Selectivity
As may be expected from equation (*19*), the quantity of carbon dioxide produced at a fixed throughput is a function of the relative quantities of air, methanol and steam. An increase in the air concentration leads to more carbon dioxide formation. Formaldehyde decomposes homogeneously at high temperature to give carbon monoxide and hydrogen, and as the reaction temperatures fall there is a rapid decrease in the rate of formation of carbon monoxide—as would be expected from the equilibrium data shown in Figure 10.15. The significance of this reaction was not fully appreciated in the earlier days of formaldehyde manufacture, and accounted for a significant loss of efficiency. It is kinetically limited, and can be minimized by rapidly quenching the effluent gases. Several cooling techniques have been adopted to do this, including forced cooling of the underside of the catalyst bed.[364] In the ICI process rapid cooling is achieved in a specially designed reactor/waste heat boiler system. Some methyl formate is also formed, and this appears to be produced in the reactor, rather than by subsequent esterification in the absorbtion section. The methyl formate level found in tail gas is typically 0.1 vol.%, and this tends to increase with increasing throughput, suggesting that the reaction in which it is formed is film diffusion controlled.

10.3.2.3. Poisoning
In the silver-catalysed process the high space velocities reflect the low catalyst volumes employed. The process is therefore sensitive to cumulative poisoning from trace levels of contaminants, and this

10.3. Methanol Oxidation

Figure 10.14. Scanning electron micrographs of granular silver methanol oxidation catalyst: (a) and (b) catalyst prior to charging ($\times$ 60 and $\times$ 175 respectively) showing faceted silver crystals; (c) and (d) same magnifications for agglomerated spent catalyst having surface contamination.

Chapter 10. Catalytic Oxidations

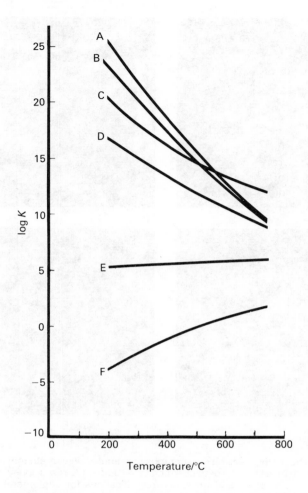

Figure 10.15. Equilibrium constants for selected reactions relevant to formaldehyde production:

A $CO + \frac{1}{2}O_2 \rightleftharpoons CO_2$;
B $2CH_2O \rightleftharpoons CO_2 + CH_4$;
C $CH_3OH + \frac{1}{2}O_2 \rightleftharpoons CH_2O + H_2O$;
D $CH_2O + \frac{1}{2}O_2 \rightleftharpoons HCOOH$;
E $CH_2O \rightleftharpoons CO + H_2$;
F $CH_3OH \rightleftharpoons CH_2O + H_2$;

appears to have been a major problem in the development of the process. It is generally accepted that halides and low levels of sulphur-containing compounds are poisons for the silver catalyst, and considerable attention has been paid to their removal, particularly from the process air. Several processes employ scrubbing devices such as packed or trayed towers with circulating dilute alkali or methanol solutions, in order to clean the process air.

It now seems likely that these effects are less severe than was thought, and the most significant catalyst poison is, in fact iron carbonyl. This may occur at trace levels in methanol, but it formerly constituted a severe source of poisoning which has now largely been eliminated. Traces of iron deposited on the surface of the silver catalyst promotes oxidation reactions leading to carbon dioxide formation, and progressive loss of selectivity and conversion to formaldehyde. Since the levels of iron carbonyl involved in the methenol were near the limits of detection this poisoning was not readily recognized, which considerably complicated the resolution of catalyst-poisoning problems. Removal of iron carbonyl has been effected by pretreating the unrefined methanol with oxidizing agents and subsequent distillation.[365, 366] The problem does not seem to occur with methanol produced using low-pressure methanol synthesis technology (see Chapter 9). The less severe cumulative effects of other trace contaminants are reflected in the longer catalyst lives currently being achieved.

An attractive advantage of the silver crystal-catalysed process is that the catalyst can be replaced at a low cost. In earlier practice formaldehyde producers regenerated catalyst themselves by acid treatment followed by alkali- and water-washing, which removed iron and other contaminants. It is now more usual to return the used catalyst to the supplier, who replaces it with fresh catalyst at a moderate cost after giving credit for the spent catalyst.

10.3.2.4. Composition of reaction gases
In the silver-catalysed process the air/methanol mixtures fed to the reactor are always maintained sufficiently rich in methanol to be non-flammable under operating conditions. This is illustrated in Figure 10.16, which shows the flammable limits over a range of temperatures and water dilutions. In the absence of water vapour flammable limits for air/methanol mixtures at 100°C lie between 7.5 and 37% of methanol in air. The effect of temperature is to widen the flammability envelope, the effect of water vapour is to close it.

Chapter 10. Catalytic Oxidations

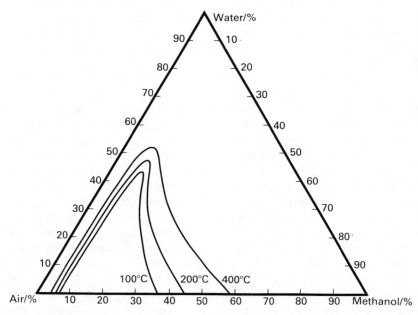

Figure 10.16. Flammable limits for air/methanol/water vapour mixtures at 100, 200 and 400°C at atmospheric pressure.

A knowledge of the flammability characteristics of the air/methanol/water system is essential to the operation and control of formaldehyde production by any method involving methanol oxidation. The silver-catalysed process operates with a methanol-rich mixture, whereas the purely oxidative metal oxide-catalysed process operates with methanol-lean mixtures (Section 10.3.3). A further constraint in this system is the effect of pressure which enhances the flammability of methanol/air mixtures, and further complicates attempts to define precise flammable limits in the air/methanol/water system under operating conditions.[367] Since the silver-catalysed process operates at about 600°C the reaction gases must at some stage become flammable. Consideration of the methanol/air mixture flame speeds shown in Figure 10.17 shows that these become progressively slower for increasingly methanol-rich mixtures, and suggest a mechanism which would contain the reaction at the catalyst surface given forward-flowing reactants. Despite the complexities of this system, operation of the process has proved readily controllable under the conditions normally used for formaldehyde production. The process is protected by the elimination

10.3. Methanol Oxidation

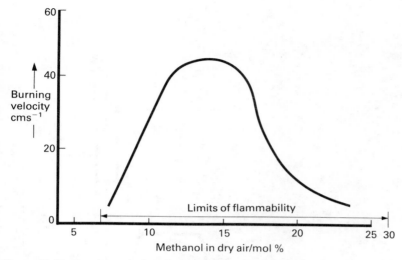

Figure 10.17. Burning velocity of methanol/air mixtures.

of ignition sources when flammable mixtures are expected; this is particularly important when there is no forward flow of reaction gases under shutdown and restart conditions.

10.3.3. The Metal Oxide-catalysed Process

This process is based on the direct oxidation of methanol to formaldehyde using a mixed metal oxide catalyst, that was originally developed in the 1930s and has been increasingly used in commercial production since the 1950s. A typical metal oxide-catalysed process is illustrated in Figure 10.18. The essential feature of the process is the tubular reactor usually containing several thousand tubes, each packed with catalyst. The tubes are contained in a shell in which a suitable heat-transfer medium is circulated in order to control the tube temperature. Formaldehyde is removed from the reacted gases in an absorption train, which normally includes refrigeration as a feature of the exhaust-gas scrubbing. As more attention is paid to environmental problems, the removal of trace levels of formaldehyde from process exhaust gas continues to assume increasing importance. Higher conversions of methanol to formaldehyde are achieved with this process than with the silver-catalysed process, and these may be as high as 97–98%. It is this feature of the process which is its greatest advantage, as products containing typically 50% formaldehyde with less than 1% methanol can be produced directly without distillation.

Chapter 10. Catalytic Oxidations

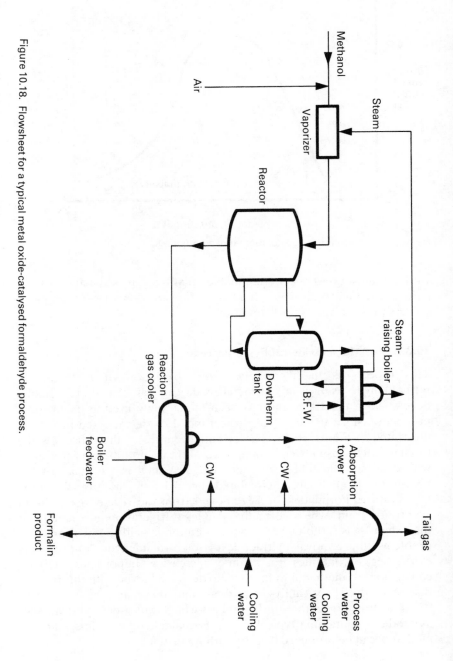

Figure 10.18. Flowsheet for a typical metal oxide-catalysed formaldehyde process.

10.3.3.1. Metal oxide-catalysed reactions

With purely the oxidative reaction a close approach to equilibrium at relatively low temperatures can be achieved, and as a result high conversion of methanol to formaldehyde is a feature of the process. A vital consideration is the removal of reaction heat in order to limit secondary reactions and to avoid overheating of the catalyst. The catalyst is contained in tubes which are cooled by a circulating heat-transfer medium, and the operating temperature is in the region 300–400°C. Under these conditions side-reactions are less important than at the higher temperatures of the silver process, and decomposition of formaldehyde as in equation (23) is considered to be the only significant secondary reaction.[368]

$$CH_2O + \tfrac{1}{2}O_2 \rightarrow CO + H_2O \qquad \triangle H = -236.5 \text{ kJ mol}^{-1} \qquad (23)$$

The most widely used metal oxide catalyst is a mixture of iron and molybdenum oxides produced by precipitation of iron and molybdenum salts which are subsequently calcined and pelleted. The pellets may be cylindrical tablets or Raschig rings, which are preferred because of their relatively low pressure-drop characteristics. The catalytically active component is claimed to be ferric molybdate $Fe_2(MeO_4)_3$. Alternative formulations include tungsten–molybdenum oxides and involve supporting the catalyst on inert refractory materials, together with the addition of other species such as transition metal oxides.

Catalyst lives are generally governed by increasing pressure across the reactor and by decreasing activity. The most significant factor in limiting these adverse effects is temperature control, which is based on the removal of heat from the tubes by the cooling medium. In systems of this type a temperature profile is established over the length of catalyst tube during operation, and a "hot spot" occurs at some point in the catalyst bed. Examination of the catalyst shows that molybdenum oxide, MoO_3, migrates from the hot spot and is deposited on the cooler catalyst. This molybdenum oxide build-up can lead to increasing pressure drop, and is the main cause in the limitation of catalyst lives, which are typically between one and two years. The hot-spot area becomes molybdenum-deficient, and this is associated with loss in overall conversion to formaldehyde. Recent investigation[369] has shown that catalyst at "hot spot" regions can exhibit improved selectivity with lowered production of carbon monoxide. This is thought to be due to the depletion of molybdenum ions in the catalyst surface, due to MoO_3 migration.

10.3.3.2. Composition of reaction gases

The stoichiometric oxygen requirement for the oxidation reaction is 2 : 1 methanol to oxygen; however, in order to avoid forming flammable reaction gas mixtures low methanol content must be maintained, and a methanol content of less than 8% is dictated by this requirement. This excess of air is a limitation of the metal oxide-catalysed process, as the requirements of process equipment and energy for compression of the process air are increased. There is also a proportionate increase in the exhaust gas volume, which increases the duty of the absorbing and scrubbing section; refrigeration may be used to assist in further removal of condensibles before final venting. As with the silver-catalysed process, a realistic definition of the process hazards associated with operation using air/methanol mixtures is difficult. With rising temperatures the flammable limits for methanol-lean mixture falls and progressively it appears that reaction gases must approach an flammable condition at reaction temperatures without propagation of combustion backwards into the feedgas stream. An early development of the process was the recycling of exhaust gas, which acts as a diluent of the reaction gas, thus enabling the requirement for excess air to be reduced.

10.3.4. Future Process Developments

Silver catalyst in various forms has been used commercially for several decades, and the limitation of both conversion and selectivity are largely imposed by process operating conditions rather than inherent catalytic properties. Given this background, the production of a significantly improved catalyst for the oxidative/dehydrogenation route is unlikely. The recent requirement for a commercial product containing more than 50% formaldehyde with a low methanol content necessitates the use of distillation, which imposes increased capital costs and energy consumption on the silver-catalyst technology. However, this disadvantage has now been offset by the recycling of process off-gas to the reactor, enabling the production of higher conversion products which do not require distillation. Similar process constraints apply to the metal oxide-catalysed process, but it is possible that some benefit could be achieved by improved selectivity arising from a better understanding of the role of MoO_3 concentration in the catalyst surface. As stated earlier, plants based on both processes are currently being constructed and operated, and it appears that the formaldehyde producer's choice of process depends on specific plant requirements and preference based on past operating experience. Possible future developments lie in further refinement of the use of recycled reaction off-gas in reaction gas mixtures, and more-refined absorber systems designed for differing product requirements. Improved integration of process heat-recovery systems may lead to lowered energy consumption, but the improvements will only be marginal. The future of formaldehyde production may lie in the attractively simple but practically difficult routes of direct methane or hydrocarbon oxidation, and these reactions are currently receiving considerable academic attention.

10.4. Sulphur Dioxide Oxidation

10.4.1. Introduction

The oxidation of sulphur dioxide was probably the first catalytic process operated on a full commercial scale. World annual production of 100% acid is more than 140 million tonnes, with 60% of this used to produce fertilizers. Among the basic chemical products sulphuric acid ranks first, above ammonia. The sulphuric acid process developed via the earliest "Chamber process" in which the reaction was catalysed by oxides of nitrogen (that is, homogeneous catalysis) to the modern "contact process". Here the process has evolved from the initial supported platinum catalyst to the modern "hybrid"-type catalyst that is a supported liquid-phase catalyst. Only the "contact process" will be considered here, emphasis being on the vanadium-based catalysts.

10.4.2. Thermodynamics

10.4.2.1. Equilibrium calculations

The thermodynamics of the oxidation of sulphur dioxide, as in equation (24), has been reviewed extensively,[370] more-recent data[371] are given in Table 10.2. Dixon and Longfield[370] expressed their equilibrium constant in the form shown in equation (25), with the same results being represented by Pearce[372] as in equation (26).

$$SO_2 + \tfrac{1}{2}O_2 \rightarrow SO_3 \qquad (24)$$

$$\log K_p = (4956/T) - 4.678 \qquad (25)$$

$$\log K_p = (5022/T) - 4.765 \qquad (26)$$

Urbanek and Trela,[373] in their comprehensive review, give the more complicated forms shown in equations (27) and (28).

$$\triangle H = 22034.3 + 5.618T - (10.4575 \times 10^{-3})T^2 + \qquad (27)$$
$$(6.4212 \times 10^{-6})T^{-3} - (1.648 \times 10^{-9})T^{-4}$$

Chapter 10. Catalytic Oxidations

Table 10.2. Thermodynamic properties of SO_2, SO_3, O_2 and data for the reaction $SO_2 + \tfrac{1}{2}O_2 \rightarrow SO_3$

Temp- erature/ K	SO_2			SO_3			O_2	$SO_2 + \tfrac{1}{2}O_2 \rightarrow SO_3$		
	$\Delta H°_f$/kcal mol^{-1}	$C°_p$/cal K^{-1} mol^{-1}	log K_p	$\Delta H°_f$/kcal mol^{-1}	$C°_p$/cal K^{-1} mol^{-1}	log K_p	$C°_p$/cal K^{-1} mol^{-1}	ΔH/kcal mol^{-1}	log K_p	K_p/atm$^{-1/2}$
500	−72.356	11.132	31.436	−96.082	15.082	36.850	7.431	−23.696 (−23.42)	5.414	25940
600	−72.824	11.723	26.143	−96.481	16.075	29.837	7.670	−23.657 (−23.42)	3.739	5843 (4180)
700	−73.206	12.180	22.342	−96.801	16.824	24.801	7.883	−23.595 (−23.27)	2.459	287.7 (257)
800	−86.593	12.532	19.825	−110.111	17.391	21.366	8.063	−23.518 (−23.08)	1.541	34.75 (32)
900	−86.577	12.806	17.197	−110.009	17.823	18.025	8.212	−23.432 (−22.87)	0.828	6.730 (6.47)
1000	−86.553	13.022	15.095	−109.891	18.157	15.354	8.336	−23.339 (−22.06)	0.259	1.816 (1.81)

Figures in parentheses due to Dixon and Longfield.[370]

10.4. Sulphur Dioxide Oxidation

$$\log K_p = (4812.3/T) - 2.8254 \log T + (2.284 \times 10^{-3})T - \\ (7.012 \times 10^{-7})T^2 + (1.197 \times 10^{-10})T^3 + 2.23 \quad (28)$$

Values of equilibrium conversions are obtained by transposing the standard equilibrium equation (29) to equation (30):

$$K_p = P_{SO_3}/P_{SO_2}P_{O_2}^{0.5} \quad (29)$$

$$K_p = [x/(1-x)]\{(100 - [SO_2]x/2)/([O_2] - [SO_2]x/2)\}^{0.5} \quad (30)$$

where x = fractional conversion of SO_2, $[SO_2]$ = initial concentration of SO_2 and $[O_2]$ = initial concentration of O_2. For initial concentrations of, say, 8% SO_2, 12% O_2, 80% N_2 this reduces to equation (31), from which values of x for different values of K_p (i.e. temperature) may be obtained. Values of conversions for selected gas mixtures are given in Table 10.3. The degree of conversion is markedly dependent on temperature, and considerable gains can be made by maintaining the reaction temperature as low as possible.

$$x^3(1 - K_p^2) - x(25 - 5K_p^2) - 7K_p^2 x + 3K_p^2 = 0 \quad (31)$$

Table 10.3. Equilibrium conversions for different sulphur dioxide containing gases at various temperatures

	Conversion/%	
Temperature/K	8% SO$_2$ 12% O$_2$ 80% N$_2$	6% SO$_2$ 10% O$_2$ 84% N$_2$
600	99.94	99.93
700	98.88	98.77
800	91.53	90.72
900	68.22	66.03
1000	37.25	34.95
Temperature/°C		
400	99.5	99.4
450	98.0	97.5
500	94.6	94.0
550	88.6	87.8
600	78.5	76.7
650	61.2	59.0
700	45.0	42.5

10.4.2.2. Application to the contact process

The effect of the strongly exothermic nature of the sulphur dioxide oxidation reaction is demonstrated by the equilibrium conversion diagram[372, 374, 375] shown in Figure 10.19. The curves are for different initial sulphur dioxide and oxygen concentrations, conversion being higher for the lower sulphur dioxide concentration. The diagonal lines, also dependent on composition, represent the adiabatic temperature rise across the catalyst bed. Slopes are directly proportional to the specific heat capacities of the process gases, which remain approximately constant for any degree of conversion. A method of calculating such systems is given by Hougen and Watson.[376] Due to the adiabatic temperature rise it is clear that conversions of only 60–70% are obtainable for a single bed, whereas for a catalyst temperature of 450°C (the actual inlet temperature) a conversion of around 98% would have been expected. In practice this is overcome by using more than one catalyst bed—usually four with inter-stage cooling by heat exchange or by direct air-quenching. Figure 10.20 shows the conversion equilibrium diagram for a typical plant, and typical conversions and temperature rises are given in Table 10.4. Conversions of the order of 98.5% are obtained. This can be further increased if the sulphur trioxide is removed from the gases entering the final catalyst bed. In such double-absorption plants conversions of 99.5% are readily obtained[375, 377] (Figure 10.21).

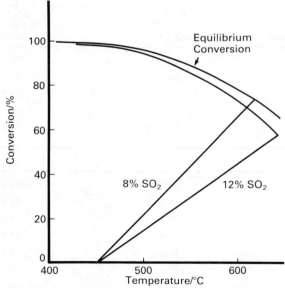

Figure 10.19. Equilibrium conversion diagram for sulphur dioxide oxidation in a single catalyst bed.

10.4. Sulphur Dioxide Oxidation

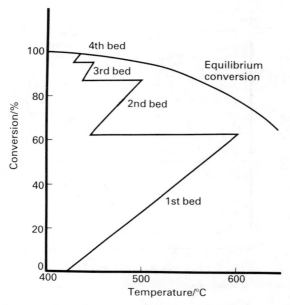

Figure 10.20. Equilibrium conversion diagram for sulphur dioxide oxidation in a four-pass intercooled system.

Table 10.4. Bed temperatures and conversions in a typical four-bed interstage-cooled sulphur dioxide oxidation converter

Catalyst Bed		Total conversion/%	Temperature/°C	Bed conversion/%
Bed 1	In	0	420	65–70
	Out	65–70	600	
Bed 2	In	70	445	18
	Out	88	500	
Bed 3	In	88	435	8
	Out	96	450	
Bed 4	In	96	428	2.5
	Out	98.5	432	

10.4.3. The Contact Process

The original patent for the oxidation of sulphur dioxide by a platinum catalyst was issued to Peregrine Phillips of Bristol in 1831. However, practical application did not follow until about 1870, when the rise in the synthetic dye industry required the availability of a cheap and reliable supply of fuming sulphuric acid. Impetus was added by various political

Chapter 10. Catalytic Oxidations

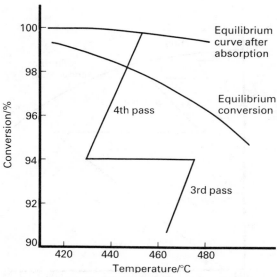

Figure 10.21. Sulphur dioxide oxidation, effect of interstage absorption on equilibrium conversion.

decisions, and until about 1925 many platinum-catalysed production units were built. After this time, due to its expense (despite its recovery value) and susceptibility to poisoning, platinum was gradually replaced by the less active but cheaper vanadium catalyst.

10.4.3.1. Vanadium catalysts

(a) Development of Commercial Catalysts

The use of vanadium catalysts for the oxidation of sulphur dioxide was reported by de Haen in 1900, but the activity of his material was low. Commercial catalysts developed by BASF were introduced around 1920.[378, 379] Many variations of this type of catalyst followed, in particular the "Seldon mass" over which a lengthy court case was fought in 1932,[374, 378] and the Monsanto catalyst[370] which eventually became the most widely used catalyst. The Seldon mass was based on a support produced by the addition of potassium silicate solution to kieselguhr, while the Monsanto catalyst was much simpler. Silica gel was precipitated from potassium silicate with hydrochloric acid in the presence of a solution of ammonium metavanadate and potassium hydroxide. The product was then dried, extruded and calcined.

Subsequently,[380] catalysts were developed containing both sodium and potassium, with increased activities at lower temperatures; for instance 31% conversion (compared with 17% for standard catalysts) at

10.4. Sulphur Dioxide Oxidation

400°C of sulphur dioxide in a gas stream containing 6% sulphur dioxide, 7% oxygen and 87% nitrogen. The typical composition of such a catalyst was: K_2O, 6.8%; Na_2O, 1.6%; V_2O_5, 5.4%; SO_3, 19.2%; SiO_2, 57.2%; Al_2O_3, 1.0%. Other variants had even better low-temperature performance as shown in Table 10.5. After heating a catalyst of this kind at 650°C in reaction gas for 24 hours, to simulate several months of full-scale operation, and then re-testing, the activities remained substantially unchanged. Many tonnes of such catalysts were manufactured and operated successfully in commercial plants. However, after about a year on-line the low-temperature activity declined and became similar to that of the conventional catalyst.

The chemistry of vanadium catalysts and their catalytic behaviour has been extensively investigated, especially in the USSR, and many excellent reviews have been published.[381, 382] Despite all of this work, however, the composition of commercial catalysts has remained relatively unchanged. The vanadium level remains normally between 5 and 9 wt% (as V_2O_5), and the V_2O_5/K_2O ratio between 2.5 and 4.5 with the support either a diatomaceous earth or precipitated silica gel.

Table 10.5. Activities of typical low-temperature sulphur dioxide oxidation catalysts

Temperature/°C	Conversion/%	
	New catalysts	Standard commercial type
380	17	5.5
400	33	11.5
420	54	22
440	72	44
460	79	65.5
480	82.5	75.5
500	83	80.5

(b) The Nature of the Active Phase
It was shown relatively early[383, 384] that the catalytically active species are liquid under reaction conditions, with the molten phase being contained in the pores of the supports. This was confirmed by others,[385, 386] and the marked promotional effect of potassium sulphate was discovered by Boreskov and Pligunov.[387] They found that the addition of silica decreased the activity of V_2O_5 to about 6% of its initial value, while addition of potassium sulphate to pure V_2O_5 lowered the reaction rate at temperatures below 490°C and increased it at higher

temperatures, but the $V_2O_5/K_2SO_4/SiO_2$ catalyst was about 20 times more active than the pure V_2O_5 catalyst over the temperature range 440–500°C. Potassium, sodium and barium vanadates with V_2O_5 were all shown to be more active than V_2O_5 alone.

Between 440 and 600°C the alkali metal sulphate/vanadium pentoxide mixtures in equilibrium with SO_2/SO_3/air mixtures are all liquid.[386] The higher the atomic weight of the alkali metal is, the lower the melting point of the mixture and the smaller the amount of V_2O_5 reduced. With a metal/vanadia ratio of 2.5, the normal pyrosulphate is formed ($M_2S_2O_7$) along with vanadyl sulphate ($VOSO_4$). There is evidence of even higher sulphates with rubidium and caesium. Topsøe and Nielsen[385] concluded that those alkali metals producing sulphates higher than pyrosulphates are the most suitable promoters; K, Rb, Cs and Tl being more effective than Na, Ba and Ag. The degree of reduction of the V_2O_5 became directly associated with the activity of the catalyst,[381] a high degree of reduction being accompanied by low activity. Some confirmation of this is given by the fact that the activity of a used catalyst can often be restored by heating in air. Boreskov and co-workers[388] identified the presence of such compounds as $6K_2O.V_2O_5.12SO_3$, $K_2O.V_2O_5.4SO_3$, $K_2O.V_2O_4.3SO_3$ and $K_2[V_2O_2.(SO_4)_3]$ in melts under reaction conditions and ESR measurements showed the existence of two V^{4+} spectral lines below 480°C, the high-temperature line disappearing below 380°C and the low temperature signal being attributed to a precipitated V^{4+} compound. Later, more-precise work[389] attributed the low-temperature line to a vanadyl sulphate ($VOSO_4$) phase and the high-temperature one to an oxygen defect surrounded by four vanadium ions in a V_2O_5 phase.

Similar examination of commercial catalysts (that is the ternary $V_2O_5/K_2SO_4/SiO_2$ systems) showed[390] that some reaction between the active layer and the support can occur, silica gel crystallizing to the cristobalite form. The rate at which this occurs depends on the surrounding gas composition and the potassium/vanadium ratio in the active layer. More recent work on the composition of the active phases has been carried out by Hansen et al.[391] and Villadsen.[392] Further evidence that activity depends on the method of manufacture and thermal treatment, and that interaction between support and melt occurs, has been furnished by Mastikhin et al.,[393] while Norwinska[394] has reported optimum compositions, K_2O/V_2O_5 ratios and V^{5+}/V^{4+} ratio for silica–alumina supported catalysts.

10.4.3.2. The modern sulphuric acid plant
One of the main problems facing designers of old sulphuric acid plants was the removal of heat to keep the temperature as low as possible, to

10.4. Sulphur Dioxide Oxidation

permit higher conversion of sulphur dioxide. With the advent of vanadium catalysts, two converters were employed, about one-third of the catalyst being placed in the first converter, where the temperature rose from ~420 to ~590°C and some 80% of the conversion was carried out. After cooling in a heat exchanger to about 425°C the gases passed to the second converter where, due to the longer contact time and much reduced temperature rise, an overall conversion of 96–97% was obtained.

The large modern plants, producing up to 2000 tonnes day^{-1} of acid have up to four or five separate layers of catalyst, totalling about 200–250 tonnes. They have cooling between each bed giving increased conversion in each bed, and more accurate temperature control. Typical data for such a plant are given in Table 10.4. Conversions to 99.5% and above can be achieved by removing much of the sulphur trioxide from the gases entering the last catalyst bed, and in the latest double-absorption plants sulphur trioxide is also removed in the earlier stages, resulting in practically complete conversion. A simplified line diagram of such a plant is shown in Figure 10.22, details are given in various reviews.[374, 378, 395–397] Gas containing between 9 and 11% sulphur dioxide and about 10% oxygen is dried, freed from dust, cooled to around 420°C and passed to the converter, with intermediate cooling between the beds. In the conventional plant the gases after the fourth pass are scrubbed with sulphuric acid to remove sulphur trioxide, and then passed through demisters to atmosphere. However, in the double-absorption process the gases from the "third bed" pass via a cold gas heat exchanger and an economizer to a primary or interpass absorption tower where the bulk of the sulphur trioxide is removed by absorption in a circulating solution of 98–99% sulphuric acid. The issuing gases are then cooled and demisted, before being reheated to ~420°C and passed through the final catalyst bed, and thence to the final absorption tower.

The life of catalysts in modern plants may be as long as 20 years, typically at least 5 years for first "passes" (bed) and at least 10–15 years for third and fourth "passes". There is a gradual deterioration of activity with time, some of which may be recovered by heating in air, but regular sieving of the catalyst, usually done annually for the first beds, to remove dust is necessary to maintain efficient operation.

10.4.4. Mechanism and Kinetics

The mechanism and kinetics of the sulphur dioxide oxidation reaction over platinum catalysts is adequately reviewed elsewhere[370, 374] and here discussion is restricted to vanadium catalysts. As would be expected

Chapter 10. Catalytic Oxidations

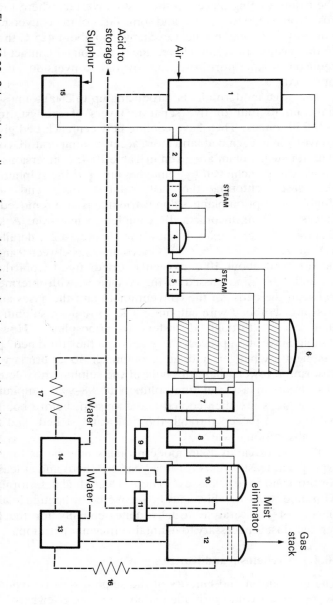

Figure 10.22. Schematic flow diagram for a typical sulphur-burning double-absorption sulphuric acid plant.
1, Drying tower; 2, sulphur burner; 3, waste heat boiler; 4, hot-gas filter; 5, waste heat boiler; 6, four-pass converter; 7, hot interpass heat exchanger; 8, cold interpass heat exchanger; 9, secondary economizer; 10, interpass absorption tower; 11, economizer superheater; 12, final absorption tower; 13, final absorption tower circulating tank; 14, drier and interpass absorption tower circulating tank; 15, sulphur melter; 16, acid cooler; 17, acid cooler.

from the complexity of the chemistry of the potassium sulphate/silica/vanadia system, interpretation of the mechanism of the reaction and derivation of suitable kinetic equations is difficult. Earlier work did not take into account the liquid nature of the catalyst, and kinetics were based on heterogeneous catalytic models. Later it became evident that the nature of the support, particularly its micromeritic properties, was extremely important and had a marked effect on catalyst performance.[370, 381, 382, 398, 399] This early work has recently been further reviewed and extended to include the mathematical modelling of catalyst beds and the optimization of adiabatic reactors.[373]

The first equation to take into account the reduction of V^{5+} to V^{4+} was due to Mars and Maessen,[400] equation (32), and this has been extensively used in later work.[382, 401, 402]

$$\text{rate} = kKp_{O_2}p_{SO_2}p_{SO_3}^{-1}/[1 + (Kp_{SO_2}p_{SO_3}^{-1})^{0.5}]^2 \quad (32)$$
$$\text{where } K = (V^{4+}/V^{5+})^2 P_{SO_3}/P_{SO_2}$$

Another well-used equation was that due to Boreskov et al.,[403] equation (33), for which Boreskov has given further support in a recent review.[404] He represents the reaction in five stages, involving such vanadium(V) sulphate complexes as $V_2(SO_3)_n SO_2$, $V_2(SO_3)_{n-1} 2SO_2$ and $V_2(SO_3)_{n+1}$, and the vanadium(IV) $V_2(SO_3)_n SO_2$. The derived kinetic equation is then brought into agreement with the earlier experimental equation of Boreskov et al.[403] by suitable manipulation of constants as indicated

$$\text{rate} = kp_{O_2}p_{SO_2}(p_{SO_2} + Ap_{SO_3})^{-1}[1 - (p_{SO_3}/Kp_{SO_2}p_{O_2}^{0.5})^2] \quad (33)$$

in equation (33). It is concluded that the rate-determining step for sulphur dioxide oxidation over vanadium catalysts is the oxygen bonding to the catalyst. The concentration of sulphur dioxide influences the amount of V^{4+} in the catalyst and hence activity.

The effects of mass and heat transfer in beds of vanadium catalysts have been extensively reviewed,[373] and some 29 earlier equations considered. However, the differences between them make them inapplicable as general rate equations. Most experiments were carried out under conditions of some diffusional effects, and no attempts have been made to include mass and heat transport effects into rate expressions. A further fact which has hitherto been ignored is the effect of inert gases which form the major part of the process gas (predominantly nitrogen). There is little doubt that the reaction mechanism is affected, but precise reasons for this are not known.[405]

The various diffusional effects may be given the following order of descending importance: internal mass transport, external mass and heat transport, internal heat transport. Improvement of catalyst activities will inevitably make these diffusional processes increasingly important.

Chapter 10. Catalytic Oxidations

In the design of fixed-bed reactors the most outstanding difficulties still existing are the lack of means of predicting linear gas velocity distributions above and within the catalyst bed, and the lack of data on the effect of axial heat distribution on the operation of the catalyst bed. A review of previous attempts to correlate data from laboratory experiments, pilot plant and full commercial plant concludes that agreement, although good for laboratory data, fails in the larger scale-up. Arbitrary excess factors are often used to convert calculated catalyst loadings to actual large-scale reactor loadings. The general effects on reactivity of various parameters have been taken from several sources and given in Figures 10.23–10.25.

Recent forms of empirical equation[375, 406] adopted the Boreskov-type kinetic equation for the liquid-phase reaction, allied with the Villadsen and Livbjerg[382, 401, 402] cluster model for the distribution of liquid in the pores. Thiele's modulus,[407] in a modified form, is applied to enable the gas phase and the liquid phase efficiency to be treated separately as in equation (34) in which r = overall reaction rate (mol SO_2 (cm^3 catalyst)$^{-1}$

$$r = Z_1 \eta_L \eta_G (1 - \xi^2) r_L \qquad (34)$$

s^{-1}); r_L = reaction rate in liquid; η_L = efficiency factor in liquid phase that is a function of Z_2, p_{O_2}, p_{SO_2}, u, T; η_G = efficiency factor in gas phase that is a function of Z_3, Z_4, p_{O_2}, p_{SO_2}, u, η_L, T; $(1 - \xi^2)$ = distance from equilibrium. The catalyst parameters Z_1–Z_4 are defined as follows:

$$Z_1 = V_p \alpha \rho_s / \rho_c, \qquad Z_2 = \sqrt{\alpha / V_p \rho_c}$$

$$Z_3 = f(R_p), \qquad Z_4 = V_c / S_c \sqrt{\alpha(1 - \alpha)}$$

$$\xi = 1/q \sqrt{p_{O_2}} K_p, \qquad q = p_{SO_2}/p_{SO_3}$$

V_p = pore volume of support (cm^3 g^{-1})

α = fraction of support pore volume filled with liquid

ρ_s, ρ_c = density of support and catalyst, respectively

R_p = mean pore radius (cm)

V_c/S_c = ratio of particle volume to geometrical surface of the particles.

10.4.5. Catalyst Poisoning

The lives of sulphur dioxide oxidation catalysts, both platinum and vanadium types, but particularly the platinum catalyst, are dependent on the purity of the reaction gases. With efficient gas purification systems platinum catalysts could last up to 10 years, although normally

10.4. Sulphur Dioxide Oxidation

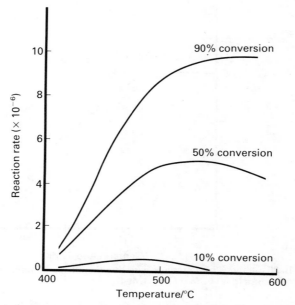

Figure 10.23. Variation of reaction rate with temperature and conversion (data from reference 373) for sulphur dioxide oxidation.

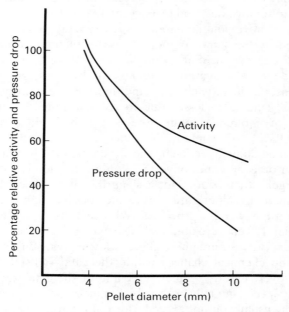

Figure 10.24. Variation of activity and pressure drop with pellet diameter (data from reference 375).

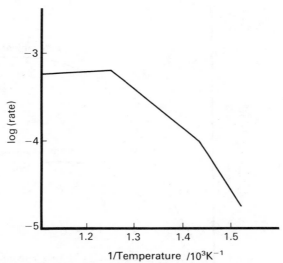

Figure 10.25. Activity of vanadium sulphur dioxide oxidation catalyst—variation with temperature (data from reference 404).

their life was much less than this. Vanadium catalysts, on the other hand, may have service lives of 20 years or more, requiring only regular sieving to remove dust and broken catalyst pellets, and occasional firing in air to remove volatile impurities. In general, vanadium catalysts are poisoned by the same compounds as are platinum catalysts, but to a much lesser extent. For example, 80 000 times as much arsenic is required to deactivate vanadium catalysts as platinum types.[408] Halogens, particularly chlorine and fluorine, are injurious to vanadium catalysts, leading to losses due to volatilization.[409] With the present supply of molten sulphur, the low-carbon grade having a low (0.03–0.05%) ash content, it is necessary to install either sulphur filtration equipment, reducing the ash content to less than 0.002%, or a gas filter preceding the converter. Wet purification is not necessary and has largely died out. Hot-gas purification where the gases are maintained at temperatures above the condensation point is now only used in a few wet-gas processes. With such treatment catalyst lives of more than 17 years are obtained. The main reasons for loss of activity of vanadium catalysts are physical breakdown, giving dust which blinds the beds, and chemical changes within the catalyst itself. The former is readily overcome by regular screening of the catalyst. First beds are usually sieved annually, other beds much less frequently. Losses in screening usually do not exceed 5–6% of the total catalyst charge. New ring-shaped pellets now being introduced, primarily to reduce pressure drop, have also reduced these losses. Much of the vanadium losses may well be due to migration of vanadium melts from the mass into the attrited dust, which is removed during sieving. Loss of activity after continuous operation at high temperatures has been partly attributed to crystallization of the silica support.[410] An extremely useful short, up-to-date review on thermal, mechanical and chemical characterization is given by Donovan *et al*.[382]

10.4.6. Disposal of Used Vanadium Catalysts

A suitable method of recovery of vanadium from spent catalyst has not yet been developed.[382] Factors against this are the relative cheapness of vanadium and the long life of the modern catalyst, there being little available material for treatment in any one area at a time. It must be borne in mind that vanadium compounds act as irritants to the conjunctive and respiratory tract. Consequently the inhalation of catalyst dust is extremely unpleasant and should be avoided with respirators being worn at all times when handling the catalyst (see Chapter 3).

10.4.7. Possible Further Developments

It has been pointed out[375] that, even with a drop to 60% of the activity of fresh catalyst, conversion can be maintained in a modern plant at 99.2% with suitable temperature adjustments. Even so, considerable effort continues towards the development of low-temperature catalysts although, to date, no major breakthrough has been achieved. However, improvements in strength and resistance to abrasion have been made which, along with the development of larger pellets, have markedly reduced the downtime needed for catalyst sieving. It should, however, be borne in mind that improvement in activity and increase in pellet size will tend to increase diffusional effects, so limiting catalyst effectiveness, and some compromise will have to be made.

Normal operating pressures are in the range 1.2–1.5 bar, increasing this should be advantageous both thermodynamically and kinetically. A medium (570 tonnes day^{-1}) double-absorption plant operating at 4–8 bar has been operated in France,[411] but so far only small units have been considered and the economics of high-pressure plants are still under review. CIL[412] in Canada has designed a pressurized single absorption plant operating at 7 bar with a three-bed system, and upwards of 99% conversion can be achieved. Any improvements in pressure drop which could accrue from such a change are being obtained by increasing the size of the catalyst pellets from 5.6 to 7.9 mm.[413] This

Chapter 10. Catalytic Oxidations

reduces the pressure drop by about 30% without affecting activity. Finally, while refinements to conventional fixed-bed systems are being continually evaluated, other processes are being devised. In particular, the development of fluid-bed processes giving much improved temperature control with reduced diffusional limitations is being actively followed, particularly in the USSR.[414] A prototype plant is being operated by Bayer,[415] using vanadium catalysts of particle size 0.3–1.0 mm. Conversions of 90% and above are achieved in one bed at a temperature of 462–502°C.

Appendix 1.

Further Reading

■ Single-volume Works

The books listed below differ in the various ways in which the subject is approached and in the depth of treatment: a guide is given in Table A1.1.

Table A1.1. Guide to further reading: single-volume works

Topics	Level of treatment	
	Introductory	Advanced
General heterogeneous catalysis	1, 3, 4, 6, 8, 9, 10, 14, 21, 23, 25a, 26, 29, 35, 36	1, 2, 3, 5, 8, 12, 13, 18, 20, 21, 22, 24, 25, 25a, 26, 29, 34, 35
Industrial catalysis	1, 3, 4, 6, 9, 10, 11, 23, 25a, 29, 33, 35	1, 3, 11, 12, 18, 19, 20, 21, 23, 25a, 29, 33, 34, 35, 37, 38
Surface science related to catalysis	2, 10, 14, 27, 31	2, 13, 27, 31, 34
Chemical aspects of catalysis	1, 2, 3, 4, 6, 8, 9, 10, 14, 16, 21, 23, 25a, 26, 35, 36	1, 2, 3, 5, 8, 12, 13, 15, 16, 17, 18, 19, 20, 21, 22, 23, 24, 25a, 26, 34, 35
Chemical engineering aspects of catalysis	1, 7, 8, 11, 21, 28, 29, 35	1, 7, 8, 11, 17, 19, 21, 29, 35
Catalyst manufacture	23, 28, 29, 32	28, 29, 30, 32

NOTES

1. The table is intended only as a guide to readers, not as a rigorous classification.

2. All of the topics listed are covered at an advanced level in serial and multi-volume works.

3. Most of the texts deal with metal, oxide and sulphide catalysts. Where the coverage is more limited, this is obvious from the title of the book.

1. *Catalyst Handbook*, Wolfe Medical, London (1970).

2. J. R. Anderson, *Structure of Metallic Catalysts*, Academic Press, New York (1975).

519

Appendices

3. J. R. Anderson and K. C. Pratt, *Introduction to Characterisation and Testing of Catalysts*, Academic Press, Sydney (1985).

4. P. G. Ashmore, *Catalysis and the Inhibition of Chemical Reactions*, Butterworth, London (1963).

5. G. C. Bond, *Catalysis by Metals*, Academic Press, London (1962).

6. G. C. Bond, *Heterogeneous Catalysis*, Second edition, Elsevier, Amsterdam (1985).

7. M. Boudart, *Kinetics of Chemical Processes*, Prentice–Hall, Englewood Cliffs, New Jersey (1968).

8. M. Boudart and G. Djegamariadassou, *Kinetics of Heterogeneous Catalytic Reactions*, (French) Masson, Paris (1982); (English) Princeton University Press, Princeton (1984).

9. I. M. Campbell, *Biomass, Catalysts and Liquid Fuels*, Technomic, Lancaster, USA (1983).

10. I. M. Campbell, *Catalysis at Surfaces*, Chapman and Hall, London (1988).

11. J. J. Carberry, *Chemical and Catalytic Reaction Engineering*, McGraw-Hill, New York (1976).

12. F. Delannay (ed.), *Characterisation of Heterogeneous Catalysts*, Dekker, New York (1984).

13. E. Drauglis and R. I. Jaffee (eds), *The Physical Basis for Heterogeneous Catalysis*, Plenum Press, New York (1975).

14. R. P. H. Gasser, *An Introduction to Chemisorption and Catalysis by Metals*, OUP, Oxford (1985).

15. J. E. Germain, *Catalytic Conversion of Hydrocarbons*, Academic Press, New York (1969).

16. J. Happel, *Isotopic Assessment of Heterogeneous Catalysis*, Academic Press, New York (1986).

17. L. L. Hegedus, *Catalyst Poisoning*, Dekker, New York (1984).

18. L. L. Hegedus (ed.), *Catalyst Design: Progress and Perspectives*, John Wiley, New York (1987).

19. R. Hughes, *Deactivation of Catalysts*, Academic Press, London (1984).

20. B. Imelik, G. A. Martin and A. J. Renouprez, *Catalysis by Metals: Fundamental and Industrial Aspects*, (French) CNRS, Paris (1984).

21. J. R. Jennings (ed.), *Catalysis, Critical Reports in Applied Chemistry, No. 12*, Society of Chemical Industry, London (1985).

22. O. V. Krylov, *Catalysis by Non-Metals*, Academic Press, New York (1970).

23. R. Pearce and W. R. Paterson, *Catalysis and Chemical Processes*, Blackie, Glasgow (1981).

24. H. Pines, *The Chemistry of Catalytic Hydrocarbon Conversions*, Academic Press, New York (1981).

25. T. S. R. Rao (ed.), *Advances in Catalysis Science and Technology*, John Wiley, New York (1985).

25a. J. T. Richardson, *Principles of Catalysis Development*, Plenum Press, New York (1989).

26. E. K. Rideal, *Concepts in Catalysis*, Academic Press, London (1968).

27. M. W. Roberts and C. S. McKee, *Chemistry of the Metal–Gas Interface*, Clarendon Press, Oxford (1978).

28. C. N. Satterfield, *Mass Transfer in Heterogeneous Catalysis*, The MIT Press, Cambridge, Massachusetts (1970).

29. C. N. Satterfield, *Heterogeneous Catalysis in Practice*, McGraw-Hill, New York (1980).

30. M. Sittig, *Handbook of Catalyst Manufacture*, Noyes, New Jersey (1978).

31. G. A. Somorjai, *Chemistry in Two Dimensions: Surfaces*, Cornell University Press, Ithaca, New York (1981).

32. A. B. Stiles, *Catalyst Manufacture: Laboratory and Commercial Preparations*, Dekker, New York (1983); *Catalyst Supports and Supported Catalysts*, Butterworths, Boston (1987).

33. C. L. Thomas, *Catalytic Processes and Proven Catalysts*, Academic Press, New York (1970).

34. J. M. Thomas and R. M. Lambert, *Characterisation of Catalysts*, John Wiley, Chichester (1980).

35. J. M. Thomas and W. J. Thomas, *Introduction to the Principles of Heterogeneous Catalysis*, Academic Press, London (1967). (Second edition in preparation.)

36. S. J. Thomson and G. Webb, *Heterogeneous Catalysis*, Oliver and Boyd, Edinburgh (1968).

37. C. A. Vancini, *Synthesis of Ammonia*, CRC/Macmillan, New York (1971).

38. O. Weisser and L. Landa, *Sulphide Catalysts. Their Properties and Applications*, Pergamon Press, Oxford (1973).

Serial Publications and Multi-volume Works

1. *Advances in Catalysis*, Academic Press. Reviews, Volume 1 1948.

2. *Applied Industrial Catalysis* (B. E. Leach, ed.), Academic Press. Treatise in progress, Volume 1 1983.

3. *Catalysis Reviews—Science and Engineering*, Marcel Dekker. Reviews, Volume 1 1968.

4. *Specialist Periodical Reports: Catalysis*, Royal Society of Chemistry. Reviews, Volume 1 1977.

5. *Catalysis* (P. H. Emmett, ed.), Reinhold. Treatise, 7 Volumes 1954–1960.

6. *Catalysis: Science and Technology* (M. Boudart and J. R. Anderson, eds), Springer. Treatise in progress, Volume 1 1981.

7. *The Chemical Physics of Solid Surfaces and Heterogeneous Catalysis* (D. A. King and D. P. Woodruffe, eds) Elsevier. Treatise in progress, Volume 1 1981.

8. *International Congress on Catalysis*, various publishers. Four-yearly intervals, first congress 1956.

9. *Studies in Surface Science and Catalysis*, Elsevier. Mostly conference proceedings, Volume 1 1976.

10. *Fundamental and Applied Catalysis*, Plenum Press, First Volume 1989.

Specialist Journals

1. *Journal of Catalysis.*

2. *Applied Catalysis.*

3. *Kinetika i Kataliz [Kinetics and Catalysis].*

4. *Journal of Molecular Catalysis.*

5. *Reaction Kinetics and Catalysis Letters.*

6. *Catalysis Today.*

7. *Catalysis Letters.*

General Journals

Papers on catalysis and related topics appear frequently in general physical chemistry and chemical engineering journals. Some other specialist journals also contain relevant material.

1. *Journal of the American Chemical Society.*

2. *Journal of Physical Chemistry.*

3. *Journal of Chemical Physics.*

4. *Journal of the Chemical Society, Faraday Transactions I.*

5. *Industrial and Engineering Chemistry Research.*

6. *Chemical Engineering Science.*

7. *American Institute of Chemical Engineers Journal (AIChE J).*

8. *Surface Science.*

9. *Applications of Surface Science.*

10. *Journal of Colloid and Interfacial Science.*

11. *Langmuir.*

12. *Zeolites.*

13. *Carbon.*

14. *Oil and Gas Journal.*

15. *Hydrocarbon Processing.*

16. *Preprints of the Petroleum Division of the American Chemical Society.*

Appendix 2.

Numerical Examples of the Use of Equations Derived in Chapter 2

Example 1. Optimum Voidage in Catalyst Bed

Cost of vessel = £1400 $(V_c + D^3)$
so, from equation (3), $C_v = C_d$ = £1400 m^{-3}
gas rate $m = 38$ kg s^{-1}
gas density $\rho = 12$ kg m^{-3}
design catalyst pressure drop is twice ideal, so $f = 2$
capitalized cost of pressure drop C_p = £300 000 bar^{-1} = £3 Pa^{-1}
equivalent diameter of pellet $d_e = 0.47$ cm $= 0.0047$ m
solid volume of catalyst $V_s = 44$ m^3

From equation (9), $a = 3.61 \times 2 \times 38^2/12 = 869$
From equation (13),
$$n = (1.89/1400)(896 \times 3/0.0047)^{1/3}(1400/44)^{2/3} = 1.11$$
From equation (12),
$$\text{optimum voidage} = \sqrt{1.11}/(1 + \sqrt{1.11}) = 0.51$$

Example 2. Optimum Catalyst Pellet Size

Voidage $e = 0.37$
catalyst cost = £5000 m^{-3}
catalyst life = 3 years
capital charge factor $\times$ $^1/_3$ year^{-1}
volume of 100% effective catalyst $V_a = 40$ m^3
Theile modulus $\phi = 270 \times d_e$

From equation (15), $b = 270$ m^{-1}
From equation (18), $c_{cat} = 5000/(3 \times 1/3) =$ £5000 m^{-3}
From equation (20),
$$m = [1.89(1 - 0.37)/(1400 + 5000)0.37] \times (1400/40)^{2/3}$$
$$\times (3 \times 869 \times 270)^{1/3}$$
$$= 0.48$$

Appendices

From equation *(21)*, approximate optimum pellet diameter
$$= (1/270)(2.5 \times 0.48)^{3/8} = 0.0040 \text{ m}$$
From equation *(22)*, optimum pellet diameter
$$= (1/270)(0.48/[0.4 + 0.0216(270 \times 0.0040)^2])^{3/8}$$
$$= 0.0039 \text{ m}$$

■ Example 3. Design Conversion of Reactor

As the design carbon monoxide slip from a LT shift reactor increases by 0.1% (dry basis), the capital cost of the plant increases by £150 000 and the running cost increases by £50 000 year^{-1}.

Capital charge factor = $^1/_3$ year^{-1}
therefore, in equation *(27)*, $C_a = [150\,000 + (3 \times 50\,000)]/0.1$
$$= £3 \text{ million } (\%CO)^{-1}$$
as in examples 1 and 2, $C_v = £1400 \text{ m}^{-3}$, $C_{cat} = £5000 \text{ m}^{-3}$
volume flow rate of gas $Q = 190\,000 \text{ N m}^3 \text{ h}^{-1}$
exit temperature of catalyst bed = 235°C
steam/dry gas ratio in LT shift = 0.33

From equation *(54)* and Table 2.5,
rate constant $k_r = \exp\{22.84 - [6600/(235 + 237)]\} = 18\,900 \text{ N m}^3 \text{ h}^{-1} \text{ m}^{-3}$

The steam/carbon ratio in the reformer is 2.75, the H_2/N_2 ratio in gas exit the methanator is 5 : 2 and the methane level in the gas exit the methanator is 1.5%. From Figure 2.17, the equilibrium level of carbon monoxide at the exit of the LT shift is:

$$A_e = 0.39 \times 0.9 \times 0.96 = 0.337\%$$

From equation *(30)*, optimum concentration of CO
$$= 0.337 + [190\,000 \times (1400 + 5000)/18\,900 \times 3 \times 10^6] = 0.358$$

■ Example 4. Calculation of Catalyst Volume

Using the data from Example 3 and with a concentration of carbon monoxide at the inlet of the LT shift of 3.5% (dry), from equation *(36)*, the active catalyst volume needed is:

$$V = (190\,000/18\,900) \ln[(3.5 - 0.337)/(0.358 - 0.337)] = 50 \text{ m}^3$$

Extra volume will be added to allow for catalyst die-off.

Example 5. Optimum Catalyst Vessel Diameter

Bulk catalyst volume $V_b = 70 \text{ m}^3$
voidage $e = 0.37$
other data as for Example 1

From equation (55), optimum vessel diameter D
$= [2 \times 3 \times 869 \times 70 \times (1 - 0.37)/1400 \times 0.37^3 \times 0.0047]^{1/9} = 4.5 \text{ m}$

Example 6. Catalyst in Vessel Dished Ends

Data as for Example 3

From equation (57), the optimum diameter that catalyst should be packed to in the dished ends D_e is:

$$4.5[2 \times 70/4.5^3 (+1)]^{-1/6} = 3.8 \text{ m}$$

Appendix 3.

ICI Catalysts for the Production of Hydrogen, Ammonia and Methanol

Process operation	Catalyst	Composition	Duty
Hydrode-sulphurization	41-6 482	Cobalt oxide/molybdenum oxide on alumina	Hydrodesulphurization for sulphur removal from hydrocarbon feedstock
	502 61-1	Nickel oxide/molybdenum oxide on alumina	
Chlorine guard	59-3	Modified alumina	Removal of chlorides from gaseous feedstocks
Desulphurization	7-1 7-2	Impregnated and activated carbon	Desulphurization of natural gas and light hydrocarbons at ambient temperatures
	32-4 75-1	Zinc oxide	Removal of hydrogen sulphide and reactive organic sulphur compounds from gas streams
Primary steam reforming	23-1 23-3 23-4 57-3 57-4	Nickel oxide supported on alumina Nickel oxide supported on calcium aluminate	Primary steam reforming of natural gas and light hydrocarbons
	25-3 25-4	Nickel oxide supported on calcium aluminate promoted with low level of potash	Primary steam reforming of natural gas-containing heavier hydrocarbons
	46-8 46-9	Nickel oxide supported on calcium aluminate promoted with low level of potash	Primary steam reforming of hydrocarbons up to C_4 feeds
	46-1 46-4	Nickel oxide on refractory support promoted with potash Nickel oxide supported on calcium aluminate	Primary steam reforming of naphtha

3. ICI Catalysts

Secondary steam reforming	23-2	Nickel oxide supported on alumina	Secondary reforming
	54-3	Nickel oxide supported on alumina	
	54-4	Nickel oxide supported on calcium aluminate	
CO conversion	15-4	Iron oxide/	High temperature CO shift
	15-5	chromium oxide	
	71-1		
	71-2		
	52-8	Copper oxide/zinc oxide/	Low temperature CO shift
	53-1	alumina	
	49-2	Cobalt oxide/molybdenum oxide on alumina	CO shift reaction under sulphiding conditions
Methanation	11-3	Supported nickel	Removal of traces of CO and CO_2 by methanation
	11-4	oxide	
Ammonia synthesis	35-4	Promoted iron oxide	Ammonia synthesis
	35-8	Pre-reduced promoted iron oxide	
Methanol synthesis	51-2	Supported copper oxide	Methanol synthesis

529

Appendix 4.

Pigtail Nipping

In the early-1960s ICI experienced problems with the premature failure of reformer tube welds. Each time this occurred it was necessary to shut down the reformer to isolate the failed tube. This was an expensive exercise because it resulted in a period of lost production. To overcome this type of problem ICI developed a mechanical device to isolate the inlet and exit pigtails by squeezing them flat. Hence, the device is known as a pigtail nipper.

Two types of pigtail nipper are now available to cope with the wide range of pigtail sizes and spacings in reformer plants. The smaller (20 ton) nipper is shown in Figure A4.1, and the more powerful (40 ton) nipper in Figure A4.2. More details of these nippers are available from ICI Catalysts at Billingham.

Further information can be obtained from:

The Catalysts & Technology Licensing Manager
ICI Chemicals & Polymers Ltd
P.O. Box 1
Billingham
Cleveland TS23 1LB
England

4. Pigtail Nipping

Figure A4.1. Photograph of 20 ton pigtail nipper.

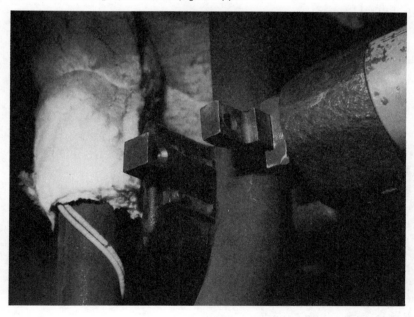

Figure A4.2. View of 40 ton pigtail nipper located on exit pigtail before pressure is applied to squeeze it flat.

Appendix 5.

ICI Technical Publications

ICI Catalysts has published a series of technical papers for the operators of ammonia, methanol and hydrogen plants. Although in the main they discuss the use and performance of catalysts in these plants, they may also consider the operation and maintenance of the plants themselves, as well as analytical methods for raw materials and process gas streams. ICI has arranged many symposia in many parts of the world for invited participants drawn from its customers. These conferences provide a unique forum for plant operators to discuss among themselves and with ICI operations staff all aspects of catalyst performance, plant operation and plant maintenance. Some of the papers presented at these gatherings are still available in limited numbers. Details of these, and copies of the freely available technical publications listed below, can be obtained from:

The Manager Catalysts & Technology Licensing
ICI Chemicals & Polymers Ltd
PO Box 1
Billingham
Cleveland TS23 1LB
England

ICI Catalysts Technical Papers

Paper 1 Catalysts used in ammonia production.

Paper 2 Purification of hydrocarbon feedstocks.

Paper 3 Steam hydrocarbon reforming.

Paper 4 Carbon monoxide conversion.

Paper 5 Ammonia synthesis catalyst.

Paper 6 The care of ammonia plant catalysts.

Paper 7 Monitoring catalyst performance.

Paper 8 Diagnostic maintenance techniques.

5. ICI Technical Publications

Paper 9 Experience with guards for low temperature shift catalysts and extended catalyst life.

Paper 10 Catalysts in ammonia plants—a replacement strategy.

Paper 11 Catalysts in industry—the steam reforming of hydrocarbons.

Paper 12 The development of copper-based catalysts for methanol synthesis and for water gas shift.

Paper 13 A comparison of laboratory and industrial reactors for heterogeneous catalysts.

Paper 14 Steam hydrocarbon reforming—processes, catalysts and feedstocks.

Paper 15 Carbon formation in industrial reactors.

Paper 16 Catalysts in coal-based ammonia plants.

Paper 17 Development in catalysts for ammonia plants.

Paper 18 Catalysts for the steam reforming of hydrocarbons.

Paper 19 Modern control rooms and associated instrumentation.

Paper 20 Catalysts—a recipe for longer life.

Paper 21 The role of promoters in the ammonia synthesis catalyst.

Paper 22 Steam reforming 1981.

Paper 23 Operation of low temperature shift catalyst.

Paper 24 Formulation and operation of methanation catalysts.

Paper 25 Vessel inspection procedure and philosophy.

Paper 26 Catalysts for the production of ammonia.

Paper 27 Methanol synthesis.

Paper 28 Decommissioning, inspection and commissioning 10 000 tons capacity atmospheric ammonia storage tank.

Paper 29 The role of catalysts in the operation of modern ammonia plants.

Paper 30 Optimising the ICI low-pressure methanol process.

Paper 31 Designing for safety and reliability in the chemical industry.

Paper 32 Stress corrosion cracking of steels in ammonia.

Appendices

Paper 33 Conversion from naphtha to natural gas—mechanical equipment changes.

Paper 34 Conversion from naphtha to natural gas—maintenance aspects.

Paper 35 Ammonia plant operations—comparison of natural gas and naphtha.

Paper 36 Recent operating & design experience on a modern gas based ammonia plant.

Paper 37 Feedstock purification catalysts.

Paper 38 Steam hydrocarbon reforming.

Paper 39 Low temperature carbon monoxide conversion.

Paper 40 An energy audit procedure for large single steam plants.

Paper 41 Radioisotope techniques for problem-solving in the chemical industry.

Paper 42 Minimum shut-down—maximum savings.

Paper 43 A new low energy ammonia process concept.

Paper 44 Reformer tube performance—analysis and prediction.

Paper 45 Factors affecting the efficiency of desulphurisation with zinc oxide.

Paper 46 Some problems handling naphtha on ammonia plants.

Paper 47 Gas desulphurisation—the consequence of moving the process offshore.

Paper 48 Steam reforming 1985.

Paper 49 Improving performance in shut-down maintenance activities.

Paper 50 Maintenance engineering: contribution to plant performance.

Paper 51 Low temperature shift catalyst operation.

Paper 52 Getting back on stream.

Paper 53 The effect of catalyst charging on catalyst life.

Paper 54 Computers for profit.

Paper 55 Energy management in large scale chemical complexes.

5. ICI Technical Publications

Paper 56 Feedstock purification for modern ammonia, methanol and hydrogen plants.

Paper 57 The desulphurisation of North Sea gas with zinc oxide absorbent.

Paper 58 Steam reforming catalysts—the importance of choosing the correct support.

Paper 59 Factors affecting the efficiency of desulphurisation with zinc oxide.

Paper 60 Problem solving on chemical process plant using radioisotopes.

Paper 61 AMV—the new ICI ammonia process.

Paper 62 Some advanced techniques in on-line maintenance.

Paper 63 The development of LTS catalysts for the modern ammonia plant.

Paper 64 Living with defects—the uses, limitations and potential of fracture mechanics.

ICI Catalysts
Analytical Methods Papers

1. Natural Gas

1.1 Determination of nitrogen and ethane in natural gas by gas chromatography.
1.2 Determination of hydrocarbons.
1.3 Determination of carbon dioxide, hydrogen, oxygen, nitrogen and carbon monoxide.
1.4 Determination of argon and helium in natural gas.
1.5 Determination of sulphur.
1.6 Determination of cyanide.
1.7 Determination of chloride.

2. Light Distillate

2.0 IP tests and procedures.
2.1 Determination of sulphur types.
2.2 Determination of total sulphur.
2.3 Determination of free sulphur.
2.4 Determination of hydrogen sulphide and mercaptans.

Appendices

2.5 Determination of disulphides.
2.6 Determination of sulphides.
2.7 Determination of total sulphur.
2.8 Determination of total chlorine.

3. De-ionized Water and Condensate

3.1 Determination of chloride.
3.2 Determination of chloride.
3.3 Determination of total alkalinity.
3.4 Determination of methanol.

4. Solids

4.1 Determination of total sulphur by gravimetry.
4.2 Determination of total sulphur iodometrically.
4.3 Determination of potassium in Catalyst 46-1.
4.4 Determination of chloride in Catalyst 59.
4.5 Determination of total chloride in catalysts.

5. Process Gas Streams

5.1 Determination of argon in ammonia plant gas streams.
5.2 Determination of nickel carbonyl in synthesis gas.
5.3 Determination of hydrogen sulphide during reduction of catalysts.
5.4 Determination of steam/gas ratio in primary reformer make gas.
5.5 Determination of benzene and toluene in the primary reformer gas stream.
5.6 Determination of naphthalenes in primary reformer gas streams.
5.7 Determination of water content during reduction of ammonia synthesis catalyst.
5.8 Determination of ammonia in ammonia plant gas streams.
5.9 Determination of hydrogen sulphide in process gas streams.
5.10 Determination of carbon dioxide, hydrogen, oxygen, nitrogen, methane and carbon monoxide.

6. Anhydrous Ammonia

6.1 Determination of water, residue and oil.
6.2 Determination of oil.
6.3 Determination of inert gases.
6.4 Determination of oxygen.
6.5 Determination of hydrocarbons.
6.6 Determination of iron.
6.7 Determination of sodium.
6.8 Determination of carbon dioxide.

Appendix 6.

Equilibrium Constants for the Methane–Steam Reaction at Various Temperatures

$$CH_4 + H_2O \rightleftharpoons CO + 3H_2$$
$$K_p = p_{CH_4} p_{H_2}/p_{CO} p_{H_2}^3 \text{ atm}^{-2}$$

The equilibrium constants tabulated below are calculated from the following equation:

$$K_p = \exp(Z(Z(Z(0.2513Z - 0.3665) - 0.58101) + 27.1337) - 3.2770)$$

where $Z = (1000/T) - 1$, with T being the absolute temperature (Kelvin). $K_p = K \times 10^n$ with K and n being listed below for temperatures over the range 200 to 1199°C.

Tempera-ture/°C	K						n
	+0°C	+1°C	+2°C	+3°C	+4°C		
200	2.1547	1.9201	1.7119	1.5269	1.3626		11
205	1.2166	1.0867	0.9712	0.8683	0.7767		11
210	6.9511	6.2236	5.5747	4.9957	4.4788		10
215	4.0172	3.6048	3.2361	2.9065	2.6115		10
220	2.3475	2.1110	1.8992	1.7094	1.5392		10
225	1.3865	1.2494	1.1264	1.0159	0.9166		10
230	8.2739	7.4713	6.7493	6.0994	5.5143		9
235	4.9872	4.5123	4.0842	3.6981	3.3498		9
240	3.0354	2.7516	2.4952	2.2636	2.0542		9
245	1.8649	1.6936	1.5387	1.3984	1.2713		9
250	1.1562	1.0519	0.9574	0.8716	0.7938		9
255	7.2322	6.5912	6.0091	5.4803	4.9996		8
260	4.5627	4.1653	3.8038	3.4748	3.1753		8

Appendices

°C	+0°C	+1°C	+2°C	+3°C	+4°C	n
265	2.9026	2.6541	2.4277	2.2213	2.0331	8
270	1.8615	1.7048	1.5619	1.4314	1.3121	8
275	1.2032	1.1037	1.0127	0.9295	0.8533	8
280	7.8369	7.1994	6.6157	6.0811	5.5913	7
285	5.1425	4.7310	4.3538	4.0077	3.6902	7
290	3.3989	3.1314	2.8857	2.6601	2.4528	7
295	2.2623	2.0871	1.9261	1.7779	1.6416	7
300	1.5161	1.4006	1.2943	1.1963	1.1060	7
305	1.0229	0.9462	0.8755	0.8103	0.7501	7
310	6.9457	6.4332	5.9600	5.5231	5.1194	6
315	4.7464	4.4017	4.0830	3.7883	3.5157	6
320	3.2635	3.0302	2.8142	2.6142	2.4290	6
325	2.2575	2.0986	1.9513	1.8147	1.6881	6
330	1.5708	1.4619	1.3608	1.2670	1.1800	6
335	1.0992	1.0241	0.9544	0.8897	0.8295	6
340	7.7351	7.2149	6.7311	6.2811	5.8625	5
345	5.4729	5.1103	4.7728	4.4585	4.1657	5
350	3.8930	3.6389	3.4021	3.1814	2.9755	5
355	2.7836	2.6046	2.4376	2.2817	2.1363	5
360	2.0005	1.8737	1.7553	1.6447	1.5414	5
365	1.4448	1.3546	1.2702	1.1913	1.1176	5
370	1.0486	0.9840	0.9236	0.8671	0.8141	5
375	7.6460	7.1820	6.7474	6.3402	5.9587	4
380	5.6012	5.2661	4.9519	4.6573	4.3811	4
385	4.1219	3.8788	3.6506	3.4365	3.2355	4
390	3.0467	2.8695	2.7031	2.5467	2.3998	4
395	2.2618	2.1321	2.0101	1.8954	1.7876	4
400	1.6862	1.5908	1.5011	1.4166	1.3371	4
405	1.2623	1.1919	1.1256	1.0631	1.0043	4
410	9.4884	8.9662	8.4740	8.0101	7.5728	3
415	7.1605	6.7717	6.4050	6.0591	5.7327	3
420	5.4248	5.1342	4.8598	4.6009	4.3564	3
425	4.1255	3.9074	3.7014	3.5068	3.3229	3
430	3.1491	2.9849	2.8296	2.6828	2.5439	3
435	2.1426	2.2884	2.1709	2.0598	1.9546	3
440	1.8550	1.7608	1.6715	1.5871	1.5071	3
445	1.4313	1.3595	1.2915	1.2271	1.1660	3
450	1.1082	1.0533	1.0013	0.9520	0.9052	3

6. Equilibrium Constants: Methane-Steam Reaction

°C	+0°C	+1°C	+2°C	+3°C	+4°C	n
455	8.6088	8.1882	7.7891	7.4104	7.0510	2
460	6.7100	6.3863	6.0789	5.7871	5.5101	2
465	5.2469	4.9970	4.7596	4.5340	4.3196	2
470	4.1159	3.9223	3.7383	3.5634	3.3970	2
475	3.2388	3.0884	2.9453	2.8092	2.6797	2
480	2.5564	2.4392	2.3276	2.2213	2.1202	2
485	2.0239	1.9322	1.8449	1.7617	1.6825	2
490	1.6070	1.5351	1.4665	1.4012	1.3390	2
495	1.2797	1.2231	1.1692	1.1177	1.0687	2
500	1.0219	0.9773	0.9347	0.8941	0.8553	2
505	8.1834	7.8303	7.4933	7.1716	6.8645	1
510	6.5712	6.2911	6.0236	5.7680	5.5239	1
515	5.2907	5.0679	4.8550	4.6515	4.4570	1
520	4.2710	4.0933	3.9234	3.7608	3.6054	1
525	3.4568	3.3146	3.1786	3.0485	2.9240	1
530	2.8049	2.6909	2.5818	2.4774	2.3774	1
535	2.2816	2.1900	2.1022	2.0182	1.9377	1
540	1.8605	1.7867	1.7159	1.6481	1.5831	1
545	1.5208	1.4611	1.4039	1.3491	1.2965	1
550	1.2461	1.1977	1.1514	1.1069	1.0643	1
555	1.0233	0.9841	0.9464	0.9103	0.8756	1
560	8.4234	8.1040	7.7974	7.5031	7.2205	0
565	6.9492	6.6886	6.4384	6.1981	5.9673	0
570	5.7456	5.5326	5.3280	5.1313	4.9424	0
575	4.7608	4.5863	4.4185	4.2573	4.1023	0
580	3.9532	3.8099	3.6721	3.5396	3.4121	0
585	3.2895	3.1716	3.0581	2.9490	2.8440	0
590	2.7429	2.6457	2.5521	2.4620	2.3753	0
595	2.2918	2.2114	2.1340	2.0595	1.9878	0
600	1.9187	1.8521	1.7880	1.7263	1.6668	0
605	1.6094	1.5542	1.5010	1.4497	1.4003	0
610	1.3527	1.3068	1.2625	1.2198	1.1787	0
615	1.1390	1.1008	1.0639	1.0283	0.9940	0
620	9.6093	9.2900	8.9821	8.6850	8.3983	−1
625	8.1216	7.8547	7.5971	7.3484	7.1084	−1
630	6.8767	6.6531	6.4372	6.2287	6.0274	−1
635	5.8330	5.6453	5.4640	5.2889	5.1198	−1
640	4.9564	4.7986	4.6461	4.4988	4.3564	−1

Appendices

°C	+0°C	+1°C	+2°C	+3°C	+4°C	n
645	4.2188	4.0859	3.9574	3.8332	3.7132	−1
650	3.5971	3.4849	3.3765	3.2716	3.1702	−1
655	3.0271	2.9773	2.8856	2.7969	2.7111	−1
660	2.6281	2.5478	2.4701	2.3950	2.3223	−1
665	2.2519	2.1838	2.1179	2.0541	1.9924	−1
670	1.9326	1.8748	1.8188	1.7646	1.7121	−1
675	1.6612	1.6120	1.5644	1.5182	1.4735	−1
680	1.4302	1.3882	1.3476	1.3082	1.2701	−1
685	1.2331	1.1973	1.1626	1.1290	1.0964	−1
690	1.0648	1.0342	1.0045	0.9758	0.9479	−1
695	9.2085	8.9464	8.6922	8.4458	8.2068	−2
700	7.9751	7.7503	7.5323	7.3209	7.1158	−2
705	6.9168	6.7238	6.5365	6.3548	6.1785	−2
710	6.0074	5.8414	5.6803	5.5240	5.3722	−2
715	5.2249	5.0819	4.9431	4.8084	4.6776	−2
720	4.5506	4.4272	4.3075	4.1912	4.0783	−2
725	3.9686	3.8621	3.7587	3.6582	3.5606	−2
730	3.4657	3.3736	3.2841	3.1971	3.1126	−2
735	3.0305	2.9508	2.8732	2.7979	2.7246	−2
740	2.6534	2.5842	2.5170	2.4516	2.3880	−2
745	2.3262	2.2661	2.2077	2.1509	2.0957	−2
750	2.0420	1.9897	1.9389	1.8895	1.8414	−2
755	1.7946	1.7492	1.7049	1.6618	1.6200	−2
760	1.5792	1.5396	1.5010	1.4634	1.4269	−2
765	1.3913	1.3567	1.3230	1.2902	1.2583	−2
770	1.2273	1.1970	1.1676	1.1389	1.1110	−2
775	1.0838	1.0573	1.0316	1.0065	0.9820	−2
780	9.5823	9.3504	9.1246	8.9047	8.6904	−3
785	8.4817	8.2784	8.0803	7.8873	7.6992	−3
790	7.5160	7.3375	7.1635	6.9939	6.8287	−3
795	6.6677	6.5107	6.3577	6.2086	6.0633	−3
800	5.9216	5.7835	5.6488	5.5175	5.3895	−3
805	5.2647	5.1430	5.0243	4.9086	4.7957	−3
810	4.6856	4.5783	4.4736	4.3715	4.2178	−3
815	4.1747	4.0799	3.9874	3.8972	3.8092	−3
820	3.7233	3.6395	3.5578	3.4780	3.4001	−3
825	3.3241	3.2500	3.1776	3.1070	3.0381	−3
830	2.9708	2.9051	2.8410	2.7784	2.7173	−3

6. Equilibrium Constants: Methane-Steam Reaction

°C	+0°C	+1°C	+2°C	+3°C	+4°C	n
835	2.6576	2.5994	2.5425	2.4870	2.4328	−3
840	2.3798	2.3281	2.2776	2.2283	2.1802	−3
845	2.1331	2.0872	2.0423	1.9985	1.9557	−3
850	1.9139	1.8730	1.8331	1.7941	1.7560	−3
855	1.7188	1.6824	1.6468	1.6121	1.5782	−3
860	1.5450	1.5126	1.4809	1.4499	1.4196	−3
865	1.3901	1.3611	1.3329	1.3052	1.2782	−3
870	1.2518	1.2260	1.2008	1.1761	1.1519	−3
875	1.1283	1.1053	1.0827	1.0606	1.0390	−3
880	1.0179	0.9973	0.9771	0.9574	0.9380	−3
885	9.1915	9.0067	8.8259	8.6490	8.4760	−4
890	8.3067	8.1410	7.9790	7.8204	7.6653	−4
895	7.5134	7.3649	7.2195	7.0772	6.9380	−4
900	6.8017	6.6683	6.5378	6.4100	6.2850	−4
905	6.1625	6.0427	5.9254	5.8105	5.6981	−4
910	5.5880	5.4803	5.3748	5.2714	5.1703	−4
915	5.0712	4.9742	4.8792	4.7862	4.6951	−4
920	4.6059	4.5185	4.4330	4.3491	4.2670	−4
925	4.1866	4.1078	4.0307	3.9551	3.8810	−4
930	3.8085	3.7374	3.6678	3.5995	3.5327	−4
935	3.4672	3.4030	3.3401	3.2785	3.2181	−4
940	3.1589	3.1009	3.0440	2.9883	2.9337	−4
945	2.8802	2.8277	2.7763	2.7259	2.6765	−4
950	2.6280	2.5805	2.5340	2.4883	2.4436	−4
955	2.3997	2.3567	2.3145	2.2732	2.2326	−4
960	2.1929	2.1539	2.1156	2.0781	2.0413	−4
965	2.0053	1.9699	1.9352	1.9012	1.8678	−4
970	1.8350	1.8029	1.7714	1.7405	1.7102	−4
975	1.6804	1.6512	1.6226	1.5945	1.5670	−4
980	1.5399	1.5134	1.4874	1.4618	1.4367	−4
985	1.4121	1.3880	1.3643	1.3411	1.3182	−4
990	1.2958	1.2739	1.2523	1.2311	1.2103	−4
995	1.1899	1.1699	1.1502	1.1309	1.1120	−4
1000	1.0934	1.0751	1.0572	1.0396	1.0223	−4
1005	1.0053	0.9887	0.9723	0.9562	0.9405	−4
1010	9.2498	9.0977	8.9483	8.8015	8.6574	−5
1015	8.5159	8.3769	8.2404	8.1063	7.9746	−5
1020	7.8453	7.7182	7.5934	7.4707	7.3503	−5

Appendices

°C	+0°C	+1°C	+2°C	+3°C	+4°C	n
1025	7.2319	7.1157	7.0015	6.8893	6.7790	−5
1030	6.6707	6.5643	6.4597	6.3570	6.2560	−5
1035	6.1568	6.0593	5.9635	5.8694	5.7769	−5
1040	5.6860	5.5966	5.5088	5.4225	5.3376	−5
1045	5.2543	5.1723	5.0917	5.0126	4.9347	−5
1050	4.8582	4.7830	4.7091	4.6364	4.5649	−5
1055	4.4946	4.4256	4.3577	4.2909	4.2253	−5
1060	4.1607	4.0972	4.0348	3.9735	3.9131	−5
1065	3.8538	3.7954	3.7381	3.6816	3.6261	−5
1070	3.5715	3.5179	3.4651	3.4131	3.3621	−5
1075	3.3118	3.2624	3.2138	3.1660	3.1189	−5
1080	3.0727	3.0272	2.9824	2.9384	2.8950	−5
1085	2.8524	2.8104	2.7692	2.7286	2.6886	−5
1090	2.6493	2.6106	2.5726	2.5351	2.4983	−5
1095	2.4620	2.4263	2.3912	2.3567	2.3227	−5
1100	2.2892	2.2563	2.2238	2.1919	2.1605	−5
1105	2.1296	2.0992	2.0693	2.0398	2.0108	−5
1110	1.9822	1.9541	1.9264	1.8992	1.8723	−5
1115	1.8459	1.8199	1.7943	1.7691	1.7443	−5
1120	1.7199	1.6959	1.6722	1.6489	1.6259	−5
1125	1.6033	1.5810	1.5591	1.5375	1.5163	−5
1130	1.4953	1.4747	1.4544	1.4344	1.4147	−5
1135	1.3953	1.3762	1.3574	1.3389	1.3206	−5
1140	1.3026	1.2849	1.2675	1.2503	1.2334	−5
1145	1.2167	1.2003	1.1841	1.1682	1.1525	−5
1150	1.1370	1.1217	1.1067	1.0919	1.0773	−5
1155	1.0630	1.0488	1.0349	1.0211	1.0076	−5
1160	9.9427	9.8112	9.6817	9.5540	9.4282	−6
1165	9.3042	9.1821	9.0617	8.9430	8.8260	−6
1170	8.7108	8.5972	8.4852	8.3749	8.2661	−6
1175	8.1589	8.0532	7.9491	7.8464	7.7452	−6
1180	7.6454	7.5471	7.4501	7.3545	7.2603	−6
1185	7.1674	7.0759	6.9856	6.8966	6.8088	−6
1190	6.7223	6.6370	6.5529	6.4699	6.3882	−6
1195	6.3075	6.2280	6.1496	6.0723	5.9961	−6

Appendix 7.

Equilibrium Constants for the CO Conversion Reaction (Shift) at Various Temperatures

$$CO + H_2O \rightleftharpoons CO_2 + H_2$$

$$K_p = p_{H_2} p_{CO_2} / p_{H_2O} p_{CO}$$

The equilibrium constants tabulated below are calculated from the following equation:

$$K_p = \exp(Z(Z(0.63508 - 0.29353Z) + 4.1778) + 0.31688)$$

where $Z = (1000/T) - 1$, with T being the absolute temperature (Kelvin). $K_p = K \times 10^n$ with K and n being listed below for temperatures over the range 200 to 1199°C.

Temperature/°C	+0°C	+1°C	+2°C	+3°C	+4°C	n
200	2.1082	2.0663	2.0254	1.9855	1.9464	2
205	1.9083	1.8711	1.8347	1.7991	1.7643	2
210	1.7304	1.6972	1.6648	1.6331	1.6021	2
215	1.5718	1.5421	1.5132	1.4849	1.4572	2
220	1.4301	1.4036	1.3777	1.3524	1.3276	2
225	1.3034	1.2797	1.2565	1.2338	1.2116	2
230	1.1899	1.1686	1.1478	1.1274	1.1075	2
235	1.0880	1.0689	1.0502	1.0319	1.0139	2
240	9.9638	9.7919	9.6236	9.4588	9.2973	1
245	9.1392	8.9843	8.8325	8.6839	8.5383	1
250	8.3956	8.2558	8.1188	7.9846	7.8530	1
255	7.7241	7.5977	7.4738	7.3524	7.2334	1
260	7.1167	7.0023	6.8901	6.7801	6.6723	1
265	6.5665	6.4628	6.3610	6.2613	6.1634	1
270	6.0674	5.9732	5.8808	5.7902	5.7012	1
275	5.6140	5.5284	5.4443	5.3619	5.2809	1

Appendices

°C	+0°C	+1°C	+2°C	+3°C	+4°C	n
280	5.2015	5.1235	5.0470	4.9719	4.8981	1
285	4.8257	4.7546	4.6848	4.6163	4.5490	1
290	4.4829	4.4180	4.3543	4.2916	4.2301	1
295	4.1697	4.1104	4.0521	3.9949	3.9386	1
300	3.8833	3.8290	3.7756	3.7232	3.6716	1
305	3.6210	3.5712	3.5222	3.4741	3.4269	1
310	3.3804	3.3347	3.2898	3.2456	3.2022	1
315	3.1595	3.1175	3.0762	3.0356	2.9957	1
320	2.9564	2.9178	2.8798	2.8424	2.8057	1
325	2.7695	2.7339	2.6989	2.6645	2.6306	1
330	2.5973	2.5645	2.5322	2.5004	2.4691	1
335	2.4384	2.4081	2.3783	2.3489	2.3200	1
340	2.2916	2.2636	2.2360	2.2089	2.1822	1
345	2.1559	2.1300	2.1045	2.0794	2.0546	1
350	2.0303	2.0063	1.9826	1.9594	1.9364	1
355	1.9139	1.8916	1.8697	1.8481	1.8268	1
360	1.8059	1.7852	1.7649	1.7448	1.7251	1
365	1.7056	1.6864	1.6675	1.6489	1.6305	1
370	1.6124	1.5945	1.5769	1.5596	1.5425	1
375	1.5257	1.5090	1.4927	1.4765	1.4606	1
380	1.4449	1.4294	1.4141	1.3991	1.3842	1
385	1.3696	1.3551	1.3409	1.3268	1.3130	1
390	1.2993	1.2858	1.2725	1.2594	1.2464	1
395	1.2337	1.2211	1.2086	1.1964	1.1842	1
400	1.1723	1.1605	1.1489	1.1374	1.1261	1
405	1.1149	1.1039	1.0930	1.0822	1.0716	1
410	1.0611	1.0508	1.0406	1.0305	1.0205	1
415	1.0107	1.0010	0.9915	0.9820	0.9727	1
420	9.6345	9.5434	9.4536	9.3648	9.2772	0
425	9.1906	9.1051	9.0207	8.9373	8.8549	0
430	8.7735	8.6931	8.6138	8.5353	8.4578	0
435	8.3813	8.3057	8.2310	8.1572	8.0843	0
440	8.0122	7.9410	7.8707	7.8012	7.7325	0
445	7.6646	7.5975	7.5312	7.4657	7.4009	0
450	7.3369	7.2737	7.2111	7.1493	7.0883	0
455	7.0279	6.9682	6.9092	6.8508	6.7932	0
460	6.7362	6.6798	6.6241	6.5690	6.5145	0
465	6.4606	6.4074	6.3547	6.3026	6.2511	0

7. Equilibrium Constants: CO Shift

°C	+0°C	+1°C	+2°C	+3°C	+4°C	n
470	6.2002	6.1498	6.1000	6.0507	6.0020	0
475	5.9538	5.9061	5.8590	5.8124	5.7662	0
480	5.7206	5.6755	5.6308	5.5867	5.5430	0
485	5.4997	5.4570	5.4147	5.3728	5.3314	0
490	5.2904	5.2499	5.2098	5.1701	5.1308	0
495	5.0919	5.0534	5.0154	4.9777	4.9404	0
500	4.9035	4.8670	4.8309	4.7951	4.7597	0
505	4.7246	4.6899	4.6556	4.6216	4.5880	0
510	4.5547	4.5217	4.4891	4.4568	4.4248	0
515	4.3931	4.3617	4.3307	4.3000	4.2695	0
520	4.2394	4.2096	4.1800	4.1508	4.1218	0
525	4.0931	4.0647	4.0366	4.0087	3.9811	0
530	3.9538	3.9268	3.9000	3.8734	3.8471	0
535	3.8211	3.7953	3.7697	3.7444	3.7194	0
540	3.6945	3.6699	3.6456	3.6214	3.5975	0
545	3.5738	3.5503	3.5271	3.5041	3.4812	0
550	3.4586	3.4362	3.4140	3.3920	3.3702	0
555	3.3486	3.3272	3.3060	3.2850	3.2642	0
560	3.2435	3.2231	3.2028	3.1827	3.1628	0
565	3.1431	3.1235	3.1041	3.0849	3.0659	0
570	3.0470	3.0283	3.0098	2.9914	2.9732	0
575	2.9551	2.9372	2.9195	2.9019	2.8844	0
580	2.8671	2.8500	2.8330	2.8162	2.7995	0
585	2.7829	2.7665	2.7502	2.7340	2.7180	0
590	2.7022	2.6864	2.6708	2.6554	2.6400	0
595	2.6248	2.6097	2.5948	2.5799	2.5652	0
600	2.5506	2.5361	2.5218	2.5075	2.4934	0
605	2.4794	2.4655	2.4518	2.4381	2.4246	0
610	2.4111	2.3978	2.3846	2.3714	2.3584	0
615	2.3455	2.3327	2.3200	2.3074	2.2949	0
620	2.2825	2.2702	2.2580	2.2459	2.2339	0
625	2.2219	2.2101	2.1984	2.1867	2.1752	0
630	2.1637	2.1524	2.1411	2.1299	2.1188	0
635	2.1077	2.0968	2.0859	2.0752	2.0645	0
640	2.0539	2.0433	2.0329	2.0225	2.0122	0
645	2.0020	1.9919	1.9818	1.9718	1.9619	0
650	1.9521	1.9423	1.9327	1.9230	1.9135	0
655	1.9040	1.8946	1.8853	1.8760	1.8668	0

Appendices

°C	+0°C	+1°C	+2°C	+3°C	+4°C	n
660	1.8577	1.8486	1.8396	1.8307	1.8218	0
665	1.8130	1.8043	1.7956	1.7870	1.7784	0
670	1.7699	1.7615	1.7532	1.7448	1.7366	0
675	1.7284	1.7203	1.7122	1.7042	1.6962	0
680	1.6883	1.6805	1.6727	1.6649	1.6573	0
685	1.6496	1.6420	1.6345	1.6270	1.6196	0
690	1.6123	1.6049	1.5977	1.5905	1.5833	0
695	1.5762	1.5691	1.5621	1.5551	1.5482	0
700	1.5413	1.5345	1.5277	1.5209	1.5143	0
705	1.5076	1.5010	1.4944	1.4879	1.4814	0
710	1.4750	1.4686	1.4623	1.4560	1.4497	0
715	1.4435	1.4373	1.4312	1.4251	1.4190	0
720	1.4130	1.4070	1.4011	1.3952	1.3893	0
725	1.3835	1.3777	1.3720	1.3663	1.3606	0
730	1.3550	1.3494	1.3438	1.3383	1.3328	0
735	1.3273	1.3219	1.3165	1.3111	1.3058	0
740	1.3005	1.2953	1.2900	1.2849	1.2797	0
745	1.2746	1.2695	1.2644	1.2594	1.2544	0
750	1.2494	1.2445	1.2396	1.2347	1.2299	0
755	1.2250	1.2203	1.2155	1.2108	1.2061	0
760	1.2014	1.1968	1.1922	1.1876	1.1830	0
765	1.1785	1.1740	1.1695	1.1651	1.1606	0
770	1.1562	1.1519	1.1475	1.1432	1.1389	0
775	1.1347	1.1304	1.1262	1.1220	1.1178	0
780	1.1137	1.1096	1.1055	1.1014	1.0974	0
785	1.0934	1.0894	1.0854	1.0814	1.0775	0
790	1.0736	1.0697	1.0659	1.0620	1.0582	0
795	1.0544	1.0506	1.0469	1.0432	1.0395	0
800	1.0358	1.0321	1.0285	1.0248	1.0212	0
805	1.0177	1.0141	1.0105	1.0070	1.0035	0
810	1.0000	0.9966	0.9931	0.9897	0.9863	0
815	9.8291	9.7954	9.7619	9.7286	9.6955	−1
820	9.6626	9.6298	9.5972	9.5648	9.5326	−1
825	9.5005	9.4686	9.4369	9.4054	9.3741	−1
830	9.3429	9.3118	9.2810	9.2503	9.2198	−1
835	9.1894	9.1592	9.1292	9.0993	9.0696	−1
840	9.0401	9.0107	8.9815	8.9524	8.9235	−1
845	8.8947	8.8661	8.8376	8.8093	8.7811	−1

7. Equilibrium Constants: CO Shift

°C	+0°C	+1°C	+2°C	+3°C	+4°C	n
850	8.7531	8.7252	8.6975	8.6699	8.6424	−1
855	8.6151	8.5880	8.5610	8.5341	8.5074	−1
860	8.4808	8.4543	8.4280	8.4018	8.3757	−1
865	8.3498	8.3240	8.2984	8.2729	8.2475	−1
870	8.2222	8.1971	8.1721	8.1472	8.1224	−1
875	8.0978	8.0733	8.0489	8.0247	8.0005	−1
880	7.9765	7.9526	7.9288	7.9052	7.8816	−1
885	7.8582	7.8349	7.8117	7.7886	7.7657	−1
890	7.7428	7.7201	7.6975	7.6750	7.6526	−1
895	7.6303	7.6081	7.5860	7.5641	7.5422	−1
900	7.5204	7.4988	7.4773	7.4558	7.4345	−1
905	7.4133	7.3921	7.3711	7.3502	7.3293	−1
910	7.3086	7.2880	7.2675	7.2470	7.2267	−1
915	7.2065	7.1863	7.1663	7.1463	7.1265	−1
920	7.1067	7.0871	7.0675	7.0480	7.0286	−1
925	7.0093	6.9901	6.9710	6.9519	6.9330	−1
630	6.9141	6.8954	6.8767	6.8581	6.8396	−1
935	6.8212	6.8028	6.7846	6.7664	6.7483	−1
940	6.7303	6.7124	6.6946	6.6768	6.6591	−1
945	6.6415	6.6240	6.6066	6.5892	6.5720	−1
950	6.5548	6.5376	6.5206	6.5036	6.4867	−1
955	6.4699	6.4532	6.4365	6.4199	6.4034	−1
960	6.3870	6.3706	6.3543	6.3381	6.3219	−1
965	6.3058	6.2898	6.2739	6.2580	6.2422	−1
970	6.2265	6.2108	6.1952	6.1797	6.1643	−1
975	6.1489	6.1335	6.1183	6.1031	6.0880	−1
980	6.0729	6.0579	6.0430	6.0281	6.0133	−1
985	5.9986	5.9839	5.9693	5.9548	5.9403	−1
990	5.9259	5.9115	5.8972	5.8830	5.8688	−1
995	5.8547	5.8406	5.8266	5.8127	5.7988	−1
1000	5.7850	5.7712	5.7575	5.7439	5.7303	−1
1005	5.7167	5.7032	5.6898	5.6765	5.6631	−1
1010	5.6499	5.6367	5.6236	5.6105	5.5974	−1
1015	5.5844	5.5715	5.5586	5.5458	5.5330	−1
1020	5.5203	5.5077	5.4950	5.4825	5.4700	−1
1025	5.4575	5.4451	5.4327	5.4204	5.4082	−1
1030	5.3960	5.3838	5.3717	5.3596	5.3476	−1
1035	5.3357	5.3237	5.3119	5.3000	5.2883	−1

Appendices

°C	+0°C	+1°C	+2°C	+3°C	+4°C	n
1040	5.2765	5.2649	5.2532	5.2416	5.2301	−1
1045	5.2186	5.2072	5.1957	5.1844	5.1731	−1
1050	5.1618	5.1506	5.1394	5.1283	5.1172	−1
1055	5.1061	5.0951	5.0841	5.0732	5.0623	−1
1060	5.0515	5.0407	5.0300	5.0192	5.0086	−1
1065	4.9979	4.9874	4.9768	4.9663	4.9558	−1
1070	4.9454	4.9350	4.9247	4.9144	4.9041	−1
1075	4.8939	4.8837	4.8735	4.8634	4.8534	−1
1080	4.8433	4.8333	4.8234	4.8134	4.8036	−1
1085	4.7937	4.7839	4.7741	4.7644	4.7547	−1
1090	4.7450	4.7354	4.7258	4.7163	4.7067	−1
1095	4.6973	4.6878	4.6784	4.6690	4.6597	−1
1100	4.6504	4.6411	4.6319	4.6226	4.6135	−1
1150	4.6043	4.5952	4.5862	4.5771	4.5681	−1
1110	4.5591	4.5502	4.5413	4.5324	4.5236	−1
1115	4.5148	4.5060	4.4972	4.4885	4.4798	−1
1120	4.4712	4.4626	4.4540	4.4454	4.4369	−1
1125	4.4284	4.4199	4.4115	4.4031	4.3947	−1
1130	4.3863	4.3780	4.3697	4.3615	4.3532	−1
1135	4.3450	4.3369	4.3287	4.3206	4.3125	−1
1140	4.3045	4.2965	4.2885	4.2805	4.2725	−1
1145	4.2646	4.2567	4.2489	4.2410	4.2332	−1
1150	4.2254	4.2177	4.2100	4.2023	4.1946	−1
1155	4.1870	4.1793	4.1717	4.1642	4.1566	−1
1160	4.1491	4.1416	4.1342	4.1267	4.1193	−1
1165	4.1119	4.1046	4.0972	4.0899	4.0826	−1
1170	4.0754	4.0681	4.0609	4.0537	4.0466	−1
1175	4.0394	4.0323	4.0252	4.0182	4.0111	−1
1180	4.0041	3.9971	3.9901	3.9832	3.9762	−1
1185	3.9693	3.9625	3.9556	3.9488	3.9420	−1
1190	3.9352	3.9284	3.9217	3.9149	3.9082	−1
1195	3.9016	3.8949	3.8883	3.8817	3.8751	−1

Reproduced from *Code of Practice for the Storage of Ammonia Under Pressure*, courtesy of Chemical Industries Association.

Appendix 8.

Nomograph of Selected Properties of Ammonia

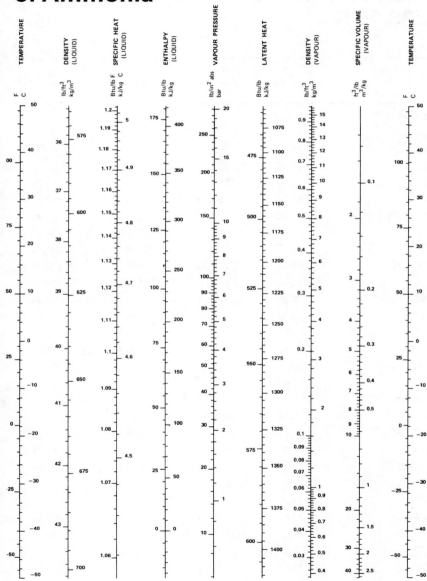

549

Appendix 9.

Thermodynamic Properties of Elements and Compounds at 298.15 K

	C_p° /J mol^{-1} K^{-1}	S° /J mol^{-1} K^{-1}	$\triangle H_f^\circ$ /kJ mol^{-1}	$\triangle G_f^\circ$ /kJ mol^{-1}
Gases				
H_2	28.84	130.57	0	0
N_2	29.12	191.50	0	0
O_2	29.37	205.03	0	0
H_2O	33.58	188.72	−241.83	−228.60
CH_4	35.64	186.15	−74.87	−50.81
Naphtha (as C_7H_{16})	165.98	427.90	−187.78	+7.99
CO	29.14	197.54	−110.53	−137.16
CO_2	37.13	219.69	−393.52	−394.40
CH_3OH	43.89	239.70	−201.17	−162.51
HCHO	35.40	218.66	−115.90	−109.90
NH_3	39.63	200.97	−45.90	−16.38
NO	29.84	210.65	+90.29	+86.60
NO_2	36.97	239.92	+33.10	+51.34
H_2S	34.18	205.77	−20.17	−33.05
SO_2	39.87	248.11	−296.85	−300.16
SO_3	50.63	256.02	−395.26	−370.37
Liquids				
H_2O	75.30	69.94	−285.84	−237.19
CH_3OH	81.6	126.78	−238.57	−166.23
HNO_3	109.87	155.60	−173.22	−79.91
H_2SO_4	138.91	156.86	−813.58	−689.65
Solids				
Al	24.29	28.41	0	0
α-Al_2O_3	79.03	50.94	−1675.27	−1581.89
γ-Al_2O_3	82.98	52.51	−1656.86	−1563.94
C (graphite)	8.53	5.69	0	0
Co	24.81	30.04	0	0

9. Thermodynamic Properties

	C_p°	S°	ΔH_f°	ΔG_f°
CoO	55.28	52.99	−237.73	−214.01
Co$_3$O$_4$	102.13	114.29	−910.02	−794.96
CoS	47.70	46.35	−84.52	−82.84
Cr	23.47	23.77	0	0
Cr$_2$O$_3$	120.36	81.15	−1134.70	−1053.12
CrO$_3$	56.02	266.06	−292.88	−273.47
Cu	24.45	33.11	0	0
Cu$_2$O	63.59	92.94	−170.29	−147.69
CuO	42.12	42.61	−155.85	−128.12
α-Fe	25.06	27.32	0	0
Fe$_{0.947}$O (Wustite)	48.12	57.59	−266.27	−245.16
Fe$_3$O$_4$	147.23	145.27	−1120.89	−1017.51
Fe$_2$O$_3$	103.75	87.40	−825.50	−743.58
Mo	23.97	28.61	0	0
MoO$_3$	74.88	77.76	−745.17	−668.13
MoS$_2$	63.47	63.18	−232.21	−225.10
Ni	25.77	29.86	0	0
NiO	44.59	38.58	−239.74	−211.79
NiS	—	—	−73.22	—
S (rhombic)	22.60	31.93	0	0
V	24.89	28.93	0	0
V$_2$O$_4$	115.40	103.52	−1427.16	−1318.52
V$_2$O$_5$	130.60	130.55	−1550.59	−1419.43
W	24.30	32.66	0	0
WO$_3$	73.14	75.91	−842.91	−764.11
WS$_2$	—	96.23	−193.72	−193.30
Zn	25.38	41.63	0	0
ZnO	40.25	43.51	−348.32	−318.40
ZnS (sphalerite)	45.19	57.74	−202.92	−198.32

Symbols Used

C_p° heat capacity at constant pressure and standard state at 298.15 K

S° entropy at the standard state, omitting contributions from isotopic mixing and nuclear spins, at 298.15 K

ΔH_f° standard enthalpy of formation with each substance in its thermodynamic standard state at 298.15 K

ΔG_f° standard free energy of formation with each substance in its thermodynamic standard state at 298.15 K

Standard state for gases = 1 atm = 101325 N m^{-2} = 1.01325 bar.

Appendices

Additional data may be obtained from the following sources:

1. Selected value of chemical thermodynamics, *Circ. NBS*, No. 500 (1952).

2. D. R. Stull and H. Prophet, *JANAF Thermochemical Tables*, 2nd edn. *NBS*, **37** (1971). Supplements in *J. Phys. Chem. Data*.

3. M. W. Chase, C. A. Davies, J. R. Downey, D. J. Frurip, R. A. McDonald and A. N. Syverud, *JANAF Thermochemical Tables*, 3rd edn. J. Phys. Chem. Reference Data, Volume 14, Supplement No. 1 (1985). Joint publication: American Chemical Society and American Institute of Physics for National Bureau of Standards.

4. D. R. Stull, E. F. Westrum and G. C. Sinke, *The Chemical Thermodynamics of Organic Compounds*, Wiley, New York (1969).

5. J. P. Coughlin, Contributions to the data on theoretical metallurgy: XII. Heats and free energies of formation of inorganic oxides, *Bur. Mines Bull.*, **542** (1954).

6. K. K. Kelley, XII High-temperature heat-content, heat-capacity and entropy data for the elements and inorganic compounds, *Bur. Mines Bull.*, **584** (1960).

7. *Bulletin of Chemical Thermodynamics* (formerly *Bulletin of Thermodynamics and Thermochemistry*), Thermochemistry, Inc., for IUPAC, Annual publication, Volume 1 (1958).

8. C. E. Wicks and F. E. Block, Thermodynamics of 65 Elements – Their Oxides, Halides, Carbides, and Nitrides, *Bur. Mines Bull.*, **605** (1963).

Appendix 10.

Physical Properties of Methanol

Boiling point at 1.013 bar (760 mmHg)/°C	64.5
Freezing point/°C	−97.5
Specific gravity/g cm^{-3}	
20/4°C	0.792
15.5/15.5°C	0.796
Vapour density at 15°C (air = 1)	1.11
Vapour pressure at 20°C/mbar	129
Latent heat of vaporization at boiling point/kJ kg^{-1}	1100
Thermal conductivity at 20°C/W m^{-1} °C^{-1}	0.2022
Coefficient of cubical expansion at 20°C/°C^{-1}	0.00119
Critical temperature/°C	240
Critical pressure/bar	79.6
Dielectric constant at 20°C	33.6
Electrical conductivity at 18°C/mhos	4.4×10^{-7}
Viscosity at 20°C/Pa s	5.88×10^{-4}
Refractive index at 20°C	1.32840
Surface tension at 20°C/mN m^{-1}	22.6
Solubility in water	Miscible in all proportions
Flash point	
Abel closed-cup/°C	∼ 9.5
Pesky–Martens open-cup/°C	∼ 16
Auto-ignition temperature/°C	∼ 470
Explosive limits of methanol in air/%(v/v)	
lower	6.0
upper	36.5

Appendix 11.

Approximate Boiling Ranges of Hydrocarbon Feedstocks

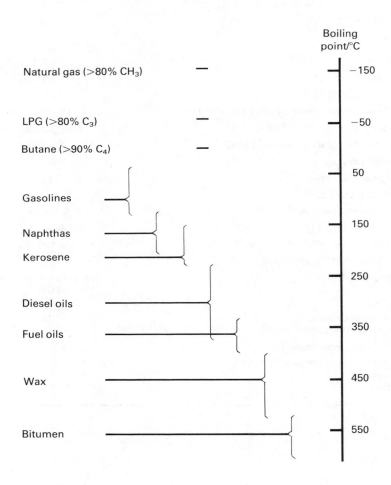

Appendix 12.

Monitoring Steam Reformer Tube Wall Temperature

Figure 5.20 in Chapter 5 shows that small changes in reformer tube wall temperature have profound effects on tube life: an increase of about 20°C above design will halve the tube life. Monitoring the maximum tube wall temperature is therefore vitally important. Temperature variation across the tubes in a reformer can occur for the following reasons.

(1) There may be a spread of flow rates through tubes because of poor catalyst packing or because of the design of the inlet header.

(2) There may be variations in the burner firing, resulting in non-uniform heat flux intensity at any level in the furnace, especially in top-fired furnaces near the top, where the influence of the flame is most pronounced.

(3) Loss of catalyst activity caused by poisoning or old age.

(4) Poor burner control due to blockage of jets or incorrect setting of combustion air register.

With most operators the tube wall temperature monitoring programme is usually based on visual examination and optical pyrometry, although occasionally some may also use direct-contact pyrometry. The methods available for measuring reformer tube wall temperature are detailed briefly below.

Visual Examination

Tubes are usually inspected every 2 hours. An experienced operator should be able to detect variations in maximum tube wall temperature of within ±30°C.

Measurement Using Optical Pyrometry

This is usually carried out weekly. The maximum tube wall temperature of all tubes in the furnace should be measured and recorded. The two types of instrument commonly used are detailed below. Both instruments suffer from the disadvantage that they are affected by radiation from furnace walls and burner flames, but with careful use they can give an accuracy of ±15–20°C.

Disappearing-filament pyrometer. With this instrument the brightness of an internal lamp is adjusted to match the brightness of the image of the tube.

Infrared sensor. The read-out from this pyrometer gives temperature directly. However, it is necessary to correct for tube-wall emissivity, and typically a value of 0.85 is used. Some models of this type of pyrometer incorporate a camera which enables a permanent record of the tube and temperature to be made.

Measurement Using Radiation Pyrometry

ICI uses a gold-cup pyrometer to check the calibration of the optical pyrometer. It is a direct-contact instrument with a useful range of 625–1200°C and an almost instantaneous response. It is mounted on a water-cooled probe to facilitate its use in the reformer furnace. The gold cup virtually eliminates errors due to emittance, and temporarily screens the target area from radiation from flames and refractory surfaces which are at a higher temperature. Properly used, this instrument has an accuracy of ±5°C.

Appendix 13.

Heat Released During Catalyst Reduction

Catalyst	ICI No.	$\triangle H_{R(298)}$/kJ kg^{-1}	T/°C	$\triangle H_R(T)$/kJ kg^{-1}
Primary reforming	46-1	+7.1	750	−26.4
Primary reforming	46-4	+3.8	750	−13.0
Primary reforming	57-3	+5.0	750	−18.0
Secondary reforming	54-3	+2.1	750	−7.1
Secondary reforming	54-4	+2.1	750	−8.4
High-temperature shift	15-4/5	−15.9 per 1% CrO_3	400	−49.0 per 1% CrO_3
Low-temperature shift	53-1	−306.8	200	−315.6
Methanation	11-3	+6.3	400	−8.4
Ammonia synthesis	35-4	+608.6	400	−487.6

The final column gives the heat of reduction at the temperature indicated in the previous column.

Appendix 14.

Heat Released During Catalyst Oxidation

Heats are calculated per kilogramme of typical catalyst in the form it is charged to the converter.

Process	Comments	$\Delta H_{R(298)}$/kJ kg^{-1}
Carbon burn-off		
$C + \frac{1}{2}O_2 \rightarrow CO$	To remove 1% C from 1 kg of charged catalyst	-92
$C + O_2 \rightarrow CO_2$	To remove 1% C from 1 kg of charged catalyst	-328
Sulphided cobalt molybdate		
$CoS + {}^3/_2O_2 \rightarrow CoO + SO_2$	To remove 1% S from 1 kg of charged catalyst	-140
Primary and secondary reforming catalysts		
$Ni + H_2O \rightarrow NiO + H_2$	Oxidation with steam (and reduction). For oxidation without steam, see methanation	Negligible
HT shift		
$2Fe_3O_4 + \frac{1}{2}O_2 \rightarrow 3Fe_2O_3$	To oxidize 1 kg of charged catalyst	-441
LT shift		
$Cu + \frac{1}{2}O_2 \rightarrow CuO$	To oxidize 1 kg of charged catalyst	-647
Methanation		
$Ni + \frac{1}{2}O_2 \rightarrow NiO$	To oxidize 1 kg of charged catalyst	-642
Ammonia synthesis		
$3Fe + 2O_2 \rightarrow Fe_3O_4$	To oxidize 1 kg of charged catalyst	-4696

Appendix 15.

Temperature Conversions

The fundamental temperature scale is the absolute, thermodynamic or Kelvin scale (K) which is based on the average kinetic energy of a perfect gas. The zero on the Kelvin scale is $-273.15°C$. The customary temperature unit in Europe is the Celsius (or Centigrade, °C) degree, which is 1/100 of the difference between the temperature of melting ice and boiling water at standard pressure. The Fahrenheit degree is 1/180 of this difference. The following relationships hold, and the accompanying chart on the next page provides a rapid means of interconverting Celsius and Fahrenheit measurements.

$0°C = 32°F = 273.15 \text{ K}$ $100°C = 212°F = 373.15 \text{ K}$

To convert °C to °F
$T°F = 32 + 1.8 t°C$

To convert °F to °C
$T°C = {}^5/_9 t°F - 17.78$

To convert °C to K
$T \text{K} = 273.15 + t°C$

To convert °F to K
$T \text{K} = 253.37 + {}^5/_9 t°F$

Appendices

Temperature Measurement
Celsius-Fahrenheit Temperature Conversion Chart

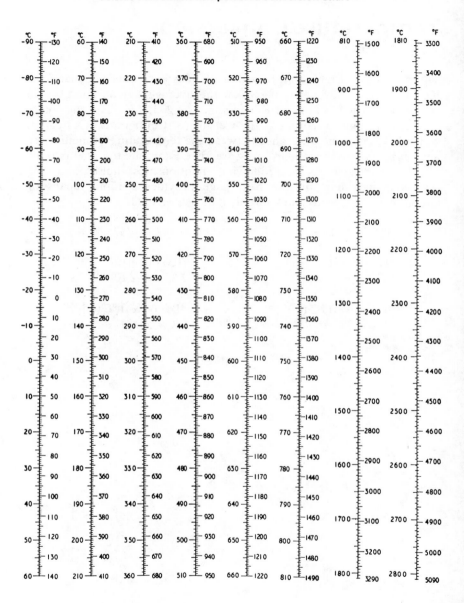

560

Appendix 16.

Specific Heats of Catalysts

The specific heat of a catalyst may be estimated as the sum of the contributions from each of its constituents. When the catalyst consists only of metal oxides the calculation is straightforward, but many catalysts contain residual carbonates or water which must be taken into account. A calculation of the heat requirement to raise the temperature of a converter requires information on the normal increase in specific heat with temperature, and also the heat required to drive off water or carbon dioxide, or the change in heat content due to chemical reactions such as reduction; there will be a decrease in both the weight and the specific heat once the catalyst has lost its water. Similarly, the specific heat of a catalyst in the reduced state will usually be slightly different from that in the oxidized state.

Fortunately, it is rarely necessary to know the specific heat of a catalyst accurately, and approximate values are normally adequate. Table A16.1 lists approximate values for the specific heat of typical catalysts used in ammonia plants; the values have been calculated for catalysts, in the manufactured state, under loss-free conditions (i.e. after water and CO_2 have been driven off).

Table A16.1. Specific heats of ammonia plant catalysts (cal g^{-1} $°C^{-1}$)

Duty	Typical Catalysts	Temperature/°C						
		20	100	200	300	400	500	800
Sulphur removal	ICI Catalyst 32-4	0.12	0.13	0.14	0.15	0.15		
Hydrodesulphurization	ICI Catalyst 41-6		0.19	0.22	0.24	0.26	0.27	0.27
Primary reforming	ICI catalysts 57-3, 46-1, 46-4, 46-9	0.18	0.20	0.21	0.22	0.23	0.24	0.25
Secondary reforming	ICI Catalyst 54-4	0.18	0.20	0.21	0.22	0.23	0.24	0.25
Methanation	ICI Catalyst 11-3	0.18	0.20	0.21	0.22	0.23	0.24	0.25
High-temperature CO conversion	ICI Catalysts 15-4/5	0.16	0.17	0.19	0.20	0.22	0.23	
Low-temperature CO conversion	ICI Catalyst 53-1	0.16	0.16	0.17	0.17			
Ammonia synthesis	ICI Catalyst 35-4	0.15	0.17	0.19	0.21	0.23	0.25	

Appendix 17.

Atomic Weights of the Common Elements[a]

Element	Symbol	Atomic number	Atomic weight
Aluminium	Al	13	26.98
Antimony	Sb	51	121.75
Argon	A	18	39.95
Arsenic	As	33	74.92
Barium	Ba	56	137.33
Beryllium	Be	4	9.01
Bismuth	Bi	83	208.98
Boron	B	5	10.81
Bromine	Br	35	79.90
Cadmium	Cd	48	112.41
Caesium	Cs	55	132.905
Calcium	Ca	20	40.08
Carbon	C	6	12.01
Cerium	Ce	58	140.12
Chlorine	Cl	17	35.45
Chromium	Cr	24	52.00
Cobalt	Co	27	58.93
Copper	Cu	29	63.55
Dysprosium	Dy	66	162.50
Erbium	Er	68	167.26
Europium	Eu	63	151.96
Fluorine	F	9	19.00
Gadolinium	Gd	64	157.25
Gallium	Ga	31	69.72
Germanium	Ge	32	72.59
Gold	Au	79	196.97
Hafnium	Hf	72	178.49
Helium	He	2	4.00
Holmium	Ho	67	164.93

17. Atomic Weights

Hydrogen	H	1	1.01
Indium	In	49	114.82
Iodine	I	53	126.90
Iridium	Ir	77	192.22
Iron	Fe	26	55.85
Krypton	Kr	36	83.80
Lanthanum	La	57	138.91
Lead	Pb	82	207.2
Lithium	Li	3	6.94
Lutetium	Lu	71	174.97
Magnesium	Mg	12	24.305
Manganese	Mn	25	54.914
Mercury	Hg	80	200.59
Molybdenum	Mo	42	95.94
Neodymium	Nd	60	144.24
Neon	Ne	10	20.18
Nickel	Ni	28	58.69
Niobium	Nb	41	92.91
Nitrogen	N	7	14.01
Osmium	Os	76	190.20
Oxygen	O	8	16.00
Palladium	Pd	46	106.42
Phosphorus	P	15	30.97
Platinum	Pt	78	195.08
Potassium	K	19	39.10
Praseodymium	Pr	59	140.91
Rhenium	Re	75	186.21
Rhodium	Rh	45	102.91
Rubidium	Rb	37	85.47
Ruthenium	Ru	44	101.07
Samarium	Sm	62	150.36
Scandium	Sc	21	44.96
Selenium	Se	34	78.96
Silicon	Si	14	28.09
Silver	Ag	47	107.87
Sodium	Na	11	22.99
Strontium	Sr	38	87.62
Sulphur	S	16	32.06
Tantalum	Ta	73	180.95

Appendices

Tellurium	Te	52	127.60
Thallium	Tl	81	204.38
Thorium	Th	90	232.04
Tin	Sn	50	118.69
Titanium	Ti	22	47.88
Tungsten	W	74	183.85
Uranium	U	92	238.03
Vanadium	V	23	50.94
Xenon	Xe	54	131.29
Yttrium	Y	39	88.91
Zinc	Zn	30	65.38
Zirconium	Zr	40	91.22

[a]Quoted to two decimal places, based on carbon-12 (IUPAC *Atomic Weights of the Elements*, 1979).

Appendix 18.

Measurement of Pressure Drop Across Steam Reformer Tubes

Each tube in a steam reformer must be packed uniformly with catalyst so that all of the tubes offer a similar resistance to the flow of process gas, otherwise it will not be evenly distributed over the reformer. If the flow is not balanced, catalyst in some tubes will tend to be overloaded with high gas flows while those tubes with low gas flows may tend to overheat. The simplest way to check the packing of a tube is to measure the pressure drop through it resulting from a fixed flow of air. The ICI pressure drop instrument shown in Figure 3.9 provides a convenient, accurate and reliable method for taking this measurement. The pressure drop across each tube in a complete furnace can then be adjusted to a suitable value and the problems of uneven flow distribution avoided.

The instrument provides a fixed flow rate of air by utilizing the phenomenon of choked flow through an orifice. Above a certain "critical" ratio of upstream to downstream pressure the mass flow rate through the orifice is a function only of the upstream pressure and the orifice diameter (at constant temperature) and can be calculated. The downstream pressure floats above atmospheric by an amount equal to the back pressure due to pressure drop, and is a measure of the pressure drop through the tube. All that is required is to load the reformer tubes with catalyst such that the instrument measures the same value across each tube. In practice there is a pressure drop range over which the catalyst may be charged. This ensures that the tubes have been vibrated sufficiently to give a stable level of catalyst, but not so excessively that a high pressure drop results and design rates cannot be achieved.

As well as establishing that catalyst packing is uniform, the inlet and exit branches of the reformer tube can also be checked to ensure that they are free from debris by comparing the expected (calculated) pressure drop with the instrument reading before the catalyst is charged. The instrument and its applications are illustrated in Figure A18.2 and typical readings expressed in psi are: exit branch 0.5; catalyst bed 10; inlet branch 0.3. The actual pressure across the catalyst depends on the tube diameter and the length packed with catalyst, as well as the size and shape of the catalyst used.

Appendices

Figure A18.1. ICI pressure drop instrument.

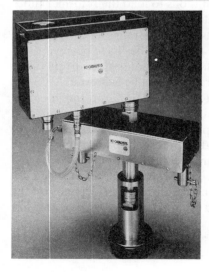

18. Pressure Drops

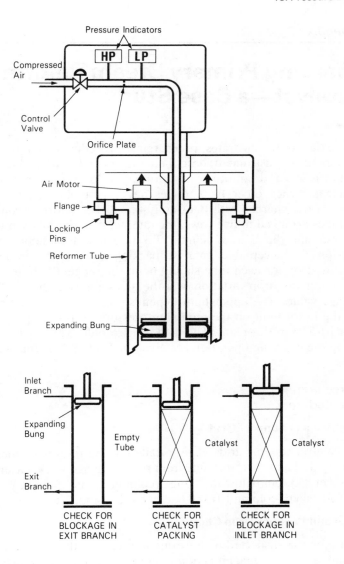

Figure A18.2. ICI pressure drop instrument and its applications.

Appendix 19.

Charging Primary Steam Reformer Catalyst—a Case Study

This case study illustrates precharging procedures and use of the pressure drop instrument described in Appendix 18 to ensure that catalyst is charged uniformly in a primary steam reformer. It is drawn from experience on a 1000 tonnes day^{-1} ammonia plant when the reformer was charged with ICI Catalyst 57-3. The reformer consisted of 225 tubes arranged in five rows, the internal diameter of each tube was 97.2 mm and the charged length 13.3 m. The criteria adopted on this occasion for acceptable catalyst packing were that the measured pressure drop for each tube should be in the range 9.5–13 psi, with an actual pressure drop variation over the tubes not exceeding ±5% of the average value. The upper limit depends on the design pressure drop, and the lower limit on the minimum pressure drop for a stable bed of Catalyst 57-3 in this plant. The data which would typically be recorded during the charging operation are given in Table A19.1 for one row of tubes.

Precharging Procedure

Installation of Support Grids

The catalyst support grids were installed in the reformer tubes over a 3-day period and their distance from the top flange was measured. The grids were 13.6 m below the flange (column 2) and three of them were badly installed (column 10) and had to be repositioned.

Visual and Pressure Drop Checks

Each tube was inspected by lowering a torch into the tube and viewing through an 8 × 30 magnification monocular. The pressure drop across each tube was also measured and typically found to be 1.7–1.8 psig (column 3). Debris found in tubes 6 and 25 by these checks was removed.

Catalyst Socking

Socking was carried out before the shutdown, by two teams of operators at the base of the reformer, using a drum hoist and sieving frame.

Weight of Catalyst Loaded

The approximate weight of catalyst to load a tube was calculated to be about 99 kg, which corresponded to 16 socks (column 4). The precise amount of catalyst charged to each tube was adjusted on the basis of pressure drop.

Charging Technique

The charging technique was as follows:

1. Charge five socks then vibrate* the tube for 10 seconds.
2. Charge another five socks and vibrate for 10 seconds.
3. Charge as many of the remaining six socks as necessary to achieve 300 mm outage in the tube then vibrate for 10 seconds.
4. Top up to 300 mm outage (column 4).
5. Record number of socks used and identify drums from which the catalyst was taken (column 5).
6. Measure the pressure drop for each tube using the ICI pressure drop instrument at a set upstream pressure of 60 psig (column 6).
7. Calculate the mean pressure drop range for a row of tubes† to check that each determination is within the accepted variation of ±5%.
8. Hammer‡ and top-up tubes with low pressure drops to raise pressure drop to within the agreed range (column 7).
9. Discharge all tubes with pressure drop values higher than the set range and record this (this amounted to nine tubes).
10. Top-up tube with additional catalyst as necessary and check the pressure drop (column 9).

FOOTNOTES

* An electrically driven vibrator was fastened to each tube in turn, approximately 1.5 m above the furnace tunnel and not less than 500 mm away from the nearest weld.

† The rest of the reformer should also be charged to this mean pressure drop range. An alternative procedure is to charge all tubes in the reformer before calculating the mean pressure drop range.

‡ Leather-faced hammers were used to strike the side wall of the top part of the tube above the furnace roof to adjust the pressure drop upwards. A total of 12 hammers were used to destruction.

Appendices

ROW NUMBER: _____B_____　　　　　　　　　　　DATE: __APRIL 1985__

1	2		3	4	5	6	7	8	9	10
	INSPECTION		EMPTY PD (psi)	No. SOCKS TO 300mm OUTAGE	DRUM NUMBER	PD (psi)	PD (psi)	RE-CHECKED PD (psi)	TOP UP / PD (psi)	REMARKS
TUBE No.	GRID DEPTH (m)	VISUAL								
1	13.6	✓	1.7	16	9 2060 / 2 2059 / 3 2123	11.2		11.2	0.3 / 12.3	
2	13.6	✓	1.7	16	8 2125 / 8 2124	12.2		12.2	0 / 12.2	
3	13.6	✓	1.8	16	5 2068 / 6 2124 / 3 2123 / 2059	12.0		11.8	0.1 / 12.1	
4	13.6	✓	1.7	16	11 2126 / 6 2125	12.5		12.5	0 / 12.5	
5	13.6	✓	1.8	16	4 2126 / 12 2098	11.9		12.1	0 / 12.1	
6	13.6	*✓	1.7	16	4 2098 / 11 2096 / 11 2097	11.9		11.9	0.1 / 12.0	*INSULATING MATERIAL FOUND IN TUBE
7	13.6	✓	1.8	16	2 2098 / 7 2096 / 7 2097	12.1		11.8	0.2 / 12.0	
8	13.6	✓	1.7	16	2 2126 / 12 2098 / 2 2096	11.9		11.8	0.2 / 12.0	
9	13.6	✓	1.7	16	1 2124 / 10 2126 / 5 2125	12.2		12.1	0 / 12.1	
10	13.6	✓	1.7	16	5 2124 / 11 2125	12.2		12.1	0 / 12.1	
11	13.6	✓	1.7	16	3 2060 / 6 2123 / 9 2124	12.5		12.3	0 / 12.3	
12	13.6	✓	1.8	16	1 2041 / 11 2123 / 4 2010	12.5		12.4	0 / 12.4	
13	13.6	✓	1.7	16	5 2041 / 10 2060 / 1 2123	11.8		11.8	0.1 / 12.0	
14	13.6	✓	1.7	16	13 2059 / 3 2041	11.3		11.7	0.2 / 12.0	
15	13.6	✓	1.7	16	12 2059 / 3 2041 / 1 2042	12.0		11.8	0.2 / 12.3	
16	13.6	✓	1.8	16	12 2115 / 3 2039 / 1 2042	12.1		11.9	0.2 / 12.3	
17	13.6	✓	1.8	16	13 2042 / 3 2040	11.8		11.6	0.3 / 12.4	
18	13.6	✓	1.7	15.5	14.5 2041 / 1 2042	11.7		11.6	0.3 / 12.2	
19	13.6	✓	1.7	16	6 2040 / 10 2042	11.8	0.4 / 12.7	12.7	0 / 12.6	
20	13.6	✓	1.8	16	6 2039 / 4 2040 / 6 2042	11.4	0.5 / 12.7	12.7	0 / 12.6	
21	13.6	✓	1.7	16	8 2039 / 8 2040	11.7	0.4 / 12.7	12.7	0 / 12.6	
22*	13.63 / 13.60	✓	1.7	15	4 2044 / 2 2045 / 9 2043	12.0	0.2 / 12.6	12.6	0 / 12.5	*GRID INCLINED – RESET
23	13.60	✓	1.8	15.5	15.5 2044	12.0	0.2 / 12.6	12.6	0 / 12.4	

Table A19.1. Typical data obtained during charging of one row of tubes of a primary reformer

19. Steam Reformer Charging

Table A19.1 continued

ROW NUMBER: __B__   DATE: __APRIL 1985__

1	2		3	4	5	6	7	8	9	10
	INSPECTION		EMPTY PD (psi)	No. SOCKS TO 300mm OUTAGE	DRUM NUMBER	PD (psi)	TOP UP / PD (psi)	RE-CHECKED PD (psi)	TOP UP / PD (psi)	REMARKS
TUBE No.	GRID DEPTH (m)	VISUAL								
24*	13.40 / 13.60	✓	1.7	15.5	3.5 / 9 / 3	2046 / 2044 / 2043	12.2	0.2 / 12.8	12.5	0 / 12.5	*GRID INCLINED - RESET
25	13.6	*/✓	1.7	16	2 / 1 / 13	2115 / 2045 / 2046	11.8	0.3 / 12.6	12.4	0 / 12.4	*INSULATING MATERIAL FOUND IN TUBE
26	13.6	✓	1.8	15.5	13.5 / 1 / 1	2040 / 2043 / 2115	11.8	0.3 / 12.7	12.5	0 / 12.5	
27	13.6	✓	1.8	15.5	13.5 / 2	2043 / 2045	12.1	0.2 / 12.6	12.4	0 / 12.4	
28	13.6	✓	1.8	15.5	6.5 / 1 / 8	2045 / 2115 / 2043	12.3	0.1 / 12.6	12.4	0 / 12.4	
29	13.6	✓	1.8	15.5	14.5 / 1	2045 / 2115	12.2	0.2 / 12.6	12.4	0 / 12.4	
30	13.6	✓	1.8	16	14 / 2	2115 / 2045	11.7	0.3 / 12.7	12.5	0 / 12.5	
31	13.6	✓	1.8	16	12 / 4	2085 / 2066	12.6	0 / 12.6	12.3	0 / 12.3	
32	13.6	✓	1.8	16	9 / 7	2066 / 2086	11.5	0.4 / 12.6	12.4	0 / 12.4	
33	13.6	✓	1.8	16	4 / 9 / 3	2066 / 2036 / 2065	11.7	0.4 / 12.6	12.4	0 / 12.4	
34	13.6	✓	1.7	16	2 / 11 / 3	2066 / 2065 / 2063	12.4	0.2 / 12.9	12.7	0 / 12.7	
35	13.6	✓	1.8	16	12 / 4	2063 / 2064	11.7	0.4 / 12.7	12.6	0 / 12.6	
36	13.6	✓	1.8	16	8 / 8	2064 / 2116	12.5	0 / 12.5	12.2	0 / 12.2	
37	13.6	✓	1.8	16	2 / 14	2116 / 2118	12.3	0.2 / 12.8	12.3	0 / 12.3	
38*	13.55 / 13.60	✓	1.8	16	3 / 7 / 2	2116 / 2118 / 2115 / 2117	12.0	0.3 / 12.8	12.6	0 / 12.6	*GRID INCLINED - RESET
39	13.6	✓	1.7	16	2 / 4 / 10	2118 / 2116 / 2117	12.3	0.2 / 12.6	12.3	0 / 12.3	
40	13.6	✓	1.8	16	3 / 9 / 4	2116 / 2118 / 2117	12.5	0 / 12.5	12.2	0 / 12.2	
41	13.6	✓	1.8	15	2 / 8 / 1	2064 / 2063 / 2116 / 2118	12.4	0.1 / 12.7	12.5	0 / 12.5	
42	13.6	✓	1.8	16	12 / 4	2064 / 2063	12.5	0 / 12.5	12.2	0 / 12.2	
43	13.6	✓	1.8	16	5 / 8 / 3	2065 / 2063 / 2064	12.7	0 / 12.7	12.4	0 / 12.4	
44	13.6	✓	1.7	16	5 / 10 / 1	2063 / 2065 / 2066	12.3	0.1 / 12.6	12.3	0 / 12.3	
45	13.6	✓	1.8	16	8 / 8	2066 / 2068	12.1	0.2 / 12.7	12.4	0 / 12.4	

Appendix 20.

Equilibrium Constants for the Reaction $ZnO + H_2S \rightleftharpoons ZnS + H_2O$

There are two forms of zinc sulphide: Wurtzite (α-ZnS) and Sphalerite (β-ZnS). Which is formed from zinc oxide by reaction with hydrogen sulphide depends on the actual operating conditions. Both forms are seen in discharged plant samples of zinc oxide. Sphalerite is the most stable form, and perhaps Wurtzite is formed first and this then converts to sphalerite. The thermodynamic data used here is based on a recent literature search, and is the average of the more reputable results. For Wurtzite $\triangle G_f$ (298K) = -184.8 kJ mol^{-1}, and for Sphalerite $\triangle G_f$ (298K) = -198.1 kJ mol^{-1}.

The calculated equilibrium constants depend on which form of zinc sulphide is assumed to be present, and two sets of values are therefore given in the following table.

Table A20.1. Equilibrium constants for formation of zinc sulphide from zinc oxide

Temperature °C	$Kp = P_{H_2O}/P_{H_2S}$	
	Wurtzite	Sphalerite
25	7.82×10^{10}	1.73×10^{13}
50	1.10×10^{10}	1.60×10^{12}
75	2.05×10^{9}	2.09×10^{11}
100	4.80×10^{8}	3.59×10^{1}
125	1.35×10^{8}	7.71×10^{9}
150	4.43×10^{7}	1.99×10^{9}
175	1.65×10^{7}	5.98×10^{8}
200	6.81×10^{6}	2.04×10^{8}
225	3.08×10^{6}	7.80×10^{7}
250	1.51×10^{6}	3.27×10^{7}
275	7.88×10^{5}	1.48×10^{7}
300	4.37×10^{5}	7.24×10^{6}

Table A20.1. *Continued*

325	2.55×10^5	3.75×10^6
350	1.55×10^5	2.05×10^6
375	9.87×10^4	1.18×10^6
400	6.49×10^4	7.08×10^5
425	4.41×10^4	4.41×10^5
450	3.08×10^4	2.85×10^5
475	2.21×10^4	1.89×10^5
500	1.62×10^4	1.30×10^5

Appendix 21.

Temperature Measurement in Catalyst Beds

Measurements of temperature gradients within catalyst beds is an important part of monitoring catalyst performance. This is commonly done using fixed thermocouples, but it is also possible to use movable thermocouples within a sheath to provide an almost continuous measurement of temperature along the path of the thermosheath. A and B in Figure A21.1 demonstrates two ways in which a long thermosheath is commonly installed in fixed bed catalyst reactors on an ammonia plant while the use of a plurality of thermosheaths is illustrated in C.

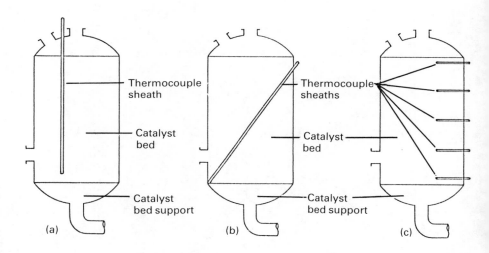

Figure A21.1. Possible arrangements of thermocouple sheaths in catalyst beds: (a) single vertical thermosheath; (b) single diagonal thermosheath; (c) multiple horizontal thermosheaths. Travelling thermocouples are often used in (a) and (b).

In ICI designs method A is normally favoured over method B. Here the thermosheath passes vertically through the bed, and advantages of this method include:

1. There is only one entry into the vessel.

2. The risk of damage to the sheath inside the vessel during catalyst charging or discharging, is less than in method B. The risk of damage on the outside of the vessel during routine maintenance is less than with method C.

3. The problem of the conduction of heat along the thermosheath is minimised compared with the other methods. This enables a truer indication of the temperatures in the reactor to be obtained.

4. A 'travelling' thermocouple can be introduced into the sheath enabling very accurate monitoring of catalyst performance to be carried out. Having measurements in a single vertical direction provides data more easily interpreted than method B.

With this method an air purge should be provided to remove water in the event of failure of the weather seal.

In method C a series of relatively short thermosheaths pass horizontally into the catalyst bed. Principal advantages of this are:

1. The position of the thermocouples is well defined, but it is unusual to be able to monitor the central regions of the bed (see 3 below).

2. There is less chance of rain water collecting in the bottom of the sheath due to failure of the weatherproof gland.

3. The thermosheath should normally extend into the vessel to within 40% of the centre line. The arrangement permits ease of access for vessel maintenance and avoids measurement errors due to heat loss to the vessel wall. However the top thermosheath should not be positioned too close to the top of the catalyst bed since normal settling and shrinkage could merely result in the duplication of the gas inlet temperature.

References

1. L. Pearce Williams, *Michael Faraday*, Chapman and Hall, London (1965), p.274.
1a. J. T. Richardson, *Principles of Catalyst Development*, Plenum Publishing Corp., New York (1989).
2. D. L. Baulch, D. D. Drysdale and D. G. Horne, *Evaluated Kinetic Data for High Temperature Reactions*, Volume 2, *Homogeneous Gas-phase reactions of the H_2–N_2–O_2 Systems*, Butterworth, London (1973).
3. D. R. Stull and H. Prophet, *JANAF Thermochemical Tables*, 2nd edition, NBS, Washington, D.C. (1971).
4. G. N. Lewis and M. Randall, *Thermodynamics and the Free Energy of Chemical Substances*, McGraw-Hill, New York (1923), p. 567.
5. N. D. Spencer, R. C. Schoonmaker and G. A. Somorjai, *J. Catal.*, **74**, 129 (1982).
6. G. Ertl, *Catal. Rev.—Sci. Eng.*, **21**, 201 (1980).
7. G. Ertl, *CRC Lit. Rev. Solid State Material Sci.*, **10**, 249 (1982).
8. G. Ertl, in: *Catalysis: Science and Technology*, Volume 4 (J. R. Anderson and M. Boudart, eds), Springer-Verlag, Berlin (1983), p. 209.
9. M. Grunze, M. Golze, J. Fuhler, M Neumann and E. Schwarz, *Proc. 8th Int. Congress Catal.*, Berlin, 1984, Volume IV, Verlag Chemie, Weinheim (1984), p. 133.
10. M. Bowker, I. B. Parker and K. C. Waugh, *Appl. Catal.*, **14**, 101 (1985).
11. D. A. Dowden, *Chem. Eng. Prog. Symposium* No. 73, **63**, 90 (1967); D. A. Dowden, C. R. Schnell and G. T. Walker, *Proc. IV Int. Congress Catal.* Moscow, 1968, Volume 2, Akademiai kiado, Budapest (1971), p. 201; D. A. Dowden, *Chim. Ind. (Milan)*, **55**, 639 (1973).
12. M. Boudart, *Proc. VI Int. Congress Catal.*, London, 1976, Volume 1, Chemical Society, London (1977), p.1; M. Boudart and M. A. McDonald, *J. Phys. Chem.*, **88**, 2185 (1984).
13. G. C. Chinchen, R. H. Logan and M. S. Spencer, *Appl. Catal.*, **12**, 69, 80 and 97 (1984).
14. L. Lloyd and M. V. Twigg, *Nitrogen*, No. 118, 30 (1979).
15. G. C. Chinchen, *Proc. Fertilizer Soc.*, No. 171 (1978).
16. S. P. S. Andrew, *Chemtech*, **9**, 180 (1979).
17. *Preparation of Catalysts I* (B. Delmon, P. A. Jacobs and G. Poucelet, eds), *Studies in Surface Science and Catalysis*, Volume 1, Elsevier, Amsterdam (1976).
18. *Preparation of Catalysts II*, (B. Delmon, P. Grange, P. A. Jacobs and G. Poucelet, eds), *Studies in Surface Science and Catalysis*, Volume 3, Elsevier, Amsterdam (1979).
19. *Preparation of Catalysts III*, G. Poucelet, P. Grange and P. A. Jacobs, eds), *Studies in Surface Science and Catalysis*, Volume 16, Elsevier, Amsterdam (1983).
20. G. W. Bridger and C. B. Snowdon, in reference 1, p. 126.
21. G. Ertl, D. Prigge, R. Schloegl and M. Weiss, *J. Catal.*, **79**, 359 (1983).
22. E. Lieber and F. L. Morritz, *Adv. Catal.*, **5**, 417 (1953).
23. G. J. K. Acres, A. J. Bird, J. W. Jenkins and F. King in: *Catalysis*, Volume 4, (C. Kemball and D. A. Dowden, eds), Specialist Periodical

Report, Royal Society of Chemistry, London (1984), p. 1.
24. G. W. Bridger, McRobert Award Lecture, Council of Engineering Institutions (1975).
25. S. P. S. Andrew, *Chem. Eng. Sci.*, **36**, 1431 (1981).
26. S. P. S. Andrew, *Ind. Eng. Chem., Prod. Res. Dev.*, **8**, 321 (1969).
27. R. M. Barrer, *Zeolites and Clay Minerals*, Academic Press, London (1978).
28. G. C. Chinchen *(ICI), Brit. Pat. No. 1,500,089; U.S. Pat. No. 4,142,988.*
29. L. H. Baekeland, *J. Ind. Eng. Chem.*, **8**, 184 (1916).
30. T. Baird in: *Catalysis*, Volume 5 Specialist Periodical Report, (G. C. Bond and G. Webb, eds), Royal Society of Chemistry, London (1982), p. 172.
31. P. Courty, D. Durand, E. Freund and A. Sugier, *J. Mol. Catal.*, **17**, 241 (1982).
32. R. W. Joyner, *Catalysis*, Volume 5 (G. C. Bond and G. Webb, eds), Specialist Periodical Report, Royal Society of Chemistry, London (1982), p.1.
33. G. A. Somorjai and S. M. Davis, *Chemtech*, 502 (1982).
34. G. A. Somorjai, *Adv. Catal.*, **26**, 1 (1977).
35. M. W. Roberts, *Adv. Catal.*, **29**, 55 (1980).
36. R. J. Madix, *Adv. Catal.*, **29**, 1 (1980).
37. P. H. Emmett, *The Physical Basis for Heterogeneous Catalysis* (E. Drauglis and R. I. Jaffee, eds), Plenum, New York (1975), p. 3.
38. A. J. Lecloux, *Catalysis: Science and Technology*, Volume 2 (J. R. Anderson and M. Boudart, eds), Springer-Verlag, Berlin (1981), p. 171.
39. ASTM D 4058-81.
40. K. S. W. Sing, *Pure Appl. Chem.*, **54**, 2201 (1982).
41. D. H. Everett, G. D. Parfitt, K. S. W. Sing and R. Wilson, *J. Appl. Chem. Biotechnol*, **24**, 199 (1974).
42. British Standard, BS 4359, Part 1 (1969); Part 3 (1970); ASTM D 3663-84.
43. Papers in *Powder Technol.*, **29**, 1–208 (1981).
44. L. L. Hegedus and E. E. Peterson, *Catal. Rev.—Sci. Eng.*, **9**, 245 (1974).
45. A. M. R. Difford and M. S. Spencer, *AIChE Symp. Ser.*, No. 143, **70**, 42 (1974).
46. J. B. Butt and V. W. Weekman, *ibid.*, p. 27.
47. L. K. Doraiswamy and D. G. Tajbl, *Catal. Rev.—Sci. Eng.*, **10**, 177 (1975).
48. J. D. Rankin and J. Plummer, *Sixth Ibero-American Catalyst Symposium*, Rio de Janeiro, p. 13 (1978).
49. E. G. Christoffel, *Catal. Rev.—Sci. Eng.*, **24**, 159 (1982).
50. J. M. Berty, *Catal. Rev.—Sci. Eng.*, **20**, 75 (1979).
51. R. J. Kokes, H. H. Tobin and P. H. Emmett, *J. Am. Chem. Soc.*, **77**, 5860 (1955); W. K. Hall and P. H. Emmett, *J. Am. Chem. Soc.*, **79**, 2091 (1957); W. K. Hall, D. S. McIver and H. P. Weber, *Ind. Eng. Chem.*, **52**, 421 (1960).
52. M. S. Spencer and T. V. Whittam, *Acta Phys. Chem.*, **24**, 307 (1978).
53. L. Gonzalez-Tejuca, K. Alka, S. Namba and J. Turkevich, *J. Phys. Chem.*, **81**, 1399 (1977).
54. M. S. Spencer, *J. Catal.*, **67**, 259 (1981); **94**, 148 (1985).
55. R. J. Madon and M. Boudart, *Ind. Eng. Chem., Fundam.*, **21**, 438 (1982).
56. S. P. S. Andrew and G. C. Chinchen, in: *Catalyst Deactivation* (B.

Delmon and G. F. Froment, eds), Elsevier, Amsterdam (1980), p. 141.
57. J. K. Musick, F. S. Thomas and J. E. Johnson, *Ind. Eng. Chem.*, *Process Des. Dev.*, **11**, 350 (1972).
58. I. Caldwell, *Appl. Catal.*, **4**, 13 (1982); **8**, 100, 293 (1983); J. M .Berty, *Appl. Catal.*, **8**, 289 (1983).
59. S. Szepe and O. Levenspiel, *Proc. European Fed.*, *4th Chem. Reaction Eng.*, Brussels, Pergamon Press, Oxford (1970); *J. Catal.*, **23**, 881 (1968).
60. O. Levenspiel, *J. Catal.*, **225**, 265 (1972).
61. J. B. Butt, C. K. Wachter and R. M. Billimoria, *Chem. Eng. Sci.*, **33**, 1321 (1978).
62. E. E. Peterson and M. S. Pachelo, *ACS Symp. Ser.*, **237**, 363 (1984).
63. I. R. Shannon, in: *Catalysis* Volume 2, (C. Kemball and D. A. Dowden, eds), Specialist Periodical Report, Chemical Society, London (1978), p. 28.
63a. S. R. Bane, D. R. Strongin and G. A. Somorjai, *J. Phys. Chem.*, **90**, 472b (1986); D. R. Strongin, J. Carrazza, S. R. Bane and G. A. Somorjai, *J. Catal.*, **103**, 213 (1987); D. R. Strongin, S. R. Bane and G. A. Somorjai, *J. Catal.*, **103**, 289 (1987).
64. J. A. Dumesic, H. Topsøe, S. Khammouma and M. Boudart, *J. Catal.*, **37**, 503 (1975); J. A. Dumesie, H. Topsøe and M. Boudart, *J. Catal.*, **37**, 513 (1975).
65. G. C. Chinchen, P. J. Denny, M. S. Spencer, K. C. Waugh and D. A. Whan, *Preprints, ACS Division of Fuel Chemistry*, **29**, 178 (1984) Annual Meeting, Philadelphia (1984).
66. G. C. Chinchen and K. C. Waugh, *J. Catal.*, **97**, 280 (1986).
66a. G. C. Chimchan, M. S. Spencer, K. C. Waugh and D. A. Whan, *J. Chem. Soc., Faraday Trans. I*, **83**, 2193 (1987).
67. M. Bowker, H. Houghton and K. C. Waugh, *J. Chem. Soc., Faraday Trans. I*, **77**, 3023 (1981); **78**, 2573 (1982); *J. Catal.*, **79**, 431 (1983); M. Bowker, H. Houghton, K. C. Waugh, T. Giddings and M. Green, *J. Catal.*, **84**, 252 (1983); M. Bowker, J. N. K. Hyland, H. D. Vandervell and K. C. Waugh, *Proc. 8th Int. Congress Catal.*, Berlin, 1984, Verlag Chemie, Weinheim (1984), Volume 2, p. 35.
68. G. C. Chinchen, in: *Catalysis*, (J. R. Jennings, ed.), Critical Reports in Applied Chemistry, Society of Chemical Industry, London (1985).
69. S. P. S. Andrew, *Post Congress Symposium, 7th Int. Congress Catal.*, Osaka, July 1980.
70. M. S. Spencer, in: *Catalysis* (J. R. Jennings, ed), Critical Reports in Applied Chemistry, Society of Chemical Industry, London (1985).
71. J. H. Sinfelt, *Catalysis: Science and Technology*, Volume 1 (J. R. Anderson and M. Boudart, eds), Springer-Verlag, Berlin (1981), p. 257.
72. *Studies in Surface Science and Catalysts*, Volume 6, *Catalyst Deactivation* (B. Delmon and G. F. Froment, eds), Elsevier, Amsterdam (1980).
73. *Studies in Surface Science and Catalysis*, Volume 4, *Growth and Properties of Metal Clusters* (J. B. Bourdon, ed.), Elsevier, Amsterdam (1979).
74. S. E. Wanke and P. C. Flynn, *Catal. Rev.—Sci. Eng.*, **12**, 93 (1975).
75. R. E. Gadsby and J. G. Livingstone, Preprint, AIChE Meeting, Denver, August 1977.
76. P. J. Denny and M. V. Twigg, in reference 72, p. 577.
77. J. B. Butt, *Chemical Reaction Engineering* (K. B. Bischoff, ed.) *Advances in Chemistry* No. 109, American Chemical Society, Washington, D.C.

(1972), p. 259.
78. J. D. Rankin and J. G. Livingstone, Preprint AIChE Meeting, Portland, Oregon, August 1980.
79. T. C. Ho, *J. Catal.*, **86**, 48 (1984).
80. J. B. Butt, in reference 72, p. 21.
81. E. B. Maxted, *Adv. Catal.*, **3**, 129 (1951).
82. C. H. Bartholomew, P. K. Agrawal and J. R. Katzer, *Adv. Catal.*, **31**, 136 (1982).
83. A. Wheeler, *Adv. Catal.*, **3**, 250 (1951); in: *Catalysis*, Volume 2 (P. H. Emmett, ed.), Reinhold, New York (1955), p.219.
84. G. J. K. Acres, A. J. Bird, J. W. Jenkins and F. King, in: *Characterisation of Catalysts* (J. M. Thomas and R. M. Lambert, eds), Wiley, Chichester (1980), p.55.
85. S. Ergun, *Chem. Eng. Prog.*, **48**(2), 89 (1952).
86. L. J. Gillespie and J. A. Beatties, *Phys. Rev.*, **36**, 743 (1930).
87. W. J. Thomas and S. Portalski, *Ind. Eng. Chem.*, **50**, 967 (1958).
88. K. Amundsen and E. Holte, *ICI Catalyst Conference, Teheran*, Paper 4 (1975).
89. A. D. Engelbrecht and L. J. Partridge, *TVA Symposium Ammonia from Coal*, p. 176 (1979).
90. P. D. Becker, *TVA Symposium Ammonia from Coal*, p. 44 (1979).
91. C. Torello, *ICI Catalyst Conference, Florence*, Paper 7 (1984).
92. G. W. Bridger, *3rd Symposium on Catalysis of The Catalyst Society of India at Dehra Dun* (1977).
93. B. W. Burkcon and R. C. Coleman, *CEP*, **55** (June 1977).
94. E. F. Need, *ICI Operating Symposium, Amsterdam*, Paper 7 (1974).
95. J. B. Gibson and D. P. Harrison, *Ind. Eng. Chem., Process Des. Dev.*, **19**, 231 (1980).
96. P. J. H. Carnell and D. P. Denny, *AICE Ammonia Safety Symposium, San Francisco*, Paper 94c (November 1984).
97. P. J. H. Carnell and P. E. Starkey, *Chem. Eng.*, **408**, 30 (1984).
98. T. Frenge, *ICI Operating Symposium, London*, Paper 11 (1978).
99. B. Nielsen and J. Villadsen, *Appl. Catal.*, **11**, 123 (1984).
100. G. J. Samuelson, *Petroleum Engr.* (December 1954).
101. T. Sandal, *ICI Catalyst Conference, Florence*, Paper 2 (1984).
102. G. W. Bridger, A. M. R. Difford and W. S. Lindsay, *ICI Operating Symposium, Amsterdam*, Paper 4 (1974).
103. E. Berecz and M. Balla-achs, *Gas Hydrates*, Elsevier, Amsterdam (1983); J. L. Cox (ed.), *Natural Gas Hydrates*, Butterworths, Boston (1983).
104. J. P. Van Hook, *Catal. Rev.—Sci. Eng.*, **21**, 1 (1981).
105. N. M. Bodrov, L. O. Apel'baum and M. I. Temkin, *Kinet. Catal.*, **5**, 614 (1964).
106. N. M. Bodrov, L. O. Apel'baum and M. I. Temkin, *Kinet. Catal.*, **8**, 696 (1967).
107. P. G. Wright, P. G. Ashmore and C. Kemball, *Trans. Faraday Soc.*, **54**, 1692 (1958).
108. C. Kemball, *Proc. Roy. Soc.*, **A207**, 529 (1951); **A217**, 376 (1953).
109. G. W. Bridger, *Catal. SPR Chem. Soc. Lond.*, **3**, 39 (1980).
110. J. R. Rostrup-Nielsen, *Catal. Sci. Technol.*, **5**, 1 (1984).
111. S. Mori and M. Uchiyana, *J. Catal.*, **42**, 323 (1976).
112. S. S. Grover, *Chim. Ind.*, **103**, 93 (1970); *Hydrocarbon Processing*, **49**(4),

References

109 (1970).
113. H. F. A. Topsøe, *J. Inst. Gas Engrs*, **6**, 401 (1966).
114. C. Raggio, E. Sebastini and C. Todini, *Chim. Ind.*, **57**, 663 (1975); E. Sebastini, C. Todini and C. Raggio, *Chim. Ind.*, **57**, 739 (1975); C. Todini, C. Raggio and E. Sebastini, *Chim. Ind.*, **57**, 818 (1975); M. H. Hyman, *Hydrocarbon Processing*, **47**, 131 (1968).
115. R. D. Arganat and A. G. Leibush, *Khim. Prom.*, **7**, 488 (1975).
116. A. A. Khomenko, L. O. Apel'baum, F. S. Shub, Y. S. Snagovskii and M. I. Tempkin, *Kinet. Catal. (USSR)*, **12**, 367 (1971).
117. R. D. Agarnat, A. G. Leibush and V. P.Semenov, *Kinet. Catal. (USSR)*, **17**, 1070 (1976).
118. T. Q. Phung Quach and D. Rouleau, *J. Appl. Chem. Biotechnol.*, **25**, 445 (1975).
119. D. W. Allen, E. R. Gerhard and M. R. Likins, *Ind. Eng. Chem. (Process Design)*, **14**, 256 (1975).
120. E. Kikuchi, S. Tanaka, Y. Yamazaki and Y. Morita, *Bull. Japan Petrol. Inst.*, **16**, 95 (1974).
121. M. Moayeri and D. L. Trimm, *J. Appl. Chem. Biotechnol.*, **26**, 419 (1976).
122. N. T. Meshenko, V. V. Veselov, F. S. Shub and M. I. Temkin, *Kinet. Catal. (USSR)*, **18**, 796 (1977).
123. E. K. Nazarov, G. S Golovin, P. M. Reshchikov and V. N. Sevrynkov, *Zh. Prikl. Khim.*, **48**, 1564 (1975).
124. E. Kikuchi, Y. Yamazaki and Y. Morita, *Bull. Japan Petrol. Inst.*, **17**, 3 (1975).
125. K. Takami, A. Ingarashi and Y. Ogino, *Bull. Japan Petrol. Inst.*, **19**, 37 (1977).
126. S. Komatsu and G. Yamaguchi, *Bull. Japan Petrol. Inst.*, **16**, 99 (1974).
127. G. Traply, G. Parlagh, G. Racz, P. Steingaszner and G. Szekely, *Acta Chim. Acad. Sci. Hung.*, **88**, 223 (1976); G. Traply, G. Parlagh, G. Racz, P. Steingaszner and G. Szekely, *Acta Chim. Acad. Sci. Hung.*, **88**, 235 (1976).
128. G. L. Rabinovich, G. N. Maslynansky and L. N. Treiger, *Symp. Mechanisms of Hydrocarbon Reactions, Siofok, Preprint 48*, 97 (1973); G. L. Rabinovich and V. N. Mozhaiko, *Neftekhimiya*, **16**, 187 (1976); G. L. Rabinovich, L. M. Treiger and G. N. Maslyanskii, *Petrol. Chem. USSR*, **13**, 199 (1973).
129. E. Kikuchi, K. Ito and Y. Morita, *Bull. Japan Petrol. Inst.*, **17**, 206 (1975).
130. D. C. Grenoble, *J. Catal.*, **51**, 203 (1978); D. C. Grenoble, *J. Catal.*, **51**, 212 (1978).
131. K. Kochloefl, *Proc. 6th Int. Conf. Catalysis, London*, Volume 2, 1122 (1977).
132. F. J. Dent, L. A. Moignard, A. H. Eastwood, W. H. Blackburn and D. Hebden, *Trans. Inst. Gas Engrs*, **95**, 604 (1945–1946).
132a. *Materials Technology in Steam Reforming Processes*, Ed. C. Edeleanu, Perganon Press, Oxford (1966).
133. J. R. Rostrup-Nielsen, *J. Catal.*, **31**, 173 (1973); **33**, 184 (1974).
134. J. R. Rostrup-Nielsen, *Ammonia Plant Safety*, **15**, 82 (1972).
135. J. R. Rostrup-Nielsen, *Steam Reforming Catalyst*, Danish Technical Press, Copenhagen (1975).
136. *Thorpes Dictionary of Applied Chemistry* (4th edn), Volume V (1941).

137. S. A. Topham, *Catalysis*, **7**, 1 (1985).
138. K. Holdermann, *Im Banne der Chemie Carl Bosch Leben und Werk*, Econ-Verlag, Dusseldorf (1953).
139. L. Mond and C. Langer, British Patent 12,608 (1888).
140. C. Bosch and W. Wild, Canadian Patent 153,379 (1914).
141. M. V. Twigg, in: *Catalysis and Chemical Processes* (R. Pearce and W. R. Patterson, eds), Leonard Hill, London (1981), p. 124.
142. M. Appl, *Nitrogen*, **100**, 47 (March/April 1976).
143. H. Bohlbro, *An Investigation on the Kinetics of the Conversion of Carbon Monoxide with Water Vapour over Iron Oxide Based Catalysts* (2nd edn), Gellerup, Copenhagen (1969).
144. D. M. Ruthven, *Can. J. Chem. Eng.*, **47**, 327 (1969).
145. P. Fott, J. Vosolsobe and V. Glaser, *Coll. Czech. Chem. Commun.*, **44**, 652 (1979).
146. W. F. Podolski and Y. G. Kim, *Ind. Eng Chem. Process, Des. Dev.*, **13**, 415 (1974).
147. G. C. Chinchen, R. H. Logan and M. S. Spencer, *Appl. Catal.*, **12**, 69 (1984).
148. G. C. Chinchen, R. H. Logan and M. S. Spencer, *Appl. Catal.*, **12**, 89 (1984).
149. G. C. Chinchen, R. H. Logan and M. S. Spencer, *Appl. Catal.*, **12**, 97 (1984).
150. G. C. Chinchen, British Patent 1,500,089 (1976); US Patent 4,142,988 (1979).
151. J. S. Campbell and S. R. Metcalfe, cited in *Catalyst Handbook*, Wolfe Publishing, London (1970) p.115.
152. Y. Kaneko and S. Oki, *J. Res. Inst. Catal. Hokkaido Univ.*, **15**, 185 (1967).
153. S. Oki, J. Happel, M. Hnatov and Y. Kaneko, *Proc. Int. Congr. Catal.*, No. 5, 173 (1973).
154. S. Oki and R. Mezaki, *J. Phys. Chem.*, **77**, 447 (1973).
155. V. Glavachek, M. Morek and M. Korzhinkova, *Kinet. Katal.*, **9**, 1107 (1968).
156. N. V. Kul'kova and M. I. Temkin, *Zh. Fiz. Chim.*, **23**, 695 (1949).
157. G. K. Boreskov, T. M. Yureva and A. S. Sergeeva, *Kinet. Katal.*, **11**, 1476 (1970).
158. J. E. Kubsh and J. A. Dumesic, *A.I.Ch.E. J.*, **28**(5), 793 (1982).
159. T. van Herwijnen and W. A. de Jong, *J. Catal.*, **63**, 83 (1980).
160. T. van Herwijnen, R. T. Guczalski and W. A. de Jong, *J. Catal.*, **63**, 94 (1980).
161. E. Fiolitakis and H. Hofmann, *J. Catal.*, **80**, 328 (1983).
162. G. C. Chinchen, M. S. Spencer, K. C. Waugh and D. A. Wahn, *J. Chem. Soc., Faraday Trans. 1*, **83**, 2193 (1987).
163. H. W. Haynes, *Chem. Eng. Sci.*, **25**, 1615 (1970).
164. D. J. Borgas and G. W. Bridger, *Chem. Ind.*, 1426 (1960).
165. P. H. Emmett and J. F. Schultz, *J. Am. Chem. Soc.*, **55**, 1376 (1933).
166. E. D. Eastman, *J. Am. Chem. Soc.*, **44**, 975 (1922).
167. D. Allen, CO conversion catalysts, in: *Ammonia Part II* 99.V. Slack and G. R. James, eds), Marcel Dekker, New York (1974), p. 3.
168. G. C. Chinchen, UK Patent 1,578,365 (1977).
169. J. S. Campbell, *Ind. Eng. Chem. Process., Des. Dev.*, **9**, 588 (1970).

170. M. S. Spencer, *Nature*, **323**, 685 (1986).
171. K. Tohji, Y. Udagawa, T. Mizushima and A. Ueno, *J. Phys. Chem.*, **89**, 5671 (1985).
172. H. M. Hulburt and C. D. S. Vasan, *A.I.Ch.E. J.*, **7**, 143 (1961).
173. S. Kodama, K. Fukui, T. Tame and M. Kinoshita, *Shokubi*, **8**, 50 (1952).
174. G. G. Shchibrya, N. M. Morozov and M. I. Temkin, *Kinet. Katal.*, **6**, 1057 (1965).
174a. C. L. Aldridge, US Patent, 3,615,216 (1968); UK Patent, 1,281,051 (1968).
174b. C. L. Aldridge and T. Kalina, UK Patent, 1,325,172 (1969); UK Patent, 1,325,173 (1969).
175. N. V. Sidgwick, *The Chemical Elements and their Compounds*, Oxford University Press, Oxford (1950), pp. 131–133.
176. H. Y. Allgood, in: *Ammonia*, Volume 2 (part 2) (A. V. Slack and G. R. James, eds), Marcel Dekker, New York (1974) pp. 289–309.
177. P. Sabatier (translated by E. E. Reid), *Catalysts in Organic Chemistry*, The Library Press, London (1923).
178. W. B. Dunwoody and J. R. Phillips, *Petrol. Refiner*, **35** (12), 169 (1956).
179. J. B. Mayland, E. A. Comley and J. C. Reynolds, *Chem. Eng. Prog.*, **50** (4), 177 (1954); D. J. Borgars, in: *Synthesis of Ammonia*, by C. A. Vancini, Macmillan, London (1970), Chap. 10.
180. P. Sabatier and C. R. Senderens, *C. R. Acad. Sci., Paris*, **134**, 514 (1902).
181. V. M. Vlasenko and G. E. Yuzefovich, *Russ. Chem. Rev.*, **38**, 728 (1969).
182. M. A. Vannice, *Catal. Rev. Sci. Eng.*, **14**, 153 (1976).
183. M. A. Vannice, *J. Catal.*, **37**, 462 (1975).
184. R. W. Joyner and M. W. Roberts, *J. Chem. Soc., Faraday Trans. I*, **70**, 1819 (1974).
185. V. Ponec, *Catal. Rev. Sci. Eng.*, **18**, 151 (1978).
186. A. T. Bell, *Catal. Rev. Sci. Eng.*, **23**, 203 (1981).
187. P. R. Wentrcek, B. J. Wood and H. Wise, *J. Catal.*, **43**, 363 (1976).
188. J. G. McCarty and H. Wise, *J. Catal.*, **57**, 406 (1979).
189. P. Winslow and A. T. Bell, *J. Catal.*, **86**, 158 (1984).
190. J. Happel, H. Cheh, M. Otarod, S. Ozawa, A. J. Sveridia, T. Yoshida and V. Fthenakis, *J. Catal.*, **75**, 317 (1982).
191. J. Happel, I. Suzuki, P. Kokayeff and V. Fthenakis, *J. Catal.*, **65**, 59 (1980).
192. M. Otarod, S. Ozawa, F. Yiu, M. Chew, H. Y. Cheh and J. Happel, *J. Catal.*, **71**, 216 (1981).
193. J. Klose and M. Baerns, *J. Catal.*, **85**, 105 (1984).
194. R. P. Underwood and C. O. Bennett, *J. Catal.*, **86**, 245 (1984).
195. P. Schoubye, *J. Catal.*, **14**, 238 (1969).
196. T. Van Herwijnen, H. Van Doesburg and W. A. de Jong, *J. Catal.*, **28**, 391 (1973).
197. Sa Van Ho and P. Harriott, *J. Catal.*, **64**, 272 (1980).
198. J. A. Dalmon and G. A. Martin, *J. Chem. Soc., Faraday Trans. I*, **75**, 1011 (1979).
199. J. L. Falconer and A. E. Zagli, *J. Catal.*, **62**, 280 (1980).
200. G. D. Weatherbee and C. H. Bartholomew, *J. Catal.*, **68**, 67 (1981).
201. V. M. Vlasenko, G. E. Yuzefovich and M. T. Rusov, *Kinet. Catal.*, **6**, 611 (1965).
202. J. L. Bousquet and S. J. Teichner, *Bull. Soc. Chim. Fr.*, 2963 (1969); 3689

(1972). J. L. Bousquet, P. Gravelle and S. J. Teichner, *Bull. Soc. Chim. Fr.*, 3693 (1972).
203. P. Schoubye, *J. Catal.*, **18**, 118 (1970).
204. W. W. Akers and R. R. White, *Chem. Eng. Prog.*, **44**, 553 (1948).
205. J. Nicolai, M. D'Hont and J. C. Jungers, *Bull. Soc. Chim. Belg.*, **55**, 160 (1946).
206. J. F. Schultz, F. S. Karn and R. B. Aderson, *US Bur. Mines Rep.*, No. 5941 (1967).
207. G. C. Binder and R. R. White, *Chem. Eng. Prog.*, **46**, 563 (1950).
208. G. C. Binder and R. R. White, *Am. Documentation Inst., (Wash.)*, Document 2834.
209. J. N. Dew, R. R. White and C. M. Sliepcevich, *Ind. Eng. Chem.*, **47**, 140 (1955).
210. M. Solc and V. Pour, *Coll. Chem. Commun. Czech.*, **29**, 857 (1964).
211. E. Sols, *Coll. Chem. Commun. Czech.*, **27**, 2621 (1962).
212. G. D. Weatherbee and C. H. Bartholomew, *J. Catal.*, **77**, 460 (1982).
213. M. E. Dry, Paper presented at Synthetic Fuels from Coal Symposium, Eindhoven, March 1977. (Published in *Energie Spectrum* (October 1977).)
214. Y. Takemura, Y. Morita and K. Yamamoto, *Bull. Japan Petrol. Inst.*, **9**, 13 (1967).
215. G. W. Bridger and C. Woodward, in: *Methanation of Synthesis Gas. Adv. Chem. Ser.*, **146**, 71 (1975).
216. J. A. Lacey, *CRG-based SNG: Principles and Process Routes*, British Gas booklet (1971).
217. A. K. Kuhn, *Chem. Eng. Prog.*, **78** (4) 64 (1982).
218. K. H. Eisenlohr, F. W. Moeller and M. Dry, in: *Methanation of Synthesis Gas. Adv. Chem. Ser.*, **146**, 113 (1975).
219. K. R. Tart and T. W. A. Rampling, *Hydrocarbon Processing*, 114 (April 1981).
220. R. L. Ensell and H. J. F. Stroud, Paper presented at 1983 International Gas Research Conference, London.
221. J. E. Franzen and F. K. Goeke, Paper presented at Sixth Synthetic Pipeline Gas Symposium, Chicago, October 1974.
222. G. A. White, T. Roszkowski and D. W. Stanbridge, in: *Methanation of Synthesis Gas. Adv. Chem. Ser.*, **146**, 138 (1975).
223. R. Harth, K. Kugeler, H. F. Niessen, U. Boltendahl and K. A. Theis, *Nucl. Technol.*, **38**, 252 (1978).
224. H. H. Gierlich, B. Hoehlein and J. Range, Paper presented at 1981 International Gas Research Conference, Los Angeles.
225. H. Harms, B. Hoehlein, E. Jorn and A. Skov, *Oil Gas J.*, 120 (14 April 1980).
226. B. Hoehlein, Paper presented at 1983 International Gas Research Conference, London.
227. B. Hoehlein, *KFA Rep., Jul-1589* (1979).
228. H. Harms, B. Hoehlein and A. Skov, *Chem.-Ing.-Tech.*, **52**, 504 (1980).
229. H. Harms, B. Hoehlein, A. Skov and M. Vorwerk, Paper presented at 17th Intersociety Energy Conversion Engineering Conference, Los Angeles, 1982.
230. B. Hoehlein, R. Menzer, H. J. R. Schiebahn, M. Vorwerk, H. Kiilerich-Hansen and A. Skov, *KFA Rep., Jul-Sept-197* (1983).
231. W. Crooks, Presidential Address to the British Association for the

Advancement of Science, September 1898.
232. S. A. Topham, *Catalysis, Sci. Technol.*, **7**, 1 (1985); A. J. Harding, *Ammonia Manufacture and Uses*, Oxford University Press, London (1959).
233. F. Haber and G. Van Oordt, *Z. Anorg. Chem.*, **43**, 111 (1904).
234. W. Nernst, W. Jost and G. Jellinek, *Z. Elektrochem.*, **13**, 521 (1915); **14**, 373 (1908); *Z. Anorg.*, **57**, 414 (1908).
235. F. Haber and R. Le Rossingnol, *Ber.*, **40**, 2144, (1907); *Elektrochemie*, **14**, 181 (1908).
236. F. Haber, S. Tamaru and C. Ponnaz, *Z. Elektrochem.*, **21**, 89 (1915).
237. A. T. Larson and R. L. Dodge, *J. Am. Chem. Soc.*, **45**, 2918 (1923).
238. A. T. Larson, *J. Am. Chem. Soc.*, **46**, 367 (1924).
239. L. J. Gillespie amd J. A. Beattie, *Phys. Rev.*, **36**, 743 (1930).
240. A. Nielsen, *An Investigation on Promoted Iron Catalysts for the Synthesis of Ammonia* (3rd Edn), Jul Gjellerups Forlag, Copenhagen (1968).
241. J. G. Livingston and A. Pinto, *AIChE Ammonia Safety Symposium*, Los Angeles, Paper 123F (1982).
242. W. Aktiebolag and H. Topsøe, US Patent 3,243,386 (1966).
243. A. M. Rubinshtein, N. A. Pribytkova, V. M. Akimov, A. L. Klyachko-Gurvich, A. A. Slinkin and I. V. Mel'nikova, *Kinet. Catal.*, **6**, 285 (1986) [Engl. Transl. pp. 243–249].
244. J. R. Jennings, European Patents, 174,078, 174,079 and 174,080.
245. A. Mittasch, *Adv. Catal.*, **2**, 81 (1950).
246. A. L. Turnock and H. P. Eugster, *J. Petrol.*, **3**, 533 (1962).
247. L. M. Dmitrenko, L. D. Kuznetsov and P. D. Rabina, *Proc. 4th Int. Congr. Catalysis*, 404 (1968).
248. W. Frankenburger, *Catalysis*, **3**, 171 (1955).
249. E. K. Jenikjev and A. V. Krylova, *Kinet. Catal. (Engl. Edn)*, **3**, 16 (1962).
250. G. Ertl, *Catal. Rev. Sci. Eng.*, **21**, 201 (1980).
251. R. Brill, E. L. Richter and E. Ruch, *Angew. Chem.*, **79**, 905 (1967).
252. N. D. Spencer, R. C. Schoonmaker and G. A. Somorjai, *J. Catal.*, **74**, 129 (1982).
253. G. Ertl, M. Weiss and S. B. Lee, *Chem. Phys. Lett.*, **60**, 391 (1979).
254. Z. Paal, G. Ertl and S. B. Lee, *Appl. Surf. Sci.*, **8**, 23 (1981).
255. S. R. Bare, D. R. Strongin and G. A. Somorjai, *J. Phys. Chem.*, **90**, 4726 (1986).
256. D. R. Strongin, J. Carrazza, S. R. Bare and G. A. Somorjai, *J. Catal.*, **103**, 213 (1987).
257. T. Rayment, R. Schlogl, J. M. Thomas and G. Ertl, *Nature*, **315**(6017), 311 (1985).
258. R. Schlogl, personal communication.
258a. J. A. Burnett, H. Y. Allgood and J. R. Hall, *Ind. Eng. Chem.*, **45**, 1678 (1953).
259. C. Bokhoven, in: *Proc. 2nd Radioisotope Conference*, Oxford, Butterworth, London (1954).
260. P. H. Emmett and S. Brunauer, *J. Am. Chem. Soc.*, **52**, 2682 (1930).
261. S. L. Kiperman, *Zh. Fiz. Khim.*, **28**, 389 (1954).
262. M. I. Temkin and V. Pyzhev, *Acta Physicochim. (USSR)*, **12**, 327 (1940).
263. D. Annable, *Chem. Eng. Sci.*, **1**, 145 (1952).
264. V. D. Livshits and I. P. Siderov, *Zh. Fiz. Khim.*, **26**, 538 (1952).
265. M. I. Temkin, N. M. Moroov and E. N. Shapatina, *Kinet. Catal. (USSR)*,

4, 260 (1963).
266. M. I. Temkin, N. M. Moroov and E. N. Shapatina, *Kinet. Catal. (USSR)*, **4**, 565 (1963).
267. P. H. Emmett and S. Brunauer, *J. Am. Chem. Soc.*, **56**, 35 (1934).
268. K. T. Kozhenova and M. Ya. Kagan, *J. Phys. Chem. USSR*, **14**, 1250 (1940).
269. F. Bozso, G. Ertl, M. Grunze and M. Weiss, *J. Catal.*, **49**, 18 (1977); G. Ertl, M. Grunze and M. Weiss, *J. Vac. Sci. Technol.*, **13**, 314 (1976).
270. F. Bozso, G. Ertl and M. Weiss, *J. Catal.*, **50**, 519 (1977).
271. K. Kishi and M. W. Roberts, *Surf. Sci.*, **62**, 252 (1977).
272. I. D. Gay, M. Textor, R. Mason and Y. Iwasawa, *Proc. Roy. Soc. (Lond.)*, **A356**, 25 (1977).
273. M. J. Grunze, M. Golze, W. Hirschwald, H.-J. Freund, H. Pulm, U. Seip, M. C. Tsai, G. Ertl and J. Kuppers, *Phys. Res. Lett.*, **53**(8), 850 (1984).
274. M. J. Grunze, J. Fuhler, M. Neumann, C. R. Brundle, D. J. Averbach and J. Behm, *Surf. Sci.*, **139**(1), 109 (1984). (Refers to precursor states on Ni, Rh and W.)
275. M. J. Grunze, F. Bozso, G. Ertl and M. Weiss, *Appl. Surf. Sci.*, **1**, 241 (1978).
276. M. Weiss, G. Ertl and F. Nitschke, *Appl. Surf. Sci.*, **2**, 614 (1979).
277. M. Dreschler, H. Hoinkes, H. Kaarman, H. Wilsch, G. Ertl and M. Weiss, *Appl. Surf. Sci.*, **3**, 217 (1979).
278. P. H. Emmett, in *The Physical Basis for Heterogeneous Catalysis* (E. Drauglis and R. I. Jaffec, eds), Plenum Press, New York (1975).
279. W. G. Frankenburg, *Catalysis*, Volume 3 (P. H. Emmett, ed.), (1955),p. 171.
280. *Nitrogen*, No. 11 (March/April) 1976.
281. M. R. Logan, J. J. McCarroll, S. R. Tennison, European Patent 0058531.
282. J. J. McCarroll, S. R. Tennison and N. P. Wilkinson, UK Patent 2,136,704.
283. C. P. van Dijk, A. Solbakken and B. G. Mandelik, US Patent 4,568,531.
284. G. S. Benner, J. R. Le Blanc, J. M. Lee, H. P. Leftin, P. J. Shires and C. P. van Dijk, US Patent 4,568,532.
285. N. D. Spencer and G. A. Somorjai, *J. Phys. Chem.*, **86**, 3493 (1982).
286. N. D. Spencer and G. A. Somorjai, *J. Catal.*, **78**, 1112 (1982).
287. M. Asscher and G. A. Somorjai, *Surf. Sci.*, **143**, L389 (1984).
288. M. J. Grunze, M. Golze, J. Fuhler, M. Neumann and E. Schwarz, *8th Int. Congr. Catalysis 1984 Proc. IV*, 133.
289. J. Chatt, A. J. Pearman and R. L. Richards, *Nature*, **243**, 39 (1975).
290. C. J. Pickett and J. Talarmin, *Nature*, **317**(6038), 652 (1985).
291. C. J. Pickett, K. S. Ryder and J. Talarmin, *J. Chem. Soc. Dalton Trans.*, 1453 (1986).
292. K. Klier, *Adv. Catal.*, **31**, 243 (1982).
293. H. H. Kung, *Catal. Rev.—Sci. Eng.*, **22**, 235 (1980).
294. P. J. Denny and D. A. Whan, in: *Catalysis*, Specialist Periodical Report, Volume 2 (C. Kemball and D. A. Dowden, eds), Chemical Society, London (1978), p. 46.
295. L. E. Wade, R. B. Gengelback, J. L. Trumbley and W. L. Hallbauer, in: *Kirk-Othmer Encyclopedia of Chemical Technology*, 3rd edn, Volume 15, Wiley, New York (1981), p. 398.
296. F. Marschner and F. W. Moeller, in: *Applied Industrial Catalysis*, Volume

2 (B. E. Leach, ed.), Academic Press, New York (1983), p. 215.
297. G. C. Chinchen, P. J. Denny, J. R. Jennings, M. S. Spencer and K. C. Waugh, *Appl. Catal.*, **36**, 1 (1988).
298. W. J. Thomas and S. Portalski, *Ind. Eng. Chem.*, **50**, 967 (1958).
299. L. Bisset, *Chem. Eng.*, **84**, 155 (24 October 1977).
300. N. J. Macnaughton, A. Pinto and P. L. Rogerson, *AIChE Spring National Mtg*, Anaheim, California, 21–23 May, 1984, Paper 23e.
301. G. Natta, *Catalysis*, Volume 3 (P. H. Emmett, ed.), Reinhold, New York (1955), p. 349.
302. M. Bowker, H. Houghton and K. C. Waugh, *J. Chem. Soc., Faraday Trans. I*, **77**, 3023 (1981).
303. M. R. Fenske and P. K. Frolich, *Ind. Eng. Chem.*, **21**, 1052 (1929).
304. P. K. Frolick and W. K. Lewis, *Ind. Eng. Chem.*, **20**, 285 (1928).
305. S. P. S. Andrew, Plenary Lecture, *Post-Congress Symposium, 7th Int. Cong. Catalysis*, Osaka, July 1980.
306. S. P. S. Andrew, in: *Catalyst Handbook*, 1st edn, Wolfe, London (1970), p. 20.
307. S. P. S. Andrew, *Chem. Eng. Sci.*, **36**, 1431 (1981).
308. M. S. Spencer, *ACS Symposium Series*, **298**, 89 (1986).
309. M. S. Spencer, *Nature*, **323**, 685 (1986).
310. G. C. Chinchen, P. J. Denny, D. G. Parker, G. D. Short, M. S. Spencer, K. C. Waugh and D. A. Whan, *Preprints, Am. Chem. Soc. Div. Fuel Chem.*, **29**(5), 178 (1984).
311. B. Denise and R. P. A. Sneeden, *Chemtech*, 108 (1982).
312. P. J. Denny, *Proc. 8th Int. Cong. Catalysis, 1984*, Volume 6, Verlag Chemie, Weinheim (1985), p. 19.
313. Yu. B. Kagan, L. G. Liberov, E. V. Slivinskii, S. M. Loktev, G. I. Lin, A. Ya. Rozovskii and A. N. Bashkirov, *Dokl. Akad. Nauk SSSR*, **221**, 1093 (1975).
314. K. Klier, V. Chatikavanu, R. G. Hermann and G. W. Simmons, *J. Catal.*, **74**, 343 (1982).
315. G. Liu, D. Willcox, M. Garland and H. H. Kung, *J. Catal.*, **90**, 139 (1984).
316. Yu. V. Lender, L. S. Parfenora and K. N. Tel'nykh, *Khim. Prom.*, **49**, 654 (1973).
317. E. Blasiak and W. Kotowski, *Prez. Chem.*, **43**, 657 (1964).
318. G. C. Chinchen, C. M. Hay, D. Vandervell and K. C. Waugh, *J. Catal.*, **103**, 79 (1987).
319. G. C. Chinchen, K. C. Waugh and D. A. Whan, *Appl. Catal.*, **25**, 101 (1986).
320. G. C. Chinchen and K. C. Waugh, *J. Catal.*, **97**, 280 (1986).
321. K. Kochloefl, *Proc. 7th Int. Cong. Catalysis, 1980*, Volume A, Elsevier, Amsterdam (1981), p. 486.
322. B. Notari, *Proc. 7th Int. Cong. Catalysis, 1980*, Volume A, Elsevier, Amsterdam (1981), p.487.
323. K. C. Waugh (to be published).
324. G. C. Chinchen, M. S. Spencer, D. A. Whan and K. C. Waugh, *J. Chem. Soc., Faraday Trans. I*, **83**, 2193 (1987).
325. G. A. Vedage. R. Pitchai, R. G. Hermann and K. Klier, *Proc. 8th Int. Cong. Catalysis, 1984*, Volume 2, Verlag Chemie, Weinheim (1984), p. 47.

326. W. L. Marsden, M. S. Wainwright and J. B. Friedrich, *Ind. Eng. Chem., Prod. Res. Dev.*, **19**, 551 (1980).
327. J. B. Friedrich, M. S. Wainwright and D. J. Young, *J. Catal.*, **80**, 1 (1983).
328. J. B. Friedrich, D. J. Young and M. S. Wainwright, *J. Catal.*, **80**, 14 (1983).
329. A. J. Bridgewater, M. S. Wainwright, D. J. Young and J. P. Orchard, *Appl. Catal.*, **7**, 369 (1983).
330. E. G. Baglin, G. B. Atkinson and L. J. Nicks, *Ind. Eng. Chem., Prod. Res. Dev.*, **20**, 87 (1981).
331. F. P. Daly, *J. Catal.*, **89**, 131 (1984).
332. G. D. Short and J. R. Jennings, European Patent 117,944 (1984).
333. R. M. Nix, T. Rayment, R. M. Lambert, J. R. Jennings and G. Owen, *J. Catal.*, **106**, 216 (1987); G. Owen, C. M. Hawkes, D. Lloyd, J. R. Jennings, R. M. Lambert and R. M. Nix, *Appl. Catal.*, **33**, 405 (1987).
334. M. L. Poutsma, L. F. Elek, P. A. Ibarbia, A. P. Risch and J. A. Rabo, *J. Catal.*, **52**, 157 (1978).
335. E. K. Poels, P. J. Mangnus, J. v. Welzen and V. Ponec, *Proc. 8th Int. Cong. Catalysis, 1984*, Volume 2, Verlag Chemie, Weinheim (1984), p. 59.
336. R. F. Hicks and A. T. Bell, *J. Catal.*, **91**, 104 (1985).
337. M. E. Frank, *Proc. Intersoc. Energy Conv. Eng. Conf., 1980*, Volume 2, p. 1567.
338. Anon., *Synfuels*, 5 (27 July 1984).
339. J. W. Mellor, *Comprehensive Treatise on Inorganic and Theoretical Chemistry*, Volume 8, Longmans and Green, London (1928).
340. T. H. Chilton, *Strong Water, Nitric Acid: Sources, Methods of Manufacture and Uses*, MIT Press, Cambridge, Massachusetts (1968), pp. 33-34.
341. W. Ostwald, US Patent 858,904.
341a. D. A. Spratt, in: *Recent Aspects of The Inorganic Chemistry of Nitrogen*, Special Publication No. 10, Chemical Society, London, (1957), pp.53-63.
342. G. C. Chinchen, UK Patent 1,342,577, (1971).
343. D. O Hughes, UK Patent 1,350,242, (1971).
344. I. R. Shannon, UK Patent 1,542,634, (1976).
345. T. H. Chilton, *Strong Water, Nitric Acid: Sources, Methods of Manufacture and Uses*, MIT Press, Cambridge, Massachusetts (1968), pp. 87–104; F. D. Miles, *Nitric Acid Manufacture and Uses*, Oxford University Press, London (1961).
346. A. E. Heywood, *Plat. Met. Rev.*, **17**(4), 118 (1973).
347. F. Sperner and W. Hohmann, *Plat. Met. Rev.*, **20**, 12 (1976); A. R. McCabe, G. D. W. Smith and A. S. Pratt, *Plat. Met. Rev.*, **30**, 54 (1986).
348. F. D. Miles, *Nitric Acid, Manufacture and Uses*, Oxford University Press, pp. 26–27, (1961).
349. R. W. McCabe, T. Piguet and L. D. Schmidt, *J. Catal.*, **32**, 114 (1974).
350. C. N. Satterfield, *Heterogeneous Catalysis in Practice*, McGraw-Hill, New York, (1980), pp. 214–221.
351. J. Pielaszek, *Plat. Met. Rev.*, **28**, 109 (1984).
352. N. H. Harbord, *Plat. Met. Rev.*, **18**, 97 (1974).
353. J. E. Philpott, *Plat. Met. Rev.*, **15**, 52 (1971).
354. N. G. Schmahl, *Z. Phys. Chem.*, **41**, 78 (1964).
355. H. Connor, *Plat. Met. Rev.*, **11**, 60 (1967).

356. G. J. K. Acres, A. J. Bird, J. W. Jenkins and F. King, in: *Characterisation of Catalysts* (J. M. Thomas and R. M. Lambert, eds), John Wiley, Chichester (1980), pp. 72–74.
357. U. Gerloff, H. Sperber and J. Zipfel, *UK Patent*, 1,217,717 (1970); *German Patent*, 1,294,360 (1969).
358. F. Brunnmüller, H. Diem, G. Lehmann and G. Matthias, *UK Patent*, 1,453,766 (1976); *German Patent*, 2,308,409 (1974).
359. A. Aicher, H. Haas, H. Diem, C. Dudeck, F. Brunnmüller and G. Lehmann, *UK Patent*, 1,526,245 (1978); *German Patent*, 2,444,586 (1976).
360. H. Diem, *Chem. Eng.*, 83–85 (27 February 1978).
361. Y. Kuwaishi and K. Yoshikawa, *Chem. Econ. Eng. Rev.*, **13**(6), 31 (1982).
362. R. P. Report 73452 (Sator), 5 March 1938; J. F. Walker, *Formaldehyde*, Reinhold, New York (1944).
363. V. L. Atroshchenko and I. P. Kushnarenko, *Int. Chem. Eng.*, 4, 581 (1964).
364. T. Eguchi, T. Yamamoto and S. Yamauchi, US Patent 2,908,715 (1959).
365. D. Ambros, W. Baumeister, G. Duembgen and U. Gerloff, *UK Patent*, 1,135,476 (1968); *German Patent*, 1,235,881 (1967).
366. U. Gerloff and O. Göhre, *UK Patent*, 1,188,215 (1970); *German Patent*, 1,277,834 (1968).
367. T. V. Ferris, *The Explosion of Methanol Air Mixtures at above Atmospheric Conditions*, Monsanto Company, Indian Orchard, Massachusetts.
368. M. Dente and I. Pasquon, *Chem. Ind. (Milan)*, **47**, 359 (1965); **47**, 821 (1965).
369. N. Burriesci, F. Garbassi, M. Petrera, G. Petrini and N. Pericone, in: *Catalyst Deactivation* (B. Delmon and F. F. Froment, eds), Elsevier, Amsterdam (1980), p. 115.
370. J. K. Dixon and J. E. Longfield, in: *Catalysis*, Volume 3 (P. H. Emmett, ed.), Reinhold, New York (1960), p. 322.
371. *Thermochemical Tables*, 2nd edn, US Department of Commerce, National Bureau of Standards, Washington (1970).
372. *Inorganic Sulphur Chemistry* (G. Nickless, ed.), Elsevier, Amsterdam (1968), p. 543.
373. A. Urbanek and M. Trela, *Catal. Rev.—Sci. Eng.*, **21**, 73 (1980).
374. W. W. Duecker and J. R. West, *The Manufacture of Sulphuric Acid*, ACS Monograph, Reinhold, New York (1959).
375. M. Appl and N. Neth, *Fertilizer Acids*, Proc. Br. Sulphur Corp. 3rd Int. Conf. on Fertilizers, Volume 1, Paper No. 20 (1979).
376. O. A. Hougen and K. M. Watson, *Chemical Process Principles*, 1st edn, Part II (1943), p. 727.
377. *Encyclopedia of Chemical Technology*, 3rd edn, Volume 22, Kirk Othmer, John Wiley, New York (1983), p. 213.
378. *Thorpes Dictionary of Applied Chemistry*, 4th edn, Volume XI, Longmans Green (1954).
379. F. Slama and H. Wolf, US Patent 1,371,004 (1921).
380. P. Davis, US Patent 3,186,794 (1965).
381. C. N. Kenney, *Specialist Periodic Report: Catalysis*, Volume 3, Chemical Society, London (1978).
382. J. Villadsen and H. Livbjerg, *Catal. Rev.—Sci. Eng.*, **17**, 203 (1978); J. R.

Donovan, R. D. Stolk and M. L. Unland, *Appl. Ind. Catal.*, **2**, 245 (1983).
383. J. H. Frazer and W. J. Kirkpatrick, *J. Am. Chem. Soc.*, **62**, 1659 (1940).
384. R. Kiyoura, *Kagaku Kogyo*, **10**, 126 (1940); *Chem. Abs.*, 7169 (1940).
385. H. F. Topsøe and A. Nielson, *Trans. Dan. Acad. Technol. Sci.*, **1**(3), 18 (1948).
386. G. H. Tandy, *J. Appl. Chem.*, **6**, 68 (1956).
387. G. K. Boreskov and V. P. Pligunov, *J. Appl. Chem. USSR*, **13**, 653 (1940).
388. H. Kuranuma, *Nippon Kagaku Kaishi*, 1078 (1972); T. Ishii, M. Aramata and R. Furvichi, *Nippon Kagaku Kaishi*, 226 (1972); G. K. Boreskov, L. P. Davydova, V. M. Mastikhin and G. M. Polyakova, *Dokl. Akad. Nauk SSSR*, **171**, 760 (1966); V. M. Mastikhin, G. M. Polyakova, Y. Zyulkovskii and G. K. Boreskov, *Kinet. Katal.*, **11**, 1219 (1970); G. K. Boreskov, G. M. Polyakova, A. A. Ivanov and V. M. Mastikhin, *Dokl. Akad. Nauk SSSR*, **210**, 626 (1973); G. K. Boreskov, V. A. Dziska, D. V. Tarasova and G. P. Balaganskaya, *Kinet. Katal.*, **11**, 144 (1970).
389. Y. Kera and K. Kuwata, *Bull. Chem. Soc., Japan*, **50**, 2831 (1977).
390. P. Putanov, D. Smiljanic, B. Djukanovic, N. Jovanovic and R. Herak, *Proc. 5th Int. Congr. Catalysis*, Miami Beach, Florida, 1972, Volume 74, 1061.
391. N. H. Hansen, R. Fehrman and N. J. Bjerrum, *Inorg. Chem.*, **21**, 744 (1982).
392. J. Villadsen, US Patent 4,193,894 (1980); *Chem. Abs.*, **93**, 10346 (1980).
393. V. M. Mastikhin, O. B. Lapina, V. F. Lyakhova and L. G. Simonova, *React. Kinet. Catal. Lett.*, **17**, 109 (1981); *Chem. Abs.*, **95**, 157347 (1981).
394. K. Norwinska, *Z. Phys. Chem.*, **126**, 117 (1981); *Chem. Abs.*, **95**, 157357 (1981).
395. A. M. Fairlie, *Sulphuric Acid Manufacture*, Reinhold, New York (1936).
396. K. Othmer, *Encyclopedia of Chemical Technology*, 2nd edn, Volume 19, (A. Standon, ed.), Interscience Publications, New York (1969), p. 441.
397. *The Modern Inorganic Chemicals Industry* (R. Thompson, ed.), Spec. Publ. No. 31, The Chemical Society, London (1977).
398. G. K. Boreskov, *Catalysis in Production of Sulphuric Acid*. Goskhimizdat M-L (1954). [In Russian.]
399. R. B. Eklund, *The Rate of Oxidation of Sulphur Dioxide with a Commercial Vanadium Catalyst*, Almquist and Wiksell, Stockholm (1956).
400. P. Mars and J. G. H. Maessen, *Proc. 3rd Int. Congr. Catalysis*, Volume 1, North–Holland, Amsterdam (1965), p. 266; *J. Catal.*, **10**, 1 (1968).
401. H. Livbjerg and J. Villadsen, *Chem. Eng. Sci.*, **27**, 21 (1972).
402. H. Livbjerg, K. F. Jensen and J. Villadsen, *J. Catal.*, **45**, 216 (1976).
403. G. K. Boreskov, R. A. Buyanov and A. A. Ivanov, *Kinet. Catal. (USSR)*, **8**, 126 (1967).
404. G. K. Boreskov, in: *Catalysis, Science and Technology*, Volume 3 (J. R. Anderson and M. Boudart, eds) (1982), p. 39.
405. A. A. Yeramian, P. L. Silveston and R. R. Hudgins, *Can. J. Chem.*, **48**, 1175 (1970).
406. M. Neth, G. Kautz, K. J. Husker and U. Wagner, *Chem. Ing. Tech.*, **51**, 825 (1979).
407. E. W. Thiele, *Ind. Eng. Chem.*, **31**, 916 (1939).
408. W. W. Duecker and J. R. West, *The Manufacture of Sulphuric Acid*, ACS

Monograph, Reinhold, New York (1959), p. 182; U. H. F. Sander, H. Fischer, U. Rothe and R. Kola, *Sulphur Dioxide, and Sulphuric Acid,* The British Sulphur Corporation Ltd, London (1984).
409. G. D. Sirotkin, *Zh. Prikl. Khim.*, **21**, 245 (1948).
410. I. E. Adadurov, *Ukrain. Khim. Zh.*, **10**, 336 (1935).
411. K. Othmer, *Encyclopedia of Chemical Technology*, 3rd edn, Volume 22, pp. 221–225; *Sulphur* No. 105, p. 51 John Wiley, New York, (1983).
412. *Sulphur*, No. 139, p. 40, (1978).
413. J. R. Donovan, R. M. Smith and J. S. Palermo, *Sulphur*, 131, 46 (1977).
414. J. P. Mukhlenov, S. V. Ikranov, V. I. Deryuzhkina, *J. Khim. Prom-st (Mosc.),* (3), 226 (1976).
415. W. Becker, *Proc. Br. Sulphur Corp. 5th Int. Conf.*, pp. 45–57, 586 (1981).

Index

Page numbers in bold type refer to figures, graphs or tables. Where subsidiary headings begin with a preposition or conjunction, ordering is according to the second word.

A

Absorption profile for poisons
  in chlorine guard 222
  sulphur on shift catalyst **326**
  on zinc oxide **212, 217**
Absorption system
  in cobalt molybdate catalysts 206
  formaldehyde plant 491
  nitric acid process, chemistry of 477-8
  sulphuric acid plant 506, 510, **512**
Acid sites, carbon formation 280
Activated alumina, for chloride 222
Activated charcoal (*see also* Charcoal)
  for single-stage sulphur removal 196-7
Activation of catalysts 69-73
  for ammonia synthesis 397
  for kinetic studies 65
  platinum/rhodium 484-7
  prereduced ammonia catalyst 403
Activation energies *for*
  deuterium exchange on nickel 239-40
  methanation 355
  methane cracking 239-40
  methane steam reforming 239
  water-gas shift reaction 288
Active sites
  cobalt molybdate catalysts 205
  iron catalysts, effect of water 407
  methanol synthesis catalyst 466
Active volume of catalyst (*see also* Activity of catalyst)
  calculation
    from gas composition 132
    from temperature profiles 133-4
  measurement of performance 128
Activity (*see also* Intrinsic activity)
  ammonia synthesis catalysts
    effect of water 401
    variation with granule size (table) **412**
  of catalysts 23-6, 89
  high-temperature shift (table) **296**
  low-temperature shift catalysts 321, **325**
  of poisoned copper catalyst **83**
  of poisoned low-temperature shift **83**
  precipitated catalysts 38
  sulphur dioxide oxidation catalysts **515**
  testing ammonia oxidation catalysts 60
ADAM-1 plant
  methanation conditions (table) **379**
  sulphur poisoning in 381
  temperature profiles **381**
Adiabatic beds
  design of reactors 90-1
  steam reforming catalyst 263
Adsorption
  on catalysts (with tables) 29-31

  on iron single crystals 67
  of nitrogen on iron surface 414, 415
  of oxygen on iron catalyst 407
  of poisons on shift catalysts 325
Ageing, life testing of catalysts 65-6
Air
  addition, in steam reforming 253
  chloride in 221
  to nitric acid plant 23
  nitric oxide equilibrium **471**
Air/methanol mixtures, flammability 497
Air mixing, in secondary reforming 125
Air separation *see* Cryogenic nitrogen unit
Air/steam, regenerating reforming catalyst 176
Alcohols
  from carbon monoxide/hydrogen 441
  higher alcohol in methanol **460**
  stability of 460
Aliphatic chlorides, in naphtha 221
Alkali
  effects on methanol synthesis **461**
  promoter in ammonia catalysts 392
  promotion of water-gas shift 338
Alkenes, addition of hydrogen sulphide 202
Alloys for steam reformer tubes 267
Alumina
  alkalized, chloride guard bed 222, 332
  catalyst, specific heat **346**
  chips, catalyst bed support 154, 291
  in methanol synthesis catalysts 457-8
  steam reforming catalyst support 249
  support for impregnated catalysts 204
  support in shift catalysts 310-12
Aluminates, in ammonia catalysts 397
Ammonia
  anhydrous, ICI papers, listing 536
  catalysts, performance of 134-7
  concentration, effect on synthesis **413**
  converters
    charging with catalyst 155
    maintenance of 164-5, 179
    temperature rise vs conversion **131**
  cracking in steam reformer 276
  deactivation of molybdate catalysts 224
  direct oxidation to nitric acid 20-1
  energy profiles for synthesis **25**
  equilibrium concentration **118**
  first plant for oxidation of 472
  heat of mixing with synthesis gas 390
  inhibition of shift reaction 330
  manufacture from natural gas (table) **22**
  nomograph of properties 549
  oxidation 19-21, 470-89
    activity of cobalt oxide catalysts 60
    adiabatic temperature rise **481**

591

Index

alternative catalysts for 476
film-diffusion control of 84
mechanisms of 477-81
modern processes for 482-4
selectivity of 19-20, 479-81
selectivity of cobalt oxide catalyst 84
use of metal gauzes 35
plants
   catalyst volumes (block diagram) **343**
   hydrodesulphurization, conditions 206
   methanation in 341-2
   process gas composition (table) **342**
production
   first use of water-gas shift 284
   growth of 386, **387**
reaction with nitric oxide 480
reactions on iron single crystals 67
synthesis of 18-19, 384-440
   catalyst volume in ammonia plant **343**
   from coal 283-4, **285**
   during catalyst reduction 401
   effects on reaction rate **413**
   energy barrier for reduction 24-6
   equilibrium constants for **110**
   heat of formation (table) **392**
   by homogeneous catalysis **20**
   kinetics of 409-13
   loop 391-2, **393**
   mechanism of 67-8
   from natural gas **285**
   optimum temperature/pressure 390
   poisoning by carbon oxides 340
   steam reforming for 256-7
   thermodynamics of 388-92
synthesis catalysts 36, 392-409, 529
   activation 397
   activity, particle size 401, 411
   aluminium in 397
   charging 155
   deactivation 178, 404-9
   discharge 186
   early research on 52
   form of surface potassium 67, 68
   heat release during oxidation 558
   heat release during reduction 557
   hydrogen formation from 185
   iron pseudomorphs in **396**
   magnesium in 397
   manufacture of 392-3
   method for stabilization 183
   nature of working surface 402
   poisoning of 404-9
   poisoning by water, carbon oxides 80
   prereduced *see* Prereduced catalysts
   promoters for 388, 392, 395-400
   reduction of 171-5, 400-4
   reduction, methods 173-5
   regeneration 178
   re-use of 188
   specific heat 561
   surface area vs position in bed **71**
   variation of activity with size **412**
synthesis reaction, in reduction 172
synthesis reactors
   commercial types 423-39
   design of 89
   operation 415-23
   performance monitoring 137

Ammonium carbamate, in loop 342
Ammonium metavanadate, 508
Ammonium nitrate, production 469
Ammonium sulphate, in fertilizers 384
Amorphous phase, ammonia catalysts 402
AMV process 88, 390, 440
Analytical methods papers, ICI, 535-6
Anhydrous ammonia, ICI papers, 536
Annual production
   ammonia 386
   formaldehyde 490
   nitric acid 469
   sulphuric acid 503
Antimony, poisoning ammonia catalysts 408
Approach to equilibrium 128-32
Aromatic compounds, feedstocks 194, 226
Aromatic disulphides, stability 199
Aromatic 'slip', from steam reformer 278
Aromatization, in desulphurization 228
Arsenic
   absorption on cobalt molybdate **224**
   avoidance in feedstock 223, 228
   poisoning of catalysts 278-80
      ammonia synthesis 404, 408
      copper 330
      high-temperature shift 305
      platinum 80
      steam reforming 220
      vanadium 516
   removal from reformer tubes 280
   transport via steam systems 279-80
Arsenic compounds, in zinc oxide 220
Associated gas, composition of 193
Atmospheric pressure nitric acid plant **473**
Atomic weights (table) 562-4
Attrition loss, testing 52
Autoradiograph, of sulphur in zinc oxide **213**

B

Baekeland, L.H., on catalyst testing 48-9
Basic carbonates, by precipitation 38, **354**
Benfield process **361**
Benzene, from steam dealkylation 240
Benzothiophenes, in naphtha 194
Berty-type reactors 63, **63, 64**
Berzelius, J.J. 17
Birkland and Eyde process, 470-1
Blanketing
   platinum/rhodium catalyst 488, 489
   pyrophoric catalyst extraction 186
   reduced catalyst 178-9
   safety precautions 189
Boiler design, and steam quality 325
Boreskov reactors 63, **63, 64**
Bosch, C.
   discovery of shift catalyst 284
   first catalytic ammonia plant 386
Brasses, in methanol catalyst 462
Breathing apparatus, vessel entry 179, 189
Brine, hydrogen from 194
Bulk density monitoring when charging 155
Bulk properties of catalysts 49-50
Burner design, secondary reforming 125
Burner efficiency, nitric acid 477, 480-1
Burners, in steam reforming furnace **265**

Index

Butylmercaptan, decomposition **199**
By-passing in catalyst bed 127, 360

C

Cadmium compounds, in zinc oxide 220
Caesium, as electronic promoter 398
Calcium aluminate, support 249
Calcium cyanamide 384
Calcium ferrites, in ammonia catalysts 398
"Calcium sulphide", on catalysts 408
Capital charge factor 103
Caprolactam plants, methanation in 352
Carberry/Brisk reactors 63, **63, 64**
Carbidic carbon, on nickel catalyst 348
Carbon-14, in methanol synthesis 68
Carbon deposition (*see also* Carbon formation) 52, 81, **82**, 84
   on ammonia synthesis catalysts 408
   promotion by acid sites 280
   in steam reformer tubes 124
Carbon dioxide
   addition in methanol synthesis 258, 453
   addition in steam reforming 259
   vs carbon monoxide, for methanol 463-4
   desulphurization by zinc oxide 219
   effect on sulphur absorption 217
   equilibrium concentrations of
     in methane steam reforming **233**
     in naphtha steam reforming **237**
   equilibrium for methanation **345**
   in feedstock for steam reforming 227-8
   heat of formation **347**
   hydrogenation of 444-5
   inhibition of methanation 350
   kinetics of methanation 349-51
   level at exit of methanator 360
   methanation of, temperature rise 346
   methanol synthesis from 442
   in natural gas 226
   poisoning ammonia catalysts 80, 404, 406
   removal processes 340
     carry-over on to methanator 220, **361**
     copper liquor scrubbing 309
     source of arsenic 408
   specific heat **346**
Carbon disulphide, presulphiding 205
Carbon formation
   avoidance in reduction 162-3
   catalyst breakage 280
   from cracking 249
   in hydrodesulphurization 208
   in methanation 344
   in naphtha steam reforming 234
   on nickel catalysts 347-8
   risk with natural gas for blanketing 179
   in steam reforming 250-3
Carbon-forming reactions, equilibria **82**
Carbon monoxide
   adsorption of 30-1
   vs carbon dioxide for methanol 463
   concentration, from shift 322
   conversion *see* Water-gas shift
   disproportionation of 81
     in high-temperature shift 298-9
     in low steam ratio operation 338-9
     in steam reforming 250, 280

   dissociation on nickel 347-8
   economics low-temperature shift **336**
   equilibrium concentrations
     for methane steam reforming **232**
     in naphtha steam reforming **236**
     in shift **116, 177**
     in steam reforming **112, 114**
   equilibrium for methanation **345**
   heat of formation **347**
   hydrogenation of 444-5
   inhibition of methanation 350
   kinetics of methanation 347-51
   level at exit of methanator 360
   levels in shift **287**
   methanation of, temperature rise 346
   in methanation reaction 31
   methanol synthesis from 442
   poisoning of ammonia catalysts 404, 406
   reduction in steam reforming 250
   removal of 286, 340
     copper liquor for 286, 309, 340
   specific heat **346**
   in water-gas 283
Carbon monoxide shift (*see* High- *and*
   Low-temperature shift)
Carbon supports 42
Carbonyl sulphide
   in coal-based plants 306
   desulphurization 203
   in natural gas 192
   poor absorption of 197
   poor removal by hydrogenation 198
Carbonylation of olefins, catalytic 259
Carrier gases for reduction 314
Carry-over
   contamination of ammonia catalyst 178
   contamination of methanator 178
   guarding methanation catalyst 220
   of solids in water gas-shift reactors 291
   on to catalyst, pressure drop 126
Casale ammonia converter 428, **431**
Catacarb process, effect on methanator **361**
Catalysts
   behaviour in use 69-84
   bibliography 519-24, 532-6
   change frequency, calculation 96-101
   charging of steam reformers 142-3, 156-60, 270, 273, 568-71
   dimensions, steam reforming 254-6
   essential properties 23-7
   first commecial use 503
   handling 140-90
   for hydrogen, ammonia, methanol 528-9
   made by impregnation processes **43**
   oxidation of, heat release (table) 558
   packing, dished end of vessel 527
   pellet shapes, in steam reforming 254-5
   pellet size
     effect on ammonia synthesis 411-2
     optimum, numerical example 525
     and shape, optimum 101-5
   performance
     calculation of 128-34
     deactivation, designing for 123-37
     effect of gas distribution 136
     measurement 126-37, **135** (table)
   poisons *see* Poisons
   reduction, heat release (table) 557

Index

sintering, metal chlorides as cause 408
specific heats 561
stages in development **55**, 56
testing of 48
volumes **343**
  calculation of 107-9
    in ammonia converters 426
    numerical example 526
Catalytic oxidations 469-518
  of ammonia 19-21, 470-89
Catalytic rich gas (CRG) (*see also* Substitute natural gas) 263
Catalytic rich gas processes 368-72
  double methanation process 368, **369**
  gas compositions (table) **370**
  hydrogasification process 370, **371, 372**
Charcoal (*see also* Activated charcoal)
  removing mercury 223
Charging catalysts 150-60
  ammonia converters 155
  devices for 153
  steam reformers 270
    case study 568-71
    facilities 273
    use of socks 143
  use of inclined screen **150**
  of water-gas shift reactors 291
Chemisorption 30, 399-400
Chile saltpetre
  deposits 384
  for nitric acid production 470
Chimney effect 179, 184
Chloride
  absorbent for 222
  contamination of LPG 193
  effect on surface area of copper 84
  hydrogenolysis 221
  limit in naphtha feedstock 228
  in natural gas 226
  poisoning of
    ammonia catalysts 399, 404, 408
    catalysts 191-2
    copper catalysts 78
    high-temperature shift catalyst 305
    low-temperature shift **221**, 328-30
      rate of reaction 289
    silver catalyst 497
    steam reforming catalyst 278
    vanadium catalyst 516
  in refinery off gases 194
  resistance of shift catalysts 312
  sources of 220-2
  stress corrosion of alloy steels 191
  washing from shift catalysts 313
Chlorine compounds (*see also* Chloride)
  effects on methanol catalysts 461
Chlorine guard 222, 528
Chromia
  in high-pressure methanol catalyst 453-5
  in high-temperature shift 33, 293, 298
  role in high-pressure methanol catalyst 457
  stabilization of magnetite 293
  as support in shift catalysts 310-11
Chromic oxide
  in high-temperature shift catalysts 294
  reduction of 295, 298, 301
  in zinc oxide/chromia catalyst 455
Claus process 198

Coal
  as feedstock in ammonia synthesis 407
  routes to substitute natural gas 372-8
  sulphur poisoning due to 293
  synthesis of ammonia from 284, 472
  synthesis gas from 225
Coal-based ammonia plants
  contamination of shift catalyst 305
  high-temperature shift catalyst in **337**
  steam/dry gas ratios 302
  sulphur compounds in 306
  sulphur-tolerant shift in 335
  use of nitrogen wash 340
Coal gasification 194, **195**
Cobalt
  catalysts for ammonia oxidation 476
  effects on methanol catalysts **461**
  as steam reforming catalyst 244
Cobalt, Raney 37
Cobalt molybdate catalysts 204-7
  absorption of arsenic and lead 224
  adsorption of hydrocarbons 202
  deactivation by ammonia 224
  discharging 208
  methanation reactions over 207
  operating conditions 205-6
  physical form 207-8
  properties (table) **204**
  sulphiding to activate 70
  for water-gas shift reaction 377-8
Cobalt oxide, ammonia oxidation 476
  loss of selectivity 84
  testing activity of 60
Cobalt spinel, in cobalt molybdate 204
Coke (*see also* Carbon deposition, Carbon formation) 52, 81, **82**, 84
  gasification for methanol synthesis 442
  in Haber Bosch process 225
  manufacture of water-gas from 283
Coke oven gas, purification 194-5
Cold shot converters, discharge of 186
Cold shot, in ammonia converters 433
Commercial catalysts, rate constants for **119**
Commercial nitrogen, oxygen content 178
Commissioning, of plant, fast 162
Composition
  of associated gas 193
  calculation of active volume 132
  of circulating gas, methanol loop 446
  of commercial zinc oxide 211
  of gas in CRG hydrogasification **372**
  of high-temperature shift catalysts 293-5
  of liquid petroleum gas 193
  of make-up gas, methanol synthesis 452-3
  of naphthas (table) **229**
  of natural gas condensates 193
  of natural gases (tables) **192, 227**
  of process gases in ammonia plant **342**
  of synthesis gas from coal (table) **195**
Compressor oils
  poisoning ammonia catalysts 407, 408
  source of sulphur poison 488
Computer programs 137-9
Condensate analysis, ICI papers 536
Condensation
  during shutdowns 175
  on shift catalyst 323-4, 332
Construction materials, for reformer 257

594

Index

Contact process 506-11
Contamination
  arsenic in feedstock 228
  of feedstock for steam reforming 227
  of LPG with chloride 193
  of methanation catalyst 177
  of platinum/rhodium catalyst 488
Continuous stirred tank reactors 62-4
Conversion
  design value 105, 526
  sulphur dioxide oxidation **506, 507**
  and temperature rise **131**
Converters *see* Reactors
Copper
  avoidance of sintering 167
  catalysts *see* Copper-based catalysts
  dispersion of crystallites 34
  enhancement of activated charcoal 196
  in low-temperature shift catalyst 33-4
  metal surface area, vs activity **465**
  oxidation state of 462, 464
  reaction with hydrogen chloride 328
  surface, mobility in catalyst 466
  surface area
    of methanol catalyst 68
    and water-gas shift activity 310
Copper-based catalysts
  activity of poisoned **83**
  chloride poisoning 192
  effect of chloride on surface area 84
  poisoning by chlorine compounds 78
Copper-based shift catalysts 287
  introduction of 310
  kinetics over 289
  reaction mechanism 290
Copper, poisoning ammonia catalysts 408
Copper halides
  melting points (table) **326**
  mobility in catalysts 325
Copper liquor 286, 309, 340
Copper(I) chlorides, melting points 328
Copper(I) sulphide, formation 326
Cracking of ammonia 276
Creep of steam reformer tubes 87
CRG *see* Catalytic rich gas
Cross-flow, ammonia reactors 424, 428, **432**
Cryogenic nitrogen unit 225, 283, 284
Cryogenic plant, hydrogen recovery 421
Cyanamide process 225, 384
Cyanide, from coal gasification 194
Cyanide removal, use of zinc oxide 219
Cyclic reformers
  contamination of shift catalyst 305
  regeneration of shift catalysts 177
Cyclohexylmercaptan, **199**
Cyclohexyphenylsulphide **199**

D

Dangers (*see also* Health hazards, Safety considerations)
  of blanketing gases 179
  of hydrogen from ammonia catalyst 185
  of nickel carbonyl formation 182, 360
DEA processes, effect on methanation **361**
Deactivation of catalysts
  ammonia synthesis 178, 404-9
  causes in shift catalyst 320
  countered by temperature-ramping 73
  high-temperature shift 304-6
  interactions in 82-4
  low-temperature shift 324-30
  methanation catalyst 77
  molybdate catalysts by ammonia 224
  platinum/rhodium 488-9
Decomposition
  of formaldehyde 501
  of hydrogen cyanide, by zinc oxide 219
Dehydrogenation, of methanol 490, 493
Demetallization
  catalysts for 208
  of feedstocks 223-4
Dentrification, of feedstocks 224
Density
  and activity of pellets 47
  ammonia (nomograph) 549
  effect of absorption by zinc oxide 212
  methanol vapour 553
  and strength of pellets 295
Dent carbon, thermodynamics of 250
Design
  of catalysts 32-4, 244-53
  of reactors 83-118
    for ammonia synthesis 423-39
    computer programs 137-9
    implications of kinetics 412-3
    process steps in **86**
    of steam reformers 240
    for water-gas shift reaction 291-2
Design conversion of reactor, example 526
Desorption
  from heterogenous catalyst 29
  from iron single crystals 67
  of surface shift intermediates 290
Desulphurization
  carbon deposition in 208
  of carbonyl sulphide 203
  catalyst volume in ammonia plant **343**
  catalysts for 528
  equilibrium constants **110**
  of feedstock 196-8, 326
  of high-temperature shift catalyst 301
  introduction for steam reforming 226
  iron oxide for 209
  of naphtha 194
  of natural gas 227-8
  performance monitoring 135
  reactor design 87
  zinc oxide for 209-20
Deuterium exchange over nickel 239-40
Deuterium exchange over iron 414
Dew-point 330
  in commissioning shift catalyst 319
  in low-temperature shift process 287
Diameter of catalyst vessel example 527
Diatomaceous earth 509
Dienes, hydrogenation of 367
Diethanolamine on methanator 178
  in sulphur removal processes 198
Diethyldisulphide, hydrogenolysis 200
Diethylsulphide 192, (table) **199**
Diffusion effects (*see also* Mass transfer limitations)
  in ammonia synthesis 411
  in hydrogenolysis 201

595

Index

limitation in poisoning 78
in methanation 350, **351**
and pellet shapes 47
and pellet size
  high-temperature shift catalysts 295-8
  low-temperature shift catalysts 312-13
in poisoning shift catalysts 313
in steam reforming 239
in sulphur dioxide oxidation 513
in water-gas shift reaction 289
Dimethyldisulphide, presulphiding 205
Dimethyl ether 455, 458, 460, **461**
Dimethylthiophene, decomposition **199**
Dinitrogen tetroxide 477, 478-9
Diphenylsulphide, decomposition **199**
Discharge of catalysts 183-7
  from cold-shot converters 186
  to correct high pressure drop 159
  high-temperature shift catalysts 330-1
  from tube-cooled converters 187
  water-gas shift reactors 291
Discharged catalysts
  low-temperature shift (table) **328**
  re-use of 187-8
  safety precautions 189
  sulphur in cobalt molybdate 205
Disposal of used catalysts 188, 517
Dissociation
  in ammonia synthesis 24-6, 67
  of carbon monoxide on nickel 31, 347-8
  of dinitrogen tetroxide 478, 479
  of sulphur compounds (table) **199**
Distribution
  and catalyst charging 152
  of gas in methanator 358
  of gas in shift reactor 322
  poor gas distribution, effect of 124
  of reactants in steam reforming 269-70
  of sulphur in zinc oxide pellets 213
Double methanation for SNG 368, **369**
Drums
  handling 141, **142**
  weight of 140
Dual pressure nitric acid plant 482, **474**
Dust exposure
  during charging 150,153
  safety precautions 190

E

Economics
  desulphurization with zinc oxide 198
  of large production units 308
  low-temperature shift catalysts 335-6
  prereduced ammonia catalysts 404
  sale of used catalysts 188
"Effective volume" of catalyst 104
Effectiveness of catalysts
  calculation 104
  high-temperature shift catalysts 295, **296**
  of low-temperature shift catalysts 313
  in steam reforming 239
Efficiency
  ammonia oxidation
    of burners 477, 480-1
    effect of gas rate 483
    effect of number of gauzes **484**

design of steam reformers for 274
effect of hydrogen recovery 422-3
nitrogen fixation **385**
zinc oxide absorbents 213-19
Ekofisk field, composition of gas **192**
Electric arc process 384, 385
Electronic promoters 395-400
Elements, atomic weights (table) **562-4**
Embrittlement of alloy steels 424
Energy barriers
  determining catalyst selectivities 26
  oxidation of methanol **28**
  reduction of 24-6
Energy profiles
  for ammonia synthesis 24-6
  for methanation 355
  for methanol oxidation **28**
  in water-gas shift reaction 288
Equilibrium (*see also* Approach to
    equilibrium)
  ammonia synthesis 389
  between haematite and magnetite 298
  of hydrogen sulphide/zinc oxide **210**
  nitric oxide/air **471**
Equilibrium constants
  ammonia synthesis 385-6, 388
  carbon-forming reactions **82**
  carbonyl sulphide (table) **306**
  hydrogenolysis 200-1, **201**
  hydrogen sulphide with zinc oxide
    (with tables) **210**, 572-3
  hydrolysis of carbonyl sulphide **203**
  magnetite and hydrogen sulphide **305**
  methanation **110**, 344-7
  methane-steam reaction (table) 537-42
  methanol synthesis 442
  oxidation of sulphur dioxide 503-6
  stages ammonia/methanol synthesis **110**
  sulphur dioxide oxidation (table) **505**
  water-gas shift reaction 285, **286**, 543-8
  for zinc and copper(I) sulphide 326
  zinc oxide/hydrogen sulphide **210**, 572-3
Equivalent diameter, of particle 101
"Ergun equation" 102-3
Ethylene, avoidance of cracking 367
Ethylmercaptan, hydrogenolysis 200
Ethylmethylsulphide, hydrogenolysis 200
EVA-ADAM project 378-83
EXAFS, of sintered copper catalysts 324
Exothermic steam reforming 234
Expansion of reformer tubes 272-3
Explosives mixtures
  ammonia and air 481
  methanol and air 497-8, **553**
Extrusion of catalysts 45, **46**, 48

F

Failure, reformer tube welds 530
Fast commissioning of plant 162
Feedstocks
  effect on heat flux profile in reformer 271
  effect on stoichiometry 258
  purification 191-223
  purification catalysts 308
Ferric-ferrous ratio, ammonia catalysts 399
Ferric molybdate 501

Index

Ferrites 397-8
Ferrous sulphide, on ammonia catalysts 408
Film diffusion, ammonia synthesis 411-2
Film diffsuion control 27-8
  in ammonia oxidation 84
  in methanation 350
  in methanol oxidation 494
Fischer-Tropsch reactions
  in high-temperature shift 338-9
  in methanol synthesis 81, 461, 441
Fischer-Tropsch synthesis
  methanation process 352
  use of steam reforming for 230
Flame lengths, in reformer furnace 271
Flammability
  ammonia/air mixtures 481
  methanol/air mixtures 497-9
Flow sheets
  atmospheric pressure nitric acid **473**
  double-absorption sulphuric acid **512**
  dual pressure nitric acid plant **474**
  fused catalyst manufacture **36**
  low-pressure methanol synthesis **448**
  manufacture, impregnated catalysts 42
  manufacture of precipitated catalysts **39**
  oxide-catalysed methanol oxidation **500**
  pulse microreactor system **57**
  single pressure nitric acid plant **475**
  stages in development of catalysts **55-6**
  water-gas shift catalysts 291
Flue gas, measuring oxygen content 274
Fluoride
  poison for vanadium catalyst 516
  removal 223
Formaldehyde
  annual production 469
  decomposition 501
  industrial uses 490
  production of 26, 490-503
Formate, in methanol synthesis 68
Fuel cell, use of steam reforming 230
Fugacity methanol synthesis kinetics 467
Fused catalysts 35-7

G

Gas by-passing, measurement of 127
Gas distribution
  effect of faulty charging 152
  effect on steam reformers 136
  maldistribution in shift reactors 292
  need for correct catalyst packing 140
  in steam reforming 269-70
Gasification
  of coal 372-8
  of oil 368-72
Gasynthane, substitute natural gas 368
Gauze
  metal 35
  platinum/rhodium catalyst
    activation 484-7
    cleaning 488-9
    deactivation 488-9
    installation **487**
  silver catalyst 490
Geometric surface area
  effect on methanation activity **351**

  effect on steam reforming catalysts 254
  low-temperature shift catalysts 312-13
  nickel and cobalt molybdate 207-8
Gold/palladium gauze, metal recovery 489
Granulation, of catalysts 45, **47**, 48
Great Plains Gasification Project 373
Groningen field, composition of gas **192**
Guard, for removing chloride 222, 528
Guard beds
  for low-temperature shift 309, 331-5
    configuration of **332**
    improvement of performance **334**
  zinc oxide 220

H

Haber, F. 386, 388, 394
Haber process 384
  ammonia synthesis loop 391
  development 225
  hydrogen for 283-4
Haematite
  in high-temperature shift 165-6, 293
  reduction 298
Health hazard (*see also* Dangers, Safety
    considerations)
  of chromic oxide 295, 455
Heat exchangers, failure of 297-8
Heat flux in steam reformer 270, 271-2
Heat of mixing, ammonia with gas 390
Heat recovery, in ammonia synthesis 425-6
Heats of formation (*see also* Energy profiles)
    550-1
  ammonia synthesis (table) **392**
  of formaldehyde 490
  in methanation (table) **347**
  of methanol 442
  of nitric acid 477
Heats of hydrogenolysis (table) **200**
Heavy metals, poisoning 192
Heterogeneous catalysis
  fundamentals 17-34
  vs homogeneous for ammonia **20**
  stages of gas-phase reaction (table) **29**
HICOM coal/SNG process 374
High-temperature methanation 376-8
High-temperature shift 293-307
  catalyst volume in ammonia plant **343**
  equilibrium of carbon monoxide **116**
  performance monitoring 136
High-temperature shift catalyst 284, 286, 529
  activity of 287
  carbon monoxide levels in **287**
  conversion of hydrogen sulphide to carbonyl
    sulphide 306
  effect of inlet temperature on life 303
  effectiveness and activity (table) **296**
  formulation 293-5
  heat release during oxidation 558
  heat release during reduction 557
  kinetics over 288-9
  low sulphur content 302
  methods for stabilization 180-1
  operation 302-4
    at low steam ratio 338-9
    in modern coal-based plant **337**
  pellet size and diffusion effects 295-8

597

Index

physical properties (table) **294**
rate constant **119**
reactions during operation (table) **308**
reduction 71, 165-6, 293, 298-302
regeneration 177, 188
specific heat 561
sulphidation 177
Higher hydrocarbons
  steam reforming catalyst reduction 276
  steam reforming of 240
Horizontal cross-flow quench 428, **432**
Hot bands in steam reformers 280-2
Hot spot, in methanol oxidation 501
Hulburt-Vasan, equations **289**
Hüttig temperature of copper 324
Hydration of magnesia 249, **250** (table)
Hydrocarbons
  hydrates 227
  in shift reaction 299
Hydrocracking, over nickel molybdate 270
Hydrodesulphurization
  aromatization in 228
  carbon deposition in 208
  catalyst in ammonia plant **343**
  catalysts for 528
    specific heat 561
  operating conditions 206
Hydroformylation 259
Hydrogasification for SNG 370, **371**, **372**
Hydrogen
  atomic, in ammonia synthesis 18
  consumption, shift catalyst reduction 167
  effect on alloy steels 424
  electrolytic, purification 194
  evolved from ammonia catalyst 187
  levels, in hydrodesulphurization 206
  methanol synthesis from 442
  from olefin plants, methanation of 367
  plants, methanation in 342
  production from coke and steam 225
  reactions on iron single crystals 67
  recovery, from loops 89, 421
  reduction of haematite 165-6
  reduction of low-temperature shift 317-19
  reduction of reforming catalyst 275
  specific heat **346**
  in water-gas 283
  from water and old catalyst 307
Hydrogen chloride
  diffusion limited poisoning 313
  effect on shift catalyst **329**
  level in process gas 329
  reaction with copper 328
  in refinery off gases 194
Hydrogen cyanide
  from coal gasification 194
  decomposition over zinc oxide 219
Hydrogen sulphide
  addition to alkenes 202
  in coal-based plants 306
  diffusion limited poisoning 313
  equilibrium constants
    conversion to carbonyl sulphide **306**
    reaction with magnetite, (table) **305**
    reaction with zinc oxide 572-3
  in natural gas 192
  in presulphiding catalysts 205
  reaction with iron oxide (table) **210**

reaction with magnetite 209
reaction with zinc oxide 209-10
release from cobalt molybdate 205
removal from feedstocks, zinc oxide 198
from sulphate 293-4, 301-2
Hydrogenation, of carbon oxides (*see also* Methanation) 444-5
Hydrogenation reactions
  carbon oxides as poisons 340
  over cobalt molybdate catalysts 206
  Raney nickel for 37
Hydrogenolysis (*see also* Desulphurization)
  of organic chlorides 221
  sulphur compounds 200-2
Hydrolysis
  of calcium cyanamide 384
  of carbonyl sulphide 203
  of potassium beta-alumina 282

I

IBC *see* Intermediate bulk containers
ICI
  500 process, for town gas 261
  high-conversion reactor 96
  high-temperature methanation 376-8
  low-pressure methanol process 26
  naphtha steam reforming 226
Impregnation processes 41-4, 245
Incipient wetness technique 43
Indonesia, natural gas (table) **192**
Inert gases
  concentration in ammonia loop 420-3
  effect on ammonia synthesis 389
  safety precautions 189
Infra red sensor (pyrometer) 556
Integral reactors 60-1, 62, 288
Inter-bed cooling
  temperature profile **93**
  types of 91
  in water-gas shift reaction 287
Intermediate bulk containers 142-3, 153
Intermetallic compounds 468
Intrinsic activity
  ammonia catalysts 401, 411
  dependence on catalyst support 32
Ion exchange, in impregnation process 44
Iridium, as steam reforming catalyst 244
IRMA (internally cooled reactor) 382-3
Iron
  catalysts
    for ammonia oxidation 476
    for methanation 341, 352
  effects on methanol synthesis **461**
  stable phases **72**
  studies on single crystals 399-400
  unpromoted, ammonia activity 394
Iron-based shift catalyst 287, 293
Iron carbides, formation 338-9
Iron/molybdenum catalysts 27
Iron ore reduction 230, 260
Iron oxide (*see also* Haematite, Magnetite)
  in catalysts for shift reaction 284
  contamination of gauze 488
  in formaldehyde catalyst 501
  for sulphur absorption 209
Iron oxide/chromia catalysts

Index

adsorption mechanism 290
kinetics over 288-9
rate equations for shift (table) **289**
Iron pentacarbonyl
  in methanol synthesis plant 461
  poisoning of silver catalyst 497
Iron pseudomorphs **396**
Iron sintering, avoidance of 401, 402
Irreversible poisoning *see* Permanent deactivation

K

Kalsilite 43, 252
Kellogg axial-flow quench converter 428, **430**
Kellogg horizontal cross-flow quench converter 428, **432**
Kinetic equations (*see also* Rate equations)
  ammonia synthesis 412
  methanation (table) **349**
  steam reforming 239, **241-3** (table)
  sulphur dioxide oxidation 513-4
  water-gas shift 287
Kinetics 61-5
  ammonia synthesis 409-13
  Berty-type vs Carberry-Brisk reactors **63**
  high-temperature shift catalyst 288-9
  hydrogenolysis, sulphur compounds 200
  over low-temperature shift catalyst 289
  steam reforming 239-43
  sulphur dioxide oxidation 511-4
Kodama model, rate equations **289**
Koppers-Totzek gasification 195, 376
Kuhlmann, C.F., ammonia oxidation 472

L

Langmuir-Hinshelwood equations 288, **289**
Larson-Miller diagram, reformer tubes **268**
le Rossignol, R., equilibrium constants 386
Lead
  absorption on cobalt molybdate 224
  in feedstocks 223
  limit in naphtha 228
  on methanol synthesis catalysts **461**
Lead compounds, in zinc oxide 220
Lean Gas reformer 261-2
Life
  of ammonia synthesis catalysts 404
  of catalysts 23, (tables: **74, 75-6**)
  of high-temperature shift catalyst 304
  of low-temperature shift catalyst 320
  of methanation catalysts 352-5, **365**
  of silver catalyst 494
  of steam reformer tubes 276, **269**
  of steam reforming catalysts 277
  of vanadium catalysts 511
Linear gas velocity, and diffusion 350
Liquefaction, of carbon monoxide 283
Liquid petroleum gas 193
  boiling range 554
  catalytic gasification 368-72
  hydrogen chloride in 221
Liquids, on methanation catalyst **361**
Location of plant, and poisoning 325-6
Low-pressure methanol synthesis

by-product formation (table) **460**
converters **449-52**
plant, flow sheet **448**
process 26
Low steam ratio
  plant operation 338-9
  for shift catalyst reduction 299
Low-temperature shift
  catalyst volume in ammonia plant **343**
  equilibrium of carbon monoxide **117**
  performance monitoring 136
  process gas in ammonia plant (table) **342**
Low-temperature shift catalysts 287, 308-35
  activities **325**
  activity of poisoned **83**
  by-pass of 360
  carbon monoxide levels in **287**
  commissioning 319-20
  development 456
  economics 335
  formulation 309-12
  heat release during oxidation 558
  heat release during reduction 557
  kinetics over 289
  manufacture 311
  method for stabilization 181-2
  monitoring performance 320-4
  once-through reduction of 317
  operation 320-4
  pellet size and diffusion effects 312-13
  poisoning 326-30, **221**
  properties (table) **312**
  protection by zinc oxide guard beds 220
  rate constants **119**
  reduction 166-71, 314-20
    arrangement of systems **315, 316**
    method 170-1
    temperature charts vs time **168, 169**
  regeneration 177
  re-use 188
  specific heat 561
  temperature profile **321**
  changes due to poisoning **80**
Lozenge cold-shot, discharge 186-7
Lozenge quench converter
  ammonia synthesis 428, **429**
  methanol synthesis 447
LPG *see* Liquid petroleum gas
Lubricating oils, poisons 407-8, 488
Lurgi coal/SNG process 195, 373

M

Magnesia
  benefit in methanation 353-8
  in steam reforming catalysts 249
Magnesium ions, in ammonia catalysts 397
Magnesium spinel support, 249, 252
Magnetite
  active phase in high-temperature shift catalyst 33, 71, 165-6, 293, 297-9
  in ammonia synthesis catalyst 388, 392, 394, 397
  equilibrium constant of reaction with hydrogen sulphide (table) **305**
  rate of oxidation by water 290
  reaction with hydrogen sulphide 209

Index

Magnetite/chromia catalyst, shift rate
 equations (table) **289**
Maldistribution of gas
 causes in charging 152
 in low-temperature shift reactor 322
 in methanator 358
 in water-gas shift reactors 292
Mass balances 129
Mass-transfer limitations (*see also* Diffusion
 effects) **30**
 in high-temperature shift reaction 295
 in water-gas shift reaction 288
Mechanical failure, heat exchangers 297-8
Mechanisms of
 ammonia synthesis reactions 413-5
 catalytic reactions 66-9
 methanation 347-51
 methanol synthesis 462-6
 rhodium oxidation 488-9
 sintering of metal catalysts 77
 sulphur dioxide oxidation 511-4
 water-gas shift reaction 288-9, 290
Mercaptans, in natural gas 192
Mercury
 in electrolytic hydrogen 194
 in feedstocks 223
 removal 223
Mesh size, platinum/rhodium catalyst 484
Metal oxide formaldehyde catalyst 499-502
Metal/zeolite catalysts **43**
Methanation (*see also* Methanation catalysts)
 340-83
 activation energy for 355
 activity of iron catalyst 406
 in ammonia plants 341-2
 avoidance in water-gas shift process 33
 benefit of magnesia 353-8
 of carbon dioxide in desulphurisation 228
 over cobalt molybdate catalysts 207
 conditions for ADAM-1 plant (table) **379**
 early use of 341
 equilibrium constants **110**, 344-7
 heat evolved during 171
 in heat transport applications 378-83
 in high-pressure methanol synthesis 461
 in high-temperature shift process 298
 in hydrogen plants 342
 kinetics and mechanisms 347-51
 in low-temperature shift process 309
 in olefin plants 367, **368** (table)
 process gas in ammonia plant **342**
 reaction 31
 for substitute natural gas 376-8
 thermodynamic data (table) **346**
Methanation catalysts 529
 calculating minimum volume 363
 contamination of 177
 deactivation 77
 formulation 352-8
 heat release during oxidation 558
 heat release during reduction 557
 heat resistance 359, 366
 life 352
 method for stablization 182
 performance data (table) **367**
 poisons 360-2
 protection by zinc oxide 220
 rate constant **119**

 reduction 171, **357**, 359-60
 re-use 188
 specific heat 561
 volume in ammonia plant **343**
Methanators
 design 89, 359
 nickel carbonyl formation in 182, 185
 in old coal-based plants 341
 performance monitoring 136
 poisoned zones in 362
 removal of carbon monoxide 286
 temperature profiles 363-4, **366** (table)
 temperature rise in 320, 345
Methane (*see also* Natural gas)
 cracking
  activation energies 239-40
  carbon formation in reforming 280-1
  in steam reforming 250
 equilibrium concentrations in steam
  reforming **112, 114, 231, 235, 537**-42
 formation in methanol synthesis 460
 heat of formation **347**
 hydrogen production from 225
 oxidation to formaldehyde 503
 specific heat **346**
Methane steam reforming
 equilibrium concentrations
  of carbon dioxide **233**
  of carbon monoxide **232**
  of methane **231**
Methanol
 catalysts, performance 134-7
 dehydrogenation 490, 493
 distillation of crude 447
 energy barriers in oxidation **28**
 flammability in mixtures with air 497-9
 oxidation 26, 490-503
  mixed metal oxide catalyst 499-502
 physical properties (table) **553**
 plants
  hydrodesulphurisation in 206
  operating conditions for reformers **258**
 steam reforming catalyst reduction 164-5, 276
Methanol synthesis 441-68
 catalysts for *see* Methanol catalysts
 effect of copper metal surface area **465**
 equilibrium constants 110
 Fischer-Tropsch reactions in 81
 flow sheet for low-pressure plant **448**
 mechanism 68-9, 462-7
 methanol concentration profiles **452**
 reaction steps 465-6
 steam reforming for 258
 thermodynamics 442-4
Methanol catalyst 40, 453-62, 529
 active components 463
 active sites 466
 effect of pressure 459
 effects of poisons (table) **461**
 high-pressure 453-5
 low-pressure 455-62
 selectivity 444-5
 stability **454** (table), 458
Methanol synthesis converters **449-51**
 design 89
 performance monitoring 137
Methanol synthesis loop 446-52
Methoxy intermediate 466

Methyl formate 493, 494
Mittasch, A., work on catalysis 388, 394
Mobility in catalysts 17
  of chloride ion 191
  of copper(I) chlorides 328
  of copper surface 466
  of surface species 325
Molybdenum oxide
  in cobalt molybdate catalysts 204, 222
  in formaldehyde catalyst 501
Molybdebum sulphide, in cobalt molybdate catalysts 205
Monticellite, for neutralization 252
MRG process for SNG 368
Multibed quench converters
  ammonia catalyst reduction in 174-5
  temperature profiles **92, 93**

N

Naphtha
  catalytic gasification 368-72, **372** (table)
  desulphurization 194
  steam reforming 81, 226
    equilibrium concentrations
      carbon dioxide **237**
      carbon monoxide **236**
      hydrogen **238**
      methane **235**
    manufacture of catalyst for 43
    scheme for production of carbon **251**
    for town gas 261
Naphthas
  boiling range 554
  for catalytic steam reforming 228
  composition (table) **229**
Naphthenes, in naphthas 228
Natural gas (see also Methane, Substitute natural gas)
  carrier gas for shift reduction 314
  chemistry of steam reforming 230
  over cobalt molybdate catalysts 206
  compositions 192, (tables: **192, 193, 227**)
  as feedstock for steam reforming 227-8
  hydrogen chloride in 221
  mercury in 223
  reduction of reforming catalyst 276
  synthesis of ammonia from, (table) **285**
  use in blanketing reduced catalyst 178-9
Natural gas condensates 193
Nernst, W.
  work on equilibrium constants 386, 388
  work on methanol synthesis 441
Netherlands, natural gas (table) **227**
Nickel
  commercial methanation catalysts 358
  crystallite size in catalysts 352
  dissociation of carbon monoxide 347-8
  effects on methanol synthesis **461**
  in feedstocks 223
  as steam reforming catalyst 244-9
  steam reforming catalysts (table) **245**
Nickel carbonyl hazard
  avoidance in gasification process 376
  in methanator 182, 185, 360
  in methanol plant **461**
  safety precautions 189

Nickel catalysts
  arsenic poisoning 278-80
  effect of potash 252
  methanation, poisoning by sulphur 340
  minimum levels of sulphur **197**
  poisoning by sulphur compounds 361
  reduction 162-3
  specific heat **346**
  stabilization 72
  sulphur poisoning 191
Nickel molybdate catalysts 207-8
  deactivation by ammonia 224
  discharging 208
  physical form 207-8
  properties (table) **207**
Nickel oxide
  calcination of **356**
  heat of formation **347**
  inhibition of sintering by magnesia 354
Nickel oxide catalyst, specific heat **346**
Nickel oxide/magnesia 353-4
Nickel spinel 353
Nickel surface area
  activity of methanation catalysts 352
  steam reforming catalysts **248**
  loss in **246** (tables), 277
Nickel tetracarbonyl see Nickel carbonyl hazard
Nitric acid
  annual production 469
  history of production 470-6
  industry, factors in expansion 472
  manufacture from natural gas (table) **22**
  oxidation of ammonia to 20-1
  plants
    concentration of nitrogen oxides 477-8
    data (table) **476**
    flow diagrams **473**
    types of 476
  theoretical reactions producing 20-3, **21**
Nitric oxide (see also Ammonia oxidation, Nitrogen oxides)
  contamination of shift catalyst 305
  gauzes in production 35
  oxidation 477-9
  from oxygen and nitrogen 385
  percentage in air at equilibrium **471**
  rates of oxidation in air **480**
  yield from ammonia oxidation 480
Nitriding
  of alloy steels 424
  of ammonia synthesis catalyst 178
Nitrogen
  carrier gas for catalyst reduction 314
  fixation via ammonia 19, **385**
  interaction with iron surface 413-5
  in natural gas 226, 228
  oxygen content of nitrogen 178
  reactions on iron single crystals 67
  safety precautions 292
  specific heat **346**
  use in blanketing catalyst 178-9
  wash, removal of carbon oxides 340
Nitrogen dioxide 477
Nitrogen-filled vessel, safety 292
Nitrogen oxides
  chemistry of absorption 477-8
  from coal gasification 194
  in nitric acid plant tail gas 477-8

601

Index

Nitrogen peroxide, from ammonia 477
Nitrous oxide, surface area determination 68
Non-ideality of gases
  in ammonia synthesis 388, 410
  in methanol synthesis 443
North Sea gas
  composition (tables) **192, 227**
  effect on town gas industry 261
  steam reforming 226
Numerical examples
  calculation of catalyst volume 526
  catalyst packing in dished end 527
  design conversion of reactor 526
  optimum catalyst pellet size 525
  optimum diameter of vessel 527
  optimum voidage 525

O

Oil refining 81, **75-6**
Olefins, hydrogenation of in hydrodesulphurization 228
Operating conditions (*see also* Operating temperatures)
  chlorine guard 222
  for steam reforming 230-8, (table) **257**
Operating temperatures
  calculation of optimum 121-3
  cobalt molybdate catalysts 205-6
  design decisions 87-9
  methanation reactors 87-9
  silver formaldehyde catalyst 491-3
Oppau, first catalytic ammonia plant 386
Optimization
  ammonia synthesis loop 417, 423
  of catalyst size 101-5, 294-5, 525
  computer programs for 138
  high-temperature shift converters 289
Optimum catalyst life, calculation 99-100
Optimum conversion, calculation 106
Optimum diameter of vessel 527
Optimum formulations, shift catalyst 312
Optimum nickel content 246
Optimum operating temperature
  calculation 121-3
  with new catalyst 122
Optimum stabilizer crystal size **457**
Optimum voidage, example 525
Organic chlorides, hydrogenolysis 221
Osmium, ammonia catalyst 386, 394
Ostwald, F. W.
  catalyst for ammonia oxidation 484
  development of process 472
Oxidation (*see also* Catalytic oxidations, Reoxidation)
  of ammonia 19-21, 470-89
  of catalysts, heat release (table) **558**
  of low-temperature shift catalysts 330-1
  of methanol 26, 490-503
  of rhodium 488-9
  of sulphided shift catalyst 307
  of sulphur dioxide 469, 503-17
Oxidation catalysts (*see also* Oxidation) 27
Oxo-alcohol 230
Oxo gas reformer **260**
Oxygen
  adsorption on iron catalyst 407
  content of commercial nitrogen 178
  coverage, on copper catalysts 290
  manufacture of nitric oxide from 385
  measuring in flue gas 274
  over cobalt molybdate catalysts 206
  poisoning of ammonia catalyst 405

P

Pakistan, natural gas (table) **227**
Palladium
  as methanol synthesis catalyst 454, 468
  as steam reforming catalyst 244
Partial oxidation plants
  low-temperature shift catalyst in 333
  sulphur-tolerant shift catalysts in 335
  use of nitrogen wash 340
Partial oxidation processes, carbon in 305
Particle size (*see also* Pellet size)
  in ammonia synthesis catalyst 411-12
  calculation of optimum 103-5
  effect on activity 401, 411
Pellet shapes (*see also* diffusion effects)
  and diffusion 47
  effect on pressure drop 296-8
  in steam reforming 254-5
Pellet size (*see also* Particle size)
  activity of sulphur dioxide catalysts **515**
  choice for shift catalysts 298
  and diffusion effects
    high-temperature shift catalysts 295-8
    low-temperature shift catalysts 312-13
  effect on ammonia synthesis 411-2
  effect on shift activity 295
  effect on pressure drop 296-8
  in guard bed 334
  high-temperature shift catalysts
    diffusion effects 295-8
    optimization 295
  optimum, numerical example 525
  and pressure drop **515**
Pellet size and shape, optimum 101-5
Pelleting, of catalysts 45-8, **46**
Performance
  ammonia synthesis loop **422**, 423
  of catalysts
    measurement (table) **135**
    tests 55-6
  of reactors 85, 123-37
Permanent deactivation of ammonia synthesis catalysts 405, 407-9
  in "temporary" poisoning 406
Phenylmercaptan, hydrogenolysis 200
Phosphorus poisoning
  of ammonia catalysts 405, 408
  of copper catalysts 330
  of iron shift catalysts 305
Physisorption 30
Pigtail nipping 530
Plant (*see under* Individual processes)
Platinum catalysts 27
  in contact process 507
  discovery for ammonia oxidation 472
  poisoning by arsenic 80
  for steam reforming 244
Platinum/rhodium catalysts (*see also* Platinum/rhodium gauze) 35

Index

for ammonia oxidation 484-9
  deactivation 488-9
  electron microscopy of 485-7
  platinum loss mechanisms 487
Platinum/rhodium gauze
  number of 483, **484**
  temperature in ammonia oxidation 482
Poisoning
  of ammonia synthesis catalysts 404-9
    resistance due to calcium 399
  effect on sintering 123
  of high-temperature shift 293, 304-6
  by impurities 77-80
  of low-temperature shift 324-30
    avoidance during catalyst reduction 301
    diffusion limitation 313
    guard beds to avoid 331-5
    resistance to 311
  by reactants and products 79-80, 81-2
  silver catalyst 494-7
  in small-scale catalyst testing 59
  of steam reforming catalysts 272
  of sulphur dioxide catalysts 514-7
  and temperature profiles **133, 134**
Poisons **79** (table)
  absence in ammonia 472
  ammonia synthesis catalysts, levels 406
  carbon oxides as 340
  low-temperature shift catalysts
    effect of temperature on 325
    rate of reaction with 289
    resistance of 312
    retention in 321
  methanation catalysts 360-2
  methanol catalysts (table) **461**
  removal from feedtsock 191-223
  steam reforming catalysts 278-80
  testing catalysts 49
  vanadium catalyst 516
  zinc oxide as absorbent 34, 209-20
  zinc-oxide/chromia catalyst 455
Poisons profiles, in shift catalyst 313
Pore-diffusion limitation (*see also* Diffusion effects) 27-8
  advantage of small pellets 298
  diminution in ammonia synthesis 433
  in high-temperature shift reaction 295
  in low-temperature shift catalyst 289, 311
Porosity
  development in ammonia catalyst 395
  effect on sulphur absorption 212
  of fused iron catalyst 36
  measurement of 54
Potash (*see also* Potassium)
  effect on nickel catalysts 252
  for neutralization of support acidity 252
  in steam reforming catalysts 281-2
Potassium
  beta-alumina 282
  depletion from reforming catalyst 252
  on iron surface 398
  promotor in ammonia catalysts 398, 399, 415
  roles in steam reforming 81
Potassium carbonate, contamination 178, 362
Precautions *see* Safety considerations
Precharging checks 150-1
  support grids in reformers 156
  use of pressure-drop device 156, **159-161**

Precious metal recovery 245
Precipitation processes 38-40
Prereduced catalysts
  ammonia synthesis 402-4
    charging 155
    economics 404
    supplied forms 172
  methanation catalysts 353
  production 72
Pressure (*see also* Operating conditions)
  dissociation of dinitrogen tetroxide **478**
  effect on methanol catalysts 459
  effect on methanol synthesis **444**
  effect on steam reforming (table) **234**
  hydrogen sulphide absorption **216**
  life of steam reformer tubes **269**
  optima for ammonia synthesis loop 418
Pressure drop
  in catalyst beds, calculation 102-3
  and catalyst shape 208
  consideration for iron ore reduction 260
  distribution in steam reformer 270
  effect of pellet size 296-8
  and height-to-diameter ratio 217, **218**
  increase, during use 152
    oxide formaldehyde catalyst 501
    reasons for 125-6
  indicating life of shift catalyst 304
  measurement
    across steam reformer tubes 565-7
    after charging reformer tubes 158
    measuring instruments 565, **566, 567**
    in precharging checks 156, **159, 161**
  over molybdate catalysts 208
  and small catalyst pellets 298
  sulphur dioxide oxidation **515**
Pressure drop, case study (table) **570-1**
Presulphiding, molybdate catalysts 205-6
Primary reformers
  performance monitoring 135-6
  top discharge 185
Primary reforming catalysts 528
  charging, case study 568-71
  heat release during oxidation 558
  heat release during reduction 557
  specific heat 561
  volume in ammonia plant **343**
Process water, chloride in 221
Producer-gas
  generation 284
  use in ammonia plants 225
Promoters 32
  for ammonia synthesis catalysts 67, 388, 392, 395-400, 415
  in Smithsonite 454
Propane
  carbon formation in reforming 253
  as feedstock for steam reforming 227
Properties of catalysts
  cobalt molybdate (table) **204**
  nickel molybdate (table) **207**
  required (with table) 23-7
Pulse microreactor system. flow sheet **57**
Purge gas
  ammonia synthesis loop 391, 420-3
  methanol synthesis loop 446
Pyrometry 555-6
Pyrophoric catalysts

603

Index

blanketing during extraction 186
discharged molybdate catalysts 208
discharging 184-5
high-temperature shift catalyst 306-7
low-temperature shift catalyst 330
safety precautions 189
sulphided zinc oxide 188

Q

Qatar, natural gas from (table) **227**
Quench converters 91
   ammonia catalyst reduction in 174-5
   for ammonia synthesis 425, 426-33
   charging with catalyst 155
   for methanol synthesis 447-52, **449**
   temperature profile of multibed **92, 93**

R

Radial-flow converters
   ammonia synthesis 424, 428, **431, 437**
   charging 152
Radioactive ^{35}S, zinc oxide 212, **213**
Radioactive isotopes
   on iron oxide/chromia catalysts 290
   in methanol synthesis 68, 464
Radioactive tracing, catalyst performance 127
Radioactivity in methanol synthesis 464
Raney
   cobalt 37
   copper 37, 468
   nickel 35, 37
Rapid commissioning of reactors 162
Rate equations
   for commercial catalysts (table) **119**
   methanation of carbon monoxide 348
   methanation (table) **349**, 364
   in reactor design 119
   for steam reforming 241
Rate-determining step
   in ammonia synthesis 67
   in methanation 348
Rating of reactors 85, 121-3, 137-9
Reactants, catalyst poisoning by 81-2
Reaction kinetics *see* Kinetics
Reaction steps
   ammonia from cyanamide 384
   high-temperature shift catalyst
      reduction 298-9, **308**
      regeneration 307
   low-temperature shift over copper 290
   methanol synthesis 465-6
   water-gas shift reaction 283
Reactivation of catalysts *see* Regeneration
Reactors
   for ammonia synthesis 415-39
   charging 151-4
   comparison of different types (table) **91**
   cost 105
   design 85-120
      for ammonia synthesis 423-39
      for desulphurization, 217-19
      for water-gas shift reaction 291-2
      dimensions, methanation 359
      height-to-diameter ratio 217, **218**
      implications of kinetics on design 412-3
      integral 60-1, 62
      performance 123-37
      predicting performance 61
      rating 85, 121-3
      single-pellet **60**
      for obtaining kinetics 62-3
      types 90-6
Recovery of hydrogen from purge 421
Rectisol process **361**
Recycling, of silver catalyst 491
Reducing gas
   for iron ore reduction 260
   operating conditions (table) **261**
Reduction
   activation of catalyst 70
   ammonia synthesis catalyst 171-5, 400-4
      effect of water during 411
   of catalysts 161-75, (table) 557
   chromate in methanol catalyst 455
   high-temperature shift catalysts 71, 165-6, 293, 298-302
   low-temperature shift catalysts 166-71, 314-20
      arrangement for **315, 316**
      once-through 317
   low-temperature shift guard beds **333**
   methanation catalysts 171, **357**, 359-60
   nickel oxide/magnesia solutions 353
   steam reforming catalysts 247, 274-7
   vanadium pentoxide catalyst 510, 513
   zinc oxide in methanol catalyst 462
Reformer tubes, charging 156-60, **157**
Reformers, top discharge 185
Reforming catalysts (*see also* Steam reforming catalysts, Primary *and* Secondary reforming catalysts)
   method for stabilization 180
   poisoning by arsenic 220
   reduction 162-3
   regeneration 176
   socks for charging 142-3
Refrigeration
   in ammonia synthesis loop 172, 418
   in formaldehyde process 499
Regeneration
   of activated charcoal 196, 197
   of catalysts 73, 176-8
      high-temperature shift 305
      in situ 81
      low-temperature shift 291
   silver catalyst 497
Regenerative mechanism, shift 290
Removal
   of arsenic 280
   of potassium carbonate 362
   of silica and fluoride 223
   of top layer of shift catalyst 304
Reorganization of iron surface 400
Reoxidation, of ammonia catalyst 178
Reoxidation, of shift catalyst 306-7
Re-use of discharged catalysts 187-8
   high-temperature shift catalyst 307
   low-temperature shift catalyst 331
Rhenium ammonia catalysts 440
Rhodium
   in ammonia oxidation catalysts 472
   as-steam reforming catalyst 240, 244

Rhodium/silica catalyst, methanol **454**
Rossignol *see* le Rossignol
Rubidium, electronic promoter 398
Ruthenium 244, 352, 440

S

Sabatier, Paul, work on methanation 340
Safety considerations 188-90
   discharging catalyst under nitrogen 186
   methanation catalyst 182
   use of blanketing gases 179, 186
Saltpetre
   for ammonia production 384
   for nitric acid production 470
SASOL
   hydrocarbon plants 352
   pilot plant for Lurgi process 373, **374**
Secondary reformers, top discharge 185
Secondary reforming (*see also* Secondary reforming catalysts) 253
   air mixing and burner design 125
   construction materials for 257
   effects of poor mixing (table) **125**
   performance monitoring 135-6
Secondary reforming catalysts 529
   heat release during oxidation 558
   heat release during reduction 557
   specific heat 561
   volume in ammonia plant **343**
Seldon mass 508
Selective chemisorption 51
Selectivity
   ammonia oxidation 19-20, 479-81
   catalysts 23, 26-7
   copper methanol catalysts 460-2
   low-temperature shift catalysts 309
   silver catalyst 494
Selexol process **361**
Self-guarding, shift catalysts 289
Shell catalyst 44
Shift converter, gas analyses (table) **285**
Shift reaction, in methanol synthesis 68
Silica
   in ammonia synthesis catalysts 397, 398
   on methanol synthesis catalysts **461**
   poisoning
      high-temperature shift catalyst 305
      low-temperature shift catalyst 329-30
   removal 223
   in steam reforming catalysts 249
Silver formaldehyde process 490-9
Silver catalysts 26, 27, 490-9
   electron microscopy **495**
   life 494
   poisoning 494-7
   regeneration 497
   selectivity 494
Single-pellet reactor **60**
Single pressure nitric acid plant **475**
Sintering
   acceleration by water 455
   of ammonia synthesis catalysts 407
   avoidance in catalyst reduction 401, 402
      low-temperature shift catalyst 167
   change in temperature profile **79**
   copper crystallites, in shift catalyst 314
   deactivation of ammonia catalysts 404
   during reduction of nickel catalysts 247
   effect of poisoning 123
   of high-temperature shift 304
   of low-temperature shift 167, 314, 324
      mechanism 328
   of nickel oxide, inhibition 354
   of supported metal catalysts 76
Smithsonite, source of zinc oxide 454
SNG *see* Substitute natural gas
Socks, catalyst charging 143, 156-7, 568
Sodium nitrate
   in fertilizers 384
   for nitric acid production 470
Specific heats
   catalysts 561
   liquid ammonia (nomograph) 549
Sphalerite 572
Spinel precursors 353
Spinels *see* Cobalt spinel, Magnesium spinel, Nickel spinel
Spinel structure, of magnetite 394-5
Stabilization
   of catalysts 179-83, **458**
   of low-temperature shift catalyst 331
   of metal catalysts, role of support 77
   of nickel catalysts 72
   of prereduced ammonia catalysts 403
Stabilizer crystal size, optimum **457**
Start-up
   fast 162
   heater in ammonia converter 434
   prereduced ammonia catalysts 403
   steam reformer 274
Steam (*see also* Water)
   addition in methanol oxidation 491
   as carrier gas for catalyst reduction 314
   injection into shift process gas 288
   manufacture of water-gas from 283
   ratios
      during steam reforming start-up 275
      effects in naphtha reforming **234**
      for steam reforming 230-8, 250, **251**
      in regenerating reforming catalyst 176
   regeneration of activated charcoal 196
   specific heat **346**
   transport of arsenic 279-80
Steam dealkylation 240
Steam-raising reactors 96
   for methanol synthesis 447-52, **451**
   temperature profiles **95**
Steam reformers
   ammonia plant (table) **257**
   design 87-8, 240
   operating conditions **88**
   performance monitoring 135-6
   schematic arrangement **262, 263**
Steam reforming 225-82
   carbon formation 250-3
   catalyst dimensions 254-6
   chemistry 230-43
   design of catalysts 244-53
   equilibrium concentration
      carbon monoxide in primary **113**
      carbon monoxide in secondary **115**
      methane in primary reformer **112**
      methane in secondary reformer **114**
   equilibrium constants **110**, 537-42

Index

feedstocks for 226-9
  purification 192-5
  kinetics 239-43, (table) **241-3**
  mechanism, higher hydrocarbons 240
  of naphtha 226
  roles of potassium 81
  throughput 265
  uses 230, 256-63
Steam reforming catalysts (*see also* Reforming catalysts, Primary *and* Secondary reforming catalysts)
  adiabatic beds 263
  nickel content (table) **245**
  nickel surface area **246** (tables), **248**
  poisons 278-80
  reduction 247, 274-7
  refractory supports 249
  sulphur poisoning 191, 278
Stenching agents in natural gas 192
Stresses, steam reformer tubes 267, 273
Structural promoters 395-8
Structural sensitivity, of ammonia synthesis 399-400, 404
Substitute natural gas (SNG) 263, 368-78
  catalytic rich gas processes (table) **372**
  gas compositions in CRG process **370**
  reactors for 90
  use of methanation 341
Sulphate, in shift catalyst 301
Sulphate of ammonia, as fertilizer 469
Sulphide ions, in zinc oxide 211
Sulphided
  cobalt molybdate, oxidation 558
  high-temperature shift catalyst
    activity 306
    oxidation 307
  magnetite 33
  zinc oxide, pyrophoricity of 188
Sulphiding
  in activation of catalysts 70
  of cobalt/molybdate catalysts 337
  high-temperature shift catalysts 177
Sulphinol process 178, **361** (with table) 362
Sulphur (*see also* Sulphur compounds)
  absorption
    iron oxide for 209
    zinc oxide for 209-20
    by zinc oxide 216-17
  absorption profiles, zinc oxide **212**
  content in discharged low-temperature shift catalyst 327, **328** (table)
  in discharged cobalt molybdate 205
  in iron ore reduction 260
  low-content shift catalysts 302
  minimum levels to poison nickel **197**
  poisoning
    of ammonia synthesis catalysts 399
    of high-temperature shift 293
    of low-temperature shift 326-7
    of nickel catalysts 191
    of platinum/rhodium catalyst 488
    of steam reforming catalysts 191, 278
  removal in early coal-based plants 294
  resistance of low-temperature shift 312
Sulphur compounds
  effects on methanol catalysts 461
  in gas from coal and oil 337
  hydrogenolysis 200-2

  in natural gas 226
  poisoning
    of ammonia catalysts 405, 407
    of methanation catalysts 361
    of nickel methanation catalysts 340
    of silver catalyst 497
  rate of reaction with shift catalysts 289
  thermal dissociation (with table) 199
Sulphur compounds, in naphtha 194
Sulphur dioxide
  from desulphurizer 207
  oxidation 469, 503-17
    kinetics and mechanism 511-4
    operating pressures 517
    thermodynamics (table) **504**
Sulphur-free high-temperature shift 294
Sulphuric acid
  catalyst for 17
  plant, amount of catalyst in 511
  process, sulphur-burning **512**
  production 503-17
    at fertilizer plants 469
    contact process 506-11
Sulphur-impregnated charcoal 223
Sulphur-tolerant shift catalysts 335-8
Sulphur trioxide, thermodynamics of **504**
Support
  refractory 32-3, 42-4
  role in methanol synthesis 465
  role in stabilization of metal catalysts 77
Surface area
  and activity of nickel 352
  of ammonia synthesis catalysts 397
    vs position in bed **71**
  of copper in methanol catalyst 456, 464
  effect of chloride in copper catalysts 84
  effect on methanation rate **351**
  effect on steam reforming catalysts 254
  of high-temperature shift catalysts 303
  of low-temperature shift catalyst 312-13
  maximizing 70
  of nickel and cobalt molybdate **246**, **248**
  and water-gas shift reaction activity 310
Surface intermediates, in shift 290
Surface-migration sintering 328
Surface roughness, ammonia catalysts 400
Swedish magnetite 394
Synthesis gas for ammonia and methanol (*see also* Steam reforming)
  use of steam reforming 230
Synthesis loops
  for ammonia 391-2, **393**, **416**
  for methanol 446-52

T

Tail gas
  concentration of nitrogen oxides 477-8
  olefin plants, methanation 367, **368**
Tail gas recycling
  in formaldehyde process 491
Temkin, M. I., ammonia kinetics 409-11
Temperature (*see also* Operating temperature)
  in commissioning shift catalyst 319
  control, of pyrophoric catalysts 184-5
  effect
    on absorption of hydrogen sulphide **215**

606

Index

on adsorption of poisons on shift 325
on dinitrogen tetroxide **478**
on equilibrium constants, methanation **345**
on heats of reaction in naphtha steam reforming (table) **234**
on life of steam reformer tubes **269**
on methanol synthesis **444**
on methanol synthesis catalysts 459
on rate of ammonia synthesis **413**
on sensitivity of steam reforming catalysts to poisons 278
increase in methanator 359-60
Temperature conversions 559, **560** (chart)
Temperature profiles
in ammonia synthesis catalyst 390
calculation of active volume 133-4
in catalyst bed after sintering **79**
during shift catalyst reduction 300
effect of poison on **97**
in high-temperature shift catalyst **303**
indirectly-cooled converter **434**
low-temperature shift catalyst **321**
changes due to poisoning **80**
measurement in catalyst beds 574-5
measurement of performance 127
methanation catalyst 363-4
in methanators 363-4, **366** (table)
multibed quench converters **92, 93**
in poisoning at top of bed only **134**
quench ammonia converter **427**
steam-raising converters **95**
tube-cooled converters **94, 439**
in uniform poisoning of catalyst **133**
in water-gas shift reactor 291-2
Temperature-ramping 73
Temperature rise
in ammonia synthesis 391
charging prereduced catalyst 155
in methanator 320, 345
Temporary deactivation 405-7
Terrace fired reforming furnace **263, 266**
Testing catalysts 48
Tetrahydrothiophene 192, 200
Thermal stability
of low-temperature shift catalysts 324
of steam reforming catalysts 244
Thermocouples 574-5
installation checking 151
in low-temperature shift reduction 167
in water-gas shift reactor 291
Thermodynamics
ammonia synthesis 388-92
of elements 550
hydrogen sulphide with zinc oxide 209-11
methanol synthesis 442-4
nickel oxide reduction **163**
oxidation of sulphur dioxide 503-6
steam reforming 230, 251
sulphur poisoning of shift catalysts 326
third law of 386
water-gas shift reaction 285-8
Thermosheaths, in shift reactors 291-2
Thiele modulus 104
for reactions of shift catalysts 313
sulphur dioxide oxidation 514
Thiophene
hydrogenolysis 200, **201**
in naphtha 194

thermal stability 199
Thiophenols, thermal dissociation 199
Top-fired steam reforming furnace (see also Steam reforming)
arrangement **265**
photograph **264**
schematic arrangement **262**
tube skin temperature profile **271**
Topsøe ammonia converter 428, **437**
Town gas (see also Substitute natural gas, Catalytic rich gas) 384
catalytic rich gas 368-72, **372** (table)
production from steam reforming 261-2
Tracer studies 290
Transition metals, activated charcoal 196
Tube-cooled reactors 96
ammonia catalyst reduction 173-4
for ammonia synthesis 425-6, 433-5, **438**
charging with catalyst 155
discharge 187
for methanol synthesis 447-52, **450**
temperature profiles **94, 439**
Tube skin temperature
measurement 555-6
modifying catalyst shape 254, 255
profile, top fired steam reformer **271**
in steam reformers **272**
Tubes (see also Steam reforming)
steam reformers 267-8
arrangement in top-fired furnace **265**
charging from socks 143
dimensions 265
effect of poisons on tube life 272
expansion and contraction 272-3
in formaldehyde production 499
Turnover number (turnover frequency) **23**
TVA ammonia converter 435

U

Ultraviolet photon spectroscopy 414
Unsupported metal catalysts 34-5, 290
Uranium, ammonia catalyst 394
Used catalysts (see also Discharged catalysts) 188

V

Vanadium catalysts 80, 510
in contact process 508-10
disposal 517
life 511
Vessel dimensions, calculation 119-20
Vetrocoke 178, **361**, 362
Vibrators 155, 160
Voidage
calculation 103
effect on pressure drop 296-8
optimum, numerical example 525
and particle size and shape 101-2
typical values (table) **102**

W

Water (see also Steam)

607

Index

ammonia synthesis catalysts
  during reduction of 172, 401
  effect on intrinsic activity 411
  poisoning of 80, 404, 406
  reaction with reduced catalyst 185
chloride in 221
heat of formation **347**
inhibition of methanation 366-7
low-temperature shift catalysts
  formed during reduction of 317
  use in wet discharge of 330
  washing of chlorides in 313
magnetite surface, oxidation of 290
methanol catalyst, effect on 455
Water-gas
  calorific value 283
  generators 225
Water-gas shift catalysts *see* High-temperature shift catalysts, Low-temperature shift catalysts
Water-gas shift process 33, 88-9
Water-gas shift reaction (*see also* High- *and* Low-temperature shift reaction) 283-339
  activation energy 288
  activity and copper surface area 310
  adjust composition, methanol synthesis 453
  during reduction of shift catalyst 314
  equilibrium constants (tables) **543-8**
  history 284
  mechanism 288-9, 290
  thermodynamics 285-8
Wurtzite 572

X

X-ray photoelectron spectroscopy 414

Z

Zeolites, X-ray diffraction of 55
Zinc carbonate formation 219
Zinc chloride, in desulphurization 222
Zinc halides, mobility of 325, 329
Zinc oxide
  absorbents 34, 209-20
  composition of commercial 211
  for desulphurization 209-20
  equilibrium constants 572-3
  formulation 211
  guard beds
    in partial oxidation plant 333
    poor performance in 332
    to protect shift catalyst 220
  hydrogen sulphide absorption **210**
  impurities 220
  in low-temperature shift catalysts 312
  in methanol synthesis catalysts 453-5
  pyrophoricity of sulphided 188
  reduction in methanol catalyst 462
  removal of hydrogen sulphide 197, 198
  role in copper methanol catalysts 456-8
  sulphur absorption 216-17
Zinc oxide bed, dimensions of **218**
Zinc oxide/chromia catalyst 27, 453-5
  by-product formation (table) **460**
  effect of adding copper 455
  for methanol synthesis 443, 452
  poison tolerance 455
Zinc oxide, free, in shift catalysts 326, **327**